Lecture Notes in Computer Science 9705

Commenced Publication in 1973
Founding and Former Series Editors:
Gerhard Goos, Juris Hartmanis, and Jan van Leeuwen

More information about this series at http://www.springer.com/series/7407

Yo-Sub Han · Kai Salomaa (Eds.)

Implementation and Application of Automata

21st International Conference, CIAA 2016
Seoul, South Korea, July 19–22, 2016
Proceedings

 Springer

Editors
Yo-Sub Han
Yonsei University
Seoul
Korea (Republic of)

Kai Salomaa
Queen's University
Kingston, ON
Canada

ISSN 0302-9743 ISSN 1611-3349 (electronic)
Lecture Notes in Computer Science
ISBN 978-3-319-40945-0 ISBN 978-3-319-40946-7 (eBook)
DOI 10.1007/978-3-319-40946-7

Library of Congress Control Number: 2016941604

LNCS Sublibrary: SL1 – Theoretical Computer Science and General Issues

This Springer imprint is published by Springer Nature
The registered company is Springer International Publishing AG Switzerland

Preface

This volume contains the papers presented at the 21st International Conference on Implementation and Application of Automata (CIAA 2016) organized by the Department of Computer Science at Yonsei University during July 19–22, 2016, in Seoul, Republic of Korea.

The CIAA conference series is a major international venue for the dissemination of new results in the implementation, application, and theory of automata. The previous 20 conferences were held in various locations all around the globe: Umeå (2015), Giessen (2014), Halifax (2013), Porto (2012), Blois (2011), Winnipeg (2010), Sydney (2009), San Francisco (2008), Prague (2007), Taipei (2006), Nice (2005), Kingston (2004), Santa Barbara (2003), Tours (2002), Pretoria (2001), London Ontario (2000), Potsdam (WIA 1999), Rouen (WIA 1998), London Ontario (WIA 1997 and WIA 1996). As for its predecessors, the theme of CIAA 2016 was the implementation of automata and applications in related fields. The topics of the presented papers include characterizations of automata, computing distances between strings and languages, implementations of automata and experiments, enhanced regular expressions, and complexity analysis.

There were 49 submissions from 34 different counties: Argentina, Australia, Belgium, Canada, China, Colombia, Czech Republic, Denmark, Finland, France, Germany, Hungary, India, Iran, Italy, Japan, Republic of Korea, Latvia, New Zealand, Poland, Portugal, Qatar, Russia, Singapore, Slovakia, South Africa, Spain, Sweden, Taiwan, Thailand, Tunisia, Turkey, the UK, and the USA. Each submission was reviewed by at least three reviewers and thoroughly discussed by the Program Committee (PC). The committee decided to accept 26 papers for oral presentation. The program also includes three invited talks by Oscar H. Ibarra, Bakhadyr Khoussainov, and Ahyoung Sung.

It is a pleasure for the PC chairs to thank the members of the PC and the external reviewers for reviewing the papers and participating in the selection process and helping to maintain the high standard of the CIAA conferences. We are especially grateful to all the invited speakers and all authors of submitted papers for making CIAA 2016 a scientific success. We appreciate the help of the EasyChair conference system for making our work of organizing CIAA 2016 much easier.

We would furthermore like to thank the editorial staff at Springer, and in particular Alfred Hofmann and Anna Kramer, for their guidance and help during the process of publishing this volume. Last but not least, we are grateful to the conference sponsors for their generous financial support and the local Organizing Committee members, Da-Jung Cho, Shin-Dong Kang, Guen-Hae Kim, and Hwee Kim, for their help.

We are all looking forward to CIAA 2017 at Université Paris-Est Marne-la-Vallée, Paris, in France.

July 2016   Yo-Sub Han
                                                                             Kai Salomaa

Organization

CIAA 2016 was organized by the Department of Computer Science at Yonsei University during July 19–22, 2016, in Seoul, Republic of Korea.

Invited Speakers

Oscar H. Ibarra	University of California, Santa Barbara, USA
Bakhadyr Khoussainov	University of Auckland, New Zealand
Ahyoung Sung	Samsung Electronics, Republic of Korea

Program Committee

Marie-Pierre Béal	Université Paris-Est Marne-la-Vallée, France
Bernd Burgstaller	Yonsei University, Republic of Korea
Cezar Câmpeanu	University of Prince Edward Island, Canada
Pascal Caron	University of Rouen, France
Jean-Marc Champarnaud	University of Rouen, France
Salimur Choudhury	Algoma University, Canada
Michael Domaratzki	University of Manitoba, Canada
Frank Drewes	Umeå University, Sweden
Dominik Freydenberger	Bayreuth University, Germany
Yo-Sub Han (Chair)	Yonsei University, Republic of Korea
Markus Holzer	University of Giessen, Germany
Oscar H. Ibarra	University of California, Santa Barbara, USA
Sang-Ki Ko	University of Liverpool, UK
Stavros Konstantinidis	Saint Mary's University, Canada
Andreas Malcher	University of Giessen, Germany
Andreas Maletti	University of Stuttgart, Germany
Sebastian Maneth	University of Edinburgh, UK
Denis Maurel	University of Tours, France
Cyril Nicaud	Université Paris-Est Marne-la-Vallée, France
Alexander Okhotin	University of Turku, Finland
Giovanni Pighizzini	University of Milan, Italy
Bala Ravikumar	Sonoma State University, USA
Daniel Reidenbach	Loughborough University, UK
Rogério Reis	University of Porto, Portugal
Michel Rigo	University of Liège, Belgium
Kai Salomaa (Co-chair)	Queen's University, Canada
Shinnosuke Seki	University of Electro-Communications, Japan
György Vaszil	University of Debrecen, Hungary
Bruce Watson	Stellenbosch University, South Africa
Hsu-Chun Yen	National Taiwan University, Taiwan

Steering Committee

Jean-Marc Champarnaud	University of Rouen, France
Markus Holzer	University of Giessen, Germany
Oscar H. Ibarra	University of California, Santa Barbara, USA
Denis Maurel	University of Tours, France
Kai Salomaa (Chair)	Queen's University, Canada
Hsu-Chun Yen	National Taiwan University, Taiwan

Additional Reviewers

Asarin, Eugene	Guingne, Franck	Ouardi, Faissal
Bedon, Nicolas	Ittoo, Ashwin	Palioudakis, Alexandros
Beier, Simon	Klimann, Ines	Patrou, Bruno
Borie, Nicolas	Kutrib, Martin	Quernheim, Daniel
Bosma, Wieb	Lavado, Giovanna	Rampersad, Narad
Bouchou, Béatrice	Lombardy, Sylvain	Reinhardt, Klaus
Broda, Sabine	Madonia, Maria	Roche, Abiel
Carayol, Arnaud	Manal, Mohammed	Rondogiannis, Panos
David, Julien	McQuillan, Ian	Santocanale, Luigi
Delgado, Manuel	Mercas, Robert	Sassolas, Mathieu
Djelloul, Ziadi	Mereghetti, Carlo	Seemann, Nina
Egecioglu, Omer	Moreira, Nelma	Serre, Olivier
Friburger, Nathalie	Ng, Timothy	Wendlandt, Matthias
Gonze, François	Nicart, Florent	Yakaryilmaz, Abuzer
Goodspeed, Ben	Nouvel, Damien	Young, Joshua

Organizing Committee

Da-Jung Cho
Yo-Sub Han (Chair)
Shin-Dong Kang
Guen-Hae Kim
Hwee Kim

Sponsoring Institutions

Department of Computer Science, Yonsei University
Yonsei Institute of Information and Communication Technology, Yonsei University
Office of Research Affairs, Yonsei University

Invited Talks

Grammatical Characterizations of NPDAs and VPDAs with Counters

Oscar H. Ibarra

Department of Computer Science, University of California,
Santa Barbara, CA 93106, USA
ibarra@cs.ucsb.edu

Abstract. We give a characterization of NPDAs with reversal-bounded counters (NPCMs) in terms of context-free grammars with monotonic counters. We show that the grammar characterization can be used to give simple proofs of previously known results such as the semilinearity of the Parikh map of any language accepted by an NPCM. We prove a Chomsky-Schutzenberger-like theorem: A language L is accepted by an NPCM if and only if there is a $k \geq 1$ and an alphabet Σ containing at least k distinguished symbols, $p_1, \ldots, p_k$, such that $L = h(D \cap E_k(R))$ for some homomorphism h, Dyck language $D \subseteq \Sigma^*$, and regular set $R \subseteq \Sigma^*$, where $E_k(R) = \{w \mid w \in R, |w|_{p_1} = \cdots = |w|_{p_k}\}$. We also give characterizations of other machine models (such as NPDAs whose stack makes only 1 reversal, NFAs, and VPDAs, i.e., visibly pushdown automata) augmented with reversal-bounded counters. Finally, we investigate the complexity of some decision problems for these grammatical models.

This research was supported, in part, by NSF Grant CCF-1117708.

Decision Problems for Finite Automata over Infinite Algebraic Structures

Bakhadyr Khoussainov and Jiamou Liu

Department of Computer Science, University of Auckland, New Zealand
bmk@cs.auckland.ac.nz
jiamou.liu@auckland.ac.nz

Abstract. We introduce the concept of finite automata over algebraic structures. We address the classical emptiness problem and its various refinements in our setting. In particular, we prove several decidability and undecidability results. We also explain the way our automata model connects with the existential first order theory of algebraic structures.

What We Experience is What You Do

Ahyoung Sung

Software Quality Assurance Group,
Division of Visual Display, Samsung Electronics
Republic of Korea
ahyoung.sung@samsung.com

1 Introduction of Embedded Systems

Embedded systems are used to control a wide variety of dynamic and complex applications, ranging from consumer products such as smart phones, smart televisions, and wearable devices to safety-critical systems such as automobiles, airplanes, and medical devices. Clearly, systems such these must be sufficiently dependable; however, there is ample evidence that often they are not. Toyota Corporation admitted that a "software glitch" was to blame for braking problems in the 2010 Prius, in which cars continued to move even when their brakes are engaged[1]. Other faults in embedded systems led to the Ariane[2], Therac-25 [1], and Mars Pathfinder [2]. Embedded systems are typically designed to perform specific tasks in particular computational environments consisting of software and hardware components. A typical system is structured in terms of four layers: the application layer, the Operating System (OS) layer, the Hardware Adaptation Layer (HAL), and the underlying hardware substrate. Interfaces between these layers provide access to, and observation points into, the internal workings of each underlying layer. Due to the layered architecture, embedded applications are more compact and portable and can easily be written to directly execute on hardware. Software development practices in the area of consumer electronics often separate developers of application software from the engineers who develop lower system layers. As an example, consider Apples App Store, which was created to allow soft-ware developers to create and market iPhone applications. Apple provided the iPhone Software Development Kit (SDK), the same tool used by Apples engineers to develop core applications, to support application development by providing basic interfaces that an application can use to obtain services from underlying software layer. There were numerous reports that some applications created using the SDK caused iPhone to exhibit unresponsiveness due to failures in the underlying layers that affected operations of the devices[3]. These failures occurred because the developers used the interfaces provided by the iPhone SDK in ways not anticipated by Apples engineers. These applications use lower-level software components (runtime services and

---

[1] CNN news report, February 4, 2010

[2] https://www.ima.umn.edu/ ~ arnold/disasters/ariane5rep.html

[3] http://arstechnica.com/apple/2008/07/iphone-app-store-problems-causing-more-than-just-headaches/

libraries) that must be developed or heavily customized by in-house software developers. In situations such as this the developers do know what scenarios underlying services will be invoked in, but they still cannot safely treat lower level components as black boxes in testing. Instead, developers and engineers need to specifically observe the execution of each component and test the scenarios in which underlying components are invoked by the application and in which tasks interact.

2   Continuous Integration and Software Testing

Continuous Integration (CI) is a development practice that requires various software developers such as user-interface (UI) developer, video application developer, kernel developer, and device-driver developer, to integrate their source codes into a shared repository or a mainstream several times a day. Each check-in follows the process of an automated build and software testing (e.g., developer's code-level testing). During the process of continuous integration, software testing is an essential activity of executing a program with the intent of finding "software defects" in product development phase; it validates and fixes potential software failure (e.g., buffer overflow) that consumers might experience in advance. When the entire integration gets done (a developer usually draws a line for the integration), now we have a software system or a product prototype that spans across from application layer to HAL. Especially, in the system deployment stage in massive productions, software test theories are rarely used for testing due to scalability challenges and environmental factors. And yet, manufacturing industry has adapted analytical testing theories such as: (a) failure analysis based on the collected defects, (b) system modeling using finite-state machine (e.g., software module analysis using state-transitions in order to verify the faulty user scenarios), (c) control-flow and its data-flow analysis for UI testing, and (d) empirical strategies for effective validation (e.g., impact analysis for regression testing in UI, test case selection for massive multi-media files, test oracle[4] design for detecting an audio-mute section). Moreover, these techniques are evolving gradually to deliver a better product for our potential customers. This talk focuses on how software engineers define a problem, prove their hypothesis, and apply their test theories to the real products that we can easily experience in our daily products

References

1. Leveson, N.G.: An investigation of the Therac-25 accidents. IEEE Comput. **26**, 18–41 (1993)
2. Muchnick, S.S., Jones, N.D.: Program Flow Analysis: Theory and Application. Prentice Hall Professional Technical Reference (1981)

---

[4] Test oracle is a mechanism for determining whether a test has passed or failed.

Contents

Invited Paper

Decision Problems for Finite Automata over Infinite Algebraic Structures

Bakhadyr Khoussainov[(✉)] and Jiamou Liu

Department of Computer Science, University of Auckland, Auckland, New Zealand
bmk@cs.auckland.ac.nz, jiamou.liu@auckland.ac.nz

Abstract. We introduce the concept of finite automata over algebraic structures. We address the classical emptiness problem and its various refinements in our setting. In particular, we prove several decidability and undecidability results. We also explain the way our automata model connects with the existential first order theory of algebraic structures.

1 Introduction

Most computer programs rely on operations and queries on a priori defined data types. An example of a such data type is the integers with the usual operations of addition, multiplication and the comparison test. Another example is the graph data type that encapsulates the operations of adding or deleting vertices and edges as well as edge and subgraph queries. Algebraically, data types are structures of the form $(D; f_1, \ldots, f_m, P_1, \ldots, P_n, \overline{c_1}, \ldots, \overline{c_\ell})$ where $f_1, \ldots, f_m$ are atomic operations, $P_1, \ldots, P_n$ are atomic relations and $\overline{c_1}, \ldots, \overline{c_\ell}$ are constants on the domain D. A program can thus be viewed as a sequence of operations and queries over a certain algebraic structure.

This view of programs motivated the definitions of many models of computation over structures. An example is the Blum-Shub-Smale (BSS) machines, where the underlying structure is the ordered ring of the reals [2]. This model builds a theoretical foundation of numerical algorithms on reals. Another example is the work of O. Bournez, et al. [5] that generalises BSS machines to models over arbitrary structures. Among several results, they prove that the set of all recursive functions over arbitrary structure S is exactly the set of decision functions computed by BSS machines over S. Other examples include various classes of counter automata that use counters in different ways [6,9,12,15,17].

In this work we introduce finite automata models over algebraic structures. The work fits into two lines of research: Firstly, one may view our models as finite automata analogues of BSS machines over arbitrary structures. These automata process finite sequences of elements from the domain of a given structure S, and accept or reject these sequences. Such an automaton is a finite state machine equipped with a fixed number of registers, a read only head that always moves to the right in the tape and transitions between the states. For the structure S, we use the term S-*automata* or *extended* S-*automata* that define two different interpretations of this computation model on S. Secondly, one may view our

© Springer International Publishing Switzerland 2016
Y.-S. Han and K. Salomaa (Eds.): CIAA 2016, LNCS 9705, pp. 3–11, 2016.
DOI: 10.1007/978-3-319-40946-7_1

automata models as automata over an infinite alphabet when the underlying structure $\mathcal{S}$ is infinite. Automata over infinite alphabet attracted considerable interests due to connections to model checking and verification [1,3,4,10,21,23]. One goal here is to extend automata-theoretic techniques to words and trees over data values. For example, Kaminsky and Francez in [13] proposed register automata. These are finite state machines equipped with a fixed number of registers which may hold values from an infinite domain D. The operations allowed by the automata are equality comparisons between the input and the register values and the copy operation. Another example is pebble automata introduced by Neven, Schwentick and Vianu [20]. Their automata use a fixed set of pebbles with a stack to keep track of values in the input data words. Operations include equality comparisons of the current pebble values, and dropping and lifting a pebble. Other examples of such automata models include Bojanczyk's data automata [3] and Alur's extended data automata [1]. While all the above automata allow only equality tests between data values, there has also been automata model for linearly ordered data domains [22]. The existence of many such models of automata calls for a general yet simple framework to formally reason about such finite state automata. We suggest one such framework.

2   Two Automata Models Over Algebraic Structures

Let $\mathcal{S} = (D; f_1, \ldots, f_m, P_1, \ldots, P_n, \overline{c_1}, \ldots, \overline{c_\ell})$ be an algebraic structure. We assume that $m = n$ and that all functions and predicate are binary. These are not restrictions. For instance, when $n < m$ we expand the structure by adding the relation P_n to its signature to ensure that in the expanded structure the number of atomic predicates and atomic operations are equal.

A D-word of length t is a sequence $d_1 \ldots d_t$ of elements of D. An automaton over $\mathcal{S}$ has k updatable registers $R_1, \ldots, R_k$; each register R_i contains an n-tuple $\overline{r_i} = (r_{i,1}, \ldots, r_{i,n})$ from D^n, and the content of the register might change. Furthermore, the automaton has ℓ constant registers containing the constants $\overline{c_1}, \ldots, \overline{c_\ell}$; these values never change. We represent these values of registers as matrices $\mathbf{R}$ and $\mathbf{C}$.

$$\mathbf{R} = \begin{pmatrix} r_{1,1} \cdots r_{1,n} \\ r_{2,1} \cdots r_{2,n} \\ \cdots \cdots \cdots \\ r_{k,1} \cdots r_{k,n} \end{pmatrix} \quad \text{and} \quad \mathbf{C} = \begin{pmatrix} c_{1,1} \cdots c_{1,n} \\ c_{2,1} \cdots c_{2,n} \\ \cdots \cdots \cdots \\ c_{\ell,1} \cdots c_{\ell,n} \end{pmatrix}$$

Let Op be the set of all atomic operations of $\mathcal{S}$. The automaton is a finite state machine where transitions are of the form (q, T_1, T_2, F, q') where q, q' are states, T_1, T_2 are a pair of $\{0,1\}$-valued matrices of sizes $k \times n$ and $\ell \times n$ respectively, and $F = (f_{i,j}) \in Op^{k \times n}$ is a $k \times n$ matrix of atomic operation of $\mathcal{S}$. Inputs to the automaton are D-words written on a one-way read-only tape. When the automaton is in state q and reads the next element x of an input D-word, the automaton proceeds with two steps:

1. (**Testing**) The automaton produces two $k \times n$ and $\ell \times n$ test matrices $Test(\mathbf{R}, x)$ and $Test(\mathbf{C}, x)$ with entries 1 (true) or 0 (false), respectively:

$$
\begin{pmatrix}
P_1(r_{1,1}, x) & \cdots & P_n(r_{1,n}, x) \\
P_1(r_{2,1}, x) & \cdots & P_n(r_{2,n}, x) \\
\cdots & \cdots & \cdots \\
P_1(r_{k,1}, x) & \cdots & P_n(r_{k,n}, x)
\end{pmatrix}, \quad
\begin{pmatrix}
P_1(c_{1,1}, x) & \cdots & P_n(c_{1,n}, x) \\
P_1(c_{2,1}, x) & \cdots & P_n(c_{2,n}, x) \\
\cdots & \cdots & \cdots \\
P_1(c_{\ell,1}, x) & \cdots & P_n(c_{\ell,n}, x)
\end{pmatrix}
$$

The automaton then makes a transition (q, T_1, T_2, F, q') where q is the current state, $T_1 = Test(\mathbf{R}, x)$ and $T_2 = Test(\mathbf{C}, x)$.

2. (**Updating**) When making the transition (q, T_1, T_2, F, q'), the automaton updates the values of registers using operations in F, transforming the matrix $\mathbf{R}$ to the matrix $F(\mathbf{R}, x)$ as presented below:

$$
\mathbf{R} = \begin{pmatrix}
r_{1,1} & \cdots & r_{1,n} \\
r_{2,1} & \cdots & r_{2,n} \\
\cdots & \cdots & \cdots \\
r_{k,1} & \cdots & r_{k,n}
\end{pmatrix}
\implies
F(\mathbf{R}, x) = \begin{pmatrix}
f_{1,1}(r_{1,1}, x) & \cdots & f_{1,n}(r_{1,n}, x) \\
f_{2,1}(r_{2,1}, x) & \cdots & f_{2,n}(r_{2,n}, x) \\
\cdots & \cdots & \cdots \\
f_{k,1}(r_{k,1}, x) & \cdots & f_{k,n}(r_{k,n}, x)
\end{pmatrix}
$$

where $f_{i,j}$ the (i,j)-entry of F for all $1 \leq i \leq k, 1 \leq j \leq n$.

After all elements on the input tape have been read, the $\mathcal{S}$-automaton stops and decides whether to accept the input depending on the last state.

In the following we put constraints on the register values $\mathbf{R}$ and the operation matrix F and introduce two finite automata models over the structure $\mathcal{S}$, namely, the $\mathcal{S}$-automata and extended $\mathcal{S}$-automata.

- $\mathcal{S}$-**automata:** We require the matrix F in each transition to be the same; furthermore, each row in F is the tuple $(f_1, \ldots, f_n) \in Op^k$ of all atomic operation of $\mathcal{S}$. Hence any transition will transform the ith register $R_i = (r_{i,1}, \ldots, r_{i,n})$ to $(f_1(r_{i,1}, x), \ldots, f_n(r_{i,n}, x))$. Thus, a transition of an $\mathcal{S}$-automaton can simply be represented as (q, T_1, T_2, q').
- **Extended $\mathcal{S}$-automata:** We require the columns in both the register matrix $\mathbf{R}$ and the operation matrix F for each transition to be the same, that is:

$$
\mathbf{R} = \begin{pmatrix}
r_1 & \cdots & r_1 \\
r_2 & \cdots & r_2 \\
\cdots & \cdots & \cdots \\
r_k & \cdots & r_k
\end{pmatrix}
\text{ and } F = \begin{pmatrix}
f_{i_1} & \cdots & f_{i_1} \\
f_{i_2} & \cdots & f_{i_2} \\
\cdots & \cdots & \cdots \\
f_{i_k} & \cdots & f_{i_k}
\end{pmatrix}
$$

Thus we can simply write $\mathbf{R}$ as a tuple of elements $(r_1, r_2, \ldots, r_k) \in D^k$ and F as a tuple $(f_{i_1}, \ldots, f_{i_k}) \in Op^k$.

Definition 1. *We define two types of automata:*

1. *An $(\mathcal{S}, k)$-automaton is a tuple $\mathcal{A} = (Q, \mathbf{R}, \Delta, I, F)$ where Q is a finite set of states, $\mathbf{R} = (\bar{r}_1, \ldots, \bar{r}_k)$ is the initial values of the registers with each $\bar{r}_i \in D^n$, $I \subseteq Q$ is the initial states set, $F \subseteq Q$ is the set of accepting states and $\Delta \subseteq Q \times \{0,1\}^{n \cdot (k+\ell)} \times Q$ is the transition relation of $\mathcal{A}$. The automaton is deterministic if Δ determines the function $\Delta : Q \times \{0,1\}^{n \cdot (k+\ell)} \to Q$. An $\mathcal{S}$-automaton is an $(\mathcal{S}, k)$-automaton for some $k \in \omega$.*

2. *An* extended (S, k)-*automaton is defined in the same way as an* (S, k)-*automaton, with the following exceptions: the register* $\mathbf{R}$ *is* $(r_1, \ldots, r_k) \in D^k$, *and* $\Delta \subseteq Q \times \{0, 1\}^{(k+\ell) \cdot n} \times Op^k \times Q$ *is the transition relation. The automaton is* deterministic *if* Δ *is a function* $\Delta : Q \times \{0, 1\}^{(k+\ell) \cdot n} \to Op^k \times Q$. *An* extended S-*automaton is an extended* (S, k)-*automaton for some* $k \in \omega$.

Let $\mathcal{A} = (Q, \mathbf{R}, \Delta, I, F)$ be an (extended) (S, k)-automaton. We define the runs of the automaton on D-words as follows. A *configuration* of the automaton is a pair $(\mathbf{R}, q)$, where q is a state of the automaton and $\mathbf{R}$ is the matrix of register values. A *run* of $\mathcal{A}$ on a D-word $d_0 \ldots d_t$ is a sequence of configurations

$$(\mathbf{R}_0, s_0), (\mathbf{R}_1, s_1), \ldots, (\mathbf{R}_{t+1}, s_{t+1})$$

such that where $\mathbf{R}_0 = \mathbf{R}$, $s_0 \in I$, the transition from s_i to s_{i+1} is labeled with the test matrices $Test(\mathbf{R}_i, d_i)$ and $Test(\mathbf{C}, d_i)$, and $\mathbf{R}_{i+1} = F(\mathbf{R}_i, d_i)$ for all i. The run is *accepting* if $s_{t+1} \in F$. We say that $\mathcal{A}$ *accepts* the D-word $d_0 \ldots d_t$ if $\mathcal{A}$ has an accepting run on $d_0 \ldots d_t$.

Definition 2. *The language* $L(\mathcal{A})$ *of the (extended)* S-*automaton* $\mathcal{A}$ *is the set of all* D-*words accepted by* $\mathcal{A}$. *We call such languages* (extended) S-*regular*.

3   Decision Problems on S-Automata

Simple Properties of S-regular Languages. The class of S-regular languages is a natural generalisation from regular languages in the following sense. Firstly, there is a natural connection between S-regular languages and regular languages. In particular, when the structure S is finite, then any S-regular language is regular. Secondly, the class of S-regular languages is closed under the Boolean operations. Thirdly, every S-regular language can be recognised by a deterministic S-automaton. Furthermore, the class of (S, k)-recognisable languages can be a proper subclass of the class of $(S, k+1)$-recognisable languages. This is true for the infinite structure $S = (D; =, \mathrm{pr}_1)$ where pr_1 is the projection: $\mathrm{pr}_1(x, y) = x$.

The Emptiness Problem for S-automata. The emptiness problem asks for an algorithm that given an S-automaton, detects if the language of the automaton is non-empty. This problem has a positive solution for regular languages and thus is decidable when S is finite. It turns out that for certain large class of structures S, the emptiness problem is decidable.

Definition 3. *An equivalence relation* $\equiv_k$ *on the set* $M_S(n, k)$ *of matrices is called* smooth *if the relation satisfies the following conditions:*

1. *The relation* $\equiv_k$ *is of finite index.*
2. *For all* $\mathbf{R}, \mathbf{R}' \in M_S(n, k)$, *matrices* X *and* Y *with* 0, 1 *entries, if* $\mathbf{R} \equiv_k \mathbf{R}'$ *then we have* $\{z \mid Test(\mathbf{R}, z) = X \ \& \ Test(\mathbf{C}, z) = Y\}$ *is the empty set if and only if* $\{z \mid Test(\mathbf{R}', z) = X \ \& \ Test(\mathbf{C}, z) = Y\}$ *is the empty set.*

As a simple example, assume $\equiv$ is an equivalence relation of finite index on the domain D such that all atomic predicates and operations are compatible with $\equiv$. This relation naturally defines the relation $\equiv_k$ on the matrices $M_S(n,k)$: Two matrices are $\equiv_k$-equivalent if the entries at the same positions of the matrices are $\equiv$-equivalent. Then the equivalence relation $\equiv_k$ is smooth.

Here is another example. Let S be a structure $(D; f_1, \ldots, f_n, =)$. On the set $M_S(n,k)$ consider the following relation $\equiv_k$: Two matrices $\mathbf{R}$ and $\mathbf{R'}$ are $\equiv_k$-equivalent if for all two positions (i,j) and (s,t) of the matrices we have $r_{i,j} = r_{s,t}$ if and only if $r'_{i,j} = r'_{s,t}$. Then the relation $\equiv_k$ is smooth.

Let $\{\equiv_k\}_{k>0}$ be a family of smooth equivalence relations on S. Assume that for each k we can effectively represent the $\equiv_k$-classes by some finite objects. For instance, when $k = 2$ and $n = 2$, the relation $\equiv_2$ considered in the paragraph above has the following representatives:

$$\begin{pmatrix} a & a \\ a & a \end{pmatrix}, \begin{pmatrix} a & b \\ a & a \end{pmatrix}, \begin{pmatrix} a & b \\ c & a \end{pmatrix}, \begin{pmatrix} a & b \\ c & d \end{pmatrix},$$

where a, b, c, d are all pairwise distinct and fixed integers. We call these representatives *types* of the equivalence classes. With this set-up, we have:

Definition 4. *The structure S is* nice *if it satisfies the following two properties:*

1. *There is an algorithm that given a type of a matrix $\mathbf{R} \in M_S(n,k)$, and given two $\{0,1\}$-valued matrices X, Y decides if the set $\{z \mid Test(\mathbf{R}, z) = X \,\&\, Test(\mathbf{C}, z) = Y\}$ is empty or not.*
2. *There is an algorithm that given a type of a matrix $\mathbf{R} \in M_S(n,k)$, computes the types of all matrices $F(\mathbf{R}, x)$ where x satisfies the equations $Test(\mathbf{R}, z) = X$ and $Test(\mathbf{R}, z) = Y$ for given X, Y.*

In particular, let S be a structure $(D; f_1, \ldots, f_n, =)$. Assume that for each f_i there is a finite set $F_i \subset D$ such that

1. For every $d \notin F_i$ the function $f_{i,d}(x) = f_i(d, x)$ is injection on D.
2. For each $d \in F_i$, the function $f_{i,d}(x) = f_i(d, x)$ is a constant function, that is, there is an $a \in D$ such that $f_{i,d}(x) = a$ for all $x \in D$.

Then the smooth equivalence relation $\equiv_k$ makes the structure S nice. For instance, the structure $(\mathbb{Z}; +, \times, =)$ satisfies the properties above.

Theorem 5. *The emptiness problem over any nice structure is decidable. More precisely, for any nice structure S over domain D, there is an algorithm that, given an S-automaton $A = (Q, \mathbf{R}, \Delta, I, F)$ and the type of $\mathbf{R}$, detects if the automaton accepts at least one D-word.*

From the theorem above we immediately get the following corollary.

Corollary 6. *The emptiness problem is decidable over the arithmetic $(\mathbb{Z}; +, \times)$, the fields of reals $(\mathbb{R}; +, \times)$ and rational numbers $(\mathbb{Q}; +, \times)$.*   □

4   Decision Problems on Extended $\mathcal{S}$-Automata

Simple Properties of Extended $\mathcal{S}$-regular Languages. Any $\mathcal{S}$-regular language is clearly extended $\mathcal{S}$-regular. On the other hand, extended $\mathcal{S}$-automata recognise larger class of languages. The limitation of $\mathcal{S}$-automata is that, when processing a D-word, the sequence of updates to the registers are the same regardless of which path the automaton take. In an extended $\mathcal{S}$-automaton, however, the operations performed on registers depends on the outcomes of the tests. This leads to a lack of some crucial properties enjoyed by $\mathcal{S}$-regular languages, such as determination. Furthermore, there exists extended $\mathcal{S}$-regular languages whose complements are not recognisable by any extended $\mathcal{S}$-automata. On the other hand, the class of languages recognised by deterministic extended $\mathcal{S}$-automata is closed under the Boolean operations.

Validation Problem for Extended $\mathcal{S}$-automata. We refine the emptiness problem for finite automata as follows. Ddesign an algorithm that, given an $\mathcal{S}$-automaton over the structure $\mathcal{S}$, and a path from an initial state to an accepting state in the automaton, builds an input sequence from the structure $\mathcal{S}$ that validates the path. We call this *the validation problem* for $\mathcal{S}$-automata. We will investigate the validation problem for extended $\mathcal{S}$-automata and connect the problem with the first order existential theory of the structure $\mathcal{S}$.

We postulate that $\mathcal{S}$ is a computable structure, i.e., its domain D and all of its atomic predicates P_1, ..., P_n and operations f_1, ..., f_n are computable. The validation problem for extended $\mathcal{S}$-automata turns out to be equivalent to deciding the existential theory (with parameters) $\mathsf{Th}_\exists(\mathcal{S})$ of the structure. For the next theorem, we use $\mathcal{S}[\mathrm{pr}_1, \mathrm{pr}_2]$ to denote the structure obtained from $\mathcal{S}$ upon expansion with two projection operations pr_1 and pr_2.

Theorem 7. *Suppose $\mathcal{S}$ is a computable structure. The validation problem for extended $\mathcal{S}[\mathrm{pr}_1, \mathrm{pr}_2]$-automata is decidable if and only if $\mathsf{Th}_\exists(\mathcal{S})$ is decidable.*

As a corollary, we see that the validation problem over computable structures with undecidable existential theory, such as the arithmetic $(\omega; +, \times, \leq,)$, is undecidable. Also, the validation problem over computable structures with decidable first order theory, such as the Presburger arithmetic, is decidable.

The Emptiness Problem for Extended $\mathcal{S}$-automata. On computable structures, the decidability of the emptiness problem implies decidability of the validation problem. The converse is not true. We discuss the emptiness problem on extended $\mathcal{S}$-automata for two special cases: the first case assumes that the transition graphs of the extended $\mathcal{S}$-automata are acyclic. The second concerns with fragments of the arithmetic $(\omega; +, \times, \leq, 0)$.

1. A state s is a *sink* if all outgoing transitions loop into s. All accepting (non-accepting) sink states can be collapsed into one (non-accepting) accepting sink state. Therefore we can always assume that every $\mathcal{S}$-automaton has at most 2 sink states. An extended $\mathcal{S}$-automaton *acyclic* if its state space without the sink states is an acyclic graph. If $\mathcal{S}$ is a computable structure, then the

emptiness problem of acyclic extended $\mathcal{S}[\mathrm{pr}_1, \mathrm{pr}_2]$-automata is equivalent to the corresponding validation problem. Hence, by Theorem 7, the emptiness problem is decidable for acyclic extended automata over such structures as $(\omega; +, \leq)$, $(\omega; \times, \leq)$, $(\mathbb{Q}; +, \leq)$ and finitely generated Abelian groups.

It is easy to find structures $\mathcal{S}$ with undecidable existential theory such that the emptiness problem for acyclic extended $(\mathcal{S}, k)$-automata is undecidable for every $k \geq 1$.

Let $\mathcal{S}_{\mathbb{Z}} = (\mathbb{Z}; +, \times, \mathrm{pr}_1, \mathrm{pr}_2, 0)$. One constructs, for any polynomial $p(\bar{x})$ in $\omega[x_1, \ldots, x_k]$, an acyclic extended $(\mathcal{S}_{\mathbb{Z}}, k+2)$-automaton $\mathcal{A}_p$ that evaluates p over a sequence $(a_1, \ldots, a_k) \in \mathbb{Z}^k$ of input values. This reduces Hilbert's tenth problem to the emptiness problem of acyclic extended $\mathcal{S}_{\mathbb{Z}}$-automata. Since Hilbert's tenth problem is undecidable for polynomials with bounded number of variables (the currently known bound that guarantees undecidability is 9 [16]), we obtain that the emptiness problem for acyclic $(\mathcal{S}_{\mathbb{Z}}, 11)$-automata is undecidable.

2. Let $\mathcal{S}$ be the following structure $(\omega; +1, \mathrm{pr}_1, 0)$ where $+1(x, y) = x + 1$. One constructs, given a 2-counter machine $\mathcal{M}$, an extended $\mathcal{S}$-automaton $\mathcal{M}'$ with 4 registers such that $\mathcal{M}$ accepts some word iff $\mathcal{M}'$ accepts some ω-word. This reduces the emptiness problem for 2-counter machines, known to be undecidable [17], to the emptiness problem for extended $(\mathcal{S}, 4)$-automata. Thus the emptiness problem for extended $(\mathcal{S}, 4)$-automata is undecidable.

The above shows for many structures $\mathcal{S}$, the emptiness problem for $\mathcal{S}$-automata is undecidable. We next present structures on which the emptiness problem is decidable. For this we use a tool similar to the notion of nice structures introduced for $\mathcal{S}$-automata; we recast Definition 4 in this setting:

Definition 8. *An equivalence relation $\equiv_k$ of finite index on the set D^k is smooth if for all $\mathbf{R}, \mathbf{R}' \in D^k$ and all $\{0, 1\}$-valued matrices X, Y, the condition $\mathbf{R} \equiv_k \mathbf{R}'$ implies that the set $\{z \mid Test(\mathbf{R}, z) = X \,\&\, Test(\mathbf{C}, z) = Y\}$ is empty iff the set $\{z \mid Test(\mathbf{R}', z) = X \,\&\, Test(\mathbf{C}, z) = Y\}$ is empty.*

Definition 9. *The structure $\mathcal{S}$ is k-nice if we have:*

(a) *There is an algorithm that given a type of a tuple $\mathbf{R} \in D^k$, and given two $\{0, 1\}$-valued matrices X, Y decides if the set*

$$\{z \mid Test(\mathbf{R}, z) = X \,\&\, Test(\mathbf{C}, z) = Y\}$$

is empty or not.

(b) *There is an algorithm that given a type of a tuple $\mathbf{R} \in D^k$, matrices X, Y, and a tuple $F \subseteq Op^k$, computes the types of all tuples $F(\mathbf{R}, x)$ where x satisfies the equation $Test(\mathbf{R}, z) = X$ and $Test(\mathbf{R}, z) = Y$.*

Theorem 10. *The emptiness problem for extended $(\mathcal{S}, k)$-automata over any k-nice structure $\mathcal{S}$ is decidable.*

Corollary 11. *The emptiness problem for extended $(\mathcal{S}, 1)$-automata is decidable for the structure $\mathcal{S} = (\omega; +, \times, \mathrm{pr}_1, \leq, c_1, \ldots, c_\ell)$.*

5   Conclusion

Our models of automata over algebraic structure provide a general framework for finite-state computation. Observe that: (1) we can vary the underlying structures S thus connecting algebraic properties of S with finite state machines, (2) in certain precise sense our machines can simulate Turing machines, (3) many known automata models (e.g., pushdown automata, Petri nets) can be simulated by our models of automata, and (4) the class of languages recognised by a S-automata is closed under the Boolean set-theoretic operations. This extends the finite automata and tree automata models of computations. However, we note that it remains to be seen whether our model of automata leads to a general framework to decidability results for various models of automata (e.g., pushdown automata, vector addition systems).

Apart from the mentioned references, we note that the current paper refines and extends the approach taken in [11]. We also mention the papers [18,19] that, motivated by the approach in [11], develop the theory of automata over the fields of reals and complex numbers. We note that the current paper also addresses some topics discussed in [14]. It could be interesting to address simulation issues for our models of automata as for finite automata, as in [7,8].

References

1. Alur, R., Černý, P., Weinstein, S.: Algorithmic analysis of array-accessing programs. In: Grädel, E., Kahle, R. (eds.) CSL 2009. LNCS, vol. 5771, pp. 86–101. Springer, Heidelberg (2009)
2. Blum, L., Shub, M., Smale, S.: On a theory of computation and complexity over the real numbers: NP-completeness, recursive functions and universal machines. Bull. Am. Math. Soc. **21**(1), 1–46 (1989)
3. Bojanczyk, M., Muscholl, A., Schwentick, T., Segoufin, L., David, C.: Two-variable logic on words with data. In: Proceedings of LICS 2006, pp. 7–16 (2006)
4. Bojanczyk, M., David, C., Muscholl, M., Schwentick, T., Segoufin, L.: Two-variablelogic on data trees and XML reasoning. In: Proceedings of PODS 2006, pp. 10–19 (2006)
5. Bournez, O., Cucker, F., de Naurois, P.J., Marion, J.-Y.: Computability over an arbitrary structure. Sequential and parallel polynomial time. In: Gordon, A.D. (ed.) FOSSACS 2003. LNCS, vol. 2620, pp. 185–199. Springer, Heidelberg (2003)
6. Bozga, M., Iosif, R., Lakhnech, Y.: Flat parametric counter automata. In: Bugliesi, M., Preneel, B., Sassone, V., Wegener, I. (eds.) ICALP 2006. LNCS, vol. 4052, pp. 577–588. Springer, Heidelberg (2006)
7. Calude, C., Calude, E., Khoussainov, B.: Deterministic automata: simulation and minimality. Ann. Pure Appl. Logic **90**(1–3), 263–276 (1997)
8. Calude, C., Calude, E., Khoussainov, B.: Finite nondeterministic automata: simulation and minimality. Theor. Comput. Sci. **242**(1–2), 219–235 (2000)
9. Comon, S., Jurski, Y.: Multiple counters automata, safety analysis and Presburger arithmetic. In: Hu, A.J., Vardi, M.Y. (eds.) CAV 1998. LNCS, vol. 1427, pp. 268–279. Springer, Heidelberg (1998)
10. Figueira, D.: Reasoning on words and trees with data. Ph.D. thesis, ENS Cachan, France (2010)

11. Gandhi, A., Khoussainov, B., Liu, J.: Finite automata over structures (Extended Abstract). In: Agrawal, M., Cooper, S.B., Li, A. (eds.) TAMC 2012. LNCS, vol. 7287, pp. 373–384. Springer, Heidelberg (2012)
12. Ibarra, O.: Reversal-bounded multicounter machines and their decision problems. J. ACM **25**(1), 116–133 (1978)
13. Kaminsky, M., Francez, N.: Finite memory automata. Theor. Comput. Sci. **134**(2), 329–363 (1994)
14. Khoussainov, B., Nerode, A.: Open questions in the theory of automatic structures. Bull. Eur. Assoc. Theor. Comput. Sci. (EATCS) (94):181–204 (2008)
15. Leroux, J., Sutre, G.: Flat counter automata almost everywhere!. In: Peled, D.A., Tsay, Y.-K. (eds.) ATVA 2005. LNCS, vol. 3707, pp. 489–503. Springer, Heidelberg (2005)
16. Matiyasevich, Y.: Hilbert's Tenth Problem. MIT Press, Massachusetts (1993)
17. Minsky, M.: Recursive unsolvability of post's problem of "Tag" and other topics in theory of turing machines. Ann. Math. **74**(3), 437–455 (1961)
18. Meer, K., Naif, A.: Generalised finite automata over real and complex numbers. Theor. Comput. Sci. **591**(C), 86–98 (2015)
19. Meer, K., Naif, A.: Periodic generalized automata over the reals. In: Dediu, A.-H., et al. (eds.) LATA 2016. LNCS, vol. 9618, pp. 168–180. Springer, Heidelberg (2016). doi:10.1007/978-3-319-30000-9_13
20. Neven, F., Schwentick, T., Vianu, V.: Finite state machines for strings over infinite alphabets. ACM Trans. Comput. Logic **15**(3), 403–435 (2004)
21. Segoufin, L.: Automata and logics for words and trees over an infinite alphabet. In: Ésik, Z. (ed.) CSL 2006. LNCS, vol. 4207, pp. 41–57. Springer, Heidelberg (2006)
22. Segoufin, L., Torunczyk, S.: Automata based verification over linearly ordered data domains. In: Proceedings of STACS, pp. 81–92. Schloss Dagstuhl - Leibniz-Zentrum fuer Informatik (2011)
23. Tan, T.: Graph reachability and pebble automata over infinite alphabets. In: Proceedings of LICS, pp. 157–166. IEEE Computer Society (2009)

Regular Papers

The Degree of Irreversibility in Deterministic Finite Automata

Holger Bock Axelsen[1], Markus Holzer[2]($\boxtimes$), and Martin Kutrib[2]

[1] Department of Computer Science, University of Copenhagen,
Universitetsparken 5, Copenhagen, Denmark
funkstar@di.ku.dk
[2] Institut für Informatik, Universität Giessen, Arndtstr. 2, 35392 Giessen, Germany
{holzer,kutrib}@informatik.uni-giessen.de

Abstract. Recently, Holzer *et al.* gave a method to decide whether the language accepted by a given deterministic finite automaton (DFA) can also be accepted by some *reversible* deterministic finite automaton (REV-DFA), and eventually proved NL-completeness. Here, we show that the corresponding problem for *nondeterministic* finite state automata (NFA) is PSPACE-complete. The recent DFA method essentially works by minimizing the DFA and inspecting it for a *forbidden pattern*. We here study the *degree of irreversibility* for a regular language, the minimal number of such forbidden patterns necessary in *any* DFA accepting the language, and show that the degree induces a strict infinite hierarchy of languages. We examine how the degree of irreversibility behaves under the usual language operations union, intersection, complement, concatenation, and Kleene star, showing tight bounds (some asymptotically) on the degree.

1 Introduction

In computation theory, reversibility is the property that computations are both forward and backward deterministic. In the context of finite state models, reversibility can usually be verified by simple inspection of the transition function, ensuring that the induced computation step relation is an injective function on configurations. Despite the apparent simplicity of reversible computations, reversibility is an interesting property that has been studied in a wide array of contexts, including the thermodynamics of computation [7], across a wide array of automata models [9], and even in robotics [10].

It is well-known that the reversibly regular languages, i.e., the languages accepted by reversible deterministic finite automata (REV-DFA), form a strict subclass of the regular languages, see, e.g., [6]. However, the exact cost of reversibility is still not well-understood: for example, changing from one-way

The authors acknowledge partial support from COST Action IC1405 *Reversible Computation*. H.B. Axelsen was supported by the Danish Council for Independent Research | Natural Sciences under the *Foundations of Reversible Computing* project, and by a COST IC1405 STSM (short-term scientific mission) grant.

Y.-S. Han and K. Salomaa (Eds.): CIAA 2016, LNCS 9705, pp. 15–26, 2016.
DOI: 10.1007/978-3-319-40946-7_2

to two-way tapes is sufficient to collapse the classes [5]. Likewise, adding a reversible transducer in front of the REV-DFA also collapses to the regular languages [1]. This motivates further study into the relationship between the regular and reversibly regular languages, and in particular towards developing methods to understand and bridge the gap in terms of the internal structure of deterministic finite automata (DFA). In this paper, we take steps in this direction.

Recently, Holzer *et al.* showed a method for deciding if the language accepted by a given DFA can also be recognized by some REV-DFA [4]. It was also shown that this is an NL-complete problem, and a decision method was given, which essentially works by minimizing the DFA and inspecting it for the presence of a *forbidden pattern*. If this pattern is present in the minimal DFA, then there is *no* REV-DFA that can accept the same language, and if not, then there *is*. What makes this particularly interesting is that the pattern is structurally more complex than the simplest violation of reversibility (see Sect. 2 for details). This suggests that the forbidden pattern captures an essential aspect of irreversibility, and offers an approach to studying the gap between the reversibly regular and regular languages based on the absence, presence, and count of occurrences, of this pattern.

Our contributions are as follows. We show that the generalization of the problem studied in [4] to *nondeterministic* finite automata (NFA), i.e., the *regular reversibility problem* of whether the language accepted by a given NFA is reversibly regular, is PSPACE-complete. Turning to DFAs, we introduce the notion of *degree of irreversibility* for DFAs, essentially the number of occurrences of the forbidden pattern in a given DFA, and extend this to (regular) languages by minimizing over all DFAs accepting the language. Finally, we show that the degree of irreversibility induces a strict, infinite hierarchy of languages. We then proceed to show exact bounds on the degree of irreversibility under the common language operations union, intersection, and complement, and asymptotically tight bounds for concatenation and Kleene star.

The paper is organized as follows. Section 2 covers the necessary preliminaries. In Sect. 3 we show that the regular reversibility problem is PSPACE-complete. Section 4 defines the degree of irreversibility, and shows the related hierarchy. We present the degree complexity results for common language operations in Sect. 5. Most proofs are omitted due to space constraints, and will be given in the full version of the paper.

2   Preliminaries

An *alphabet* Σ is a non-empty finite set, its elements are called *letters* or *symbols*. We write Σ^* for the *set of all words* over the finite alphabet Σ.

We recall some definitions on finite automata as contained, for example, in [3]. A *deterministic finite automaton* (DFA) is a 5-tuple $A = (Q, \Sigma, \delta, q_0, F)$, where Q is the finite set of *internal states*, Σ is the alphabet of *input symbols*, $q_0 \in S$ is the *initial state*, $F \subseteq Q$ is the set of *accepting states*, and $\delta \colon Q \times \Sigma \to Q$ is the partial *transition function*. Note that here the transition function is not required to be *total*. The *language accepted* by A is $L(A) = \{\, w \in \Sigma^* \mid \delta(q_0, w) \in F \,\}$, where the transition function is recursively extended to $\delta \colon Q \times \Sigma^* \to Q$. By $\delta^R \colon Q \times \Sigma \to 2^Q$, with $\delta^R(q, a) = \{\, p \in S \mid \delta(p, a) = q \,\}$, we denote the *reverse* transition function of δ. Similarly, also δ^R can be extended to words instead of symbols. Two devices A and A' are said to be *equivalent* if they accept the same language, that is, $L(A) = L(A')$.

Let $A = (Q, \Sigma, \delta, q_0, F)$ be a DFA accepting the language L. The set of words $R_{A,q} = \{\, w \in \Sigma^* \mid \delta(q, w) \in F \,\}$ refers to the *right language* of the state q in A. In case $R_{A,p} = R_{A,q}$, for some states $p, q \in Q$, we say that p and q are *equivalent* and write $p \equiv_A q$. The equivalence relation $\equiv_A$ partitions the state set Q of A into equivalence classes, and we denote the equivalence class of $q \in S$ by $[q] = \{\, p \in S \mid p \equiv_A q \,\}$. Equivalence can also be defined between states of different automata: a state p of DFA A and a state q of DFA A' are *equivalent*, denoted by $p \equiv q$, if $R_{A,p} = R_{A',q}$.

A state $p \in Q$ is *accessible* in A if there is a word $w \in \Sigma^*$ such that $\delta(q_0, w) = p$, and it is *productive* if there is a word $w \in \Sigma^*$ such that $\delta(p, w) \in F$. If p is both accessible and productive then we say that p is *useful*. In this paper we only consider automata with all states useful. Let A and A' be two equivalent DFAs. Observe that if p is a useful state in A, then there exists a useful state p' in A', with $p \equiv p'$. A DFA is *minimal* (among all DFAs) if there does not exist an equivalent DFA with fewer states. It is well known that a DFA is minimal if and only if all its states are useful and inequivalent.

Next we define reversible DFAs. Let $A = (Q, \Sigma, \delta, q_0, F)$ be a DFA. A state $r \in Q$ is said to be *irreversible* if there are two distinct states p and q in Q and a letter $a \in \Sigma$ such that $\delta(p, a) = r = \delta(q, a)$. Then a DFA is *reversible* if it does not contain any irreversible state. In this case the automaton is said to be a *reversible* DFA (REV-DFA). Equivalently the DFA A is reversible, if every letter $a \in \Sigma$ induces an *injective partial mapping* from Q to itself *via* the mapping $\delta_a \colon Q \to Q$ with $p \mapsto \delta(p, a)$. In this case, the reverse transition function δ^R can be seen as a (partial) injective function $\delta^R \colon Q \times \Sigma \to Q$. Notice that if p and q are two distinct states in a REV-DFA, then $\delta(p, w) \neq \delta(q, w)$, for all words $w \in \Sigma^*$. Finally, a REV-DFA is *minimal* (among all REV-DFAs) if there is no equivalent REV-DFA with a smaller number of states.

In [4] the following structural characterization of regular languages that can be accepted by REV-DFAs in terms of their minimal DFAs is given. The conditions of the characterization are illustrated in Fig. 1.

Theorem 1. *Let $A = (Q, \Sigma, \delta, q_0, F)$ be a* minimal *deterministic finite automaton. The language $L(A)$ can be accepted by a reversible deterministic finite automaton if and only if there do not exist useful states $p, q \in Q$, a letter $a \in \Sigma$, and a word $w \in \Sigma^*$ such that $p \neq q$, $\delta(p, a) = \delta(q, a)$, and $\delta(q, aw) = q$.*

Finally we need some notations on computational complexity theory. We classify problems on DFAs with respect to their computational complexity. Consider the complexity class NL (PSPACE, respectively) which refers to the set

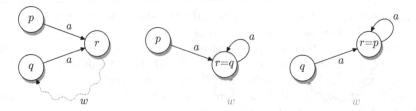

Fig. 1. The "forbidden pattern" of Theorem 1: the language accepted by a minimal DFA A can be accepted by a REV-DFA if and only if A does not contain the structure depicted on the left. Here the states p and q must be distinct, but state r could be equal to state p or state q. The situations where $r = q$ or $r = p$ are shown in the middle and on the right, respectively—here the word w and its corresponding path are grayed out because they are not relevant: in the middle, the word w that leads from r to q is not relevant since it can be identified with the a-loop on state $r = q$. Also on the right hand side, word w is not important because we can simply interchange the roles of the states q and $r = p$.

of problems accepted by nondeterministic logspace bounded (polynomial space, respectively) Turing machines. Further, hardness and completeness is always meant with respect to deterministic logspace bounded reducibility, unless otherwise stated.

3   Complexity of the Regular Reversibility Problem

In [4] it was shown that the regular language reversibility problem—given a DFA A, decide whether $L(A)$ is accepted by any REV-DFA—is NL-complete. If the regular language is given by an NFA or a regular expression, the problem becomes intractable.

Theorem 2. *The regular language reversibility problem is* PSPACE-*complete, if the language is given as a nondeterministic finite automaton or a regular expression.*

Before we can prove this result we need a technical lemma, which will be used in the PSPACE-hardness argument later.

Lemma 3. *Let $A = (Q, \Sigma, \delta, q_0, F)$ be a minimal DFA. If there is a state $q \in Q$, other than the initial state, such that $R_{A,q} = \Sigma^*$, then $L(A)$ is irreversible.* □

Let $L \subseteq \Sigma^*$. Then the *left derivative* of L with respect to the letter a in Σ is the set $a^{-1} \cdot L = \{ w \mid aw \in L \}$. This notation generalizes to words. By this definition, there is an obvious relation between these left derivative set and the states of the minimal finite automaton A accepting L. To be more precise, the set $u^{-1} \cdot L$, for $u \in \Sigma^*$, is a description of the state $q_u = \delta(q_0, u)$, where $A = (Q, \Sigma, \delta, q_0, F)$, and *vice versa*. Now we are ready to proof Theorem 2 in a convenient way.

Proof (of Theorem 2). The containment within PSPACE is easily seen. For the hardness we reduce the PSPACE-complete universality problem for regular expressions [8] to the reversibility problem for NFAs or regular expressions. Let the regular expression r be an instance of the universality problem. We may assume that r is an expression over the alphabet $\Sigma = \{a, b\}$. Then we construct the expression

$$s = a \cdot r + b \cdot \Sigma^* + \lambda$$

or equivalently the NFA depicted in Fig. 2 in deterministic logspace. Now assume that $L(r) = \Sigma^*$. Then it is easy to see that $L(s) = \Sigma^*$, too, and therefore a reversible language. On the other hand, if $L(r) \neq \Sigma^*$, then there is a word $u \notin L(r)$. From this it follows that $au \notin L(s)$ but $bu \in L(s)$. Thus we conclude that the states $a^{-1} \cdot L(s)$ and $b^{-1} \cdot L(s)$ are not equivalent in the DFA accepting $L(s)$. Moreover, in that DFA states $L(s)$ and $b^{-1} \cdot L(s)$ are not equivalent, too. Note that the former state is the initial state of the DFA that accepts $L(s)$. Since the right language of the state $b^{-1} \cdot L(s)$ is equal to Σ^* and it is not equal to the initial state, Lemma 3 applies, and the language $L(s)$ is *not* reversible. This proves PSPACE-hardness. □

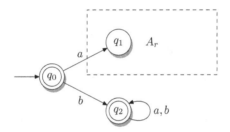

Fig. 2. Finite automaton that accepts the language $L(s)$. It is built from the regular expression r, where A_r is an NFA with initial state q_1 that accepts the language $L(r)$.

4   On the Degree of Irreversibility

For an automaton A we define its *degree of irreversibility* $d(A)$ as the number of irreversible states that are part of one of the forbidden patterns shown in Fig. 1. Observe, that since our DFAs need not to be complete and only contain useful states, the non-accepting sink state does not count for the degree of irreversibility. This notation is generalized to languages in the usual way. This means, for a regular language $L \subseteq \Sigma^*$ we define its *degree of irreversibility* $d(L)$ as the minimum degree of irreversibility among all equivalent DFAs A, that is,

$$d(L) = \min\{ d(A) \mid A \text{ is a DFA with } L(A) = L \}.$$

The next example explains our notation.

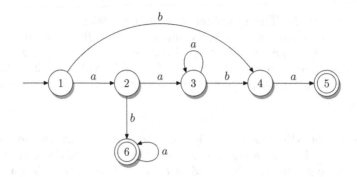

Fig. 3. DFA which accepts $aba^* + a^*ba$ that has irreversibility degree one.

Example 4. Consider the following DFA depicted in Fig. 3, which accepts the union of aba^* and a^*ba. This automaton has irreversibility degree one by state 3. Note that although state 4 has two ingoing b-transitions, this state does not yield a forbidden pattern as shown in Fig. 1. There is no word that leads from state 4 to either state 1 or 3. Moreover, the language accepted by this automaton, which is $aba^* + a^*ba$ is also of irreversibility degree one, since it is *not* reversible by Theorem 1.   □

Next we consider the hierarchy on regular languages that is induced by the irreversibility degree. To this end let

$$\text{IREV}_k\text{-DFA} = \{\, A \mid A \text{ is a DFA and } d(A) \leq k \,\}.$$

We have $\text{IREV}_0\text{-DFA} = \{\, A \mid A \text{ is a reversible DFA} \,\}$ and thus the equality $\mathscr{L}(\text{IREV}_0\text{-DFA}) = \mathscr{L}(\text{REV-DFA})$ holds, where the family of all languages accepted by an automaton of some type X is denoted by $\mathscr{L}(X)$. Moreover, by definition the inclusion $\text{IREV}_k\text{-DFA} \subseteq \text{IREV}_{k+1}\text{-DFA}$ follows and therefore the corresponding language classes satisfy $\mathscr{L}(\text{IREV}_k\text{-DFA}) \subseteq \mathscr{L}(\text{IREV}_{k+1}\text{-DFA})$, for $k \geq 0$. By the example above we have

$$\mathscr{L}(\text{REV-DFA}) = \mathscr{L}(\text{IREV}_0\text{-DFA}) \subset \mathscr{L}(\text{IREV}_1\text{-DFA}).$$

Before we show that the degree of irreversibility induces an infinite strict hierarchy we need some tool that allows us to determine the irreversibility degree for an arbitrary regular language. Since for the degree of irreversibility of a language L we quantify over all equivalent DFAs we have to show that we cannot trade more states for less irreversibility. The following example shows that this is in fact not the case in general.

Example 5. Consider the substructure of a DFA as depicted in Fig. 4. It is not hard to see that this pattern may appear in a *minimal* DFA. Both states r_1 and r_2 in the substructure are irreversible. By splitting both of these states, we obtain a connecting structure as shown in Fig. 5. The structure has *one* irreversible state only. Thus, the irreversibility degree of a *minimal* DFA is not necessarily the irreversibility degree of the language under consideration.   □

The Degree of Irreversibility in Deterministic Finite Automata

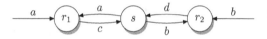

Fig. 4. Substructure of a DFA containing two irreversible states r_1 and r_2.

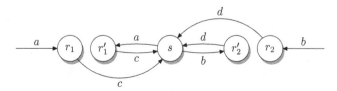

Fig. 5. Substructure of a DFA with just one irreversible state s obtained after splitting both irreversible states.

For a special class of finite automata, we can show that the minimal DFA already gives the degree of irreversibility. A DFA is *simply-irreversible* if all irreversible states are of the form depicted in the middle and right drawing shown in Fig. 1. That is, the irreversibility state is entered by an a-transition and has an a-self-loop, which is the simplest form of irreversibility. For the languages accepted by these automata we can prove the next result.

Theorem 6. *Let L be a regular language and A be its minimal deterministic finite automaton. If A is simply-irreversible, then the degree of irreversibility of A is equal to the irreversibility degree for L. That is $d(L) = d(A)$.* □

Now we are ready to show that the strict hierarchy on regular languages induced by the irreversibility degree is tight and infinite.

Theorem 7. *For all $k \geq 0$, $\mathscr{L}(IREV_k\text{-}DFA) \subset \mathscr{L}(IREV_{k+1}\text{-}DFA)$.*

Proof. Consider the languages L_k over the alphabet $\{a, b\}$ defined as follows: for $k \geq 0$ set

$$L_{2k} = (aa^*bb^*)^k \quad \text{and} \quad L_{2k+1} = (aa^*bb^*)^k aa^*.$$

The language L_k, for $k \geq 0$, is accepted by the DFA $A_k = (Q_k, \{a, b\}, \delta_k, q_0, F_k)$ with $Q_k = \{1, 2, \ldots, k+1\}$, $q_0 = 1$, $F_k = \{k+1\}$, and

$$\delta(i, a) = \begin{cases} i+1 & \text{if } i \text{ is odd and } 1 \leq i < k+1 \\ i & \text{if } i \text{ is even and } 1 < i \leq k+1 \end{cases}$$

and

$$\delta(i, b) = \begin{cases} i+1 & \text{if } i \text{ is even and } 1 \leq i < k+1 \\ i & \text{if } i \text{ is odd and } 1 < i \leq k+1. \end{cases}$$

By construction the DFA A_k is minimal and simply-irreversible. Thus, by the previous theorem the degree of irreversibility of A_k is equal to the irreversibility degree of the language L_k. Since A_k contains exactly k irreversible states, we have $d(L_k) = k$. This shows that $L_k \in \mathscr{L}(\text{IREV}_k\text{-DFA}) \setminus \mathscr{L}(\text{IREV}_{k-1}\text{-DFA})$, for $k \geq 1$.   □

Finally, we consider unary regular languages and their irreversibility degree. It is not difficult to see that a unary *complete* DFA consists of a path, which starts from the initial state, followed by a cycle of one or more states. Thus the irreversibility degree of any unary DFA is at most one. Thus, the hierarchy on the irreversibility degree collapses to its second level and $\mathscr{L}(\text{IREV}_1\text{-DFA}) \cap 2^{\{a\}^*}$ is already equal to the class of all unary regular languages. Moreover, we conclude that $\mathscr{L}(\text{IREV}_0\text{-DFA}) \cap 2^{\{a\}^*}$ is the class of languages that contains only finite or cyclic unary regular languages. Here a unary regular language is *cyclic* if it is accepted by a unary DFA which is a cycle of one or more states.

5   Operations on Languages and Degree of Irreversibility

In this section we study the descriptional complexity of the operation problem for reversible languages. We start with the Boolean operations and continue with the concatenation and Kleene star operation.

First we consider the union operation. For the union of two reversible language, the increase of the degree of irreversibility is linear in the sum of the number of states of the involved automata. This can be seen in the next theorem.

Theorem 8. *Let $m, n \geq 1$ be two integers, A be an m-state and B be an n-state reversible deterministic finite automaton. Then the degree $m+n$ of irreversibility for the language $L(A) \cup L(B)$ is sufficient and necessary in the worst case.*

Proof. Let $A = (Q_A, \Sigma, \delta_A, q_{0,A}, F_A)$ and $B = (Q_B, \Sigma, \delta_B, q_{0,B}, F_B)$. In order to accept the union of $L(A)$ and $L(B)$ we apply the standard cross-product construction. To this end define $C = (Q_C, \Sigma, \delta_C, q_{0,C}, F_C)$, where

$$Q_C = (Q_A \times Q_B) \cup (Q_A \times \{-\}) \cup (\{ \ \} \times Q_B),$$

$q_{0,C} = (q_{0,A}, Q_{0,B})$, and $F_C = (Q_A \times F_B) \cup (F_A \times Q_B) \cup (F_A \times \{-\}) \cup (\{-\} \times F_B)$. The transition function δ_C is set to

$$\delta_C((p,q), a) =$$
$$\begin{cases} (\delta_A(p,a), \delta_B(q,a)) & \text{if both } \delta_A(p,a) \text{ and } \delta_B(q,a) \text{ are defined} \\ (\delta_A(p,a), -) & \text{if } \delta_A(p,a) \text{ is defined and } \delta_B(q,a) \text{ is undefined} \\ (-, \delta_B(q,a)) & \text{if } \delta_A(p,a) \text{ is undefined and } \delta_B(q,a) \text{ is defined} \end{cases}$$

and furthermore $\delta_C((p,-), a) = (\delta_A(p,a), -)$, if $\delta_A(p,a)$ is defined, as well as $\delta_C((-,q), a) = (-, \delta_B(q,a))$, if $\delta_B(q,a)$ is defined, for $a \in \Sigma$. So we have

$L(C) = L(A) \cup L(B)$. From the $m \cdot n + m + n$ states of C at most $m + n$ are irreversible. To be more precise, none of the states from $Q_A \times Q_B$ are irreversible. This is seen as follows: consider a state $(r, r') \in Q_A \times Q_B$. Assume to the contrary that (r, r') is irreversible. Then there are different states (p, p') and (q, q') with $\delta_C((p, p'), a) = (r, r') = \delta_C((q, q'), a)$, for some $a \in \Sigma$. Since (p, p') is not equal to (q, q') we have $p \neq q$ or $p' \neq q'$. We only consider the case $p \neq q$ by symmetric reasons. But then we find that r is an irreversible state, because $\delta_A(p, a) = r = \delta_A(q, a)$, for the letter a from above. This is a contradiction, because automaton A is a reversible DFA. It is worth mentioning that a similar argumentation does not apply to states of the form $(r, -)$ or $(-, r)$. This is seen by the counterexample $\delta_C((r, p), a) = (r, -) = \delta_C((r, -), a)$, for some $a \in \Sigma$, which induces only $\delta_A(r, a) = r$ and $\delta_B(p, a)$ is undefined—an analogous example can be given for state of the form $(-, r)$. Hence, this does not contradict the irreversibility of either A or B.

It remains to be shown that the bound $m + n$ is tight. Define the reversible DFA $A = (Q_A, \{a, b\}, \delta_A, q_0, F)$ with $Q_A = \{1, 2, \ldots, m\}$, $q_0 = 1$, $F = \{m\}$, and the transition function is given by $\delta(i, a) = i + 1$, for $1 \leq i < m$, and $\delta(i, b) = i$, for $1 \leq i \leq m$. The automaton B is the same as A, but with n states, and where the letters a and b are interchanged. The automaton C constructed above is easily seen to be minimal.

Finally we show that all states of the form $(i, -)$ and $(-, j)$, for $1 \leq i \leq m$ and $1 \leq j \leq n$, are irreversible and yield a forbidden pattern as shown in Fig. 1. The below given argument shows even more, namely that the automaton C is simply-irreversible. We have already argued that the state (i, n) is accessible. Then it is easy to see that from state (i, n) reading a b the automaton C enters state $(i, -)$, which has a b loop. Therefore state $(i, -)$ is simply-irreversible. A similar argument shows that state $(-, j)$ is simply-irreversible as well. By Theorem 6 the stated claim follows.   □

A careful inspection of the previous proof reveals that we can use parts of it for the intersection of two reversible languages. For two reversible automata A and B we construct an automaton C by the cross-product construction described in the proof of Theorem 8 but only using states of the form $Q_A \times Q_B$ and by altering the set of accepting states to be $F = F_A \times F_B$. Then $L(C) = L(A) \cap L(B)$. It was shown that none of the states from $Q_A \times Q_B$ are irreversible. Hence C does not contain any irreversible state. Thus, we have shown the following result.

Theorem 9. *Let $m, n \geq 1$ be two integers, A be an m-state and B be an n-state reversible deterministic finite automaton. Then the language $L(A) \cap L(B)$ is accepted by a reversible deterministic finite automaton.*   □

Next we deal with the complementation operation, and show that the degree of irreversibility can be increased by one.

Theorem 10. *Let $n \geq 1$ be an integers and A be an n-state reversible deterministic finite automaton. Then the degree 1 of irreversibility for the complement of $L(A)$ is sufficient and necessary in the worst case.*   □

In the remainder of this section we investigate the effect of the concatenation and the Kleene star operation on the degree of irreversibility. First we recall the construction of DFAs for the concatenation [11]. Let $A = (Q_A, \Sigma, \delta_A, q_{0,A}, F_A)$ and $B = (Q_B, \Sigma, \delta_B, q_{0,B}, F_B)$ be two DFAs. As in [11] we construct the DFA $C = (Q_C, \Sigma, \delta_C, q_{0,C}, F_C)$, where

$$Q_C = (Q_A \times 2^{Q_B}) \setminus (F_A \times 2^{Q_B \setminus \{q_{0,B}\}}),$$

the initial state is

$$q_{0,C} = \begin{cases} (q_{0,A}, \emptyset) & \text{if } q_{0,A} \notin F_A \\ (q_{0,A}, \{q_{0,B}\}) & \text{otherwise,} \end{cases}$$

the final states are

$$F_C = \{ (p, P) \mid (p, P) \in Q_C \text{ and } P \cap F_B \neq \emptyset \},$$

and the transition function is defined by $\delta_C((p, P), a) = (q, Q)$, for $a \in \Sigma$, where $q = \delta_A(p, a)$ and

$$Q = \begin{cases} \delta_B(P, a) \cup \{q_{0,B}\} & \text{if } q \in F_A \\ \delta_B(P, a) & \text{otherwise.} \end{cases}$$

Clearly, automaton C accepts $L(A) \cdot L(B)$ and has at most $m \cdot 2^n - 2^{n-1}$ states. Thus, the construction gives rise to an exponential upper bound on the number of irreversible states.

Theorem 11. *Let $m, n \geq 2$ be two integers, A be an m-state and B be an n-state reversible deterministic finite automaton. Then the degree $m \cdot 2^n - 2^{n-1}$ of irreversibility is sufficient for a deterministic finite automaton to accept the language $L(A) \cdot L(B)$.* $\quad\square$

The next theorem gives an exponential lower bound on the degree of irreversibility for the concatenation operation.

Theorem 12. *Let $m, n \geq 2$ be two integers. There are a reversible m-state deterministic finite automaton A and a reversible n-state deterministic finite automaton B such that any deterministic finite automaton accepting $L(A) \cdot L(B)$ has at least the degree $(3m - 2) \cdot 2^{n-2}$ of irreversibility.*

Proof. Define the left automaton to be $A = (Q_A, \{a, b, c, d\}, \delta_A, q_{0,A}, F_A)$ with $Q_A = \{0, 1, \ldots, m - 1\}$, initial state $q_{0,A} = 0$, final states $F_A = \{m - 1\}$, and the transition function

$$\delta_A(i, a) = \begin{cases} i + 1 & \text{if } 0 \leq i < m - 1 \\ 0 & \text{otherwise} \end{cases} \quad \text{and} \quad \delta_A(i, b) = \delta_A(i, c) = \delta_A(i, d) = i$$

for $0 \leq i \leq m - 1$. The right automaton is $B = (Q_B, \{a, b, c, d\}, \delta_B, q_{0,B}, F_B)$ with $Q_B = \{0, 1, \ldots, n - 1\}$, initial state $q_{0,B} = 0$, final states $F_A = \{0\}$, and the transition function

$$\delta_B(i, a) = i, \quad \text{for } 0 \le i \le n - 1, \text{ and } \quad \delta_B(i, b) = \begin{cases} i+1 & \text{if } 0 \le i < n - 1 \\ 0 & \text{otherwise,} \end{cases}$$

and

$$\delta_B(i, c) = i, \quad \text{for } 0 < i \le n - 1, \text{ and } \quad \delta_B(i, d) = i, \quad \text{for } i = 0 \text{ or } 2 \le i \le n - 1.$$

Both reversible automata are depicted in Fig. 6.

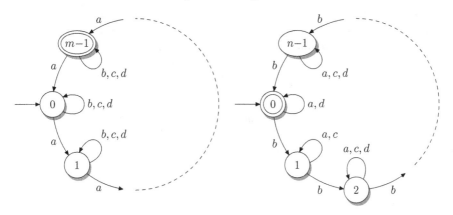

Fig. 6. The reversible automata A (left) and B (right) with m and n states, respectively, that witness the irreversibility degree lower bound for the concatenation operation.

We construct the DFA C for the concatenation $L(A) \cdot L(B)$ as described above. In order to apply Theorem 6 we need to show that C is minimal. Thus, one has to verify that every state in C is useful and defines a distinct equivalence class.

Finally, it remains to determine the lower bound on the irreversibility degree of C. We show that all states of C whose second component does not contain 0 and 1 at the same time are simply-irreversible. We have already seen that all states of the form $(p, P \cup \{0\})$ and $(p, P \cup \{1\})$ are reachable in C. We distinguish two cases:

1. Assume $p = m - 1$. Then $0 \in P$, but then by assumption $1 \notin P$. We have $\delta_C((p, P \cup \{1\}), d) = (p, P)$ and $\delta_C((p, P), d) = (p, P)$. Thus (p, P) is simply-irreversible.
2. Let $p = i$ with $0 \le i < m - 1$. If $0 \notin P$, then $\delta_C((p, P \cup \{0\}), c) = (p, P)$ and $\delta_C((p, P), c) = (p, P)$. Also in the case $1 \notin P$, the two transitions $\delta_C((p, P \cup \{1\}), d) = (p, P)$ and $\delta_C((p, P), d) = (p, P)$ follow. In both cases the state (p, P) is simply-irreversible.

Next we count the number of simply-irreversible states. The first item above induces 2^{n-2} possibilities, and the second item $3(m-1) \cdot 2^{n-2}$. There are $(m-1)$ choices for p and the number of different sets P that do not contain 0 or 1 is $3 \cdot 2^{n-2}$. For each of the cases (i) both 0 and 1 are not in P, (ii) element 0 is in P but 1 is not, and (iii) element 0 is not in P but 1 is member of P, there are 2^{n-2} possibilities. This results in $3(m-1) \cdot 2^{n-2}$ sets for the second item above. Putting things together results in in at least $(3m-2) \cdot 2^{n-2}$ simply-irreversible states in C. By Theorem 6 the stated claim follows.   □

Finally, we consider the Kleene star operation. From [11] the tight worst case bound for a DFA to accept the Kleene closure of an n-state DFA language is $2^{n-1} + 2^{n-2}$. Thus, the upper bound for the irreversibility degree for the Kleene closure is exponential.

Theorem 13. *Let $n \geq 2$ be an integers and A be an n-state reversible deterministic finite automaton. Then the degree $2^{n-1} + 2^{n-2}$ of irreversibility is sufficient for a deterministic finite automaton to accept the language $L(A)^*$.*   □

As in the case of the concatenation operation we can provide an exponential lower bound.

Theorem 14. *Let $n \geq 3$ be an integer. There is a reversible n-state deterministic finite automaton A such that any deterministic finite automaton accepting $L(A)^*$ has at least the degree $3 \cdot 2^{n-3} - 1$ of irreversibility.*   □

References

1. Axelsen, H.B., Kutrib, M., Malcher, A., Wendlandt, M.: Boosting reversible pushdown machines by preprocessing. In: RC 2016, LNCS. Springer (2016)
2. Glushkov, V.M.: The abstract theory of automata. Russ. Math. Surv. **16**, 1–53 (1961)
3. Harrison, M.A.: Introduction to Formal Language Theory. Addison-Wesley, Boston (1978)
4. Holzer, M., Jakobi, S., Kutrib, M.: Minimal reversible deterministic finite automata. In: Potapov, I. (ed.) DLT 2015. LNCS, vol. 9168, pp. 276–287. Springer, Heidelberg (2015)
5. Kondacs, A., Watrous, J.: On the power of quantum finite state automata. In: FOCS 1997, pp. 66–75. IEEE (1997)
6. Kutrib, M.: Aspects of reversibility for classical automata. In: Calude, C.S., Freivalds, R., Kazuo, I. (eds.) Gruska Festschrift. LNCS, vol. 8808, pp. 83–98. Springer, Heidelberg (2014)
7. Landauer, R.: Irreversibility and heat generation in the computing process. IBM J. Res. Dev. **3**, 183–191 (1961)
8. Meyer, A.R., Stockmeyer, L.J.: The equivalence problem for regular expressions with squaring requires exponential time. In: SWAT 1972, pp. 125–129. IEEE (1972)
9. Morita, K.: Reversible computing and cellular automata–a survey. Theoret. Comput. Sci. **395**(1), 101–131 (2008)
10. Schultz, U.P., Laursen, J.S., Ellekilde, L., Axelsen, H.B.: Towards a domain-specific language for reversible assembly sequences. In: Krivine, J., Stefani, J.-B. (eds.) RC 2015. LNCS, vol. 9138, pp. 111–126. Springer, Switzerland (2015)
11. Yu, S., Zhuang, Q., Salomaa, K.: The state complexity of some basic operations on regular languages. Theoret. Comput. Sci. **125**, 315–328 (1994)

Deterministic Stack Transducers

Suna Bensch[1], Johanna Björklund[1], and Martin Kutrib[2(✉)]

[1] Department of Computing Science, Umeå University, 90187 Umeå, Sweden
{suna,johanna}@cs.umu.se
[2] Institut für Informatik, Universität Giessen, Arndtstr. 2, 35392 Giessen, Germany
kutrib@informatik.uni-giessen.de

Abstract. We introduce and investigate stack transducers, which are one-way stack automata with an output tape. A one-way stack automaton is a classical pushdown automaton with the additional ability to move the stack head inside the stack without altering the contents. For stack transducers, we distinguish between a digging and a non-digging mode. In digging mode, the stack transducer can write on the output tape when its stack head is inside the stack, whereas in non-digging mode, the stack transducer is only allowed to emit symbols when its stack head is at the top of the stack. These stack transducers have a motivation from natural language interface applications, as they capture long-distance dependencies in syntactic, semantic, and discourse structures. We study the computational capacity for deterministic digging and non-digging stack transducers, as well as for their non-erasing and checking versions. We finally show that even for the strongest variant of stack transducers the stack languages are regular.

1 Introduction

Natural language interfaces are prevalent. We encounter them as automated booking services, as question-answering systems, and as intelligent personal assistants (Apple's Siri and Microsoft's Cortana belong to this category). As of recent, Google can support natural-language queries and exploratory dialogues. If the search engine is asked, in sequence, "Who is the president of the US?", "Where was he born?", "Who is his wife?", and finally "Where was she born?", it will interpret the questions as intended and perform the required anaphora resolution. For example, it will understand that the subject of the last question is the same entity as the second-to-last answer [13].

Natural language interfaces (NLI) have several advantages. They are fast and intuitive to use, and inclusive for social groups such as children, illiterates, and dyslectics. They allow for different modalities to input and output data, for example, microphones, speakers, keyboards, and terminals. For this reason, NLIs are accessible also while performing manual tasks, and open new possibilities for the disabled. On the downside, more is required on the side of the computer to process and represent natural language. In particular, efficient and reliable methods are needed to translate between NL sentences and structured data.

© Springer International Publishing Switzerland 2016
Y.-S. Han and K. Salomaa (Eds.): CIAA 2016, LNCS 9705, pp. 27–38, 2016.
DOI: 10.1007/978-3-319-40946-7_3

In natural language processing, translations are often done by transducers. These are abstract devices that map input strings, trees, or graphs to some target output domain. We find them in, for example, speech processing [12], machine translation [3], and increasingly in dialog systems [9]. A disadvantage of the currently used devices is that they cannot capture long-distance dependencies, as they interpret input words in the context of a very restricted history. However, the dependency structure is a determinative factor for syntactic, semantic and discourse analyses. In response, we introduce what we believe is a promising alternative, namely finite-state transducers with stacks that can be read, but not written, in their entirety throughout the execution. The aim is a balance between expressive power on the one hand, in particular the ability to model long-distance dependencies, and computational efficiency on the other.

This paper initiates the investigation of stack transducers. We begin with stack transducers in their unweighted and deterministic form, though as the reader will see, this also produces results for more general devices in the passing.

Stack automata were introduced in [5] as a mathematical model of compilation, with a computational power in between that of pushdown automata and Turing machines. The stack automaton in [5] is a generalization of a pushdown automaton, as its input pointer can move to the right or left while reading input symbols, and its stack pointer can move *inside* the stack while reading stack symbols. The interior part of the stack cannot be altered, the operations push and pop are only allowed at the top of the stack. In [6] the authors restrict the stack automaton model to a *one-way* automaton that moves only to the right while reading input symbols. One-way nondeterministic stack automata can be simulated by deterministic linear-bounded automata, so the accepted languages are deterministic context sensitive [8]. Although compilation is a translation process from source code to object code, the authors of [5] focus on the acceptance of the input language.

We introduce stack transducers that are one-way stack automata with an output tape, to compute relations between input and output words. Like in [5], our devices are allowed to read information from the entire stack, but the operations push and pop are only allowed at the top of the stack. The stack pointer can thus move inside the stack, but the interior stack content cannot be altered. If the stack pointer is inside the stack, we say that the stack transducer is in *internal mode*. If the stack pointers scans the top most stack symbol, the transducer is said to be in *external mode*. In external mode the stack transducers can push or pop symbols from the top of the stack, or leave the stack unchanged, and are also allowed to write on the output tape with each operation. For stack transducers in internal mode, we distinguish between a digging and a non-digging[1] mode: In the former, the stack transducer can write on the output tape, whereas in the latter, the stack transducer is not allowed to output symbols. We believe that making the interior of the stack available as a read-only memory improves the expressiveness and space-efficiency of the transduction model, at a relatively low cost in terms of computational complexity.

---

[1] The term 'digging' refers to the intuition of digging up soil from a deep hole.

2   Definitions and Preliminaries

Let Σ^* denote the set of all words over the finite alphabet Σ. The *empty word* is denoted by λ, and we set $\Sigma^+ = \Sigma^* \setminus \{\lambda\}$. The *reversal* of a word w is denoted by w^R. For the *length* of w we write $|w|$. We denote the powerset of a set S by 2^S. We use $\subseteq$ for inclusion, and $\subset$ for proper inclusion.

A one-way stack automaton is a classical pushdown automaton with the additional ability to move the stack head inside the stack without altering the contents. In this way, it is possible to read but not to change the stored information. Well-known variants are so-called non-erasing stack automata, that are not allowed to pop from the stack, and checking stack automata, that are non-erasing stack automata which cannot push any symbols once the head has moved into the stack, even if it has then returned to the top. The devices are called 'transducers' if they are equipped with an output tape, to which they can append symbols during the course of the computation. More formally:

A *deterministic one-way stack transducer (abbreviated 1DSaT)* is a system $M = \langle Q, \Sigma, \Gamma, \Delta, \delta_{ext}, \delta_{int}, q_0, \bot, F \rangle$, where Q is the finite set of *internal states*, Σ is the finite set of *input symbols*, Γ is the finite set of *stack symbols*, Δ is the finite set of *output symbols*, $q_0 \in Q$ is the *initial state*, $\bot \in \Gamma$ is the *initial stack* or *bottom-of-stack symbol*, $F \subseteq Q$ is the set of *accepting states*, δ_{ext} is the *external transition function* mapping $Q \times (\Sigma \cup \{\lambda\}) \times \Gamma$ to $Q \times (\Gamma^* \cup \{-1\}) \times \Delta^*$, ($\delta_{ext}$ is the next move mapping when the stack head is at the top of the stack. Here, -1 refers to the stack head moving down one symbol), δ_{int} is the *internal transition function* mapping $Q \times (\Sigma \cup \{\lambda\}) \times \Gamma$ to $Q \times \{-1, 0, +1\} \times \Delta^*$, ($\delta_{int}$ is the next move mapping when the stack head is inside the stack. Here, 0 means that the stack head does not move, and $+1$ means the stack pointer moves up one cell).

A *configuration* of a stack transducer $M = \langle Q, \Sigma, \Gamma, \Delta, \delta_{ext}, \delta_{int}, q_0, \bot, F \rangle$ at some time $t \geq 0$ is a quintuple (q, w, s, p, u) where $q \in Q$ is the current state, $w \in \Sigma^*$ is the unread part of the input, $s \in \Gamma^*$ gives the current stack content with the topmost symbol left, $1 \leq p \leq |s|$ gives the current position of the stack pointer from the top of the stack, and $u \in \Delta^*$ gives the content of the current output tape. The *initial configuration* for input w is set to $(q_0, w, \bot, 1, \lambda)$.

During the course of its computation, M runs through a sequence of configurations. One step from a configuration to its successor configuration is denoted by $\vdash$, and the reflexive and transitive (resp., transitive) closure of $\vdash$ by $\vdash^*$ (resp., $\vdash^+$). Let $a \in \Sigma \cup \{\lambda\}$, $x \in \Sigma^*$, $Z \in \Gamma$, $z, y \in \Gamma^*$, $u, v \in \Delta^*$, and $p, q \in Q$. We set

1. $(q, ax, Zy, u, 1) \vdash (p, x, zy, uv, 1)$ if $\delta_{ext}(q, a, Z) = (p, z, v)$, (push/pop, stack head on top),
2. $(q, ax, Zy, u, 1) \vdash (p, x, Zy, uv, 2)$ if $\delta_{ext}(q, a, Z) = (p, -1, v)$, (go inside the stack from ext mode),
3. $(q, ax, Zy, u, i) \vdash (p, x, Zy, uv, i - d)$ if $\delta_{int}(q, a, Z) = (p, d, v)$, $2 \leq i \leq |Zy|$, $d \in \{-1, 0, +1\}$, (inside the stack, up, down, stay).

To simplify matters, we require that the bottom-of-stack symbol $\bot$ can neither be pushed onto nor be popped from the stack, and that the stack head never moves below the bottom of the stack. This normal form is always available through effective constructions.

In accordance with the language acceptors, a stack transducer is said to be *non-erasing* (1DNESaT) if it is not allowed to pop from the stack, that is, δ_{ext} maps $Q \times (\Sigma \cup \{\lambda\}) \times \Gamma$ to $Q \times (\Gamma^+ \cup \{-1\}) \times \Delta^*$. A non-erasing stack transducer is *checking* (1DCSaT) if it cannot push any further symbol once the head has moved into the storage. In order to formalize this property, it is sufficient to partition the state set into $Q_1 \cup Q_2$ with $q_0 \in Q_1$ so that once the stack head is moved down, a state from Q_2 is entered and there is no transition from a state in Q_2 to a state in Q_1. That is, δ_{ext} maps $Q_1 \times (\Sigma \cup \{\lambda\}) \times \Gamma$ to $Q_1 \times \Gamma^+ \times \Delta^*$ or to $Q_2 \times \{-1\} \times \Delta^*$, and it maps $Q_2 \times (\Sigma \cup \{\lambda\}) \times \Gamma$ to $Q_2 \times \{-1, 0\} \times \Delta^*$, while δ_{int} is defined only for states from Q_2.

Finally, we distinguish two modes of writing to the output tape. So far, the stack transducers are allowed to write in any step, even if the stack head is not at the top. These transducers are called *digging* stack transducer (or simply, digger). In *non-digging* mode a stack transducer may only write when the stack head is at the top. Formally, this means that δ_{int} maps to $Q \times \{-1, 0, +1\}$. Non-digging stack transducers and their non-erasing and checking versions are abbreviated as ndi-1DSaT, ndi-1DNESaT, and ndi-1DCSaT.

A stack transducer *halts* if the transition function is not defined for the current configuration. It *transforms* an input word $w \in \Sigma^*$ into an output word $v \in \Delta^*$. For a successful computation M has to halt in an accepting state after having read the whole input, otherwise the output is not recorded: $M(w) = v$, where $(q_0, w, \perp, 1, \lambda) \vdash^* (p, \lambda, y, i, v)$ with $p \in F$, $1 \leq i \leq |y|$, and M halts in configuration (p, λ, y, i, v). The *transduction realized by* M, denoted by $\tau(M)$, is the set of pairs $(w, v) \in \Sigma^* \times \Delta^*$ such that $v = M(w)$. If we build the projection on the first components of $\tau(M)$, denoted by $L(M)$, then the transducer degenerates to a deterministic language acceptor.

The family of transductions realized by a device of type X is denoted by $\mathcal{T}(\mathsf{X})$.

In order to clarify our notion we continue with an example.

Example 1. The length-preserving transduction

$$\tau_1 = \{(a^n \$ a^n \$ a^m \$, a^n \$ a^m \$ a^n \$) \mid m \geq 0, n \geq 1\}$$

is realized by the non-digging stack transducer

$$M = \langle \{q_0, q_1, q_2, q_3, q_+\}, \{a, \$\}, \{A, \perp\}, \{a, \$\}, \delta_{\text{ext}}, \delta_{\text{int}}, q_0, \perp, \{q_+\} \rangle,$$

where the transition functions are as follows.

(1) $\delta_{\text{ext}}(q_0, a, \perp) = (q_0, A\perp, a)$

(2) $\delta_{\text{ext}}(q_0, a, A) = (q_0, AA, a)$

(3) $\delta_{\text{ext}}(q_0, \$, A) = (q_1, A, \$)$

(4) $\delta_{\text{ext}}(q_1, a, A) = (q_1, -1, \lambda)$

(5) $\delta_{\text{int}}(q_1, a, A) = (q_1, -1, \lambda)$

(6) $\delta_{\text{int}}(q_1, \$, \perp) = (q_2, +1, \lambda)$

(7) $\delta_{\text{int}}(q_2, \lambda, A) = (q_2, +1, \lambda)$

(8) $\delta_{\text{ext}}(q_2, a, A) = (q_2, A, a)$

(9) $\delta_{\text{ext}}(q_2, \$, A) = (q_3, A, \$)$

(10) $\delta_{\text{ext}}(q_3, \lambda, A) = (q_3, \lambda, a)$

(11) $\delta_{\text{ext}}(q_3, \lambda, \perp) = (q_+, \perp, \$)$

Since δ_{int} never emits a symbol, M is non-digging.

Given an input $a^n\$a^n\$a^m\$$, the ndi-1DSaT M starts to read the prefix a^n with Transitions (1) and (2) whereby A^n is successively pushed onto the stack and a^n is emitted. Then Transition (3) reads and writes the first $\$$ and sends M into state q_1 without changing the stack. State q_1 is used to move the stack head to the bottom of the stack while the next sequence of a's is read (Transitions (4) and (5)). Nothing is written during this phase. If the next $\$$ appears in the input exactly when the stack head reaches the bottom, the input prefix is $a^n\$a^n\$$ and M enters state q_2 with Transition (6). In state q_2 the stack head is moved to the top again (Transition (7)) whereby nothing is written to the output. At the top of the stack transition function δ_{ext} is applied again and M reads and emits the suffix $a^m\$$ with Transitions (8) and (9). The stack content is not changed in this phase. Finally, in state q_3 the stack is successively emptied with λ-moves and a^n is appended to the output tape (Transition (10)). The last λ-move at the bottom of the stack drives M into the accepting state q_+ while the concluding $\$$ is written (Transition (11)). ∎

3 Computational Capacity

We turn to consider the computational capacity of the stack transducers. In particular, whenever two devices have different language acceptance power, then the simple identity transduction applied to a language from their symmetric difference would be a witness for separating also the power of the transducers. However, in the following we consider transductions of languages that are accepted by both types of devices in question. In this way, we are separating in fact the capabilities of computing transductions. First the relation with pushdown transducers (cf. [1]) is studied. A *pushdown transducer* (PDT) is a pushdown automaton equipped with a one-way write-only output tape. In our terms this is a stack automaton whose internal transition function δ_{int} is completely undefined. Our first result shows that pushdown transducers are strictly weaker than ndi-1DSaT, even if the language transformed is deterministic context free.

Theorem 2. *The length-preserving transduction*

$$\tau_1 = \{(a^n\$a^n\$a^m\$, a^n\$a^m\$a^n\$) \mid m \geq 0, n \geq 1\}$$

is a witness for the strictness of the inclusion $\mathscr{T}(\textsf{PDT}) \subset \mathscr{T}(\textsf{ndi-1DSaT})$.

The situation changes when the non-digging stack transducers are non-erasing. Clearly, the deterministic context-free language $\{a^m\$a^n\$ \mid m \geq n \geq 1\}$ is also accepted by a deterministic one-way checking stack automaton.

Lemma 3. *The transduction $\tau_2 = \{(a^m\$a^n\$, a^{m-n}) \mid m \geq n \geq 1\}$ belongs to the difference $\mathscr{T}(\textsf{PDT}) \setminus \mathscr{T}(\textsf{ndi-1DNESaT})$.*

Proof. A PDT realizing τ_2 is constructed from a real-time deterministic pushdown automaton that accepts the language $\{\,a^m\$a^n\$ \mid m \geq n \geq 1\,\}$. First the leading a's are read and pushed on the stack. When the first $\$$ appears, for every further input symbol a, one symbol is popped. Finally, the remaining symbols are popped and emitted.

In order to show that τ_2 is not realized by any ndi-1DNESaT we contrarily assume that it is realized by the ndi-1DNESaT $M = \langle Q, \Sigma, \Gamma, \Delta, \delta_{\text{ext}}, \delta_{\text{int}}, q_0, \bot, F \rangle$.

We consider the situation when M has processed an input prefix $a^m\$$, for m large enough. Up to that time nothing can have been written on the output tape. Otherwise, assume M has already written some word a^i with $i \geq 1$. Then the accepting computation on input $a^m\$a^m\$$ would produce a pair $(a^m\$a^m\$, a^j)$, for some $j \geq 1$, belonging to the transduction realized, but not to τ_2. Furthermore, by the same argumentation it follows that M cannot emit anything until the second $\$$ appears in the input, that is, until the input has been read entirely. Since M is non-erasing and non-digging, it has to write a^{m-n} on the tape with λ-moves and with the stack head on top of the stack. In between several write operations the stack head may move into the stack and back. The behavior of M in these phases can entirely be described by a table that lists for every state in which M moves the stack head into the stack what happens. This can either be halting or moving the state head back to the top in some state. Altogether there are only finitely many of such tables. We conclude that there are two numbers n_1 and n_2 so that $a^m\$a^{n_1}$ and $a^m\$a^{n_2}$ drive M into the same state, with the same topmost stack symbol, having the stack head on top, and the same table describing the behavior while the head is in the stack. So, if $a^m\$a^{n_1}\$$ is transformed into a^{m-n_1}, then so is $a^m\$a^{n_2}\$$, a contradiction.   □

Since ndi-1DCSaT accept non-context-free languages the incomparabilities of the next corollary follow in general.

Corollary 4. *The family $\mathscr{T}(PDT)$ is incomparable with each of the families $\mathscr{T}(ndi\text{-}1DNESaT)$ and $\mathscr{T}(ndi\text{-}1DCSaT)$.*

Moreover, the inclusion shown in Theorem 2 together with the transduction τ_2 belonging to the difference $\mathscr{T}(PDT) \setminus \mathscr{T}(ndi\text{-}1DNESaT)$ by Lemma 3 reveals the strictness of the following inclusions. The inclusions themselves follows for structural reasons.

Corollary 5. *The family $\mathscr{T}(ndi\text{-}1DSaT)$ properly contains the two families $\mathscr{T}(ndi\text{-}1DNESaT)$ and $\mathscr{T}(ndi\text{-}1DCSaT)$.*

Since the language recognition power of ndi-1DNESaT are stronger than that of ndi-1DCSaT there is a proper inclusion between the corresponding families of transductions. However, it is currently an open problem whether there is a ndi-1DNESaT M so that $L(M)$ is accepted by some ndi-1DCSaT as well, but $\tau(M)$ cannot be realized by any ndi-1DCSaT.

3.1   Digging Versus Non-digging

We turn to show that all types of stack transducers that are able to write to the output tape while the stack head is inside the stack are strictly stronger than their corresponding non-digging variant. To this end, the witness transduction $\tau_3 = \{\,(a^n\$b^m\$, b^m\$a^n\$a^n\$) \mid m \geq 0, n \geq 1\,\}$ is exploited.

Example 6. The transduction τ_3 is realized by the checking stack transducer

$$M = \langle\{q_0, q_1, q_2, q_3, q_+\}, \{a, b, \$\}, \{A, \perp\}, \{a, b, \$\}, \delta_{\text{ext}}, \delta_{\text{int}}, q_0, \perp, \{q_+\}\rangle,$$

where the transition functions are as follows.

(1) $\delta_{\text{ext}}(q_0, a, \perp) = (q_0, A\perp, \lambda)$ (7) $\delta_{\text{int}}(q_2, \lambda, A) = (q_2, -1, a)$

(2) $\delta_{\text{ext}}(q_0, a, A) = (q_0, AA, \lambda)$ (8) $\delta_{\text{int}}(q_2, \lambda, \perp) = (q_3, 0, \$)$

(3) $\delta_{\text{ext}}(q_0, \$, A) = (q_1, A, \lambda)$ (9) $\delta_{\text{int}}(q_3, \lambda, \perp) = (q_3, +1, a)$

(4) $\delta_{\text{ext}}(q_1, b, A) = (q_1, A, b)$ (10) $\delta_{\text{int}}(q_3, \lambda, A) = (q_3, +1, a)$

(5) $\delta_{\text{ext}}(q_1, \$, A) = (q_2, A, \$)$

(6) $\delta_{\text{ext}}(q_2, \lambda, A) = (q_2, -1, a)$ (11) $\delta_{\text{ext}}(q_3, \lambda, A) = (q_+, A, \$)$ ∎

So, transduction τ_3 is realized by the weakest type of digging stack transducers, where the language $L(M)$ is regular. On the other hand, the next result shows that τ_3 is not even realized by the strongest type of non-digging stack transducers.

Lemma 7. *Transduction τ_3 does not belong to the family $\mathscr{T}(\text{ndi-1DSaT})$.*

Example 6 and Lemma 7 show the following proper inclusions.

Theorem 8.

1. $\mathscr{T}(\text{ndi-1DCSaT}) \subset \mathscr{T}(\text{1DCSaT})$
2. $\mathscr{T}(\text{ndi-1DNESaT}) \subset \mathscr{T}(\text{1DNESaT})$
3. $\mathscr{T}(\text{ndi-1DSaT}) \subset \mathscr{T}(\text{1DSaT})$

3.2   Relations Between Diggers

Here we compare the capacities of the three different types of stack transducers that may emit symbols while the stack head is inside the stack. Our first result separates the restricted families of non-erasing and checking transducers. Again, the witness transduction relies on an input language that is accepted by the weaker devices. We define $\tau_4 = \{\,(a^n\$a^n\$v\$, v^R\$a^n\$a^n\$) \mid n \geq 1, v \in \{a, b\}^*\,\}$. Transduction τ_4 is realized by some non-erasing stack transducer.

Theorem 9. *The length-preserving transduction τ_4 is a witness for the strictness of the inclusion $\mathscr{T}(\text{1DCSaT}) \subset \mathscr{T}(\text{1DNESaT})$.*

With almost literally the same proof as in the previous theorem the next corollary can be shown.

Corollary 10. *The transductions* $\{ (a^n \$ a^n \$ v \$, v^R \$ a^n \$) \mid n \geq 1, v \in \{a, b\}^* \}$ *and* $\{ (a^m \$ a^n \$ v \$, v^R \$ a^{m-n} \$) \mid m \geq 1, n \geq 0, v \in \{a, b\}^* \}$ *do not belong to the family* $\mathscr{T}(1DCSaT)$. □

The final comparison separates the most general family $\mathscr{T}(1DSaT)$ of transductions considered here from the 'next' weaker family of transductions realized by non-erasing transducers. As before, the witness transduction relies on an input language that is accepted by the weaker devices. We define the transduction $\tau_5 = \{ (a^m \$ a^n \$ v \$, v^R \$ a^{m-n} \$) \mid m \geq 1, n \geq 0, v \in \{a, b\}^* \}$.

Example 11. The transduction τ_5 is realized by the stack transducer

$$M = \langle \{q_0, q_1, \ldots, q_4, q_+\}, \{a, b, \$\}, \{A, B, \$, \bot\}, \{a, b, \$\}, \delta_{\text{ext}}, \delta_{\text{int}}, q_0, \bot, \{q_+\}\rangle,$$

where the transition functions are as follows. Let $X \in \{A, B, \$\}$.

(1) $\delta_{\text{ext}}(q_0, a, \bot) = (q_0, A\bot, \lambda)$

(2) $\delta_{\text{ext}}(q_0, a, A) = (q_0, AA, \lambda)$

(3) $\delta_{\text{ext}}(q_0, \$, A) = (q_1, A, \lambda)$

(4) $\delta_{\text{ext}}(q_1, a, A) = (q_1, \lambda, \lambda)$

(5) $\delta_{\text{ext}}(q_1, \$, \bot) = (q_2, \$\bot, \lambda)$

(6) $\delta_{\text{ext}}(q_1, \$, A) = (q_2, \$A, \lambda)$

(7) $\delta_{\text{ext}}(q_2, a, X) = (q_2, AX, \lambda)$

(8) $\delta_{\text{ext}}(q_2, b, X) = (q_2, BX, \lambda)$

(9) $\delta_{\text{ext}}(q_2, \$, \$) = (q_4, \lambda, \$)$

(10) $\delta_{\text{ext}}(q_2, \$, A) = (q_3, \lambda, a)$

(11) $\delta_{\text{ext}}(q_2, \$, B) = (q_3, \lambda, b)$

(12) $\delta_{\text{ext}}(q_3, \lambda, A) = (q_3, \lambda, a)$

(13) $\delta_{\text{ext}}(q_3, \lambda, B) = (q_3, \lambda, b)$

(14) $\delta_{\text{ext}}(q_3, \lambda, \$) = (q_4, \lambda, \$)$

(15) $\delta_{\text{ext}}(q_4, \lambda, A) = (q_4, \lambda, a)$

(16) $\delta_{\text{ext}}(q_4, \lambda, \bot) = (q_+, \bot, \$)$

Given an input $a^m \$ a^n \$ v \$$, the *1DSaT* M starts to read the prefix a^m with the Transitions (1)–(3) whereby A^m is successively pushed onto the stack and nothing is emitted. Then Transition (4) reads the following a's as long as $n \leq m$, whereby as many stack symbols are popped as input symbols are read. If $n > m$, the computation blocks when the bottom-of-stack symbol appears. Otherwise, Transitions (5) and (6) read the next $\$$ and push it on the stack. Now the stack content is $\$ A^{m-n}$. Next, state q_2 is used to read and push the input factor v by Transitions (7)–(11). When the last $\$$ appears in the input with a $\$$ at the top of the stack, then v is empty (Transition (9)). Finally, the stack content, that is, $v^R \$ A^{m-n}$ is emitted by Transitions (12)–(15). In the last step, the closing $\$$ is emitted by Transition (16) that drives M into the accepting state q_+. ∎

Based on transduction τ_5 the next separation is shown.

Theorem 12. *The transduction τ_5 is a witness for the strictness of the inclusion* $\mathscr{T}(1DNESaT) \subset \mathscr{T}(1DSaT)$.

Finally, we can derive the relationships between the family $\mathscr{T}(PDT)$ with all families of stack automata transductions considered. Since $\mathscr{T}(PDT)$ is properly included in $\mathscr{T}(\text{ndi-}1DSaT)$ (Theorem 2), it is properly included in $\mathscr{T}(1DSaT)$. With all other families we obtain incomparabilities as follows. By Corollary 4 there is a transduction in $\mathscr{T}(\text{ndi-}1DCSaT)$ not belonging to $\mathscr{T}(PDT)$. On the other hand, the stack transducer of Example 11 is in fact a pushdown transducer, since δ_{int} is completely undefined. So, transduction τ_5 belongs to $\mathcal{T}(\text{PDT})$. But by Theorem 12 it does not belong even to $\mathcal{T}(\text{1DNESaT})$. This implies the following corollary.

Corollary 13. *The family $\mathcal{T}(PDT)$ is incomparable with each of the families $\mathcal{T}(1DNESaT)$, $\mathcal{T}(1DCSaT)$, $\mathcal{T}(ndi\text{-}1DNESaT)$, and $\mathcal{T}(ndi\text{-}1DCSaT)$.*

The inclusion structure of the families in question are depicted in Fig. 1.

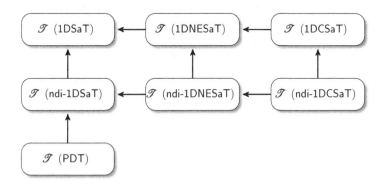

Fig. 1. Inclusion structure of transduction families realized by stack automata with different properties. The solid arrows indicate strict inclusions. The family $\mathcal{T}(\text{PDT})$ is incomparable with all families to which it is not connected by a path.

4 Regularity of Stack Languages

It is well known that the set of reachable pushdown contents in a pushdown automaton is a regular language. Here we generalize this result to even the strongest type of stack transducer in question. Clearly, this implies the same result for stack automata as language acceptors.

The *stack language* of a stack transducer M is the set of all stack contents that occur in some configuration of a successful computation of M.

Before we turn to the proof of the main result in this section, we consider the notion of *stack trees* to model how the stack transducer transition function interacts with the stack. Intuitively, a stack tree t stores the stack contents as they appear in a computation, organized so that the right-most path from the root to a leaf of t holds the current stack. Without loss of generality, we restrict our attention to stack transducers that halt with an empty stack.

Definition 14 (Stack Tree). *Let M be a stack transducer with stack alphabet Γ, and ρ be a computation of M on some input string w. The stack tree t_ρ of ρ is created as follows. At the start of the computation, the tree consists of a* *single node labeled ⊥, and we place a pointer p at this node. This is the base case. Assume now that we have a stack tree t for the prefix ρ' of ρ, and that p marks one of its nodes. Depending on the next operation in ρ, the tree t is updated accordingly (see Figs. 2 and 3):*

- *If M pushes the symbol a ∈ Γ onto the stack, then*
  - *if p points to a leaf v, then a new leaf v' labeled a is created below it, an p is set to point to v', but*
  - *if p points to a non-leaf v at which a tree t' is rooted, then a new node v' is created below v, marked with the auxiliary symbol ◇, t' is moved down and placed as the left child of v', and a new leaf v̂ labeled a is added and placed as the right child of v'. The pointer p is set to v̂.*
- *If M pops a symbol, then p is moved towards the root of t, to the closest ancestor node that is not an auxiliary node.*

Since ρ is a valid computation, t_ρ is well defined, and from the construction we know that it is binary.

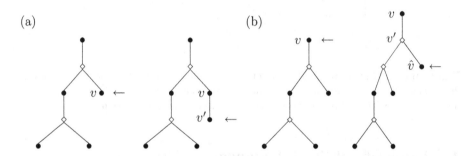

Fig. 2. Stack trees associated to a push operation of the stack transducer. Figure(a) shows a stack tree whose tree pointer points at a leaf node v, and after a push operation on the stack, a new leaf v' is created and p is set to point at v'. Figure(b) shows a stack tree whose tree pointer points at a non-leaf v, and after a push operation on the stack, a new node v' is created and marked with ◇. The subtree that was rooted at v becomes the left child of v'.

From here on, we denote by S_Γ the set of all stack trees over the alphabet $\Gamma \cup \{◇\}$.

Definition 15 (Composition Operators). *Given $t, s \in S_\Gamma$, we denote by*

- *$t \otimes s$ the stack tree obtained by adding the root of s as a child of the right-most leaf of t, and*
- *$t \oplus s$ the stack tree obtained by creating a new node labeled with ◇ and adding t and s as its left and right subtree, respectively.*

We denote by $\Gamma_{\oplus,\otimes}$ the set of stack trees that can be built from the symbols in Γ, seen as trees of height 0, and the operators $\oplus$ and $\otimes$.

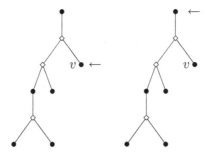

Fig. 3. Stack trees associated to a pop operation of the stack transducer. The figure shows a stack tree whose tree pointer points at a leaf node v, and after a pop operation on the stack, the tree pointer moves towards the root and to the closest ancestor node that is not an auxiliary node $\diamond$.

Lemma 16. *For every stack alphabet Γ, we have $S_\Gamma = \Gamma_{\oplus,\otimes}$.*

Proof. It is easy to see that every tree in $\Gamma_{\oplus,\otimes}$ is the stack tree for some choice of M, w, and ρ. So $\Gamma_{\oplus,\otimes} \subseteq S_\Gamma$.

The fact that every tree in S_Γ is in $\Gamma_{\oplus,\otimes}$ can be shown by induction on the height of the tree. The statement is true for all trees of height zero, that is, for Γ. Assume then that the inclusion holds for all trees of height n or less. The root of a tree t of height $n+1$ is either $\diamond$ or in Γ. In the first case, t can be constructed by applying the $\oplus$ operator to two trees of height at most n. In the second case, t can be constructed by applying the $\otimes$ operator to one tree in Γ and one tree of height at most n. By the induction hypotheses, all trees of height at most n can be constructed from Γ using $\oplus$ and $\otimes$. □

With these considerations the main result of this section can be shown.

Theorem 17. *The stack language of any stack transducer is regular.*

Theorem 17 shows that a stack transducer M with a one-way read-only input tape cannot be simulated by any stack transducer M' that receives its input directly on the stack. This holds even if M is deterministic, but M' is allowed to be nondeterministic. Moreover, since the intersection of regular languages is regular, any of the following ways of providing input to a stack transducer in place of the tape will cause the domain to be regular:

- the input is given on the stack,
- the machine guesses the input string and verifies its guess,
- the machine computes the input string on the stack, in such a way that the entire string is on the stack at once.

References

1. Aho, A.V., Ullman, J.D.: The Theory of Parsing, Translation, and Compiling, vol. I: Parsing. Prentice-Hall, Upper Saddle River (1972)

2. Bojanczyk, M.: Transducers with origin information. In: Invited Talk at the 3rd International Workshop on Trends in Tree Automata and Tree Transducers, The Queen Mary University of London (2015)
3. Braune, F., Seemann, N., Quernheim, D., Maletti, A.: Shallow local multi-bottom-up tree transducers in statistical machine translation. In: Association for Computational Linguistics (ACL 2013), vol. 1, pp. 811–821. The Association for Computer Linguistics (2013)
4. Drewes, F., van der Merwe, B.: Path languages of random permitting context tree grammars are regular. Fundam. Inform. **82**, 47–60 (2008)
5. Ginsburg, S., Greibach, S.A., Harrison, M.A.: Stack automata and compiling. J. ACM **14**, 172–201 (1967)
6. Ginsburg, S., Greibach, S.A., Harrison, M.A.: One-way stack automata. J. ACM **14**, 389–418 (1967)
7. Hibbard, T.N.: A generalization of context-free determinism. Inform. Control **11**, 196–238 (1967)
8. Hopcroft, J.E., Ullman, J.D.: Sets accepted by one-way stack automata are context sensitive. Inform. Control **13**, 114–133 (1968)
9. Hori, C., Ohtake, K., Misu, T., Kashioka, H., Nakamura, S.: Weighted finite state transducer based statistical dialog management. In: Automatic Speech Recognition and Understanding (ASRU 2009), pp. 490–495. IEEE (2009)
10. Kutrib, M., Wendlandt, M.: On simulation cost of unary limited automata. In: Shallit, J., Okhotin, A. (eds.) DCFS 2015. LNCS, vol. 9118, pp. 153–164. Springer, Heidelberg (2015)
11. Kutrib, M., Wendlandt, M.: Reversible limited automata. In: Durand-Lose, J., Nagy, B. (eds.) MCU 2015. LNCS, vol. 9288, pp. 113–128. Springer, Switzerland (2015)
12. Mohri, M., Pereira, F.C.N., Riley, M.: Speech recognition with weighted finite-state transducers. In: Benesty, J., Sondhi, M.M., Huang, Y. (eds.) Handbook on Speech Processing and Speech Communication, Part E: Speech Recognition, pp. 559–584. Springer, Heidelberg (2008)
13. Petrov, S.: Towards Universal Syntactic and Semantic Processing of Natural Language. Invited talk at SLTC 2016, Uppsala University (2014)
14. Pighizzini, G., Pisoni, A.: Limited automata and regular languages. Int. J. Found. Comput. Sci. **25**, 897–916 (2014)
15. Pighizzini, G., Pisoni, A.: Limited automata and context-free languages. Fundamenta Inform. **136**, 157–176 (2015)
16. Wagner, K., Wechsung, G.: Computational Complexity. Reidel, Dordrecht (1986)

Computing the Expected Edit Distance from a String to a PFA

Jorge Calvo-Zaragoza[1], Colin de la Higuera[2(✉)], and Jose Oncina[1]

[1] DLSI, University of Alicante, Alicante, Spain
{jcalvo,oncina}@dlsi.ua.es
[2] LINA Lab, UMR 6241, University of Nantes, Nantes, France
cdlh@univ-nantes.fr

Abstract. In a number of fields one is to compare a witness string with a distribution. One possibility is to compute the probability of the string for that distribution. Another, giving a more global view, is to compute the expected edit distance from a string randomly drawn to the witness string. This number is often used to measure the performance of a prediction, the goal then being to return the *median* string, or the string with smallest expected distance.

To be able to measure this, computing the distance between a hypothesis and that distribution is necessary. This paper proposes two solutions for computing this value, when the distribution is defined with a probabilistic finite state automaton. The first is exact but has a cost which can be exponential in the length of the input string, whereas the second is a FPRAS.

Keywords: Edit distance · Probabilistic finite state automata

1 Introduction

The edit or Levenshtein distance is often used to measure how close one string is to another [14]. This distance has given rise to many questions: if one is given a set instead of a string, the question may be to compute rapidly the distance between the set and a string or between two sets [17,28]. In turn, a set defines an empirical distribution which can be represented by a probabilistic finite state automaton (PFA), a hidden Markov model or a weighted automaton [17,27].

If the set is used as a learning sample, the distribution may be more general, again represented by the above machines, but these may now contain cycles and therefore define a distribution over all possible strings.

The following questions are then posed: what is the expected distance between a given *witness* string and such a distribution? What could a representative string be for this distribution? One possible answer to the second question is the *most probable string* [10,11]. Another is the *median string* which is the string minimizing the expected distance to the distribution, which in turn contributes to make the first question relevant. These questions have not only a

© Springer International Publishing Switzerland 2016
Y.-S. Han and K. Salomaa (Eds.): CIAA 2016, LNCS 9705, pp. 39–50, 2016.
DOI: 10.1007/978-3-319-40946-7_4

precise mathematical interest, but they have been posed in very different settings like bio-informatics [7], pattern recognition [19] or computational linguistics [24].

Alternative distances have been studied, such as the minimum cost obtaining by summing the weight of a string and its distance to the witness string [2]. Balls of strings, Levenshtein automata and other finite state machines linking regular languages and the edit distance have been introduced, discussed and studied [4,16,17,23].

We prove in this paper two results.

The first is that the expected edit distance can be computed, and that if the weights of the PFA are rational, then the result itself is rational. The construction involves building a multiplicity automaton which can be of size exponential in the length of the string w, but only increases polynomially with the number of states of the PFA or the size of the alphabet.

The second result is that the problem admits a *fully polynomial time randomized schema* (FPRAS), that is, a randomized algorithm which will return a probably approximatively correct value in time polynomial with the length of the string, the size of the automaton representing the distribution, and the inverse of the accepted error.

Each algorithm has its advantages and inconveniences, as we will show in the experimental section: the method involving the multiplicity automaton will give an exact result, but only for small witness strings. The FPRAS, on the other hand, can only build an approximate result, but there is a guarantee on the error bound and the method can handle long witness strings and large PFA.

After introducing notations and definitions (Sect. 2), we prove in Sect. 3 that the problem is decidable and provide an algorithm which gives the correct result; Sect. 4 presents a polynomial randomized computation whose result is probably approximately correct. Our experiments, described in Sect. 5, empirically confirm the bounds in both error and complexity of the proposed strategies. Section 6 concludes the present work.

2   Preliminaries

2.1   Basic Notations

An *alphabet* Σ is a finite non-empty set of symbols called *letters*. A *string* w over Σ is a finite sequence $w = w_1 \ldots w_m$ of letters. Letters will be indicated by $a, b, c, \ldots$, and strings by $u, v, \ldots, z$. Let $|w|$ denote the length of w. In this case we have $|w| = |w_1 \ldots w_m| = m$. The *empty string* is denoted by λ.

We denote by Σ^* the set of all strings, by Σ^m the set of those of length m.

A *probabilistic language* $\mathcal{D}$ is a probability distribution over Σ^*. The probability of a string $w \in \Sigma^*$ under the distribution $\mathcal{D}$ is denoted as $Pr_{\mathcal{D}}(w)$. The distribution must satisfy $\sum_{w \in \Sigma^*} Pr_{\mathcal{D}}(w) = 1$.

If the distribution is modelled by some syntactic machine $\mathcal{M}$, the probability of x according to the probability distribution defined by $\mathcal{M}$ is denoted by $Pr_{x \sim \mathcal{M}}(x)$ or simply $Pr_{\mathcal{M}}(x)$.

2.2   Multiplicity Automata

An n-state Multiplicity Automata (MA) $\mathcal{M}$ (also known as *recognizable series* [5] or *Stochastic Sequential Machines* [20]) can be defined by a 4-tuple $\langle \Sigma, \mathbf{S}, \mathbf{M}, \mathbf{F} \rangle$ here: Σ is the alphabet, $\mathbf{S} \in \mathbb{Q}^{1 \times n}$, $\mathbf{M} = \{\mathbf{M}_a \in \mathbb{Q}^{n \times n} : a \in \Sigma\}$, and $\mathbf{F} \in \mathbb{Q}^{n \times 1}$. $\mathcal{M}$ realizes a function from Σ^* to $\mathbb{Q}$ such that:

$$\mathcal{M}(x_1 \cdots x_k) = \mathbf{S} \sum_{i=1}^{k} \mathbf{M}_{x_i} \mathbf{F}$$

This machine can also be defined from a graph point of view as an n-state machine $\langle \Sigma, Q, \mathbb{S}, \mathbb{F}, \delta \rangle$ where $Q = \{q_0, \cdots, q_{n-1}\}$, $\mathbb{S} : Q \to \mathbb{Q}$ are the initial weights ($\mathbb{S}(q_i) = \mathbf{S}[i]$), $\mathbb{F} : Q \to \mathbb{Q}$ are the final weights ($\mathbb{F}(q_i) = \mathbf{F}[i]$), and $\delta : Q \times \Sigma \times Q \to \mathbb{Q}$ are the transition weights ($\delta(q_i, a, q_j) = [\mathbf{M}_a]_{i,j}$).

Given $x \in \Sigma^*$, $\Pi_{\mathcal{M}}(x)$ is the set of all paths accepting x: an *accepting x-path* is a sequence $\pi = q_{i_0} x_1 q_{i_1} x_2 \ldots x_k q_{i_k}$ where $x = x_1 \cdots x_k$, $a_i \in \Sigma$, and $\forall j \in [1, k]$ such that $\delta(q_{i_{j-1}}, a_j, q_{i_j}) \neq 0$. Let $\pi = q_{i_0} x_1 q_{i_1} x_2 \ldots x_k q_{i_k}$, we denote by $\delta(\pi) = \prod_{j=1}^{k} \delta(q_{i_{i-1}}, a_j, q_{i_j})$, $\alpha(\pi) = q_{i_0}$ and $\omega(\pi) = q_{i_k}$.

$$\mathcal{M}(x) = \sum_{\pi \in \Pi_{\mathcal{M}}(x)} \mathbb{S}(\alpha(\pi)) \delta(\pi) \mathbb{F}(\omega(\pi))$$

This can be computed efficiently using the Forward (or Backwards) algorithm. Obviously, the two ways to compute $\mathcal{M}(x)$ are equivalent.

Probabilistic Finite Automata (PFA) can be viewed as a special type of MA that are restricted to describe probability distributions over sets of strings. Then further restrictions should be applied. Let $\mathbf{1} \in \mathbb{Q}^{n \times 1} : \mathbf{1}[i] = 1 \forall i$, $I \in \mathbb{Q}^{n \times n}$ be the identity matrix and $\mathbf{M}_\Sigma = \sum_{a \in \Sigma} \mathbf{M}_a$, then:

- the components of $\mathbf{S}$, $\mathbf{M}$ and $\mathbf{F}$ are interpreted as probabilities, that is, they should be in $[0, 1]$
- $\mathbf{S1} = 1$: the sum of the starting probabilities should add one
- $\mathbf{M}_\Sigma \mathbf{1} + \mathbf{F} = 1$: for any state, the sum of the outgoing probability plus the ending probability should add one
- $(I - \mathbf{M}_\Sigma)$ should be non-singular: this is a sufficient condition to assure the non existence of absorbing states (or set of states).

2.3   The Edit Distance

The edit distance between two strings $d_e(x, y)$ is the minimum number of edition operations needed to transform x into y [14].

We will make use of the following generous bounds for the edit distance:

$$d_e(x, y) \leq \max\{x, y\} \leq |x| + |y| \qquad (1)$$

The *relative edit distance* from x to y is $d_r(x, y) = d_e(x, y) - |y|$. Notice that this is not a metric.

It follows from (1) that for a fixed string x the set of values that $d_r(x, y)$ can take is finite, with values ranging from $-|x|$ to $|x|$, even though the set of strings from which y is chosen is infinite.

We extend the definitions to distributions over strings (*string-distribution edit distance*):

$$d_e(w, \mathcal{D}) = \sum_{y \in \Sigma^*} d_e(w, y) Pr_{\mathcal{D}}(y) = \sum_{y \in \Sigma^*} d_r(w, y) Pr_{\mathcal{D}}(y) + \sum_{y \in \Sigma^*} |y| Pr_{\mathcal{D}}(y) \quad (2)$$

When $\mathcal{D}$ is given by a PFA $\mathcal{A}$, we can also write $d_e(w, \mathcal{A})$.

2.4 Complexity Issues

Let us recall that a decision problem is one for which the possible answers are **true** and **false**. Such a problem is in class **P** if there is a deterministic Turing machine solving any instance in polynomial time, in **NP** if this machine is non-deterministic, **NP**-complete if it as hard as any of the other **NP**-complete problems.

An optimization problem asks for a numerical value to be computed given an instance. This value can sometimes be approximated by a polynomial-time approximation scheme (PTAS) which can compute a value within a factor $1+\epsilon$ of the optimum in time polynomial in the size of the approximation scheme. If the runtime also depends polynomially of $1/\epsilon$, the scheme is called a fully polynomial-time approximation scheme or FPTAS. For more about approximation algorithms, see [26].

Sometimes, deterministic algorithms are unable to approximate, but randomized algorithms [18] can solve the problem in the following sense: an algorithm **A** is a *fully polynomial time randomized schema* or FPRAS if it can return a solution which is at distance ϵ of the optimum, with confidence at least $1 - \delta$ and runs in time polynomial in the size of the instance, $1/\epsilon$ and $1/\delta$.

The key problem in this work is called EDD:

Name: Computing the edit distance to a distribution (EDD)
Instance: A distribution $\mathcal{D}$ over an alphabet Σ. A string w over Σ.
Question: Compute $d_e(w, \mathcal{D})$.

If we need to only consider the decision problem we will be also taking a rational input r and asking if $d_e(w, \mathcal{D}) \leq r$. And the associated approximation problem consists in computing a value x such $|x - d_e(w, \mathcal{D})| < \epsilon$.

The exact status of EDD is an open question. We conjecture it is **NP**-hard.

3 EDD Is Decidable

We first prove that there exists an algorithm which takes a string w and a PFA $\mathcal{A}_{\mathcal{D}}$ and computes $d_e(w, \mathcal{A}_{\mathcal{D}})$. The computation cannot be bounded by a polynomial, but it terminates. The construction we propose follows three steps:

1. We first (Sect. 3.1) build from w an MA $\mathcal{A}_w$ which can compute $d_r(w, x)$.
2. We next (Sect. 3.2) build from $\mathcal{A}_\mathcal{D}$ and $\mathcal{A}_w$ an MA $\mathcal{A}_{\mathcal{D},w}$ which computes the product of the relative edit distance and the probability of the string.
3. Using the matrix representation of $\mathcal{A}_{\mathcal{D},w}$ and $\mathcal{A}_\mathcal{D}$ we are able to compute the values of the infinite series $\sum_{x \in \Sigma^*} |x| Pr_{\mathcal{A}_\mathcal{D}}(x)$ and $\sum_{x \in \Sigma^*} d_r(w, x) Pr_{\mathcal{A}_\mathcal{D}}(x)$.

3.1 Building a Multiplicity Automaton Computing the Edit Distance to a String (Step 1)

Given a string w, we build (with Algorithm 1 MA_BUILD) an MA, $\mathcal{A}_w$, which will allow to parse any other string x and in linear time obtain $d_r(w, x)$.

The states of the MA are the different columns one may obtain when running the classical edit distance algorithm for strings w (used to index the lines) and u (used to index the columns), and subtracting, in each cell, the length u, i.e., computing $d_r(w, u)$, with w fixed and u being any string. The number of states is finite, because $d_r(w, \cdot) \in [-|w|, |w|]$, so the number of possible columns is bounded by $(2|w|)^{|w|}$. Moreover, if we take into account that the difference between two consecutive elements in a column is in $\{-1, 0, 1\}$, the number of different columns, hence of states, is bounded by $3^{|w|}$.

There is a transition in the MA labelled by symbol a between the state corresponding to the last column of $d_r(w, u)$ to the state corresponding to the last column of $d_r(w, ua)$, for some string u.

There is no guarantee that the construction terminates in polynomial time. We give an example of this construction in Appendix A of [6] (Fig. 4) and in Appendix B of [6] we provide a counter-example, i.e. a parameterized string such that the size of $\mathcal{A}_w$ increases faster than any polynomial in $|w|$.

Yet even when exponential, the construction does terminate, and the following result can be given:

Proposition 1. *Given any string x, $d_r(w, x) = \mathcal{A}_w(x)$.*

3.2 Computing the Product Automaton (Step 2)

We are now given a PFA $\mathcal{A}_\mathcal{D} = \langle \Sigma, Q_\mathcal{D}, \mathbb{S}_\mathcal{D}, \mathbb{F}_\mathcal{D}, \delta_\mathcal{D} \rangle$ and a multiplicity automaton $\mathcal{A}_w = \langle \Sigma, Q_w, \mathbb{S}_w, \mathbb{F}_w, \delta_w \rangle$.

The new machine, denoted by $\mathcal{A}_{\mathcal{D},w}$ has as states pairs $\langle q, q' \rangle$ with $q \in Q_\mathcal{D}$, $q' \in Q_w$. $\mathcal{A}_{\mathcal{D},w} = \langle \Sigma, Q_{\mathcal{D},w}, \mathbb{S}_{\mathcal{D},w}, \mathbb{F}_{\mathcal{D},w}, \delta_{\mathcal{D},w} \rangle$:

- $\delta_{\mathcal{D},w}(\langle q, q' \rangle, a, \langle s, s' \rangle) = \delta_\mathcal{D}(q, a, s) \delta_w(q', a, s')$,
- $\mathbb{S}_{\mathcal{D},w}(\langle q, q' \rangle) = \mathbb{S}_\mathcal{D}(q) \mathbb{S}_w(q')$,
- $\mathbb{F}_{\mathcal{D},w}(\langle q, q' \rangle) = \mathbb{F}_\mathcal{D}(q) \mathbb{F}_\mathcal{D}(q')$.

By construction, $\mathcal{A}_{\mathcal{D},w}(x) = d_r(w, x) Pr_\mathcal{A}(x)$.
An example is proposed in Appendix A, Fig. 6 of [6].

Algorithm MA_BUILD(w)
   Data: $w = w_1 \ldots w_m$ of length m
   Result: a multiplicity automaton $\mathcal{A}_w = \langle \Sigma, Q, \mathbb{S}, \mathbb{F}, \delta \rangle$

   $q_0 \leftarrow [0, 1, 2, \ldots, m]$; $Q \leftarrow \{q_0\}$; $\mathbb{S}(q_0) \leftarrow 1$; $\mathbb{F}(q_0) \leftarrow m$;
   unmarked $\leftarrow \{q_0\}$;
   while unmarked $\neq \emptyset$ **do**
      Choose q in unmarked ;
      unmarked $\leftarrow$ unmarked $- \{q\}$;
      for $a \in \Sigma$ **do**
         $q'[0] \leftarrow 0$;
         for $i = 1$ **to** m **do**
            if $w_i = a$ **then** $x \leftarrow 0$;
            else $x \leftarrow 1$;
            $q'[i] \leftarrow \min\{q[i], q[i-1] + x - 1, q'[i-1]\}$;
         if $q' \in Q$ **then** $\delta(q, a, q') \leftarrow 1$;
         else
            $Q \leftarrow Q \cup \{q'\}$; $\delta(q, a, q') \leftarrow 1$; $\mathbb{F}(q')$; $\leftarrow q'[m]$; $\mathbb{S}(q') \leftarrow 0$;
            unmarked $\leftarrow$ unmarked $\cup \{q'\}$;

Algorithm 1. Algorithm MA_BUILD(w) computing, given a string w, the deterministic MA $\mathcal{A}_w$ such that on input x, $d_r(w, x)$ is computed as $\mathcal{A}_w(x)$.

3.3 Computing the Distance (Step 3)

We have to compute $d_e(w, \mathcal{A}_\mathcal{D}) = \sum_{x \in \Sigma^*} |x| Pr_\mathcal{A}(x) + \sum_{x \in \Sigma^*} d_r(w, x) Pr_\mathcal{A}(x)$.

Let $(\Sigma, \overset{\mathcal{D}}{\mathbf{S}}, \overset{\mathcal{D}}{\mathbf{M}}, \overset{\mathcal{D}}{\mathbf{F}})$ be the matrix representation of the PFA $\mathcal{A}_\mathcal{D}$. Since $(I - \overset{\mathcal{D}}{\mathbf{M}})$ is non-singular by definition of PFA, the average length of the strings generated by $\mathcal{A}_\mathcal{D}$ can be computed as in [11]:

$$\sum_{x \in \Sigma^*} |x| Pr_\mathcal{A}(x) = \sum_{i=0}^{\infty} i \, Pr_\mathcal{A}(\Sigma^i) = \sum_{i=0}^{\infty} i \, \overset{\mathcal{D}}{\mathbf{S}} \overset{\mathcal{D}}{\mathbf{M}}_\Sigma{}^i \overset{\mathcal{D}}{\mathbf{F}} = \overset{\mathcal{D}}{\mathbf{S}} \overset{\mathcal{D}}{\mathbf{M}}_\Sigma (I - \overset{\mathcal{D}}{\mathbf{M}}_\Sigma)^{-2} \overset{\mathcal{D}}{\mathbf{F}}$$

Let $(\Sigma, \overset{w}{\mathbf{S}}, \overset{w}{\mathbf{M}}, \overset{w}{\mathbf{F}})$ be the matrix representation of $\mathcal{A}_{D,w}$. Each addend of the series $\sum_{x \in \Sigma^*} d_r(w, x) Pr_\mathcal{A}(x)$ can be computed as:

$$\sum_{x \in \Sigma^*} d_r(w, x) Pr_\mathcal{A}(x) = \sum_{x \in \Sigma^*} \overset{w}{\mathbf{S}} \overset{w}{\mathbf{M}}_x \overset{w}{\mathbf{F}} = \sum_{i=0}^{\infty} \overset{w}{\mathbf{S}} \overset{w}{\mathbf{M}}{}^i \overset{w}{\mathbf{F}} = \overset{w}{\mathbf{S}} (I - \overset{w}{\mathbf{M}})^{-1} \overset{w}{\mathbf{F}}$$

One point to check is that the matrix $(I - \overset{w}{\mathbf{M}})$ is non-singular.

By construction, $[\overset{w}{\mathbf{M}}]_{i,j} \geq 0$. Moreover, in any adjacency matrix, $[\mathbf{M}^k]_{i,j}$ is the sum of the weights of all the paths of length exactly n that goes from node i to node j. In our case, by construction, $[\overset{\mathcal{D}}{\mathbf{M}^k}]_{i,j} = \sum_{q,s} [\overset{w}{\mathbf{M}^k}]_{<i,q>,<j,s>}$ hence $[\overset{\mathcal{D}}{\mathbf{M}^k}]_{i,j} \geq [\overset{w}{\mathbf{M}^k}]_{<i,q>,<j,s>}$. We also know that $(I - \overset{\mathcal{D}}{\mathbf{M}})$ is non-singular so

$\lim_{k\to\infty}[\overset{\mathcal{D}}{\mathbf{M}}^k]_{i,j} = 0$. Summarising, we have that, $0 \leq \lim_{k\to\infty}[\overset{w}{\mathbf{M}}^k]_{<i,q>,<j,s>} \leq \lim_{k\to\infty}[\overset{\mathcal{D}}{\mathbf{M}}^k]_{i,j} = 0$, so $\lim_{k\to\infty}[\overset{w}{\mathbf{M}}^k]_{i,j} = 0$ and then, $(I - \overset{w}{\mathbf{M}})$ is non-singular.

Therefore, $d_e(w, \mathcal{A}_{\mathcal{D}}) = \overset{\mathcal{D}}{\mathbf{S}}\overset{\mathcal{D}}{\mathbf{M}}_{\Sigma}(I - \overset{\mathcal{D}}{\mathbf{M}}_{\Sigma})^{-2}\mathbf{F} + \overset{w}{\mathbf{S}}(I - \overset{w}{\mathbf{M}})^{-1}\mathbf{F}$. It follows:

Theorem 1. *EDD is decidable and the edit distance between a witness string and a* PFA *with rational weights is rational.*

The construction described here is not polynomially bounded. The final computation is (with arbitrary precision and unit computation time for all arithmetic operations) cubic in the size of the product finite state machine. In turn, the size of this machine essentially depends on the length of the input string.

4   An FPRAS for EDD

As can be seen in the experiments (or in the theoretical analysis from Appendix B of [6]), the method described in Sect. 3 may lead to a combinatorial explosion during the construction of $\mathcal{A}_w$. In this section we propose an FPRAS to approximate the value of $d_e(w, \mathcal{A}_{\mathcal{D}})$.

Alternatively, the result can be seen as a Probably Approximate Correct (PAC) algorithm [25]. The goal of this framework is to learn (in this case, to compute) a concept for which, with high probability, we obtain a sufficiently good approximation of it.

We are given a PFA $\mathcal{A}_{\mathcal{D}}$, a string w and two values $\epsilon > 0$, $\delta > 0$.

An FPRAS would be an algorithm which, in time polynomial in $|\mathcal{A}_{\mathcal{D}}|, |w|, \frac{1}{\epsilon}, \frac{1}{\delta}$ computes a value v such that, with probability at least $1 - \delta$,

$$\left| v - d_e(w, \mathcal{A}_{\mathcal{D}}) \right| \leq \epsilon$$

Theorem 2. *There exists an* FPRAS *computing the expected distance between a string and a distribution given by a* PFA.

Proof. A full description of Algorithm COMPUTE_BOUNDS is given in Appendix C of [6]. This algorithm returns L which is the length at which the generation process of the PFA should be stopped. The goal is to have a polynomial limit to the length of the strings without this impacting the quality of the result. Then, for this L a value N, also polynomial, is computed. These numbers are used by Algorithm BUILD_SAMPLE which with high probability and complexity in $O(NL)$ is going to return a correct sample. The main Algorithm EXPECTED_DISTANCE uses this sample and computes the distance.

The complexity of Algorithm BUILD_SAMPLE is in $O(NL)$. There is a (non null, but lower than $\frac{\delta}{2}$) probability that the number of generated samples is less than N. □

Algorithm BUILD_SAMPLE($\mathcal{A}_\mathcal{D}$, L, N)

   Data: a PFA $\mathcal{A}_\mathcal{D}$

   Result: a sample S which, with probability $> 1 - \frac{\delta}{2}$, contains N strings

   $S \leftarrow \emptyset$;

   for $i : 1 \leq i \leq N$ **do**

        generate a string of length at most L, using $\mathcal{A}_\mathcal{D}$ and add it to S. If
        during the generation the string becomes too long, generate nothing

   return S

Algorithm 2. Algorithm BUILD_SAMPLE($\mathcal{A}_\mathcal{D}$)

Algorithm EXPECTED_DISTANCE(w, $\mathcal{A}_\mathcal{D}$, ϵ, δ)

   Data: a string w, a PFA $\mathcal{A}_\mathcal{D}$, ϵ, δ

   Result: the expected distance between w and $\mathcal{A}_\mathcal{D}$

   $\langle N, L \rangle \leftarrow$ COMPUTE_BOUNDS($\mathcal{A}_\mathcal{D}, \epsilon, \delta, w$) ;

   $S \leftarrow$ BUILD_SAMPLE($\mathcal{A}_\mathcal{D}$, L, N);

   $Res \leftarrow 0$;

   for $x \in S$ **do** $Res \leftarrow Res + d_e(w, x)$;

   return Res/N

Algorithm 3. Algorithm EXPECTED_DISTANCE(w, $\mathcal{A}_\mathcal{D}$, ϵ, δ)

5   Experiments

As a preliminary evaluation, we ran our FPRAS with a fixed value of $\delta = 0.01$ and varying values of ϵ on 100 pairs of PFA and strings w. In all cases, the difference between the real value and the one computed with the FPRAS was always less than ϵ, which confirms that the values computed by COMPUTE_BOUNDS represent a pessimistic lower bound.

In the series of experiments we want to empirically confirm the time complexity of the algorithms. We showed that the MA-based method grows with the size of the witness string, whereas the FPRAS is bounded by N, the number of necessary samples, which is closely related to the expected length of a string from the PFA. In order to focus these experiments on the most relevant issues, we are using the small PFA shown in Fig. 1. Parameter $p_f \in (0, 0.9)$ allows us to tune the expected lengths nicely: the lower p_f, the higher the length.

The first experiment examines the time complexity using the method described in Sect. 3. Parameter p_f varies so that the expected lengths of the strings are 6.22, 7.75, 9.71, 12.33, and 16. For this experiment, we generate strings randomly and uniformly of lengths ranging from 1 to 13 from an alphabet of size 2. Then, we measure the execution time consumed to compute the distance between the string and the PFA, including the construction of the $\mathcal{A}_w$. The experiment is repeated 100 times for each PFA and each witness string length considered. Average results are shown in Fig. 2.

As expected, the complexity of the procedure grows very fast as the length of the witness string is increased. Note that the y-axis is shown in logarithmic

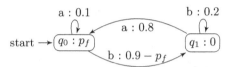

Fig. 1. Parametrizable PFA used in the experiments.

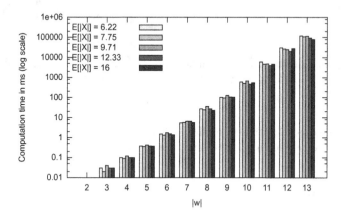

Fig. 2. Average execution time in milliseconds needed for the method based on a MA to compute the distance between a witness string w and a PFA.

scale, so the curve suggests an exponential growth. According to the empirical curve shown in these experiments, computing the distance of a witness string of length 50 would mean an execution time in the order of 10^7 years. Another thing to remark is that this method is not dependent upon the configuration of the PFA, as long as its number of states does not change.

On the other hand, the same experiments is repeated using the FPRAS for ϵ and δ fixed to 0.01. The length of the witness strings are 4, 7, 10, 13 and 16. The parameter p_f is configured so that the expected length of the strings of the PFA are 7.75, 12.33, 16, 21.5, 30.66, 36.31, 44, and 49. Figure 3 illustrates the average results of these experiments.

It can be noticed that the configuration of the PFA is the most relevant factor for the FPRAS. As might be expected by the relationship between the size of the PFA measured by $c_\mathcal{A}$ and the expected length of the strings, the empirical growth seems to be polynomial [3]. It is also observed that the length of the witness string is a factor that can vary the time complexity since the distances to compute are more expensive. Nevertheless, it is important to emphasize that the FPRAS scales relatively well: for a PFA whose expected length of strings is 100 and a witness string of length 100, the result (89.3235) was computed in around 25 min fixing both ϵ and δ to 0.01.

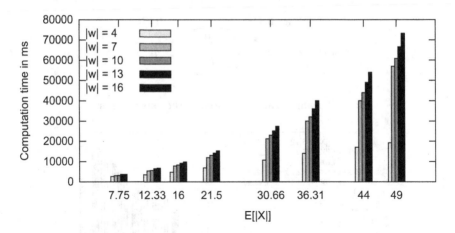

Fig. 3. Average execution time in milliseconds needed for the FPRAS method to compute the distance between a witness string w and a PFA.

6   Conclusion

Two algorithms have been proposed to deal with the question of computing the expected edit distance from a witness string to a distribution given by a PFA. Whereas one is able to compute this value exactly, it is limited to cases where the length of the witness string is short, as the construction involves building a multiplicity whose size can increase in an exponential way with the length of this string. The first one also shows that the question is decidable and that the solution can be expressed with rational weights.

On the other hand we have a FPRAS which will return with high confidence a value (ϵ)-close to the correct result. It has been shown that its complexity essentially depends on the expected length of the strings of the distribution.

The above results raise several extra questions:

- Computing the expected edit distance between two PFA. In [17] a technique is proposed for the special cases where these PFA correspond to finite languages, or can be determinized. A randomized technique in which strings are drawn from both distributions is likely to work.
- The exact status of EDD remains unclear. The (decision) problem is decidable, as witnessed by Theorem 1. But is it in **NP**?
- A more technical puzzling question concerns the size of the multiplicity automaton built in Sect. 3. The experiments and the construction proposed in Appendix B of [6] shows that polynomial bounds are not going to be met. But the proof relies on an alphabet whose size increases with the length of string w. Having a construction with a fixed size (ideally 2) is an open question.

More importantly, the really crucial question is that of computing the median string, given a PFA.

When given a distribution, a prediction system will often attempt to return the most probable string in order to minimize the empirical risk by following a *maximum a posteriori probability* (MAP) criterion.

Nevertheless, while this may be applicable in a large number of applications, other loss functions can be better suited than the 0/1 loss. For instance, very often the final goal is to reduce the number of post-processing corrections required to transform a hypothesis. This is usually counted by means of the Levenshtein or edit distance (d_e), or a related metric like the Word Error Rate (WER). Then, the empirical risk becomes

$$R(w|x) = \sum_{v \in \Sigma^*} Pr(v|x) d_e(w,v)$$

In which case, the optimum string is the *median string*. Yet most often the most probable string (or an approximation of it) is proposed instead of the median string, whose search is related to a $\mathcal{NP}$-hard problem [9] even in a finite case. This inconsistency is well known [12], and there have been a number of studies addressing this issue [8,21], with recently a specific analysis of the relationship between 0/1 loss functions and other discrete loss functions [22]. Other approaches include the introduction of heuristics to approximate the median string [1,13,15].

This constitutes of course a real challenge.

Acknowledgements. The authors wish to acknowledge the help of Borja Balle in establishing the proof of Theorem 2, and the comments of the 3 anonymous reviewers of this paper. Also the financial help of the Spanish Ministerio de Educación, Cultura y Deporte through a FPU grant (Ref. AP2012-0939) and the Spanish Ministerio de Economía y Competitividad through Project No. TIN2013-48152-C2-1-R (supported by UE FEDER funds). This work was partly done while the second author was supported by the University of Kyoto.

References

1. Abreu, J., Rico-Juan, J.R.: A new iterative algorithm for computing a quality approximate median of strings based on edit operations. Patt. Rec. Lett. **36**, 74–80 (2014)
2. Allauzen, C., Mohri, M.: Linear-Space Computation of the Edit-Distance Between a String and a Finite Automaton (2009). CoRR abs/0904.4686
3. Balle, B.: Learning finite-state machines: algorithmic and statistical aspects. Ph.D. thesis, Universitat Politécnica de Catalunya (2013)
4. Becerra-Bonache, L., de la Higuera, C., Janodet, J.C., Tantini, F.: Learning balls of strings from edit corrections. J. Mach. Learn. Res. **9**, 1841–1870 (2008)
5. Berstel, J., Reutenauer, C.: Rational Series and Their Languages. Springer, Heidelberg (1988)
6. Calvo-Zaragoza, J., de la Higuera, C., Oncina, J.: Computing the expected edit distance from a string to a PFA. Complete Version with Appendices (2016). https://hal.archives-ouvertes.fr/hal-01308549

7. Durbin, R., Eddy, S., Krogh, A., Mitchison, G.: Biological Sequence Analysis: Probalistic Models of Proteins and Nucleic Acids. Cambridge University Press, Cambridge (1998)
8. Ehling, N., Zens, R., Ney, H.: Minimum Bayes risk decoding for BLEU. In: Proceedings of 45th Annual Meeting of the Association for Computational Linguistics (ACL) (2007)
9. de la Higuera, C., Casacuberta, F.: Topology of strings: median string is NP-complete. Theor. Comput. Sci. **230**, 39–48 (2000)
10. de la Higuera, C., Oncina, J.: Computing the most probable string with a probabilistic finite state machine. In: Proceedings of FSMNLP (2013)
11. de la Higuera, C., Oncina, J.: The most probable string: an algorithmic study. J. Logic Comput. **24**(2), 311–330 (2014)
12. Jelinek, F.: Statistical Methods for Speech Recognition. MIT Press, Cambridge (1997)
13. Kruzslicz, F.: Improved greedy algorithm for computing approximate median strings. Acta Cybernetica **14**(2), 331–339 (1999)
14. Levenshtein, V.I.: Binary codes capable of correcting deletions, insertions, and reversals. Dokl. Akad. Nauk SSSR **163**(4), 845–848 (1965)
15. Martínez-Hinarejos, C.D., Juan, A., Casacuberta, F.: Median strings for k-nearest neighbour classification. Pattern Rec. Lett. **24**(1–3), 173–181 (2003)
16. Mihov, S., Schulz, K.U.: Fast approximate search in large dictionaries. Comput. Linguist. **30**(4), 451–477 (2004)
17. Mohri, M.: Edit-distance of weighted automata: general definitions and algorithms. Int. J. Found. Comput. Sci. **14**(6), 957–982 (2003)
18. Motwani, R., Raghavan, P.: Randomized Algorithm. Springer, Berlin (1995)
19. Navarro, G., Raffinot, M.: Flexible Pattern Matching. Cambridge University Press, Cambridge (2002)
20. Paz, A.: Introduction to Probabilistic Automata. Academic Press, New York (1971)
21. Schluter, R., Nussbaum-Thom, M., Ney, H.: On the relationship between Bayes risk and word error rate in ASR. IEEE Trans. Audio Speech Lang. Process. **19**(5), 1103–1112 (2011)
22. Schluter, R., Nussbaum-Thom, M., Ney, H.: Does the cost function matter in Bayes decision rule? IEEE Trans. Pattern Anal. Mach. Intell. **34**(2), 292–301 (2012)
23. Schulz, K.U., Mihov, S.: Fast string correction with Levenshtein automata. IJDAR **5**(1), 67–85 (2002)
24. Stolcke, A., Konig, Y., Weintraub, M.: Explicit word error minimization in n-best list rescoring. In: 5th European Conference on Speech Communication and Technology (1997)
25. Valiant, L.: A theory of the learnable. Commun. ACM **27**(11), 1134–1142 (1984)
26. Vazirani, V.: Approximation Algorithms. Springer, Berlin (2003)
27. Vidal, E., Thollard, F., de la Higuera, C., Casacuberta, F., Carrasco, R.C.: Probabilistic finite state automata - part I and II. Pattern Anal. Mach. Intell. **27**(7), 1013–1039 (2005)
28. Wagner, R.A.: Order-n correction for regular languages. Commun. ACM **17**(5), 265–268 (1974)

Derived-Term Automata of Multitape Rational Expressions

Akim Demaille[✉]

EPITA Research and Development Laboratory (LRDE),
14-16, rue Voltaire, 94276 Le Kremlin-Bicêtre, France
akim@lrde.epita.fr

Abstract. We introduce (weighted) rational expressions to denote series over Cartesian products of monoids. To this end, we propose the operator | to build multitape expressions such as $(a^+ \mid x + b^+ \mid y)^*$. We define expansions, which generalize the concept of derivative of a rational expression, but relieved from the need of a free monoid. We propose an algorithm based on expansions to build multitape automata from multitape expressions.

1 Introduction

Automata and rational (or regular) expressions share the same expressive power, with algorithms going from one to the other. This fact made rational expressions an extremely handy practical tool to specify some rational languages in a concise way, from which acceptors (automata) are built. There are many largely used implementations, probably starting with Thompson [15], the creator of Unix, grep, etc.

There are numerous algorithms to build an automaton from an expression. We are particularly interested in the derivative-based family of algorithms [3–5,7,10], because they offer a very natural interpretation to states (they are labeled by an expression that denotes the future of the states, i.e., the language/series accepted from this state). This allowed to support several extensions: extended operators (intersection, complement) [4,5], weights [10], additional products (shuffle, infiltration), etc.

Multitape automata, including transducers, share many properties with "single-tape" automata, in particular the Fundamental Theorem [14, Theorem 2.1, p. 409]: under appropriate conditions, multitape automata and rational (multitape) series share the same expressive power. However, as far as the author knows, there is no definition of multitape rational expressions that allows expressions such as $E_2 := (a^+ \mid x + b^+ \mid y)^*$ (Example 5). To denote such a binary relation between words, one had to build a (usual) rational expression in "normal form", without tupling of expressions but only tuples of letters such as a set of generators. So for instance instead of E_2, one must use $E_2' := ((a\mid\varepsilon)^+(\varepsilon\mid x) + (b\mid\varepsilon)^+(\varepsilon\mid y))^*$, which is larger, as is its derived-term automaton.

The contributions of this paper are twofold: we define (weighted) multitape rational expressions featuring a | operator, and we provide an algorithm to build

Y.-S. Han and K. Salomaa (Eds.): CIAA 2016, LNCS 9705, pp. 51–63, 2016.
DOI: 10.1007/978-3-319-40946-7_5

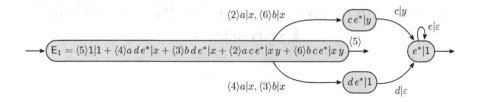

Fig. 1. The derived-term automaton of E_1 (see Examples 1 to 3) with $E_1 := \langle 5\rangle 1|1 + \langle 4\rangle a\,d\,e^*|x + \langle 3\rangle b\,d\,e^*|x + \langle 2\rangle a\,c\,e^*|x\,y + \langle 6\rangle b\,c\,e^*|x\,y$.

an equivalent automaton. This algorithm is a generalization of the derived-term based algorithms, freed from the requirement that the monoid is free.

We first settle the notations in Sect. 2, provide an algorithm to compute the expansion of an expression in Sect. 3, which is used in Sect. 4 to propose an alternative construction of the derived-term automaton.

The constructs exposed in this paper are implemented in Vcsn.[1] Vcsn is a free-software platform dedicated to weighted automata and rational expressions [8]; its lowest layer is a C++ library, on top of which Python/IPython bindings provide an interactive graphical environment.

2 Notations

Our purpose is to define (weighted) multitape rational expressions, such as $E_1 := \langle 5\rangle 1|1 + \langle 4\rangle a\,d\,e^*|x + \langle 3\rangle b\,d\,e^*|x + \langle 2\rangle a\,c\,e^*|x\,y + \langle 6\rangle b\,c\,e^*|x\,y$ (weights are written in angle brackets). It relates ade with x, with weight 4. We introduce an algorithm to build a multitape automaton (aka *transducer*) from such an expression, e.g., Fig. 1. This algorithm relies on *rational expansions*. They are to the derivatives of rational expressions what differential forms are to the derivatives of functions. Defining expansions requires several concepts, defined bottom-up in this section. The following figure presents these different entities, how they relate to each other, and where we are heading to: given a weighted multitape rational expression such as E_1, compute *its* expansion:

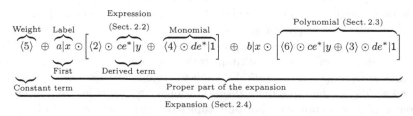

[1] See the interactive environment, http://vcsn-sandbox.lrde.epita.fr, or its documentation, http://vcsn.lrde.epita.fr/dload/2.3/notebooks/expression.derived_term.html, or this paper's companion notebook, http://vcsn.lrde.epita.fr/dload/2.3/notebooks/CIAA-2016.html.

from which we build its derived-term automaton (Fig. 1).

It is helpful to think of expansions as a normal form for expressions.

2.1 Rational Series

Series will be used to define the semantics of the forthcoming structures: they are to weighted automata what languages are to Boolean automata. Not all languages are rational (denoted by an expression), and similarly, not all series are rational (denoted by a weighted expression). We follow Sakarovitch [14, Chap. III].

In order to cope with (possibly) several tapes, we cannot rely on the traditional definitions based on the free monoid A^* for some alphabet A.

Labels. Let M be a monoid (e.g., A^* or $A^* \times B^*$), whose neutral element is denoted ε_M, or ε when clear from the context. For consistency with the way transducers are usually represented, we use $m \mid n$ rather than (m, n) to denote the pair of m and n. For instance $\varepsilon_{A^* \times B^*} = \varepsilon_{A^*} \mid \varepsilon_{B^*}$, and $\varepsilon_M \mid a \in M \times \{a\}^*$. A *set of generators* G of M is a subset of M such that $G^* = M$. A monoid M is of *finite type* (or *finitely generated*) if it admits a finite set of generators. A monoid M is *graded* if it admits a *gradation* function $|\cdot| \in M \to \mathbb{N}$ such that $\forall m, n \in M$, $|m| = 0$ iff $m = \varepsilon$, and $|mn| = |m| + |n|$. Cartesian products of graded monoids are graded, and Cartesian products of finitely generated monoids are finitely generated. Free monoids and Cartesian products of free monoids are graded and finitely generated.

Weights. Let $\langle \mathbb{K}, +, \cdot, 0_{\mathbb{K}}, 1_{\mathbb{K}} \rangle$ (or $\mathbb{K}$ for short) be a semiring whose (possibly non commutative) multiplication will be denoted by juxtaposition. $\mathbb{K}$ is *commutative* if its multiplication is. $\mathbb{K}$ is a *topological semiring* if it is equipped with a topology, and both addition and multiplication are continuous. It is *strong* if the product of two summable families is summable.

Series. A (formal power) *series* over M with *weights* (or *multiplicities*) in $\mathbb{K}$ is a map from M to $\mathbb{K}$. The weight of $m \in M$ in a series s is denoted $s(m)$. The *null* series, $m \mapsto 0_{\mathbb{K}}$, is denoted 0; for any $m \in M$ (including ε_M), m denotes the series $u \mapsto 1_{\mathbb{K}}$ if $u = m, 0_{\mathbb{K}}$ otherwise. If M is of finite type, then we can define the Cauchy product of series. $s \cdot t := m \mapsto \sum_{u,v \in M \mid uv=m} s(u) \cdot t(v)$. Equipped with the pointwise addition $(s + t := m \mapsto s(m) + t(m))$ and $\cdot$ as multiplication, the set of these series forms a semiring denoted $\langle \mathbb{K} \langle\!\langle M \rangle\!\rangle, +, \cdot, 0, \varepsilon \rangle$.

The *constant term* of a series s, denoted s_ε, is $s(\varepsilon)$, the weight of the empty word. A series s is *proper* if $s_\varepsilon = 0_{\mathbb{K}}$. The *proper part* of s is the proper series s_p such that $s = s_\varepsilon + s_p$.

Star. The *star* of a series is an infinite sum: $s^* := \sum_{n \in \mathbb{N}} s^n$. To ensure semantic soundness, we need M to be graded monoid and $\mathbb{K}$ to be a strong topological semiring.

Proposition 1. *Let M be a graded monoid and $\mathbb{K}$ a strong topological semiring. Let $s \in \mathbb{K}\langle\!\langle M \rangle\!\rangle$, s^* is defined iff s_ε^* is defined and then $s^* = s_\varepsilon^* + s_\varepsilon^* s_p s^*$.*

Proof. By [14, Proposition 2.6, p. 396] s^* is defined iff s_ε^* is defined and then $s^* = (s_\varepsilon^* s_p)^* s_\varepsilon^* = s_\varepsilon^* (s_p s_\varepsilon^*)^*$. The result then follows directly from $s^* = \varepsilon + ss^*$: $s^* = s_\varepsilon^* (s_p s_\varepsilon^*)^* = s_\varepsilon^* (\varepsilon + (s_p s_\varepsilon^*)(s_p s_\varepsilon^*)^*) = s_\varepsilon^* + s_\varepsilon^* s_p (s_\varepsilon^* (s_p s_\varepsilon^*)^*) = s_\varepsilon^* + s_\varepsilon^* s_p s^*$. □

Tuple. We suppose $\mathbb{K}$ is commutative. The *tupling* of two series $s \in \mathbb{K}\langle\!\langle M \rangle\!\rangle, t \in \mathbb{K}\langle\!\langle N \rangle\!\rangle$, is the series $s \mid t := m \mid n \in M \times N \mapsto s(m)t(n)$. It is a member of $\mathbb{K}\langle\!\langle M \times N \rangle\!\rangle$.

Proposition 2. *For all series* $s, s' \in \mathbb{K}\langle\!\langle M \rangle\!\rangle$ *and* $t, t' \in \mathbb{K}\langle\!\langle N \rangle\!\rangle$, $(s + s') \mid t = s \mid t + s' \mid t$ *and* $s \mid (t + t') = s \mid t + s \mid t'$.

Proof. Let $m \mid n \in M \times N$. $((s+s') \mid t)(m \mid n) = (s+s')(m) \cdot t(n) = (s(m) + s'(m)) \cdot t(n) = s(m) \cdot t(n) + s'(m) \cdot t(n) = (s \mid t)(m \mid n) \cdot (s' \mid t)(m \mid n) = (s \mid t + s' \mid t)(m \mid n)$. Likewise for right distributivity.   □

From now on, M is a graded monoid of finite type, and $\mathbb{K}$ a commutative strong topological semiring.

2.2   Weighted Rational Expressions

Contrary to the usual definition, we do not require a finite alphabet: any set of generators $G \subseteq M$ will do. For expressions with more than one tape, we required $\mathbb{K}$ to be commutative; however, for single tape expressions, our results apply to non-commutative semirings, hence there are two exterior products.

Definition 1 (Expression). *A rational expression* E *over* G *is a term built from the following grammar, where* $a \in G$ *denotes any non empty label, and* $k \in \mathbb{K}$ *any weight:* E ::= 0 $\mid$ 1 $\mid$ a $\mid$ E + E $\mid$ $\langle k \rangle$E $\mid$ E$\langle k \rangle$ $\mid$ E $\cdot$ E $\mid$ E* $\mid$ E $\mid$ E.

Expressions are syntactic; they are finite notations for (some) series.

Definition 2 (Series Denoted by an Expression). *Let* E *be an expression. The series denoted by* E, *noted* $[\![E]\!]$, *is defined by induction on* E:

$$[\![0]\!] := 0 \qquad [\![1]\!] := \varepsilon \qquad [\![a]\!] := a \qquad [\![E + F]\!] := [\![E]\!] + [\![F]\!] \qquad [\![\langle k \rangle E]\!] := k[\![E]\!]$$

$$[\![E\langle k \rangle]\!] := [\![E]\!]k \quad [\![E \cdot F]\!] := [\![E]\!] \cdot [\![F]\!] \quad [\![E^*]\!] := [\![E]\!]^* \quad [\![E \mid F]\!] := [\![E]\!] \mid [\![F]\!]$$

An expression is *valid* if it denotes a series. More specifically, there are two requirements. First, the expression must be well-formed, i.e., concatenation and disjunction must be applied to expressions of appropriate number of tapes. For instance, $a + b|c$ and $a(b|c)$ are ill-formed, $(a \mid b)^* \mid c + a \mid (b \mid c)^*$ is well-formed. Second, to ensure that $[\![F]\!]^*$ is well defined for each subexpression of the form F*, the constant term of $[\![F]\!]$ must be *starrable* in $\mathbb{K}$ (Proposition 1). This definition, which involves series (semantics) to define a property of expressions (syntax), will be made effective (syntactic) with the appropriate definition of the constant term $d_\varepsilon(E)$ *of an expression* E (Definition 6).

Let $[n]$ denote $\{1, \ldots, n\}$). The *size* (aka *length*) of a (valid) expression E, $|E|$, is its total number of symbols, not counting parenthesis; for a given tape number $i \in [k]$ the *width on tape* i, $\|E\|_i$, is the number of occurrences of labels on the tape i, the *width* of E (aka *literal length*), $\|E\| := \sum_{i \in [k]} \|E\|_i$ is the total number of occurrences of labels.

Two expressions E and F are *equivalent* iff $[\![E]\!] = [\![F]\!]$. Some expressions are "trivially equivalent"; any candidate expression will be rewritten via the following *trivial identities*. Any subexpression of a form listed to the left of a '$\Rightarrow$' is rewritten as indicated on the right.

$$E + 0 \Rightarrow E \qquad 0 + E \Rightarrow E$$

$$\langle 0_{\mathbb{K}} \rangle E \Rightarrow 0 \quad \langle 1_{\mathbb{K}} \rangle E \Rightarrow E \quad \langle k \rangle 0 \Rightarrow 0 \quad \langle k \rangle \langle h \rangle E \Rightarrow \langle kh \rangle E$$

$$E \langle 0_{\mathbb{K}} \rangle \Rightarrow 0 \quad E \langle 1_{\mathbb{K}} \rangle \Rightarrow E \quad 0 \langle k \rangle \Rightarrow 0 \quad E \langle k \rangle \langle h \rangle \Rightarrow E \langle kh \rangle$$

$$(\langle k \rangle E) \langle h \rangle \Rightarrow \langle k \rangle (E \langle h \rangle) \qquad \ell \langle k \rangle \Rightarrow \langle k \rangle \ell$$

$$E \cdot 0 \Rightarrow 0 \qquad 0 \cdot E \Rightarrow 0$$

$$(\langle k \rangle^? 1) \cdot E \Rightarrow \langle k \rangle^? E \qquad E \cdot (\langle k \rangle^? 1) \Rightarrow E \langle k \rangle^?$$

$$0^* \Rightarrow 1$$

$$(\langle k \rangle^? E) \mid (\langle h \rangle^? F) \Rightarrow \langle kh \rangle^? E \mid F$$

where E is a rational expression, $\ell \in G \cup \{1\}$ a label, $k, h \in \mathbb{K}$ weights, and $\langle k \rangle^? \ell$ denotes either $\langle k \rangle \ell$, or ℓ in which case $k = 1_{\mathbb{K}}$ in the right-hand side of $\Rightarrow$. The choice of these identities is beyond the scope of this paper (see [14]), however note that they are limited to trivial properties; in particular *linearity* ("weighted ACI": associativity, commutativity and $\langle k \rangle E + \langle h \rangle E \Rightarrow \langle k+h \rangle E$) is not enforced. In practice, additional identities help reducing the automaton size [12].

2.3   Rational Polynomials

At the core of the idea of "partial derivatives" introduced by Antimirov [3], is that of *sets* of rational expressions, later generalized in *weighted sets* by Lombardy and Sakarovitch [10], i.e., functions (partial, with finite domain) from the set of rational expressions into $\mathbb{K} \setminus \{0_{\mathbb{K}}\}$. It proves useful to view such structures as "polynomials of expressions". In essence, they capture the linearity of addition.

Definition 3 (Rational Polynomial). *A polynomial (of rational expressions) is a finite (left) linear combination of expressions. Syntactically it is a term built from the grammar* $P ::= 0 \mid \langle k_1 \rangle \odot E_1 \oplus \cdots \oplus \langle k_n \rangle \odot E_n$ *where* $k_i \in \mathbb{K} \setminus \{0_{\mathbb{K}}\}$ *denote non-null weights, and* E_i *denote non-null expressions. Expressions may not appear more than once in a polynomial. A monomial is a pair* $\langle k_i \rangle \odot E_i$.

We use specific symbols ($\odot$ and $\oplus$) to clearly separate the outer polynomial layer from the inner expression layer. Let $P = \bigoplus_{i \in [n]} \langle k_i \rangle \odot E_i$ be a polynomial of expressions. The *"projection"* of P is the expression $\mathsf{expr}(P) := \langle k_1 \rangle E_1 + \cdots + \langle k_n \rangle E_n$ (or 0 if P is null); this operation is performed on a canonical form of the polynomial (expressions are sorted in a well defined order). Polynomials denote series: $[\![P]\!] := [\![\mathsf{expr}(P)]\!]$. The *terms* of P is the set $\mathsf{exprs}(P) := \{E_1, \ldots, E_n\}$.

Example 1. Let $E_1 := \langle 5 \rangle 1 | 1 + \langle 4 \rangle a\,d\,e^* | x + \langle 3 \rangle b\,d\,e^* | x + \langle 2 \rangle a\,c\,e^* | x\,y + \langle 6 \rangle b\,c\,e^* | x\,y$. Polynomial '$P_{1,a|x} := \langle 2 \rangle \odot ce^* | y \oplus \langle 4 \rangle \odot de^* | 1$' has two monomials: '$\langle 2 \rangle \odot ce^* | y$' and '$\langle 4 \rangle \odot de^* | 1$'. It denotes the (left) quotient of $[\![E_1]\!]$ by $a | x$, and '$P_{1,b|x} := \langle 6 \rangle \odot ce^* | y \oplus \langle 3 \rangle \odot de^* | 1$' the quotient by $b | x$.

Let $P = \bigoplus_{i \in [n]} \langle k_i \rangle \odot E_i, Q = \bigoplus_{j \in [m]} \langle h_i \rangle \odot F_i$ be polynomials, k a weight and F an expression, all possibly null, we introduce the following operations:

$$P \cdot F := \bigoplus_{i \in [n]} \langle k_i \rangle \odot (E_i \cdot F) \quad \langle k \rangle P := \bigoplus_{i \in [n]} \langle k k_i \rangle \odot E_i \quad P \langle k \rangle := \bigoplus_{i \in [n]} \langle k_i \rangle \odot (E_i \langle k \rangle)$$

$$P | 1 := \bigoplus_{i \in [n]} \langle k_i \rangle \odot E_i | 1 \quad 1 | P := \bigoplus_{i \in [n]} \langle k_i \rangle \odot 1 | E_i$$

$$P | Q := \bigoplus_{(i,j) \in [n] \times [m]} \langle k_i \cdot h_j \rangle \odot E_i | F_j$$

Trivial identities might simplify the result. Note the asymmetry between left and right exterior products. The addition of polynomials is commutative, multiplication by zero (be it an expression or a weight) evaluates to the null polynomial, and the left-multiplication by a weight is distributive.

Lemma 1. $[\![P \cdot F]\!] = [\![P]\!] \cdot [\![F]\!]$ $[\![\langle k \rangle P]\!] = \langle k \rangle [\![P]\!]$ $[\![P \langle k \rangle]\!] = [\![P]\!] \langle k \rangle$ $[\![P | Q]\!] = [\![P]\!] | [\![Q]\!]$.

2.4   Rational Expansions

Definition 4 (Rational Expansion). *A* rational expansion X *is a term* $X ::= \langle X_\varepsilon \rangle \oplus a_1 \odot [X_{a_1}] \oplus \cdots \oplus a_n \odot [X_{a_n}]$ *where* $X_\varepsilon \in \mathbb{K}$ *is a weight (possibly null),* $a_i \in G \setminus \{\varepsilon\}$ *non-empty labels (occurring at most once), and* X_{a_i} *non-null polynomials. The* constant term *is* X_ε, *the* proper part *is* $X_p := a_1 \odot [X_{a_1}] \oplus \cdots \oplus a_n \odot [X_{a_n}]$, *the* firsts *is* $f(X) := \{a_1, \ldots, a_n\}$ *(possibly empty) and the* terms $\mathsf{exprs}(X) := \bigcup_{i \in [n]} \mathsf{exprs}(X_{a_i})$.

To ease reading, polynomials are written in square brackets. Contrary to expressions and polynomials, there is no specific term for the null expansion: it is represented by $\langle 0_\mathbb{K} \rangle$, the null weight. Except for this case, null constant terms are left implicit. Expansions will be written: $X = \langle X_\varepsilon \rangle \oplus \bigoplus_{a \in f(X)} a \odot [X_a]$. When more convenient, we write $X(\ell)$ instead of X_ℓ for $\ell \in f(X) \cup \{\varepsilon\}$.

An expansion X can be "projected" as a rational expression $\mathsf{expr}(X)$ by mapping weights, labels and polynomials to their corresponding rational expressions, and $\oplus / \odot$ to the sum/concatenation of expressions. Again, this is performed on a canonical form of the expansion: labels are sorted. Expansions also denote series: $[\![X]\!] := [\![\mathsf{expr}(X)]\!]$. An expansion X is *equivalent* to an expression E iff $[\![X]\!] = [\![E]\!]$.

Example 2 (Example 1 continued). Expansion $X_1 := \langle 5 \rangle \oplus a|x \odot [P_{1,a|x}] \oplus b|x \odot [P_{1,b|x}]$ has $X_1(\varepsilon) = \langle 5 \rangle$ as constant term, and maps the generator $a|x$ (resp. $b|x$) to the polynomial $X_1(a|x) = P_{1,a|x}$ (resp. $X_1(b|x) = P_{1,b|x}$). X_1 can be proved to be equivalent to E_1.

Let X, Y be expansions, k a weight, and E an expression (all possibly null):

$$X \oplus Y := \langle X_\varepsilon + Y_\varepsilon \rangle \oplus \bigoplus_{a \in f(X) \cup f(Y)} a \odot [X_a \oplus Y_a] \tag{1}$$

$$\langle k \rangle X := \langle k X_\varepsilon \rangle \oplus \bigoplus_{a \in f(X)} a \odot [\langle k \rangle X_a] \qquad X \langle k \rangle := \langle X_\varepsilon k \rangle \oplus \bigoplus_{a \in f(X)} a \odot [X_a \langle k \rangle] \tag{2}$$

$$X \cdot E := \bigoplus_{a \in f(X)} a \odot [X_a \cdot E] \qquad \text{with } X \text{ proper: } X_\varepsilon = 0_\mathbb{K} \tag{3}$$

$$X \mid Y := \langle X_\varepsilon Y_\varepsilon \rangle \oplus \langle X_\varepsilon \rangle \bigoplus_{b \in f(Y)} (\varepsilon|b) \odot (1 \mid Y_b) \oplus \langle Y_\varepsilon \rangle \bigoplus_{a \in f(X)} (a|\varepsilon) \odot (X_a \mid 1)$$
$$\oplus \bigoplus_{a|b \in f(X) \times f(Y)} (a|b) \odot (X_a \mid Y_b) \tag{4}$$

Since by definition expansions never map to null polynomials, some firsts might be smaller that suggested by these equations. For instance in $\mathbb{Z}$ the sum of $\langle 1 \rangle \oplus a \odot [\langle 1 \rangle \odot b]$ and $\langle 1 \rangle \oplus a \odot [\langle -1 \rangle \odot b]$ is $\langle 2 \rangle$.

The following lemma is simple to establish: lift semantic equivalences, such as Proposition 2, to syntax, using Lemma 1.

Lemma 2. $[\![X \oplus Y]\!] = [\![X]\!] + [\![Y]\!] \qquad [\![\langle k \rangle X]\!] = \langle k \rangle [\![X]\!] \qquad [\![X \langle k \rangle]\!] = [\![X]\!] \langle k \rangle$
$[\![X \cdot E]\!] = [\![X]\!] \cdot [\![E]\!] \qquad [\![X \mid Y]\!] = [\![X]\!] \mid [\![Y]\!]$

2.5 Finite Weighted Automata

Definition 5 (Weighted Automaton). *A weighted automaton $\mathcal{A}$ is a tuple $\langle M, G, \mathbb{K}, Q, E, I, T \rangle$ where:*

- *M is a monoid,*
- *G (the labels) is a set of generators of M,*
- *$\mathbb{K}$ (the set of weights) is a semiring,*
- *Q is a finite set of states,*
- *I and T are the initial and final functions from Q into $\mathbb{K}$,*
- *E is a (partial) function from $Q \times G \times Q$ into $\mathbb{K} \setminus \{0_\mathbb{K}\}$; its domain represents the transitions: (source, label, destination).*

An automaton is proper if no label is ε_M.

A *computation* $p = (q_0, a_0, q_1)(q_1, a_1, q_2) \cdots (q_n, a_n, q_{n+1})$ in an automaton is a sequence of transitions where the source of each is the destination of the previous one; its *label* is $a_0 a_1 \cdots a_n \in M$, its *weight* is $I(q_0) \otimes E(q_0, a_0, q_1) \otimes \cdots \otimes E(q_n, a_n, q_{n+1}) \otimes T(q_{n+1}) \in \mathbb{K}$. The *evaluation* of word u by $\mathcal{A}$, $\mathcal{A}(u)$, is the sum of the weights of all the computations labeled by u, or $0_\mathbb{K}$ if there are none.

The *behavior* of an automaton $\mathcal{A}$ is the series $[\![\mathcal{A}]\!] := m \mapsto \mathcal{A}(m)$. A state q is *initial* if $I(q) \neq 0_{\mathbb{K}}$. A state q is *accessible* if there is a computation from an initial state to q. The *accessible* part of an automaton $\mathcal{A}$ is the subautomaton whose states are the accessible states of $\mathcal{A}$. The size of an automaton, $|\mathcal{A}|$, is its number of states.

We are interested, given an expression E, in an algorithm to compute an automaton $\mathcal{A}_E$ such that $[\![\mathcal{A}_E]\!] = [\![E]\!]$ (Definition 7). To this end, we first introduce a simple recursive procedure to compute *the* expansion of an expression.

3   Expansion of a Rational Expression

Definition 6 (Expansion of a Rational Expression). *The* expansion of a rational expression E, *written $d(E)$, is defined inductively as follows:*

$$d(0) := \langle 0_{\mathbb{K}} \rangle \qquad d(1) := \langle 1_{\mathbb{K}} \rangle \qquad d(a) := a \odot [\langle 1_{\mathbb{K}} \rangle \odot 1] \tag{5}$$

$$d(\mathsf{E} + \mathsf{F}) := d(\mathsf{E}) \oplus d(\mathsf{F}) \tag{6}$$

$$d(\langle k \rangle \mathsf{E}) := \langle k \rangle d(\mathsf{E}) \qquad d(\mathsf{E}\langle k \rangle) := d(\mathsf{E})\langle k \rangle \tag{7}$$

$$d(\mathsf{E} \cdot \mathsf{F}) := d_p(\mathsf{E}) \cdot \mathsf{F} \oplus \langle d_\varepsilon(\mathsf{E}) \rangle d(\mathsf{F}) \tag{8}$$

$$d(\mathsf{E}^*) := \langle d_\varepsilon(\mathsf{E})^* \rangle \oplus \langle d_\varepsilon(\mathsf{E})^* \rangle d_p(\mathsf{E}) \cdot \mathsf{E}^* \tag{9}$$

$$d(\mathsf{E} \mid \mathsf{F}) := d(\mathsf{E}) \mid d(\mathsf{F}) \tag{10}$$

where $d_\varepsilon(\mathsf{E}) := d(\mathsf{E})_\varepsilon, d_p(\mathsf{E}) := d(\mathsf{E})_p$ are the constant term/proper part of $d(\mathsf{E})$.

The right-hand sides are indeed expansions. The computation trivially terminates: induction is performed on strictly smaller subexpressions. These formulas are enough to compute the expansion of an expression; there is no secondary process to compute the firsts — indeed $d(a) := a \odot [\langle 1_{\mathbb{K}} \rangle \odot 1]$ suffices and every other case simply propagates or assembles the firsts — or the constant terms.

Note that the firsts are a subset of the labels of the expression, hence of $G \setminus \{\varepsilon\}$. In particular, no first includes ε.

Proposition 3. *The expansion of a rational expression is equivalent to the expression.*

Proof. We prove that $[\![d(\mathsf{E})]\!] = [\![\mathsf{E}]\!]$ by induction on the expression. The equivalence is straightforward for (5) to (7) and (10), viz., $[\![d(\mathsf{E} \mid \mathsf{F})]\!] = [\![d(\mathsf{E}) \mid d(\mathsf{F})]\!]$ (by (10)) $= [\![d(\mathsf{E})]\!] \mid [\![d(\mathsf{F})]\!]$ (by Lemma 2) $= [\![\mathsf{E}]\!] \mid [\![\mathsf{F}]\!]$ (by induction hypothesis) $= [\![\mathsf{E} \mid \mathsf{F}]\!]$ (by Lemma 2). The case of multiplication, (8), follows from:

$$\begin{aligned}
[\![d(\mathsf{E} \cdot \mathsf{F})]\!] &= [\![d_p(\mathsf{E}) \cdot \mathsf{F} \oplus \langle d_\varepsilon(\mathsf{E}) \rangle \cdot d(\mathsf{F})]\!] & &= [\![d_p(\mathsf{E})]\!] \cdot [\![\mathsf{F}]\!] + \langle d_\varepsilon(\mathsf{E}) \rangle \cdot [\![d(\mathsf{F})]\!] \\
&= [\![d_p(\mathsf{E})]\!] \cdot [\![\mathsf{F}]\!] + \langle d_\varepsilon(\mathsf{E}) \rangle \cdot [\![\mathsf{F}]\!] & &= ([\![\langle d_\varepsilon(\mathsf{E}) \rangle]\!] + [\![d_p(\mathsf{E})]\!]) \cdot [\![\mathsf{F}]\!] \\
&= [\![\langle d_\varepsilon(\mathsf{E}) \rangle + d_p(\mathsf{E})]\!] \cdot [\![\mathsf{F}]\!] & &= [\![d(\mathsf{E})]\!] \cdot [\![\mathsf{F}]\!] \\
&= [\![\mathsf{E}]\!] \cdot [\![\mathsf{F}]\!] & &= [\![\mathsf{E} \cdot \mathsf{F}]\!]
\end{aligned}$$

It might seem more natural to exchange the two terms (i.e., $\langle d_\varepsilon(\mathsf{E}) \rangle \cdot d(\mathsf{F}) \oplus d_p(\mathsf{E}) \cdot \mathsf{F}$), but an implementation first computes $d(\mathsf{E})$ and then computes $d(\mathsf{F})$ *only if $d_\varepsilon(\mathsf{E}) \neq 0_{\mathbb{K}}$*. The case of Kleene star, (9), follows from Proposition 1. □

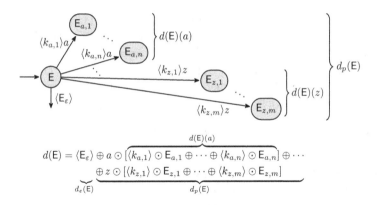

$$d(\mathsf{E}) = \langle \mathsf{E}_\varepsilon \rangle \oplus a \odot \overbrace{[\langle k_{a,1} \rangle \odot \mathsf{E}_{a,1} \oplus \cdots \oplus \langle k_{a,n} \rangle \odot \mathsf{E}_{a,n}]}^{d(\mathsf{E})(a)} \oplus \cdots$$
$$\underbrace{\oplus z \odot [\langle k_{z,1} \rangle \odot \mathsf{E}_{z,1} \oplus \cdots \oplus \langle k_{z,m} \rangle \odot \mathsf{E}_{z,m}]}_{d_p(\mathsf{E})}$$

$\underbrace{\phantom{d(\mathsf{E}) = \langle \mathsf{E}_\varepsilon \rangle}}_{d_\varepsilon(\mathsf{E})}$

Fig. 2. Initial part of $\mathcal{A}_\mathsf{E}$, the derived-term automaton of E. This figure is somewhat misleading in that some $\mathsf{E}_{a,i}$ might be equal to an $\mathsf{E}_{z,j}$, or E (but never another $\mathsf{E}_{a,j}$).

4   Expansion-Based Derived-Term Automaton

Definition 7 (Expansion-Based Derived-Term Automaton). *The* derived-term automaton *of an expression* E *over* G *is the* accessible part *of the automaton* $\mathcal{A}_\mathsf{E} := \langle M, G, \mathbb{K}, Q, E, I, T \rangle$ *defined as follows:*

– *Q is the set of rational expressions on alphabet A with weights in $\mathbb{K}$,*
– *$I = \mathsf{E} \mapsto 1_\mathbb{K}$,*
– *$E(\mathsf{F}, a, \mathsf{F}') = k$ iff $a \in f(d(\mathsf{F}))$ and $\langle k \rangle \mathsf{F}' \in d(\mathsf{F})(a)$,*
– *$T(\mathsf{F}) = k$ iff $\langle k \rangle = d(\mathsf{F})(\varepsilon)$.*

Since the firsts exclude ε, this automaton is proper. It is straightforward to extract an algorithm from Definition 7, using a work-list of states whose outgoing transitions to compute. The Fig. 2 illustrates the process. This approach admits a natural lazy implementation: the whole automaton is not computed at once, but rather, states and transitions are computed on-the-fly, on demand, for instance when evaluating a word [7]. However, we must justify Definition 7 by proving that this automaton is finite (Theorem 1).

Example 3 (Examples 1 and 2 continued). With $\mathsf{E}_1 := \langle 5 \rangle 1|1 + \langle 4 \rangle a\,d\,e^*|x + \langle 3 \rangle b\,d\,e^*|x + \langle 2 \rangle a\,c\,e^*|x\,y + \langle 6 \rangle b\,c\,e^*|x\,y$, one has:

$$d(\mathsf{E}_1) = \langle 5 \rangle \oplus a|x \odot [\langle 2 \rangle \odot ce^* \mid y \oplus \langle 4 \rangle \odot de^* \mid \varepsilon] \oplus b|x \odot [\langle 6 \rangle \odot ce^* \mid y \oplus \langle 3 \rangle \odot de^* \mid \varepsilon]$$
$$= \mathsf{X}_1 \quad \text{(from Example 2)}$$

Figure 1 shows the resulting derived-term automaton.

Theorem 1. *For any k-tape expression E, $|\mathcal{A}_\mathsf{E}| \leq \prod_{i \in [k]}(\|\mathsf{E}\|_i + 1) + 1$.*

Proof. The proof goes in several steps. First introduce the *true derived terms* of E, a set of expressions noted $\mathrm{TD}(\mathsf{E})$, and the *derived terms* of E, $\mathrm{D}(\mathsf{E}) := \mathrm{TD}(\mathsf{E}) \cup \{\mathsf{E}\}$. $\mathrm{TD}(\mathsf{E})$ admits a simple inductive definition similar to [2,

Def. 3], to which we add $TD(E\,|\,F) := (TD(E)\,|\,TD(F)) \cup (\{1\}\,|\,TD(F)) \cup (TD(E)\,|\,\{1\})$, where for two sets of expressions E, F we introduce $E|F := \{E|F\}_{(E,F)\in E\times F}$. Second, verify that $|TD(E)| \le \prod_{i\in[k]}(\|E\|_i + 1)$ (hence finite). Third, prove that $D(E)$ is "stable by expansion", i.e., $\forall F \in D(E), \text{exprs}(d(F)) \subseteq D(E)$. Finally, observe that the states of $\mathcal{A}_E$ are therefore members of $D(E)$, whose size is less than or equal to $1 + |TD(E)|$. □

Theorem 2. *Any expression* E *and its expansion-based derived-term automaton* $\mathcal{A}_E$ *denote the same series, i.e.,* $[\![\mathcal{A}_E]\!] = [\![E]\!]$.

Example 4. Let $\mathcal{A}_k$ be the derived-term automaton of the k-tape expression $a_1^* \,|\, \cdots \,|\, a_k^*$. The states of $\mathcal{A}_k$ are all the possible expressions where the tape i features 1 or a_i^*, except $1|\cdots|1$. Therefore $|\mathcal{A}_k| = 2^k - 1$, and $\prod_{i\in[k]}(\|E\|_i + 1) = 2^k$.

$\mathcal{A}_3$, the derived-term automaton of $a^* \,|\, b^* \,|\, c^*$, is depicted on the right.

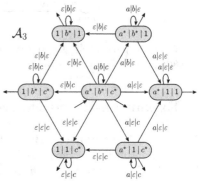

Proof (Theorem 2). We will prove $[\![\mathcal{A}_E]\!](m) = [\![E]\!](m)$ by induction on $m \in M$. If $m = \varepsilon$, then $[\![\mathcal{A}_E]\!](m) = E_\varepsilon = d(E)(\varepsilon) = [\![d(E)]\!](\varepsilon) = [\![E]\!](\varepsilon)$.

If m is not ε, then it can be generated in a (finite) number of ways: let $F(E, m) := \{(a, m_a) \in f(d(E)) \times M \mid m = am_a\}$. $F(E, m)$ is a function: for a given a, there is at most one m_a such that $(a, m_a) \in F(E, m)$. Figure 2 is helpful.

$$
\begin{aligned}
[\![\mathcal{A}_E]\!](m) &= \sum_{(a,m_a)\in F(E,m)} \sum_{i\in[n_a]} \langle k_{a,i}\rangle [\![\mathcal{A}_{E_{a,i}}]\!](m_a) && \text{by definition of } \mathcal{A}_E \\
&= \sum_{(a,m_a)\in F(E,m)} \sum_{i\in[n_a]} \langle k_{a,i}\rangle [\![E_{a,i}]\!](m_a) && \text{by induction hypothesis} \\
&= \sum_{(a,m_a)\in F(E,m)} \Big[\!\!\sum_{i\in[n_a]} \langle k_{a,i}\rangle E_{a,i}\Big]\!\!\Big](m_a) && \text{by Lemma 1} \\
&= \sum_{(a,m_a)\in F(E,m)} [\![d(E)(a)]\!](m_a) = \sum_{(a,m_a)\in F(E,m)} [\![a \odot d(E)(a)]\!](am_a) && \\
&= \sum_{a\in f(d(E))} [\![a \odot d(E)(a)]\!](m) && F(E, m) \text{ is a function} \\
&= \Big[\!\!\sum_{a\in f(d(E))} a \odot d(E)(a)\Big]\!\!\Big](m) && \text{by Lemma 2} \\
&= [\![d_\varepsilon(E)]\!](m) && \text{by definition} \\
&= [\![d(E)]\!](m) && \text{since } m \ne \varepsilon \\
&= [\![E]\!](m) && \text{by Proposition 3 □}
\end{aligned}
$$

Example 5. Let $\mathsf{E}_2 := (a^+ \mid x + b^+ \mid y)^*$, where $\mathsf{E}^+ := \mathsf{E}\mathsf{E}^*$. Its expansion is

$$d(\mathsf{E}_2) = \langle 1 \rangle \oplus a|x \odot \left[(a^* \mid 1)(a^+ \mid x + b^+ \mid y)^* \right] \oplus b|y \odot \left[(b^* \mid 1)(a^+ \mid x + b^+ \mid y)^* \right]$$
$$= \langle 1 \rangle \oplus a|x \odot \left[(a^* \mid 1)\mathsf{E}_2 \right] \oplus b|y \odot \left[(b^* \mid 1)\mathsf{E}_2 \right]$$

Its derived-term automaton is:

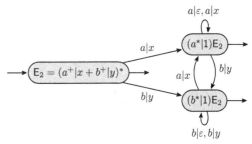

5 Related Work

Multitape rational expressions have been considered early [11], but "an n-way regular expression is simply a regular expression whose terms are n-tuples of alphabetic symbols or ε" [9]. However, Kaplan and Kay [9] do consider the full generality of the semantics of operations on rational languages and rational relations, including $\times$, the Cartesian product of languages, and even use rational expressions more general than their definition. They do not, however, provide an explicit automaton construction algorithm, apparently relying on the simple inductive construction (using the Cartesian product between automata). Our $|$ operator on series was defined as the *tensor product*, denoted $\otimes$, by Sakarovitch [14, Sect. III.3.2.5], but without equivalent for expressions.

Brzozowski [4] introduced the idea of derivatives of expressions as a means to construct an equivalent automaton. The method applies to extended (unweighted) rational expressions, and constructs a deterministic automaton. Antimirov [3] modified the computation to rely on parts of the derivatives ("partial derivatives"), which results in nondeterministic automata. Lombardy and Sakarovitch [10] extended this approach to support weighted expressions; independently, and with completely different foundations, Rutten [13] proposed a similar construction. Caron et al. [5] introduced support for (unweighted) extended expressions. Demaille [7] provides support for weighted extended expressions; expansions, originally mentioned by Brzozowski [4], are placed at the center of the construct, replacing derivatives, to gain independence with respect to the size of the alphabet, and efficiency. However, the proofs still relied on derivatives, contrary to the present work.

Based on (10) one could attempt to define a derivative-based version, with $\partial_{a|b}(\mathsf{E} \mid \mathsf{F}) := \partial_a \mathsf{E} \mid \partial_b \mathsf{F}$, however this is troublesome on several regards. First, it would also require $\partial_{a|\varepsilon}$ and $\partial_{\varepsilon|b}$, whose semantics is dubious. Second, from an implementation point of view, that would lead to repeated computations of $\partial_a \mathsf{E}$ and of $\partial_b \mathsf{F}$, unless one would cache them, but that's exactly what expansions do. And finally observe that in the derived-term automaton in Example 5, the state $(a^* \mid 1)(a^+ \mid x + b^+ \mid y)^*$ accepts words starting with a on the first tape, and y on the second, yet an outgoing transition on $a|y$ would result in a more complex automaton. Alternative definitions of derivatives may exist[2], but anyway they would no longer be equivalent to taking the left-quotient of the corresponding language: $a|y$ *is* a viable prefix from this state.

Different constructions of the derived-term automaton have been discovered [1,6]. They do not rely on derivatives at all. It is an open question whether these approaches can be adapted to support a tuple operator.

6 Conclusion

Our work is in the continuation of derivative-based computations of the derived-term automaton [3–5,10]. However, we replaced the derivatives by expansions, which lifted the requirement for the monoid of labels to be free.

In order to support k-tape (weighted) rational expressions, we introduced a tupling operator, which is more compact and readable than simple expressions on k-tape letters. We demonstrated how to build the derived-term automaton for any such expressions.

Vcsn (see footnote 1) implements the techniques exposed in this paper. Our future work aims at other operators, and studying more closely the complexity of the algorithm.

Acknowledgments. The author thanks the anonymous reviewers for their constructive comments, and A. Duret-Lutz, S. Lombardy, L. Saiu and J. Sakarovitch for their feedback during this work.

References

1. Allauzen, C., Mohri, M.: A unified construction of the Glushkov, follow, and Antimirov automata. In: Královič, R., Urzyczyn, P. (eds.) MFCS 2006. LNCS, vol. 4162, pp. 110–121. Springer, Heidelberg (2006)
2. Angrand, P.-Y., Lombardy, S., Sakarovitch, J.: On the number of broken derived terms of a rational expression. J. Autom. Lang. Comb. **15**(1/2), 27–51 (2010)
3. Antimirov, V.: Partial derivatives of regular expressions and finite automaton constructions. TCS **155**(2), 291–319 (1996)
4. Brzozowski, J.A.: Derivatives of regular expressions. J. ACM **11**(4), 481–494 (1964)
5. Caron, P., Champarnaud, J.-M., Mignot, L.: Partial derivatives of an extended regular expression. In: Dediu, A.-H., Inenaga, S., Martín-Vide, C. (eds.) LATA 2011. LNCS, vol. 6638, pp. 179–191. Springer, Heidelberg (2011)
6. Champarnaud, J.-M., Ouardi, F., Ziadi, D.: An efficient computation of the equation $\mathbb{K}$-automaton of a regular $\mathbb{K}$-expression. In: Harju, T., Karhumäki, J., Lepistö, A. (eds.) DLT 2007. LNCS, vol. 4588, pp. 145–156. Springer, Heidelberg (2007)
7. Demaille, A.: Derived-term automata for extended weighted rational expressions. Technical report 1605.01530, arXiv, May 2016. http://arxiv.org/abs/1605.01530

[2] Makarevskii and Stotskaya [11] define derivatives, but (i) in the case of expressions over tuples of letters, and (ii) only when in so-called "standard form", for which he notes "no method of constructing [an] n-expression in standard form for a regular n-expression is known".

8. Demaille, A., Duret-Lutz, A., Lombardy, S., Sakarovitch, J.: Implementation concepts in Vaucanson 2. In: Konstantinidis, S. (ed.) CIAA 2013. LNCS, vol. 7982, pp. 122–133. Springer, Heidelberg (2013)
9. Kaplan, R.M., Kay, M.: Regular models of phonological rule systems. Comput. Linguist. **20**(3), 331–378 (1994)
10. Lombardy, S., Sakarovitch, J.: Derivatives of rational expressions with multiplicity. TCS **332**(1–3), 141–177 (2005)
11. Makarevskii, A.Y., Stotskaya, E.D.: Representability in deterministic multi-tape automata. Cybern. Syst. Anal. **5**(4), 390–399 (1969)
12. Owens, S., Reppy, J., Turon, A.: Regular-expression derivatives re-examined. J. Funct. Program. **19**(2), 173–190 (2009)
13. Rutten, J.J.M.M.: Behavioural differential equations: a coinductive calculus of streams, automata, and power series. TCS **308**(1–3), 1–53 (2003)
14. Sakarovitch, J.: Elements of Automata Theory. Cambridge University Press, Cambridge (2009). Corrected English translation of Éléments de théorie des automates, Vuibert (2003)
15. Thompson, K.: Programming techniques: regular expression search algorithm. Commun. ACM **11**(6), 419–422 (1968)

Solving Parity Games Using
an Automata-Based Algorithm

Antonio Di Stasio[1], Aniello Murano[1], Giuseppe Perelli[2(✉)],
and Moshe Y. Vardi[3]

[1] Università di Napoli Federico II, Naples, Italy
[2] University of Oxford, Oxford, UK
giuseppe.perelli@cs.ac.ox.uk
[3] Rice University, Houston, USA

Abstract. *Parity games* are abstract infinite-round games that take an important role in formal verification. In the basic setting, these games are two-player, turn-based, and played under perfect information on directed graphs, whose nodes are labeled with priorities. The winner of a play is determined according to the parities (even or odd) of the minimal priority occurring infinitely often in that play. The problem of finding a winning strategy in parity games is known to be in UPTIME ∩ CoUP-TIME and deciding whether a polynomial time solution exists is a long-standing open question. In the last two decades, a variety of algorithms have been proposed. Many of them have been also implemented in a platform named `PGSolver`. This has enabled an empirical evaluation of these algorithms and a better understanding of their relative merits.

In this paper, we further contribute to this subject by implementing, for the first time, an algorithm based on alternating automata. More precisely, we consider an algorithm introduced by Kupferman and Vardi that solves a parity game by solving the emptiness problem of a corresponding alternating parity automaton. Our empirical evaluation demonstrates that this algorithm outperforms other algorithms when the game has a small number of priorities relative to the size of the game. In many concrete applications, we do indeed end up with parity games where the number of priorities is relatively small. This makes the new algorithm quite useful in practice.

1   Introduction

Parity games [11,31] are abstract infinite-duration games that represent a powerful mathematical framework to address fundamental questions in computer science. They are intimately related to other infinite-round games, such as *mean* and *discounted* payoff, *stochastic*, and *multi-agent* games [3,4,6,7].

Work supported by NSF grants CCF-1319459 and IIS-1527668, NSF Expeditions in Computing project "ExCAPE: Expeditions in Computer Augmented Program Engineering", BSF grant 9800096, ERC Advanced Investigator Grant 291528 ("Race") at Oxford and GNCS 2016: Logica, Automi e Giochi per Sistemi Auto-adattivi.
G. Perelli—Part of the work has been done while visiting Rice University.

© Springer International Publishing Switzerland 2016
Y.-S. Han and K. Salomaa (Eds.): CIAA 2016, LNCS 9705, pp. 64–76, 2016.
DOI: 10.1007/978-3-319-40946-7_6

In the basic setting, parity games are two-player, turn-based, played on directed graphs whose nodes are labeled with priorities (also called, *colors*) and players have perfect information about the adversary moves. The two players, Player 0 and Player 1, take turns moving a token along the edges of the graph starting from a designated initial node. Thus, a play induces an infinite path and Player 0 wins the play if the smallest priority visited infinitely often is even; otherwise, Player 1 wins the play. The problem of deciding if Player 1 has a winning strategy (i.e., can induce a winning play) in a given parity game is known to be in UPTIME ∩ CoUPTIME [15]; whether a polynomial time solution exists is a long-standing open question [30].

Several algorithms for solving parity games have been proposed in the last two decades, aiming to tighten the known complexity bounds for the problem, as well as come out with solutions that work well in practice. Among the latter, we recall the recursive algorithm (RE) proposed by Zielonka [31], the Jurdziński's small-progress measures algorithm [16] (SP), the strategy-improvement algorithm by Jurdziński and Vöge [28], the (subexponential) algorithm by Jurdzinki et al. [17], and the big-step algorithm by Schewe [25]. These algorithms have been implemented in the platform `PGSolver`, and extensively investigated experimentally [12,13]. This study has also led to a few key optimizations, such as the decomposition into strongly connected components, the removal of self-cycles on nodes, and the application of a priority compression [2,16]. Specifically, the latter allows to reduce a game to an equivalent game where the priorities are replaced in such a way they form a dense sequence of natural numbers, $1, 2, \ldots, d$, for a minimal possible d. Table 1 summarizes the mentioned algorithms along with their known worst-case complexity, where the parameters n, e, and d denote the number of nodes, edges, and priorities, respectively (see [12,13], for more).

Table 1. Parity algorithms along with their computational complexities.

Algorithm	Computational complexity
Recursive (RE) [31]	$O(e \cdot n^d)$
Small Progress Measures (SP) [16]	$O(d \cdot e \cdot (\frac{n}{d})^{\frac{d}{2}})$
Strategy Improvement (SI) [28]	$O(2^e \cdot n \cdot e)$
Dominion Decomposition (DD) [17]	$O(n^{\sqrt{n}})$
Big Step (BS) [25]	$O(e \cdot n^{\frac{1}{3}d})$

In formal system design [8,9,21,24], parity games arise as a natural evaluation machinery for the automatic synthesis and verification of distributed and reactive systems [1,19,27], as they allow to express liveness and safety properties in a very elegant and powerful way [22]. Specifically, in model-checking, one can check the correctness of a system with respect to a desired behavior, by checking whether a model of the system, that is, a *Kripke structure*, is correct with respect to a formal specification of its behavior, usually described in terms of a modal logic formula. In case the specification is given as a μ-calculus formula [18], the model checking question can be rephrased, in linear-time, as a parity game [11]. So, a parity game solver can be used as a model checker for a μ-calculus specification (and vice-versa), as well as for fragments such as CTL, CTL*, and the like.

In the automata-theoretic approach to μ-calculus model checking, under a linear-time translation, one can also reduce the verification problem to a question about automata. More precisely, one can take the product of the model and an alternating tree automaton accepting all tree models of the specification. This product can be defined as an alternating word parity automaton over a singleton alphabet, and the system is correct with respect to the specification iff this automaton is nonempty [21]. It has been proved there that the nonemptiness problems for nondeterministic tree parity automata and alternating word parity automata over a singleton alphabet are equivalent and that their complexities coincide. For this reason, in the sequel we refer to these two kinds of automata just as parity automata. Hence, algorithms for the solution of the μ-calculus model checking problem, parity games, and the emptiness problem for parity automata can be interchangeably used to solve any of these problems, as they are linear-time equivalent. Several algorithms have been proposed in the literature to solve the non-emptiness problem of parity automata, but none of them has been ever implemented under the purpose of solving parity games.

In this paper, we study and implement an algorithm, which we call APT, introduced by Kupferman and Vardi in [20], for solving parity games via emptiness checking of alternating parity automata, and evaluate its performance over the PGSolver platform. This algorithm has been sketched in [20], but not spelled out in detail and without a correctness proof, two major gaps that we fill here. The core idea of the APT algorithm is an efficient translation to *weak alternating automata* [23]. These are a special case of Büchi automata in which the set of states is partitioned into partially ordered sets. Each set is classified as accepting or rejecting. The transition function is restricted so that the automaton either stays at the same set or moves to a smaller set in the partial order. Thus, each run of a weak automaton eventually gets trapped in some set in the partition. The special structure of weak automata is reflected in their attractive computational properties. In particular, the nonemptiness problem for weak automata can be solved in linear time [21], while the best known upper bound for the nonemptiness problem for Büchi automata is quadratic [5]. Given an alternating parity word automaton with n states and d colors, the APT algorithm checks the emptiness of an equivalent weak alternating word automaton with $O(n^d)$ states. The construction goes through a sequence of d intermediate automata. Each automaton in the sequence refines the state space of its predecessor and has one less color to check in its parity condition. Since one can check in linear time the emptiness of such an automaton, we get an $O(n^d)$ overall complexity for the addressed problem. APT does not construct the equivalent weak automaton directly, but applies the emptiness test directly, constructing the equivalent weak automaton on the fly.

We evaluated our implementation of the APT algorithm over several random game instances, comparing it with RE and SP algorithms. Our main finding is that when the number of the priority in a game is significantly smaller (specifically, logarithmically) than the number of nodes in the game graph, the APT algorithm significantly outperform the other algorithms. We take this as an important development since in many real applications of parity games we do get game instances where the number of priorities is indeed very small compared to the size of the game graph. For example, coming back to the automata-theoretic approach to μ-calculus model checking [21], the translation usually results in a parity automaton (and thus in a parity game) with few priorities, but with a huge number of nodes. This is due to the fact that usually specification formulas are small, while the system is big. A similar phenomenon occurs in the application of parity games to reactive synthesis [27].

Outline. The sequel of the paper is as follows. Section 2 gives preliminary concepts on parity games. Section 3 introduces extended parity games and describes the APT algorithm in detail, including a proof of correctness. Section 4 describes the implementation of the APT algorithm in the tool PGSolver. Section 5 contains the experimental results on runtime for APT over random benchmarks. Finally, Sect. 6 gives some conclusions.

2 Preliminaries

In this section, we briefly recall some basic concepts regarding parity games. A *Parity Game* (PG, for short) is a tuple $\mathcal{G} \triangleq \langle \mathrm{Ps}_0, \mathrm{Ps}_1, Mv, \mathrm{p} \rangle$, where Ps_0 and Ps_1 are two finite disjoint sets of nodes for Player 0 and Player 1, respectively, with $\mathrm{Ps} = \mathrm{Ps}_0 \cup \mathrm{Ps}_1$, $Mv \subseteq \mathrm{Ps} \times \mathrm{Ps}$, is the left-total binary relation of moves, and $\mathrm{p} : \mathrm{Ps} \to \mathbb{N}$ is the priority function[1]. Each player moves a token along nodes by means of the relation Mv. By $Mv(q) \triangleq \{q' \in \mathrm{Ps} : (q, q') \in Mv\}$ we denote the set of nodes to which the token can be moved, starting from node q.

As a running example, consider the PG depicted in Fig. 1. The set of players's nodes is $\mathrm{Ps}_0 = \{q_0, q_3, q_4, q_5\}$ and $\mathrm{Ps}_1 = \{q_1, q_2, q_6\}$; we use circles to denote nodes belonging to Player 0 and squares for those belonging to Player 1. Mv is described by arrows. Finally, the priority function p is given by $\mathrm{p}(q_1) = 1$, $\mathrm{p}(q_3) = \mathrm{p}(q_4) = \mathrm{p}(q_6) = 2$, $\mathrm{p}(q_0) = 3$, and $\mathrm{p}(q_2) = \mathrm{p}(q_5) = 5$.

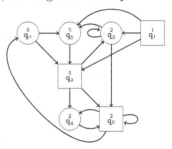

Fig. 1. A parity game.

A *play* (resp., *history*) over $\mathcal{G}$ is an infinite (resp., finite) sequence $\pi = q_1 \cdot q_2 \cdot \ldots \in \mathrm{Pth} \subseteq \mathrm{Ps}^\omega$ (resp., $\rho = q_1 \cdot \ldots \cdot q_n \in \mathrm{Hst} \subseteq \mathrm{Ps}^*$) of nodes that agree with Mv, i.e., $(\pi_i, \pi_{i+1}) \in Mv$, for each natural number $i \in \mathbb{N}$ (resp., $i \in [1, n-1]$). In the PG in Fig. 1, a possible play is $\overline{\pi} = q_1 \cdot q_5 \cdot q_2 \cdot (q_3)^\omega$, while a possible history is given by $\overline{\rho} = q_1 \cdot q_5 \cdot q_2 \cdot q_3$.

For a given play $\pi = q_1 \cdot q_2 \cdot \ldots$, by $\mathrm{p}(\pi) = \mathrm{p}(q_1) \cdot \mathrm{p}(q_2) \cdot \ldots \in \mathbb{N}^\omega$ we denote the associated priority sequence. As an example, the associated priority sequence to $\overline{\pi}$ is given by $\mathrm{p}(\overline{\pi}) = 1 \cdot 5 \cdot 5 \cdot (2)^\omega$.

For a given history $\rho = q_1 \cdot \ldots \cdot q_n$, by $\mathsf{fst}(\rho) \triangleq q_1$ and $\mathsf{lst}(\rho) \triangleq q_n$ we denote the first and last node occurring in ρ, respectively. For the example history, we have that $\mathsf{fst}(\overline{\rho}) = q_1$ and $\mathsf{lst}(\overline{\rho}) = q_3$. By Hst_0 (resp., Hst_1) we denote the set of

---

[1] Here, we mean the set of non-negative integers, excluding zero.

histories ρ such that $\mathsf{lst}(\rho) \in \mathrm{Ps}_0$ (*resp.*, $\mathsf{lst}(\rho) \in \mathrm{Ps}_1$). Moreover, by $\mathsf{Inf}(\pi)$ and $\mathsf{Inf}(\mathsf{p}(\pi))$ we denote the set of nodes and priorities that occur infinitely often in π and $\mathsf{p}(\pi)$, respectively. Finally, a play π is winning for Player 0 (*resp.*, Player 1) if $\min(\mathsf{Inf}(\mathsf{p}(\pi)))$ is even (*resp.*, odd). In the running example, we have that $\mathsf{Inf}(\overline{\pi}) = \{\mathsf{q}_3\}$ and $\mathsf{Inf}(\mathsf{p}(\overline{\pi})) = \{2\}$ and so, π is winning for Player 0.

A Player 0 (*resp.*, Player 1) strategy is a function $\mathsf{str}_0 : \mathrm{Hst}_0 \to \mathrm{Ps}$ (*resp.*, $\mathsf{str}_1 : \mathrm{Hst}_1 \to \mathrm{Ps}$) such that, for all $\rho \in \mathrm{Hst}_0$ (*resp.*, $\rho \in \mathrm{Hst}_1$), it holds that $(\mathsf{lst}(\rho), \mathsf{str}_0(\rho)) \in Mv$ (*resp.*, $\mathsf{lst}(\rho), \mathsf{str}_1(\rho)) \in Mv$).

Given a node q, Player 0 and a Player 1 strategies str_0 and str_1, the play of these two strategies, denoted by $\mathsf{play}(q, \mathsf{str}_0, \mathsf{str}_1)$, is the only play π in the game that starts in q and agrees with both Player 0 and Player 1 strategies, *i.e.*, for all $i \in \mathbb{N}$, if $\pi_i \in \mathrm{Ps}_0$, then $\pi_{i+1} = \mathsf{str}_0(\pi_i)$, and $\pi_{i+1} = \mathsf{str}_1(\pi_i)$, otherwise.

A strategy str_0 (*resp.*, str_1) is *memoryless* if, for all $\rho_1, \rho_2 \in \mathrm{Hst}_0$ (*resp.*, $\rho_1, \rho_2 \in \mathrm{Hst}_1$), with $\mathsf{lst}(\rho_1) = \mathsf{lst}(\rho_2)$, it holds that $\mathsf{str}_0(\rho_1) = \mathsf{str}_0(\rho_2)$ (*resp.*, $\mathsf{str}_0(\rho_1) = \mathsf{str}_1(\rho_2)$). Note that a memoryless strategy can be defined on the set of nodes, instead of the set of histories. Thus we have that they are of the form $\mathsf{str}_0 : \mathrm{Ps}_0 \to \mathrm{Ps}$ and $\mathsf{str}_1 : \mathrm{Ps}_1 \to \mathrm{Ps}$.

We say that Player 0 (*resp.*, Player 1) *wins* the game $\mathcal{G}$ from node q if there exists a Player 0 (*resp.*, Player 1) strategy str_0 (*resp.*, str_1) such that, for all Player 1 (*resp.*, Player 0) strategies str_1 (*resp.*, str_0) it holds that $\mathsf{play}(q, \mathsf{str}_0, \mathsf{str}_1)$ is winning for Player 0 (*resp.*, Player 1).

A node q is *winning* for Player 0 (*resp.*, Player 1) if Player 0 (*resp.*, Player 1) wins the game from q. By $\mathrm{Win}_0(\mathcal{G})$ (*resp.*, $\mathrm{Win}_1(\mathcal{G})$) we denote the set of winning nodes in $\mathcal{G}$ for Player 0 (*resp.*, Player 1). Parity games enjoy determinacy, meaning that, for every node q, either $q \in \mathrm{Win}_0(\mathcal{G})$ or $q \in \mathrm{Win}_1(\mathcal{G})$ [11]. Moreover, it can be proved that, if Player 0 (resp., Player 1) has a winning strategy from node q, then it has a memoryless winning strategy from the same node [31].

3   Extended Parity Games

In this section we recall the APT algorithm, introduced by Kupferman and Vardi in [20], to solve parity games via emptiness checking of parity automata. More important, we fill two major gaps from [20] which is to spell out in details the definition of the APT algorithm as well as to give a correctness proof. The APT algorithm makes use of two special (incomparable) sets of nodes, denoted by V and A, and called set of *Visiting* and *Avoiding*, respectively. Intuitively, a node is declared visiting for a player at the stage in which it is clear that, by reaching that node, he can surely induce a winning play and thus winning the game. Conversely, a node is declared avoiding for a player whenever it is clear that, by reaching that node, he is not able to induce any winning play and thus losing the game. The algorithm, in turns, tries to partition all nodes of the game into these two sets. The formal definition of the sets V and A follows.

An *Extended Parity Game*, (EPG, for short) is a tuple $\langle \mathrm{Ps}_0, \mathrm{Ps}_1, \mathrm{V}, \mathrm{A}, Mv, \mathsf{p} \rangle$ where $\mathrm{Ps}_0, \mathrm{Ps}_1, Mv$ are as in PG. The subsets of nodes $\mathrm{V}, \mathrm{A} \subseteq \mathrm{Ps} = \mathrm{Ps}_0 \cup \mathrm{Ps}_1$ are two disjoint sets of *Visiting* and *Avoiding* nodes, respectively. Finally, $\mathsf{p} : \mathrm{Ps} \to \mathbb{N}$ is a parity function mapping every non-visiting and non-avoiding set to a color.

The notions of histories and plays are equivalent to the ones given for PG. Moreover, as far as the definition of strategies is concerned, we say that a play π that is in $Ps \cdot (Ps \setminus (V \cup A))^* \cdot V \cdot Ps^\omega$ is winning for Player 0, while a play π that is in $Ps \cdot (Ps \setminus (V \cup A))^* \cdot A \cdot Ps^\omega$ is winning for Player 1. For a play π that never hits either V or A, we say that it is winning for Player 0 iff it satisfies the parity condition, $i.e.,\min(\mathsf{Inf}(\mathsf{p}(\pi)))$ is even, otherwise it is winning for Player 1.

Clearly, PGs are special cases of EPGs in which $V = A = \emptyset$. Conversely, one can transform an EPG into an equivalent PG with the same winning set by simply replacing every outgoing edge with loop to every node in $V \cup A$ and then relabeling each node in V and A with an even and an odd number, respectively.

In order to describe how to solve EPGs, we introduce some notation. By $F_i = \mathsf{p}^{-1}(i)$ we denote the set of all nodes labeled with i. Doing that, the parity condition can be described as a finite sequence $\alpha = F_1 \cdot \ldots \cdot F_k$ of sets, which alternates from sets of nodes with even priorities to sets of nodes with odd priorities and the other way round, forming a partition of the set of nodes, ordered by the priority assigned by the parity function. We call the set of nodes F_i an even (resp., odd) parity set if i is even (resp., odd).

For a given set $X \subseteq Ps$, by $\mathsf{force}_0(X) = \{q \in Ps_0 : X \cap Mv(q) \neq \emptyset\} \cup \{q \in Ps_1 : X \subseteq Mv(q)\}$ we denote the set of nodes from which Player 0 can force to move in the set X. Analogously, by $\mathsf{force}_1(X) = \{q \in Ps_1 : X \cap Mv(q) \neq \emptyset\} \cup \{q \in Ps_0 : X \subseteq Mv(q)\}$ we denote the set of nodes from which Player 1 can force to move in the set X. For example, in the PG in Fig. 1, $\mathsf{force}_1(\{q_6\}) = \{q_2, q_4, q_6\}$.

We now introduce two functions that are co-inductively defined that will be used to compute the winning sets of Player 0 and Player 1, respectively.

For a given EPG $\mathcal{G}$ with α being the representation of its parity condition, V its visiting set, and A its avoiding set, we define the functions $\mathsf{Win}_0(\alpha, V, A)$ and $\mathsf{Win}_1(\alpha, A, V)$. Informally, $\mathsf{Win}_0(\alpha, V, A)$ computes the set of nodes from which the player 0 has a strategy that avoids A and either force a visit in V or he wins the parity condition. The definition is symmetric for the function $\mathsf{Win}_1(\alpha, A, V)$. Formally, we define $\mathsf{Win}_0(\alpha, V, A)$ and $\mathsf{Win}_1(\alpha, A, V)$ as follows.

If $\alpha = \varepsilon$ is the empty sequence, then

- $\mathsf{Win}_0(\varepsilon, V, A) = \mathsf{force}_0(V)$ and
- $\mathsf{Win}_1(\varepsilon, A, V) = \mathsf{force}_1(A)$.

Otherwise, if $\alpha = F \cdot \alpha'$, for some set F, then

- $\mathsf{Win}_0(F \cdot \alpha', V, A) = Ps \setminus \mu Y(\mathsf{Win}_1(\alpha', A \cup (F \setminus Y), V \cup (F \cap Y)))$ and
- $\mathsf{Win}_1(F \cdot \alpha', A, V) = Ps \setminus \mu Y(\mathsf{Win}_0(\alpha', V \cup (F \setminus Y), A \cup (F \cap Y)))$,

where μ is the least fixed-point operator[2].

To better understand how APT solves a parity game we show a simple piece of execution on the example in Fig. 1. It is easy to see that such parity game is

---

[2] The unravellings of Win_0 and Win_1 have some analogies with the fixed-point formula introduced in [29] also used to solve parity games. Unlike our work, however, the formula presented there is just a translation of the Zielonka algorithm [31].

won by Player 0 in all the possible starting nodes. Then, the fixpoint returns the entire set Ps. The parity condition is given by $\alpha = F_1 \cdot F_2 \cdot F_3 \cdot F_4 \cdot F_5$, where $F_1 = \{q_1\}$, $F_2 = \{q_3, q_4, q_6\}$, $F_3 = \{q_0\}$, $F_4 = \emptyset$, $F_5 = \{q_5, q_6\}$. The repeated application of functions $\text{Win}_0(\alpha, V, A)$ and $\text{Win}_1(\alpha, A, V)$ returns:

$$\text{Win}_0(\alpha, \emptyset, \emptyset) = \text{Ps} \backslash \mu Y^1(\text{Ps} \backslash \mu Y^2(\text{Ps} \backslash \mu Y^3(\text{Ps} \backslash \mu Y^4(\text{Ps} \backslash \mu Y^5(\text{Ps} \backslash \text{force}_1(V^6))))))$$

in which the sets Y^i are the nested fixpoint of the formula, while the set V^6 is obtained by recursively applying the following:

- $V^1 = \emptyset$, $V^{i+1} = A^i \cup (F_i \setminus Y^i)$, and
- $A^1 = \emptyset$, $A^{i+1} = V^i \cup (F_i \cap Y^i)$.

As a first step of the fixpoint computation, we have that $Y^1 = Y^2 = Y^3 = Y^4 = Y^5 = \emptyset$. Then, by following the two iterations above for the example in Fig. 1, we obtain that $V^6 = \{q_0, q_1, q_2, q_5\}$.

At this point we have that $\text{force}_1(V^6) = \{q_0, q_1, q_5, q_6\} \neq \emptyset = Y^5$. This means that the fixpoint for Y^5 has not been reached yet. Then, we update the set Y^5 with the new value and compute again V^6. This procedure is repeated up to the point in which $\text{force}_1(V^6) = Y^5$, which means that the fixpoint for Y^5 has been reached. Then we iteratively proceed to compute $Y^4 = \text{Ps} \setminus Y^5$ until a fixpoint for Y^4 is reached. Note that the sets A^i and V^i depends on the Y^i and so they need to be updated step by step. As soon as a fixpoint for Y_1 is reached, the algorithm returns the set $\text{Ps} \setminus Y_1$. As a fundamental observation, note that, due to the fact that the fixpoint operations are nested one to the next, updating the value of Y^i implies that every Y^j, with $j > i$, needs to be reset to the empty set.

We now prove the correctness of this procedure. Note that the algorithm is an adaptation of the one provided by Kupferman and Vardi in [20], for which a proof of correctness has never been shown.

Theorem 1. *Let $\mathcal{G} = \langle \text{Ps}_0, \text{Ps}_1, V, A, Mv, p \rangle$ be an EPG with α being the parity sequence condition. Then, the following properties hold.*

1. *If $\alpha = \varepsilon$ then $\text{Win}_0(\mathcal{G}) = \text{Win}_0(\alpha, V, A)$ and $\text{Win}_1(\mathcal{G}) = \text{Win}_1(\alpha, V, A)$;*
2. *If α starts with an odd parity set, it holds that $\text{Win}_0(\mathcal{G}) = \text{Win}_0(\alpha, V, A)$;*
3. *If α starts with an even parity set, it holds that $\text{Win}_1(\mathcal{G}) = \text{Win}_1(\alpha, V, A)$.*

Proof. The proof of Item 1 follows immediately by definition, as $\alpha = \epsilon$ forces the two players to reach their respective winning sets in one step.

For Item 2 and 3, we need to find a partition of F into a winning set for Player 0 and a winning set for Player 1 such that the game is invariant *w.r.t.* the winning sets, once they are moved to visiting and avoiding, respectively. We proceed by mutual induction on the length of the sequence α. As base case, assume $\alpha = F$ and F to be an odd parity set. Then, first observe that Player 0 can win only by eventually hitting the set V, as the parity condition is made by only odd numbers. We have that $\text{Win}_0(F, V, A) = \mu Y(\text{Ps} \setminus \text{Win}_1(\varepsilon, A \cup (F \setminus Y), V \cup (F \cap (Y)))) = \mu Y(\text{Ps} \setminus \text{force}_1(A \cup (F \setminus Y)))$ that, by definition, computes the set from which Player 1 cannot avoid a visit to V, hence the winning set for Player 0. In the case the set F is an even parity set the reasoning is symmetric.

As an inductive step, assume that Items 2 and 3 hold for sequences α of length n, we prove that it holds also for sequences of the form $F \cdot \alpha$ of length $n+1$. Suppose that F is a set of odd priority. Then, we have that, by induction hypothesis, the formula $Win_1(\alpha, A \cup (F \setminus Y), V \cup (F \cap Y))$ computes the winning set for Player 1 for the game in which the nodes in $F \cap Y$ are visiting, while the nodes in $F \setminus Y$ are avoiding. Thus, its complement $Ps \setminus Win_1(\alpha, A \cup (F \setminus Y), V \cup (F \cap Y))$ returns the winning set for Player 0 in the same game. Now, observe that, if a set Y' is bigger than Y, then $Ps \setminus Win_1(\alpha, A \cup (F \setminus Y'), V \cup (F \cap Y'))$ is the winning set for Player 0 in which some node in $F \setminus Y$ has been moved from avoiding to visiting. Thus we have that $Ps \setminus Win_1(\alpha, A \cup (F \setminus Y), V \cup (F \cap Y)) \subseteq Ps \setminus Win_1(\alpha, A \cup (F \setminus Y'), V \cup (F \cap Y'))$. Moreover, observe that, if a node $q \in F \cup A$ is winning for Player 0, then it can be avoided in all possible winning plays, and so it is winning also in the case q is only in F. It is not hard to see that, after the last iteration of the fixpoint operator, the two sets $F \setminus Y$ and $F \cap Y$ can be considered in avoiding and winning, respectively, in a way that the winning sets of the game are invariant under this update, which concludes the proof of Item 2.

Also in the inductive case, the reasoning for Item 3 is perfectly symmetric to the one for Item 2.   □

4   Implementation of APT in PGSolver

In this section we describe the implementation of APT in the well-known platform PGSolver developed in OCaml by Friedman and Lange [13], which collects the large majority of the algorithms introduced in the literature to solve parity games [14,16,17,25,26, 28,31].

We briefly recall the main aspects of this platform. The graph data structure is represented as a fixed length array of tuples. Every tuple has all information that a node needs, such as the owner player, the assigned priority and the adjacency list of nodes. The platform implements a collection of tools to generate and solve parity games, as well as compare the performance of different algorithms. The purpose of this platform is not just that of making available an environment to deploy and test a generic solution algorithm, but also to investigate the practical aspects of the different

```
fun win i G Ps α V A =
  if (α ≠ ε) then
    W := Ps \( min_fp (1−i) G Ps α V A );
  else
    W := force_i(V);

  return W;;

fun min_fp i G Ps α V A =
  Y₁ := ∅;
  Y₂ := ∅;
  F := head [α];
  α' := tail [α];

  A' := A ∪ (F ∩ Y₁);
  V' := V ∪ (F ∩ Y₁);

  Y₂ := win i G Ps α' A' V' ;

  while ( Y₂ ≠ Y₁ ) do
  (
    Y₁ := Y₂ ;

    A' := A ∪ (F ∩ Y₁);
    V' := V \ (F ∩ Y₁);

    Y₂ := win i G Ps α' A' V' ;
  )
  done

  return Y₂;;
```

Fig. 2. APT algorithm

algorithms on the different classes of parity games. Moreover, `PGSolver` implements optimizations that can be applied to all algorithms in order to improve their performance. The most useful optimizations in practice are decomposition into strongly connected components, removal of self-cycles on nodes, and priority compression.

We have added to `PGSolver` an implementation of the `APT` algorithm introduced in Sect. 3. Our procedure applies the fixpoint algorithm to compute the set of winning positions in the game by means of two principal functions that implement the two functions of the algorithm core processes, i.e., function force_i and the recursive function $\mathrm{Win}_i(\alpha, V, A)$. The pseudocode of the `APT` algorithm implementation is reported in Fig. 2. It takes six parameters: the Player (0 or 1), the game, the set of nodes, the condition α, the set of visiting and avoiding. Moreover, we define the function min_fp for the calculation of the fixed point. The whole procedure makes use of Set and List data structures, which are available in the OCaml's standard library, for the manipulation of the sets visiting and avoiding, and the accepting condition α. The tool along with the implementation of the `APT` algorithm is available for download from https://github.com/antoniodistasio/pgsolver-APT.

For the sake of clarity, we report that in `PGSolver` it is used the maximal priority to decide who wins a given parity game. Conversely, the `APT` algorithm uses the minimal priority. However, these two conditions are well known to be equivalent and, in order to compare instances of the same game on different implementations of parity games algorithms in `PGSolver`, we simply convert the game to the specific algorithm accordingly. For the conversion, we simply use a suitable permutation of the priorities.

5   Experiments

In this section, we report the experimental results on evaluating the performance for the `APT` algorithm implemented in `PGSolver` over the random benchmarks generated in the platform. We have compared the performance of the implementation of `APT` with those of `RE` and `SP`. We have chosen these two algorithms as they have been proved to be the best-performing in practice [13].

All tests have been run on an AMD Opteron 6308 @2.40 GHz, with 224 GB of RAM and 128 GB of swap running Ubuntu 14.04. We note that `APT` has been executed without applying any optimization implemented in `PGSolver` [13], while `SP` and `RE` are run with such optimizations. Applying these optimization on `APT` is a topic of further research.

We evaluated the performance of the three algorithms over a set of games that are randomly generated by `PGSolver`, in which it is possible to give the number n of states and the number k of priority as parameters. We have taken 20 different game instances for each set of parameters and used the average time among them returned by the tool. For each game, the generator works as follows. For each node q in the graph-game, the priority $\mathsf{p}(q)$ is chosen uniformly between 0 and $k - 1$, while its ownership is assigned to Player 0 with probability $\frac{1}{2}$, and to Player 1 with probability $\frac{1}{2}$. Then, for each node q, a number d from 1 to n is chosen uniformly and d distinct successors of q are randomly selected.

Table 2. Runtime executions with fixed priorities 2, 3 and 5

n	2 Pr			3 Pr			5 Pr		
	RE	SP	APT	RE	SP	APT	RE	SP	APT
2000	4.94	5.05	0.10	4.85	5.20	0.15	4.47	4.75	0.42
4000	31.91	32.92	0.17	31.63	31.74	0.22	31.13	32.02	0.82
6000	107.06	108.67	0.29	100.61	102.87	0.35	100.81	101.04	1.39
8000	229.70	239.83	0.44	242.24	253.16	0.5	228.48	245.24	2.73
10000	429.24	443.42	0.61	482.27	501.20	0.85	449.26	464.36	3.61
12000	772.60	773.76	0.87	797.07	808.96	0.98	762.89	782.53	6.81
14000	1185.81	1242.56	1.09	1227.34	1245.39	1.15	1256.32	1292.80	10.02

5.1 Experimental Results

We ran two experiments. First, we tested games with 2, 3, and 5 priorities, where for each of them we measured runtime performance for different state-space sizes, ranging in $\{2000, 4000, 6000, 8000, 10000, 12000, 14000\}$. The results are in Table 2, in which the number of states is reported in column 1, the number of colors is reported in the macro-column 2, 3, and 5, each of them containing the runtime executions, expressed in seconds, for the three algorithms. Second, we evaluated the algorithms on games with an exponential number of nodes $w.r.t.$ the number of priorities. More precisely, we ran experiments for $n = 2^k$, $n = e^k$ and $n = 10^k$, where n is the number of states and k is the number of priorities.

The experiment results are reported in Table 3. By abort_T, we denote that the execution has been aborted due to time-out (greater of one hour), while by abort_M we denote that the execution has been aborted due to mem-out.

The first experiment shows that with a fixed number of priorities (2, 3, and 5) APT significantly outperforms the other algorithms, showing excellent runtime execution

Table 3. Runtime executions with $n = e^k$ and $n = 2^k$ and $n = 10^k$

n	Pr	RE	SP	APT
$n = 2^k$				
1024	10	1.25	1.25	8.58
2048	11	7.90	8.21	71.08
4096	12	52.29	52.32	1505.75
8192	13	359.29	372.16	abort_T
16384	14	2605.04	2609.29	abort_T
32768	15	abort_T	abort_T	abort_T
$n = e^k$				
21	3	0	0	0
55	4	0	0	0.02
149	5	0.01	0.01	0.08
404	6	0.14	0.14	0.19
1097	7	1.72	1.72	0.62
2981	8	24.71	24.46	7.88
8104	9	413.2.34	414.65	35.78
22027	10	abort_T	abort_T	311.87
$n = 10^k$				
10	1	0	0	0
100	2	0	0	0
1000	3	1.3	1.3	0.04
10000	4	738.86	718.24	4.91
100000	5	abort_M	abort_M	66.4

even on fairly large instances. For example, for $n = 14000$, the running time for both RE and SP is about 20 min, while for APT it is less than a minute.

The results of the exponential-scaling experiments, shown in Table 3, give more nuanced results. Here, APT is the best performing algorithm for $n = e^k$ and $n = 10^k$. For example, when $n = 100000$ and $k = 5$, both RE and SP memout, while APT completes in just over one minute. That is, the efficiency of APT is notable also in terms of memory usage. At the same APT underperforms for $n = 2^k$. Our conclusion is that APT has superior performance when the number of priorities is logarithmic in the number of game-graph nodes, but the base of the logarithm has to be large enough. As we see experimentally, e is sufficiently large base, but 2 is not. This point deserve further study, which we leave to future work. In Fig. 3 we just report graphically the benchmarks in the case $n = e^k$. An interested reader can find more detailed experiment results at https://github.com/antoniodistasio/pgsolver-APT.

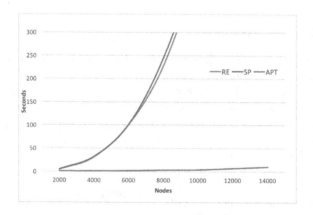

Fig. 3. Runtime executions with $n = e^k$ (Color figure online)

6   Conclusion

The APT algorithm, an automata-theoretic technique to solve parity games, has been designed two decades ago by Kupferman and Vardi [20], but never considered to be useful in practice [12]. In this paper, for the first time, we fill missing gaps and implement this algorithm. By means of benchmarks based on random games, we show that it is the best performing algorithm for solving parity games when the number of priorities is very small w.r.t. the number of states. We believe that this is a significant result as several applications of parity games to formal verification and synthesis do yield games with a very small number of priorities.

The specific setting of a small number of priorities opens up opportunities for specialized optimization technique, which we aim to investigate in future work. This is closely related to the issue of accelerated algorithms for three-color parity games [10]. We also plan to study why the performance of the APT algorithm is so sensitive to the relative number of priorities, as shown in Table 3.

References

1. Aminof, B., Kupferman, O., Murano, A.: Improved model checking of hierarchical systems. Inf. Comput. **210**, 68–86 (2012)
2. Antonik, A., Charlton, N., Huth, M.: Polynomial-time under-approximation of winning regions in parity games. ENTCS **225**, 115–139 (2009)
3. Berwanger, D.: Admissibility in infinite games. In: Thomas, W., Weil, P. (eds.) STACS 2007. LNCS, vol. 4393, pp. 188–199. Springer, Heidelberg (2007)
4. Chatterjee, K., Doyen, L., Henzinger, T.A., Raskin, J.-F.: Generalized mean-payoff and energy games. In: FSTTCS 2010. LIPIcs, vol. 8, pp. 505–516 (2010)
5. Chatterjee, K., Henzinger, M.: An $O(n^2)$ time algorithm for alternating Büchi games. In: SODA 2012, pp. 1386–1399 (2012)
6. Chatterjee, K., Henzinger, T.A., Jurdzinski, M.: Mean-payoff parity games. In: LICS 2005, pp. 178–187 (2005)
7. Chatterjee, K., Jurdzinski, M., Henzinger, T.A.: Quantitative stochastic parity games. In: SODA 2004, pp. 121–130 (2004)
8. Clarke, E.M., Emerson, E.A.: Design and synthesis of synchronization skeletons using branching time temporal logic. In: Kozen, D. (ed.) LP 1981. LNCS, vol. 131, pp. 52–71. Springer, Heidelberg (1982)
9. Clarke, E.M., Grumberg, O., Peled, D.A.: Model Checking (2002)
10. de Alfaro, L., Faella, M.: An accelerated algorithm for 3-color parity games with an application to timed games. In: Damm, W., Hermanns, H. (eds.) CAV 2007. LNCS, vol. 4590, pp. 108–120. Springer, Heidelberg (2007)
11. Emerson, E.A., Jutla, C.: Tree automata, μ-calculus and determinacy. In: FOCS 1991, pp. 368–377 (1991)
12. Friedmann, O., Lange, M.: The PGSolver collection of parity game solvers. University of Munich (2009)
13. Friedmann, O., Lange, M.: Solving parity games in practice. In: Liu, Z., Ravn, A.P. (eds.) ATVA 2009. LNCS, vol. 5799, pp. 182–196. Springer, Heidelberg (2009)
14. Heljanko, K., Keinänen, M., Lange, M., Niemelä, I.: Solving parity games by a reduction to SAT. J. Comput. Syst. Sci. **78**(2), 430–440 (2012)
15. Jurdzinski, M.: Deciding the winner in parity games is in UP ∩ co-Up. Inf. Process. Lett. **68**(3), 119–124 (1998)
16. Jurdziński, M.: Small progress measures for solving parity games. In: Reichel, H., Tison, S. (eds.) STACS 2000. LNCS, vol. 1770, p. 290. Springer, Heidelberg (2000)
17. Jurdzinski, M., Paterson, M., Zwick, U.: A deterministic subexponential algorithm for solving parity games. SIAM J. Comput. **38**(4), 1519–1532 (2008)
18. Kozen, D.: Results on the propositional μ-calculus. TCS **27**(3), 333–354 (1983)
19. Kupferman, O., Vardi, M., Wolper, P.: Module checking. Inf. Comput. **164**(2), 322–344 (2001)
20. Kupferman, O., Vardi, M.Y.: Weak alternating automata and tree automata emptiness. In: STOC, pp. 224–233 (1998)
21. Kupferman, O., Vardi, M.Y., Wolper, P.: An automata theoretic approach to branching-time model checking. J. ACM **47**(2), 312–360 (2000)
22. Mogavero, F., Murano, A., Sorrentino, L.: On promptness in parity games. In: McMillan, K., Middeldorp, A., Voronkov, A. (eds.) LPAR-19 2013. LNCS, vol. 8312, pp. 601–618. Springer, Heidelberg (2013)
23. Muller, D.E., Saoudi, A., Schupp, P.E.: Weak alternating automata give a simple explanation of why most temporal and dynamic logics are decidable in exponential time. In: LICS 1988, pp. 422–427 (1988)

24. Queille, J.P., Sifakis, J.: Specification and verification of concurrent systems in CESAR. In: Dezani-Ciancaglini, M., Montanari, U. (eds.) Programming 1982. LNCS, vol. 137, pp. 337–351. Springer, Heidelberg (1982)
25. Schewe, S.: Solving parity games in big steps. In: Arvind, V., Prasad, S. (eds.) FSTTCS 2007. LNCS, vol. 4855, pp. 449–460. Springer, Heidelberg (2007)
26. Schewe, S.: An optimal strategy improvement algorithm for solving parity and payoff games. In: Kaminski, M., Martini, S. (eds.) CSL 2008. LNCS, vol. 5213, pp. 369–384. Springer, Heidelberg (2008)
27. Thomas, W.: Facets of synthesis: revisiting church's problem. In: de Alfaro, L. (ed.) FOSSACS 2009. LNCS, vol. 5504, pp. 1–14. Springer, Heidelberg (2009)
28. Vöge, J., Jurdziński, M.: A discrete strategy improvement algorithm for solving parity games. In: Emerson, E.A., Sistla, A.P. (eds.) CAV 2000. LNCS, vol. 1855, pp. 202–215. Springer, Heidelberg (2000)
29. Walukiewicz, I.: Pushdown processes: games and model checking. In: Alur, R., Henzinger, T.A. (eds.) CAV 1996. LNCS, vol. 1102, pp. 62–74. Springer, Heidelberg (1996)
30. Wilke, T.: Alternating tree automata, parity games, and modal μ-calculus. Bull. Belg. Math. Soc. Simon Stevin 8(2), 359 (2001)
31. Zielonka, W.: Infinite games on finitely coloured graphs with applications to automata on infinite trees. Theor. Comput. Sci. 200(1–2), 135–183 (1998)

Ternary Equational Languages

Zoltán Ésik[(✉)]

Department of Computer Science, University of Szeged, Szeged 6720, Hungary
`ze@inf.u-szeged.hu`

Abstract. We present a general method for solving fixed point equations involving pseudo-complementation over complete ternary algebras satisfying some infinite distributivity conditions as generalized truth values, and finitely decomposable algebras as data domains. When the algebra of data domains is a word or a tree algebra, fixed point equations may be interpreted as grammars and we obtain wide classes of (fuzzy) languages and tree languages.

1 Introduction

A ternary algebra, introduced in [6], is a De Morgan algebra [2] equipped with an additional constant satisfying certain natural equational laws. Ternary algebras form a variety [13,14] and are closely related to Kleene algebras [2]. Moreover, every ternary algebra is a normal distributive i-lattice as defined in [15]. The initial (and only subdirectly irreducible, cf. [15]) ternary algebra is Kleene's 3-element lattice of the classical truth values and an intermediate truth value which is the interpretation of the additional constant. In addition to the binary infimum and supremum operations, it is equipped with the unary operation of pseudo-complementation, or involution, which interchanges the classical truth values and maps the intermediate truth value to itself.

Systems of fixed point equations over the 3-element Kleene algebra, called stipulations, were considered already in [17]. In this paper, our aim is to describe a method for solving systems of fixed point equations over fuzzy sets in T^A, where T is a complete ternary algebra of truth values satisfying certain infinite distributivity conditions, and A is a finitely decomposable algebra (defined in the paper) serving as data domain. The fixed point equations may involve the anti-monotonic pseudo-complementation operation of the ternary algebra.

In order to obtain canonical solutions to fixed point equations, we show that every complete ternary algebra satisfying the infinite distributivity conditions may naturally be equipped with a new ordering (the uncertainty or information order) turning the algebra into a complete partial order (cf. Theorem 2). Our proof of this fact uses a representation theorem for complete ternary algebras satisfying the distributivity conditions (Theorem 1). We then show that the structure map associated with a system of fixed point equations over a complete

---

Z. Ésik—Partially supported by the NKFI grant no. 108488.

Y.-S. Han and K. Salomaa (Eds.): CIAA 2016, LNCS 9705, pp. 77–88, 2016.
DOI: 10.1007/978-3-319-40946-7_7

ternary algebra T satisfying the distributivity conditions and a finitely decomposable algebra A is continuous with respect to the uncertainty order and thus has a least fixed point which can be constructed in ω iterations (Theorem 3). Our algebraic theorems are necessary for this result. The canonical solution of a system is defined as this least fixed point. A ternary equational (fuzzy) set is a component of the canonical solution of a finite system. Each finite system of equations over T^A (in normal form) can be seen as a sort of generalized grammar. The canonical solution of the system provides the semantics of the grammar. By specializing the algebra A to be a monoid of finite words or an algebra of trees, we obtain various notions of grammars defining ternary equational (fuzzy) sets, or languages, of words or trees over T.

2   Ternary Algebras and Symmetric Complete Lattices

Regarding partially ordered sets and lattices, we will use standard terminology, see e.g. [9]. It is well-known that a lattice L can be viewed as a partially ordered set such that the supremum $x \vee y$ and the infimum $x \wedge y$ exist for all $x, y \in L$. Alternatively, a lattice can be seen as an abstract algebra, equipped with binary operations $\vee$ and $\wedge$, subject to certain equational axioms. A lattice homomorphism preserves these operations and is (thus) monotonic, i.e., it preserves the order relation. Moreover, a bijective function between lattices is a lattice isomorphism iff it is monotonic and its inverse is also monotonic, i.e., when it is an order isomorphism. A lattice is bounded if it is equipped with distinguished constants $\bot$ and $\top$ which are the least and the greatest elements, respectively. A bounded lattice homomorphism also preserves these distinguished constants. A lattice L is called complete if each subset X of L has a supremum denoted $\bigvee X$. It then follows that each $X \subseteq L$ also has an infimum $\bigwedge X$. In particular, a complete lattice has a least and a greatest element and thus gives rise to a bounded lattice.

An (order reversing) involution of a partially ordered set $P = (P, \leq)$ is a function $\sim: P \to P$ such that for all $x, y \in P$,

$$x \leq y \ \to \ \sim x \geq \sim y \quad \text{and} \quad \sim\sim x = x.$$

Note that an involution of P is a bijective function, in fact an anti-automorphism which is its own inverse. Moreover, for any $X \subseteq P$, $\bigvee X$ exists iff $\bigwedge \sim X$ does, and symmetrically, $\bigwedge X$ exists iff $\bigvee \sim X$ does, and if they exist, then

$$\sim \bigvee X = \bigwedge \sim X \quad \text{and} \quad \sim \bigwedge X = \bigvee \sim X.$$

Suppose that D is a bounded distributive lattice [2] with least and greatest elements $\bot$ and $\top$, respectively. When D is equipped with an involution $\sim$, also called pseudo-complementation, then it is a De Morgan algebra. A ternary algebra T [6] is a De Morgan algebra endowed with a constant Φ which is a fixed point of the function $\sim$ and satisfies

$$x \wedge \sim x \leq \Phi \tag{1}$$

for all $x \in T$. Dually, it follows that $\Phi \leq y \vee \sim y$ for all $y \in L$. Hence every ternary algebra T satisfies Kleene's law $x \wedge \sim x \leq y \vee \sim y$ for all $x, y \in L$ and is thus a Kleene algebra [2], and by (1), also a centered Kleene algebra. Conversely, every centered Kleene algebra is clearly a ternary algebra. Also, if T is a ternary algebra (or equivalently, a centered Kleene algebra), then Φ is the only fixed point of $\sim$, as if Φ' is another fixed point, then $\Phi' = \Phi' \wedge \sim \Phi' \leq \Phi \leq \Phi' \vee \sim \Phi' = \Phi'$. For more information about ternary algebras, we refer to [1,5,10,13].

Homomorphisms of De Morgan algebras are bounded lattice homomorphisms preserving the involution. A homomorphism of ternary algebras is a De Morgan algebra homomorphism which preserves the constant Φ.

Example 1. Examples of ternary algebras include the 3-element ternary algebra $\{\perp < \Phi < \top\}$ or $\{\text{ff} < \Phi < \text{tt}\}$ (mentioned in the introduction), equipped with the (unique) involution interchanging the extremal elements and fixing Φ. The extremal elements represent the classical truth values while Φ is an intermediate truth value expressing uncertainty. Another example is the closed interval $[0, 1]$, ordered as usual, with involution $\sim x = 1 - x$. The constant Φ is represented by $1/2$. This algebra is of course isomorphic to the ternary algebra given by the closed interval $[-1, 1]$, equipped with the usual order, and the function $x \mapsto -x$ as involution. The constant Φ is 0. Here, a point $p \in [-1, 1]$ may represent the level of confidence that a proposition holds or not. We denote this ternary algebra by $\mathbb{Q}$.

We now describe a general construction originating from [15], see also [8]. Suppose that L is a bounded distributive lattice. Then we can form another bounded distributive lattice on the set $L \times L$ by defining $(x_1, x_2) \leq (y_1, y_2)$ iff $x_1 \geq y_1$ and $x_2 \leq y_2$. Note that

$$(x_1, x_2) \vee (y_1, y_2) = (x_1 \wedge y_1, x_2 \vee y_2) \quad \perp = (\top, \perp)$$
$$(x_1, x_2) \wedge (y_1, y_2) = (x_1 \vee y_1, x_2 \wedge y_2) \quad \top = (\perp, \top)$$

By defining $\sim (x_1, x_2) = (x_2, x_1)$, we obtain a De Morgan algebra, denoted $D(L)$. Let $T(L)$ consist of those elements $(x_1, x_2) \in L \times L$ with $x_1 \wedge x_2 = \perp$. Since $T(L)$ is closed with respect to $\vee, \wedge$ and $\sim$ and contains $(\top, \perp)$ and $(\perp, \top)$, it is also a De Morgan algebra. Moreover, equipped with the constant $(\perp, \perp)$ as Φ, it is a ternary algebra.

Example 2. When L is the lattice $P(A)$ of all subsets of a set A, then $T(L)$ is called a subset pair algebra [5].[1] The elements of $T(L)$ are all ordered pairs (X_1, X_2) of disjoint sets $X_1, X_2 \subseteq A$. The operations are $(X_1, X_2) \cup (Y_1, Y_2) = (X_1 \cap X_2, Y_1 \cup Y_2)$ etc. The intuition for this algebra is that an ordered pair (X_1, X_2) specifies those facts X_1 that falsify a certain proposition P, while X_2 specifies those which justify P.

Proposition 1. *For every ternary algebra T there is a bounded distributive lattice L such that T can be embedded into $T(L)$ by an injective ternary algebra homomorphism.*

---

[1] Actually [5] uses a dual construction.

Proof. For later reference, we outline the proof. Given T, let $L = \{x \in T : \Phi \leq x\}$. Then L is a bounded distributive lattice equipped with the $\vee$ and $\wedge$ operations of T restricted to L, and with the elements Φ and $\top$ as distinguished constants. The ternary algebra T can be embedded in $T(L)$ by mapping each $x \in T$ to $h(x) = (\Phi \vee \sim x, \Phi \vee x)$. This function is injective and preserves the operations and constants. Also, $h(x) \in T(L)$ for all $x \in T$. $\qquad\square$

Remark 1. Since every bounded distributive lattice can be embedded by an injective homomorphism in some $P(A)$, it follows that each ternary algebra can be embedded in a subset pair algebra by some injective ternary algebra homomorphism. See [10], or [5] for the finite case.

We say that a ternary algebra is complete if it is a complete lattice. The ternary algebras described above in Examples 1 and 2 are all complete. In fact, all finite ternary algebras are complete.

Proposition 2. *Suppose that L is a complete lattice satisfying the infinite distributivity conditions*

$$x \vee \bigwedge_{i \in I} y_i = \bigwedge_{i \in I}(x \vee y_i) \quad \text{and} \quad x \wedge \bigvee_{i \in I} y_i = \bigvee_{i \in I}(x \wedge y_i), \tag{2}$$

where I is any infinite set. Then $T(L)$ is a complete ternary algebra satisfying these conditions.

Proof. Indeed, if $(x_{i1}, x_{i2}) \in T(L)$ for all $i \in I$, then $\bigvee_{i \in I} x_{i1} \wedge \bigwedge_{i \in I} x_{i2} = \bigvee_{i \in I}(x_{i1} \wedge \bigwedge_{j \in I} x_{j2}) = \bigvee_{i \in I} \bigwedge_{j \in J}(x_{i1} \wedge x_{j2}) \leq \bigvee_{i \in I}(x_{i1} \wedge x_{i2}) = \perp$. Hence, if $(x_{i1}, x_{i2}) \in T(L)$ for all $i \in I$, then $\bigwedge_{i \in I}(x_{i1}, x_{i2}) = (\bigvee_{i \in I} x_{i1}, \bigwedge_{i \in I} x_{i2})$ exists in $T(L)$. Similarly, $\bigvee_{i \in I}(x_{i1}, x_{i2}) = (\bigwedge_{i \in I} x_{i1}, \bigvee_{i \in I} x_{i2})$ also exists. Since the infinite distributivity conditions hold in L, they also hold in $T(L)$. To prove this, suppose that $y_i = (y_{i1}, y_{i2}) \in T(L)$ for all $i \in I$ and $x = (x_1, x_2) \in T(L)$, where I is infinite. Then

$$x \vee \bigwedge_{i \in I} y_i = (x_1, x_2) \vee \left(\bigvee_{i \in I} y_{i1}, \bigwedge_{i \in I} y_{i2} \right)$$

$$= (x_1 \wedge \bigvee_{i \in I} y_{i1}, x_2 \vee \bigwedge_{i \in I} y_{i2})$$

$$= \left(\bigvee_{i \in I}(x_1 \wedge y_{i1}), \bigwedge_{i \in I}(x_2 \vee y_{i2}) \right)$$

$$= \bigwedge_{i \in I}(x_1 \wedge y_{i1}, x_2 \vee y_{i2}) = \bigwedge_{i \in I}(x \vee y_i),$$

proving the first equation in (2). Since $\sim$ is an involution, the second holds also. $\qquad\square$

Theorem 1. *For any complete ternary algebra T satisfying the infinite distributivity conditions there is a complete lattice L which also satisfies the infinite distributivity conditions such that T can be embedded in $T(L)$ by an injective ternary algebra homomorphism which preserves all suprema and infima.*

Proof. When T is complete and satisfies the infinite distributivity conditions, then $L = \{x \in T : \Phi \leq x\}$ is also complete and satisfies the infinite distributivity conditions. Consider the ternary algebra embedding h of T into $T(L)$ defined by $x \mapsto (\Phi \vee \sim x, \Phi \vee x)$ as above. If $x_i \in T$ for all $i \in I$, where I is infinite, then

$$h(\bigvee_{i \in I} x_i) = (\Phi \vee \sim \bigvee_{i \in I} x_i, \Phi \vee \bigvee_{i \in I} x_i)$$

$$= (\Phi \vee \bigwedge_{i \in I} \sim x_i, \Phi \vee \bigvee_{i \in I} x_i)$$

$$= (\bigwedge_{i \in I}(\Phi \vee \sim x_i), \bigvee_{i \in I}(\Phi \vee x_i))$$

$$= \bigvee_{i \in I}(\Phi \vee \sim x_i, \Phi \vee x_i) = \bigvee_{i \in} h(x_i).$$

Hence, h preserves all suprema. Since h preserves $\sim$, it also preserves all infima. $\square$

In [6], the 3-element ternary algebra $\{\mathsf{ff} < \Phi < \mathsf{tt}\}$ was equipped with another partial order $\sqsubseteq$ given by $\Phi \sqsubseteq \mathsf{ff}$ and $\Phi \sqsubseteq \mathsf{tt}$. The partial order was then extended to direct powers of the 3-element ternary algebra in a pointwise manner. But actually the partial order $\sqsubseteq$ can be defined on all ternary algebras.

Suppose that T is a ternary algebra. Following [4][2], we define, for all $x, y \in T$,

$$x \sqcap y = (x \vee y) \wedge ((x \wedge y) \vee \Phi)$$

and $x \sqsubseteq y \Leftrightarrow x \sqcap y = x$. The following fact is stated in [4] in an equivalent form. We give a simple proof based on Theorem 1.

Proposition 3 ([4]). *Let T be a ternary algebra. Then the relation $\sqsubseteq$ is a partial order on T, and for each $x, y \in T$, $x \sqcap y$ is the infimum of $\{x, y\}$ w.r.t. this partial order. Moreover, Φ is the least element of T w.r.t. $\sqsubseteq$.*

Proof. By Proposition 1, it suffices to prove our claim when T is a subalgebra of $T(L)$, where L is a bounded distributive lattice (with the extremal elements $\bot$ and $\top$). Assuming this, let $x = (x_1, x_2)$ and $y = (y_1, y_2)$ in T. Then

$$x \sqcap y = (x \vee y) \wedge ((x \wedge y) \vee \Phi)$$

$$= (x_1 \wedge y_1, x_2 \vee y_2) \wedge ((x_1 \vee y_1, x_2 \wedge y_2) \vee (\bot, \bot))$$

$$= (x_1 \wedge y_1, x_2 \vee y_2) \wedge (\bot, x_2 \wedge y_2) = (x_1 \wedge y_1, x_2 \wedge y_2)$$

is in T. Hence $(x_1, x_2) \sqsubseteq (y_1, y_2)$ iff $x_1 \leq y_1$ and $x_2 \leq y_2$, so that $\sqsubseteq$ is just the component-wise order on $T(L)$, restricted to T. Moreover, $x \sqcap y$ is the infimum of x and y w.r.t. this order relation. The last claim is obvious. $\square$

Theorem 2. *Suppose that T is a complete ternary algebra satisfying the infinite distributivity conditions. Then the infimum $\sqcap X$ of X w.r.t. $\sqsubseteq$ exists for all nonempty subsets X of T. Moreover, with respect to the ordering $\sqsubseteq$, every chain has a supremum.*

---

[2] Actually a dual operation and a dual ordering are defined in [4].

Proof. Indeed, by Theorem 1, it suffices to prove the claim when T is a ternary subalgebra of $T(L)$ for some complete lattice L satisfying the infinite distributivity conditions, moreover, T is closed under arbitrary infima and suprema. Let $X = \{(x_{i1}, x_{i2}) : i \in I\} \subseteq T$, where I not empty. Then, denoting the least element of L by $\bot$,

$$\bigvee X \wedge \left(\bigwedge X \vee \Phi\right) = \left(\bigwedge_{i \in I} x_{i1}, \bigvee_{i \in I} x_{i2}\right) \wedge \left(\left(\bigvee_{i \in I} x_{i1}, \bigwedge_{i \in I} x_{i2}\right) \vee (\bot, \bot)\right)$$

$$= \left(\bigwedge_{i \in I} x_{i1}, \bigvee_{i \in I} x_{i2}\right) \wedge \left(\bot, \bigwedge_{i \in I} x_{i2}\right) = \left(\bigwedge_{i \in I} x_{i1}, \bigwedge_{i \in I} x_{i2}\right),$$

showing that $(\bigwedge_{i \in I} x_{i1}, \bigwedge_{i \in I} x_{i2})$ is in T. But then $(\bigwedge_{i \in I} x_{i1}, \bigwedge_{i \in I} x_{i2})$ is necessarily $\bigsqcap X$ in T (as in $T(L)$).

We now show that, with respect to the ordering $\sqsubseteq$, every chain has a supremum. So suppose that I is a linearly ordered set and $(x_{i1}, x_{i2}) \in T$ for all $i \in I$ such that $(x_{i1}, x_{i2}) \sqsubseteq (x_{j1}, x_{j2})$ whenever $i \leq j$. Let $x_{\infty 1} = \bigvee_{i \in I} x_{i1}$ and $x_{\infty 2} = \bigvee_{i \in I} x_{i2}$. We prove that $(x_{\infty 1}, x_{\infty 2}) \in T$. First, note that for all $i \leq j$, $(x_{i1}, x_{j2}) = (x_{i1}, x_{i2}) \vee (x_{j1}, x_{j2}) \in T$. Hence for each fixed $i \in I$, $(x_{i1}, x_{\infty 2}) = (x_{i1}, \bigvee_{i \leq j} x_{j2}) = \bigvee_{i \leq j} (x_{i1}, x_{j2})$ is in T. But then, $(x_{\infty 1}, x_{\infty 2}) = (\bigvee_{i \in I} x_{i1}, x_{\infty 2}) = \bigvee_{i \in I} (x_{i1}, x_{\infty 2})$ is also in T. It is now clear that $(x_{\infty 1}, x_{\infty 2})$ is the least upper bound $\bigsqcup_{i \in I} (x_{i1}, x_{i2})$ in T (as in $T(L)$). $\square$

Recall from [9] that a partial order P is called complete if each chain in P has a supremum. Since this also applies for the empty chain, each complete partial order has a least element. (It is known that every directed set has a supremum in a complete partial order.) We have thus proved that if T is a complete ternary algebra satisfying the infinite distributivity conditions, then $(T, \sqsubseteq)$ is a complete partial order. We can strengthen Theorem 1.

Corollary 1. *Suppose that T is a complete ternary algebra satisfying the infinite distributivity conditions. Then there is a complete lattice L satisfying the infinite distributivity conditions such that T can be embedded in $T(L)$ by a ternary algebra homomorphism which preserves all suprema and infima with respect to the ordering $\leq$, all infima with respect to the ordering $\sqsubseteq$ and all suprema of chains with respect to $\sqsubseteq$.*

Proposition 4. *Suppose that T is a ternary algebra. Then all distributive laws hold for the operations $\vee, \wedge$ and $\sqcap$. And when T is a complete ternary algebra satisfying the infinite distributivity conditions, then all infinite distributivity conditions hold in T for the operations $\vee, \wedge, \sqcap$ and $\bigvee, \bigwedge, \bigsqcap$.*

Proof. We only prove the second claim. By Theorem 1, without loss of generality we may assume that there is a complete lattice L satisfying the infinite distributivity conditions such that T is a subalgebra of $T(L)$ closed under all infima and suprema.

Suppose that $x_i \in T$ for all $i \in I$ and $y \in T$, where I is infinite. Then we can write each x_i and y in the form $x_i = (x_{i1}, x_{i2})$ and $y = (y_1, y_2)$. Using Theorem 2 we have:

$$y \sqcap \bigvee_{i \in I} x_i = (y_1, y_2) \sqcap \bigvee_{i \in I} (x_{i1}, x_{i2}) = (y_1, y_2) \sqcap (\bigwedge_{i \in I} x_{i1}, \bigvee_{i \in I} x_{i2})$$

$$= (y_1 \wedge \bigwedge_{i \in I} x_{i1}, y_2 \wedge \bigvee_{i \in I} x_{i2}) = (\bigwedge_{i \in I} (y_1 \wedge x_{i1}), \bigvee_{i \in I} (y_2 \wedge x_{i2}))$$

$$= \bigvee_{i \in I} (y_1 \wedge x_{i1}, y_2 \wedge x_{i2}) = \bigvee_{i \in I} ((y_1, y_2) \sqcap (x_{i1}, x_{i2})) = \bigvee_{i \in I} (y \sqcap x_i).$$

This proves the infinite distributivity condition for $\sqcap$ and $\bigvee$. Similar arguments prove the infinite distibutivity condition for the other cases. For $\vee$ and $\wedge$ and for $\wedge$ and $\bigvee$, see also Proposition 2. $\square$

We now recall from [9] that a function $f : P \to Q$ between partially ordered sets P and Q is continuous if it preserves the supremum of nonempty chains (or equivalently, the supremum of nonempty directed sets).

Proposition 5. *Suppose that T is a complete ternary algebra satisfying the infinite distributivity conditions. Then the operations $\vee, \wedge, \sqcap$ and $\sim$ are continuous with respect to the partial order $\sqsubseteq$.*

Proof. It again suffices to prove our claim when T is a subalgebra of a ternary algebra $T(L)$, where L is a complete lattice satisfying the distributivity conditions, which is closed under arbitrary suprema $\bigvee X$ and infima $\bigwedge X$. We already know that suprema $\bigsqcup X$ exist if X is a chain w.r.t. $\sqsubseteq$. Moreover, all these suprema are given as in $T(L)$.

We prove that $\vee$ is continuous w.r.t. $\sqsubseteq$ in the second argument. To this end, suppose that I is a nonempty chain ordered by the relation $\leq$ and $(x_{i1}, x_{i2}) \in T$ for all $i \in I$. Moreover, suppose that $(x_{i1}, x_{i2}) \sqsubseteq (x_{j1}, x_{j2})$ whenever $i \leq j$ in I. Then for all $(y_1, y_2) \in T$,

$$(y_1, y_2) \vee \bigsqcup_{i \in I} (x_{i1}, x_{i2}) = (y_1, y_2) \vee (\bigvee_{i \in I} x_{i1}, \bigvee_{i \in I} x_{i2})$$

$$= (y_1 \wedge \bigvee_{i \in I} x_{i1}, y_2 \vee \bigvee_{i \in I} x_{i2})$$

$$= (\bigvee_{i \in I} (y_1 \wedge x_{i1}), \bigvee_{i \in I} (y_2 \vee x_{i2}))$$

$$= \bigsqcup_{i \in I} (y_1 \wedge x_{i1}, y_2 \vee x_{i2}) = \bigsqcup_{i \in I} ((y_1, y_2) \vee (x_{i1}, x_{i2}))$$

proving that $\vee$ is continuous w.r.t. $\sqsubseteq$. Since $\vee$ is commutative, it is also continuous in the first argument and hence continuous. The facts that $\wedge$ and $\sqcap$ are also continuous can be proved similarly. Finally, $\sim$ is also continuous which can be seen easily using the above representation of T. $\square$

3 Solving Systems of Fixed Point Equations

Let Σ be a ranked alphabet. A Σ-algebra A consists of a nonempty set, also denoted A, and an operation $\sigma^A : A^n \to A$ for each symbol σ of rank n.

Call a Σ-algebra A finitely decomposable if for each $a \in A$ and $\sigma \in \Sigma$ of rank n, there are only a finite number of decompositions of a in the form $a = \sigma^A(a_1, \ldots, a_n)$, where $a_1, \ldots, a_n \in A$.

For the rest of this section, suppose that T is a complete ternary algebra satisfying the infinite distributivity conditions and A is a finitely decomposable Σ-algebra. For each $n \geq 0$, let Σ_n denote the set of symbols of rank n. We introduce systems of fixed point equations over the complete ternary algebra T of generalized truth values and the finitely decomposable algebra A as data domain.

Since T is a complete ternary algebra satisfying the infinite distributivity conditions, so is the set T^A of all functions $A \to T$ w.r.t. the pointwise order, given for $f, g \in L^A$ by $f \leq g$ iff $f(a) \leq g(a)$ for all $a \in A$. (A function $f : A \to T$ is usually called a fuzzy set.) The supremum $\bigvee F$ and infimum $\bigwedge F$ of a subset F of T^A are formed pointwise, so that $(\bigvee F)(a) = \bigvee_{f \in F} f(a)$ for all $a \in A$, and similarly for $\bigwedge F$. The operation $\sim$ is also defined pointwise. It follows that the partial order $\sqsubseteq$ on T^A is also the pointwise order, so that for all $f, g \in T^A$, $f \sqsubseteq g$ iff $f(a) \sqsubseteq g(a)$ for all $a \in A$. Moreover, for any nonempty $X \subseteq T^A$, $\bigsqcap X$ can be computed pointwise, and similarly for $\bigsqcup X$ whenever X is a chain w.r.t. $\sqsubseteq$. The least and greatest elements of T^A are the constant functions with value $\bot$ and $\top$, respectively. The additional constant is the function mapping each element of A to Φ. Below we will denote these functions by $\bot, \top$ and Φ as well.

We now turn T^A into a Σ-algebra. Let $\sigma^{T^A}(f_1, \ldots, f_n)(a) = \bigvee\{\bigwedge_{i=1}^n f_i(a_i) : a = \sigma^A(a_1, \ldots, a_n)\}$, for all $\sigma \in \Sigma_n$, $n \geq 0$, $f_1, \ldots, f_n \in T^A$ and $a \in A$. Note that when $n = 0$, σ^{T^A} maps σ^A to $\top$ and all other elements of A to $\bot$.

Proposition 6. *For each $\sigma \in \Sigma$, σ^{T^A} is continuous w.r.t. the ordering $\sqsubseteq$.*

Proof. For simplicity we assume that σ is of rank 2. (When the rank is 0, our claim is obvious.) It suffices to prove that σ^{T^A} is separately continuous w.r.t. $\sqsubseteq$ in each argument. We prove it is continuous in the first argument. Suppose that $f_i \in T^A$ for all $i \in I$, where I is a nonempty linearly ordered set, such that $f_i \sqsubseteq f_j$ whenever $i \leq j$ in I. Let $g \in T^A$. We need to show that $\sigma^{T^A}(\bigsqcup_{i \in I} f_i, g) = \bigsqcup_{i \in I} \sigma^{T^A}(f_i, g)$. But for all $a \in A$, $(\sigma^{T^A}(\bigsqcup_{i \in I} f_i, g))(a) =$

$$= \bigvee\{(\bigsqcup_{i \in I} f_i)(a_1) \wedge g(a_2) : a = \sigma^A(a_1, a_2)\}$$

$$= \bigvee\{(\bigsqcup_{i \in I} f_i(a_1)) \wedge g(a_2) : a = \sigma^A(a_1, a_2)\}$$

$$= \bigvee\{\bigsqcup_{i \in I}(f_i(a_1) \wedge g(a_2)) : a = \sigma^A(a_1, a_2)\}$$

$$= \bigsqcup_{i \in I} \bigvee\{f_i(a_1) \wedge g(a_2) : a = \sigma^A(a_1, a_2)\}$$

$$= \bigsqcup_{i \in I}(\sigma^{T^A}(f_i, g)(a)) = (\bigsqcup_{i \in I} \sigma^{T^A}(f_i, g))(a).$$

In the above argument, we used that by finite decomposability, the supremum $\bigvee$ is over a finite set, and moreover, $\vee$ and $\wedge$ are continuous w.r.t. $\sqsubseteq$ as shown above. □

Our aim is to provide a method for solving systems of fixed point equations over T^A involving the Σ-operations, the binary $\vee$ and $\wedge$ operations, involution $\sim$, the elements of T including $\bot$, $\top$ and Φ as constants, and possibly other functions $T \to T$, $T \times T \to T$, etc. which are continuous w.r.t. $\sqsubseteq$. Let X be a set of variables. A system of fixed point equations over T^A is of the form

$$x = t_x, \quad x \in X \qquad (E)$$

where each t_x is a well formed term composed of the variables in X, the symbols in Σ and $\{\vee, \wedge, \sim\}$ and the elements of T as constants.

Since T^A is both a Σ-algebra and a complete ternary algebra, each term t induces a function $t^{T^A} : (T^A)^X \to T^A$. This is defined by induction on the structure of the term. Let $t, t_1, t_2, \ldots$ be terms and $\sigma \in \Sigma_n$.

- For each $x \in X$, x^{T^A} is the corresponding projection $(T^A)^X \to T^A$, so that $(x^{T^A})(u)(a) = u(x)(a)$ for all $u \in (T^A)^X$ and $a \in A$.
- $(t_1 \vee t_1)^{T^A} = \vee \circ \langle t_1^{T^A}, t_2^{T^A} \rangle$, so that $(t_1 \vee t_2)^{T^A}(u)(a) = t_1^{T^A}(u)(a) \vee t_2^{T^A}(u)(a)$ for all $u \in (T^A)^X$ and $a \in A$.
- $(t_1 \wedge t_1)^{T^A} = \wedge \circ \langle t_1^{T^A}, t_2^{T^A} \rangle$, so that $(t_1 \wedge t_2)^{T^A}(u)(a) = t_1^{T^A}(u)(a) \wedge t_2^{T^A}(u)(a)$ for all $u \in (T^A)^X$ and $a \in A$.
- $(\sim t)^{T^A} = \sim \circ t^{T^A}$, so that $(\sim t)^{T^A}(u)(a) = \sim (t^{T^A}(u)(a))$, for all $u \in (T^A)^X$ and $a \in A$.
- $(\sigma(t_1, \ldots, t_n))^{T^A} = \sigma^{T^A} \circ \langle t_1^{T^A}, \ldots, t_n^{T^A} \rangle$, i.e.,

$$(\sigma(t_1, \ldots, t_n))^{T^A}(u)(a) = \bigvee_{a = \sigma^A(a_1, \ldots, a_n)} \bigwedge_{i=1}^{n} t_i^{T^A}(u)(a_i)$$

for all $\sigma \in \Sigma_n$, $u \in (T^A)^X$ and $a \in A$.
- When $t = s \in T$, $t^{T^A}(u)(a) = s$ for all $u \in (T^A)^X$ and $a \in A$.

Since T^A is a complete ternary algebra satisfying the infinite distributivity conditions, so is $(T^A)^X$. In particular, $(T^A)^X$ is a complete partial order with respect to the ordering $\sqsubseteq$ (which agrees with the pointwise $\sqsubseteq$ ordering).

Lemma 1. *For each term t, $t^A : (T^A)^X \to t^A$ is continuous with respect to the the partial order $\sqsubseteq$.*

Proof. This follows by a straightforward induction using that the projection functions $(T^A)^X \to T^A$ are continuous as are the operations σ^{T^A} and the pointwise extensions of the functions $\vee, \wedge, \sim$ and the constant functions associated with elements of T, together with the fact that any composition or target tupling of continuous functions is continuous. □

Consider again the system of fixed point equations (E). The target tupling $E^{T^A} = \langle t_x^{T^A} \rangle_{x \in X}$ maps $(T^A)^X$ to itself, and solutions of (E) correspond to fixed points of E^{T^A}. We call $(E^A)^X$ the structure map associated with E. From Lemma 1 we immediately have:

Corollary 2. *For any system of fixed point equations* (E) *as above,* E^{T^A} : $(T^A)^X \to (T^A)^X$ *is continuous w.r.t.* $\sqsubseteq$.

It is well-known that a continuous endofunction of a complete partial order has a least fixed point which can be obtained in at most ω iterations. We thus have:

Theorem 3. *Any system of fixed point equations* (E) *has a unique least solution w.r.t.* $\sqsubseteq$ *which can be obtained as* $\bigsqcup_{n \geq 0} u_n$, *where* $u_0 = \Phi$ *and* $u_{n+1} = E^{T^A}(u_n)$ *for all* $n \geq 0$.

We call the least solution the canonical solution of E over T^A. Moreover, we call a function (or fuzzy set) in T^A equational, if it is equal to some component of the canonical solution of a finite system. Since ternary equational sets are defined by least fixed points, the following result can be proved by standard techniques [3]:

Proposition 7. *Suppose that* T *is a complete ternary algebra satisfying the distributivity conditions and* A *is a finitely decomposable algebra. Then the equational sets in* T^A *are closed under the operations* $\vee, \wedge, \sim$ *and* σ^{T^A}, *where* $\sigma \in \Sigma$.

Remark 2. There are several extensions. Since the operation $\sqcap$ is also continuous, we may also allow this operation (or the corresponding symbol) in terms. In a similar vein, we can introduce additional $\sqsubseteq$-continuous functions $T \to T$, $T \times T \to T$ etc. and use them in terms.

Certain systems of equations in normal form can be considered as grammars, see the example below. As usual, the language generated by the grammar is then defined as a selected component of the canonical solution of the system. We can associate a $\sqsubseteq$-continuous function $T \to T$ with each rule as the 'contribution' of the application of the rule. This will be illustrated by an example below.

Remark 3. Every continuous function between partially ordered sets is monotonic, and when P is complete and $f : P \to P$, then for the existence of the least fixed point of f it suffices that the function f is monotonic. (However, usually more than ω iterations are needed to reach the least fixed point.) Now the structure map $(E^A)^X$ is monotonic w.r.t. $\sqsubseteq$ without the assumption that A is finitely decomposable, hence we can define the canonical solution as the least solution of (E) for all algebras A.

Example 3. In this example, the underlying algebra is the free monoid Z^* over the alphabet $Z = \{a, b, c\}$ equipped with the operation of concatenation and constants denoting the letters of Z and the empty word ϵ. Consider the following grammar from [21]:

$$S \to AB \& \neg DC \quad A \to aA|\epsilon \quad B \to bBc|\epsilon$$
$$C \to cC|\epsilon \qquad D \to aDb|\epsilon$$

As a system of equations, it is

$$S = AB \wedge \sim (DC) \quad A = aA \vee \epsilon \quad B = bBc \vee \epsilon$$
$$C = cC \vee \epsilon \qquad\quad D = aDb \vee \epsilon$$

Its semantics in [21] is given by $L = L(S) = \{a^m b^n c^n : m, n \geq 0, \ m \neq n\}$. Using the 3-element ternary algebra, it is the function that maps each word in L to $\mathfrak{tt}$ and all other words to $\mathfrak{ff}$, since this function is the canonical solution of the corresponding system of equations. (This is actually the only solution.)

Let us now move to the ternary algebra $\mathbb{Q}$. Let r be a constant $0 < r < 1$, and let f_r map $q \in [-1, 1]$ to $rq \in [-1, 1]$. Let us attach f_r to the rules for A and B. Technically this means that the equation corresponding to A becomes $A = f_r(aA) \vee f_r(\epsilon)$, and similarly for B. Let f denote the component of the canonical solution of the system of equations corresponding to S, i.e. the ternary equational set specified by the grammar. Then $f(w) = r^{\max\{m,n\}+1}$ for all $w = a^m b^n c^n$ with $m \neq n$. For all other words, $f(w)$ is negative.

4 Related Work

Fixed point equations over languages of words and trees and other structures have been studied since the 1960's. When the functions involved are all monotonic or continuous with respect to the subset ordering of languages, the canonical solution of a system is usually defined as the least solution. In some cases, the least solution is also the unique solution. In the presence of non-monotonic operations such as complementation, a system of fixed point equations may not have a solution, or several minimal solutions may exist. Such systems of fixed point equations for word languages have been extensively studied in [18], and more recently in [20]. A closely related concept is that of Boolean grammars, introduced in [19]. For a recent survey on Boolean grammars, see [21].

Fixed point equations involving non-monotonic operations have traditionally been considered in logic programming, see [12] for a survey. There is now a common agreement that the meaning of a logic program involving negation can be best described using 3-valued logic. Using methods from logic programming, a novel 3-valued semantics to Boolean grammars is given in [16]. In order to specify the meaning of a grammar, a system of fixed point equations over valuations in the 3-element Kleene, or ternary algebra, is associated with the grammar, and the language generated by the grammar is described as the unique 3-valued well-founded fixed point solution of the associated system. An attempt to extend the well-founded approach to a more general setting has been made in [11]. However, as shown recently [7], the well-founded fixed point solution does not behave in the expected way. For example, one cannot freely substitute the term t_x on the right hand side of an equation $x = t_x$ for some, or for all occurrences of x. (In systems of equations corresponding to grammars, the terms t_x are in a certain normal form, this property is not preserved by arbitrary substitution.) Motivated in part by this reason, in this paper we investigated another, more classical way of defining the meaning of grammars, this time taking least fixed points of the associated system of equations w.r.t. the information ordering. It has the advantage that systems of equations can be manipulated as expected. Also, the semantics is symmetric, whereas the well-founded semantics is not, since the preferred default truth value is ff. But the well-founded semantics is more precise, since the $\sqsubseteq$-least fixed point is less than or equal to the well-founded fixed point in the ordering $\sqsubseteq$.

Acknowledgment. The author would like to thank the referees for useful suggestions.

References

1. Balbes, R.: Free ternary algebras. IJAC **10**(6), 739–749 (2000)
2. Balbes, R., Dwinger, P.: Distributive Lattices. University of Missouri Press, Columbia (1974)
3. Bloom, S.L., Ésik, Z.: Iteration Theories. Springer, Heidelberg (1993)
4. Brzozowski, J.A.: Involuted semilattices and uncertainty in ternary algebras. IJAC **14**(3), 295–310 (2004)
5. Brzozowski, J.A., Lou, J.J., Negulescu, R.: A characterization of finite ternary algebras. IJAC **7**(6), 713–722 (1997)
6. Brzozowski, J.A., Seeger, C.J.: Asynchronous Circuits. Springer, Berlin (1995)
7. Carayol, A., Ésik, Z.: An analysis of the equational properties of the well-founded fixed point, arXiv:1511.09423 (2015)
8. Cignoli, R.: The class of Kleene algebras satisfying the interpolation property and Nelson algebras. Alg. Univ. **23**, 262–292 (1986)
9. Davey, B.A., Priestley, H.A.: Introduction to Lattices and Order, 2nd edn. Cambridge University Press, Cambridge (2002)
10. Ésik, Z.: A Cayley theorem for ternary algebras. IJAC **8**(3), 311–316 (1998)
11. Ésik, Z., Kuich, W.: Boolean fuzzy sets. IJFCS **18**(6), 1197–1207 (2007)
12. Fitting, M.: Fixpoint semantics for logic programming–a survey. TCS **278**(1–2), 25–51 (2002)
13. Figallo, A.V., Gomes, C.M., Sarmiento, L.S., Videla, M.E.: Notes on the variety of ternary algebras. Adv. Pure Math. **4**, 506–512 (2014)
14. Jezek, J.: Universal Algebra, 1st edn. (2008). http://www.karlin.mff.cuni.cz/~jezek/
15. Kalman, J.A.: Lattices with involution. Trans. Amer. Math. Soc. **87**, 485–491 (1958)
16. Kountouriotis, V., Nomikos, C., Rondogiannis, P.: Well-founded semantics for Boolean grammars. Inf. Comput. **207**(9), 945–967 (2009)
17. Kripke, S.: Outline of a theory of truth. J. Philos. **72**, 690–716 (1975)
18. Leiss, E.L.: Language Equations. Springer, New York (1999)
19. Okhotin, A.: Boolean grammars. Inf. Comput. **194**(1), 19–48 (2004)
20. Okhotin, A., Yakimova, O.: Language equations with complementation: expressive power. TCS **416**, 71–86 (2012)
21. Okhotin, A.: Conjunctive and Boolean grammars: the true general case of the context-free grammars. Comput. Sci. Rev. **9**, 27–59 (2013)

Problems on Finite Automata
and the Exponential Time Hypothesis

Henning Fernau[1]($\boxtimes$) and Andreas Krebs[2]($\boxtimes$)

[1] Fachbereich 4 – Abteilung Informatik, Universität Trier, 54286 Trier, Germany
fernau@uni-trier.de
[2] Wilhelm-Schickard-Institut für Informatik,
Universität Tübingen, Sand 13, 72076 Tübingen, Germany
krebs@informatik.uni-tuebingen.de

Abstract. We study several classical decision problems on finite automata under the (Strong) Exponential Time Hypothesis. We focus on three types of problems: universality, equivalence, and emptiness of intersection. All these problems are known to be CoNP-hard for nondeterministic finite automata, even when restricted to unary input alphabets. A different type of problems on finite automata relates to aperiodicity and to synchronizing words. We also consider finite automata that work on commutative alphabets and those working on two-dimensional words.

Keywords: Finite automata · Exponential Time Hypothesis · Universality problem · Equivalence problem · Emptiness of intersection

1 Introduction

Many computer science students will get the impression, at least when taught about the Chomsky hierarchy, that finite automata are fairly simple devices, and hence it is expected that typical decidability questions on finite automata are easy ones. In fact, for instance, non-emptiness for finite automata is solvable in polynomial time, as well as the uniform word problem.[1]

However, this impression is somewhat misled. Finite automata can be also viewed as edge-labeled directed graphs, and as many combinatorial problems are harder on directed graphs compared to undirected ones, it should not come as such a surprise that many interesting questions are NP-hard for finite automata.

We study hard problems for finite automata under the perspective of the Exponential Time Hypothesis (ETH) and variants thereof, as surveyed in [17]. In particular, using the famous sparsification lemma [13], ETH implies that there is no $O(2^{o(n+m)})$ algorithm for SATISFIABILITY (SAT) of m-clause 3CNF formulae with n variables, or 3SAT for short. Occasionally, we will also use SETH (Strong

---

[1] Even tighter descriptions of the complexities can be given within classical complexity theory, but this is not so important for our presentation here, as we mostly focus on polynomial versus exponential time.

© Springer International Publishing Switzerland 2016
Y.-S. Han and K. Salomaa (Eds.): CIAA 2016, LNCS 9705, pp. 89–100, 2016.
DOI: 10.1007/978-3-319-40946-7_8

Table 1. Universality/Equivalence; functions refer to exponents of bounding functions

Universality/Equivalence			
Σ	Lower bound	Upper bound	
Unary	$o(\sqrt[3]{q})$	$O(\sqrt{q \log q})$	Theorem 6
Binary	$o(q)$	q	Theorem 10
Unbounded	$o(q)$	q	Theorem 10

ETH); this hypothesis implies that there is no $O((2 - \varepsilon)^n)$ algorithm for solving (CNF-)SAT with n variables for any $\varepsilon > 0$.

Let us now discuss the objects and questions that we are going to study in the following. Mostly, we consider finite-state automata that read input words over the input alphabet Σ one-way, from left to right, and they accept when entering a final state upon reading the last letter of the word. We only consider deterministic finite automata (DFAs) and nondeterministic finite automata (NFAs). The language (set of words) accepted by a given automaton A is denoted by $L(A)$. We are going to study the following three problems.

Problem 1 (Universality). *Given an automaton A with input alphabet Σ, is $L(A) = \Sigma^*$? Parameters are the number q of states of A and the size of Σ.*

Problem 2 (Equivalence). *Given two automata A_1, A_2 with input alphabet Σ, is $L(A_1) = L(A_2)$? Parameters are an upper bound q on the number of states of A_1, A_2 and the size of Σ.*

Clearly, UNIVERSALITY reduces to EQUIVALENCE by choosing the automaton A_2 such that $L(A_2) = \Sigma^*$. Also, all these problems can be solved by computing the equivalent (minimal) deterministic automata, which requires time $O^*(2^q)$.[2] Our results on these problems for NFAs are summarized in Table 1. The functions refer to the exponents, so, e.g., in the first row, there is no $2^{o(\sqrt[3]{q})}$ algorithm for UNIVERSALITY for q-state NFAs with unary input alphabets.

Problem 3 (Intersection). *Given k automata $A_1, \ldots, A_k$, each with input alphabet Σ, is $\bigcap_{i=1}^{k} L(A_i) = \varnothing$? Parameters are the number of automata k, an upper bound q on the number of states of the automata A_i, and the size of Σ.*

For (EMPTINESS OF) INTERSECTION, our results are summarized in Table 2.

All these problems are already CoNP-hard for NFAs on unary inputs. Hence, we will study these first, before turning towards larger input alphabets. The classical complexity status of these and many more related problems is nicely surveyed in [12].

In the second part of the paper, we are extending our research in two directions: we consider further hard problems on finite automata, more specifically,

---

[2] Recall that the O^* notation suppresses polynomial factors.

Table 2. Intersection; functions refer to exponents of bounding functions

Intersection				
# of states	Σ	Lower bound	Upper bound	
2			O(1), i.e., in P	
3	Unary		O(1), i.e., in P	
3	Binary		O(1), i.e., in P	
3	Unbounded	$o(k)$	k	Proposition 11
q	Unary	$o(\min(k, \sqrt{q}/\log q))$	$\min(k \log q, q)$	Theorem 8 & Proposition 9
q	Binary	$o(\min(k, 2^q))$	$\min(k \log q, 2^{2q} \log q)$	Propositions 12 & 13
q	Unbounded	$o(k \log q)$	$k \log q$	Proposition 11

the question of whether a given DFA accepts an aperiodic language, and questions related to synchronizing words, and we also look at finite automata that work on objects different from strings.

2   Universality, Equivalence, Intersection: Unary Inputs

We first study UNIVERSALITY. Given an NFA A with input alphabet $\{a\}$, is $L(A) = \{a\}^*$? In [23], the corresponding problem for regular expressions was examined and shown to be CoNP-complete. As the reduction given in [23] starts off from 3SAT, we can easily analyze the proof to obtain the following result.

Theorem 4. *Unless ETH fails, for any $\epsilon > 0$, there is no $O^*(2^{o(q^{1/4-\epsilon})})$-time algorithm for deciding, given a tally NFA A on q states, whether $L(A) = \{a\}^*$.*

We are now trying to strengthen the assertion of the previous theorem. There are actually two weak spots in the mentioned reduction: (a) The ϵ-term in the statement of the theorem is due to logarithmic factors introduced by encodings with prime numbers; however, the encodings suggested in [23] leave rather big gaps of numbers that are not coding any useful information. (b) The $\sqrt[4]{\cdot}$-term is due to writing down all possible reasons for not satisfying any clause, which needs about $\tilde{O}(mn^3)$ many states (ignoring logarithmic terms) on its own; so, we are looking for a problem that allows for cheaper encodings of conflicts. To achieve our goals, we need the following theorem, see [5, Theorem 14.6].

Theorem 5. *Unless ETH fails, there is no $O^*(2^{o(m+n)})$-time algorithm for deciding if a given m-edge n-vertex graph has a (proper) 3-coloring.*

The previous result can be used to prove the following theorem. What is important to know is that the NFA A_G that can be associated to a graph instance G of 3-COLORING, with n vertices and m edges, decomposes into $3m+1$ components A_i, each with $O(n^2)$ many states, one taking care of unary words that do not encode graph colorings, and $3m$ expressing that one edge has two endpoints with the same color. So, $L(A_G) = \bigcup_{i=1}^{3m+1} L_i$ (L_i corresponding to A_i).

Theorem 6. *Unless ETH fails, there is no $O^*(2^{o(q^{1/3})})$-time algorithm for deciding, given an NFA A on q states, whether $L(A) = \{a\}^*$.*

How good is this? For the conversion of a q-state NFA on unary inputs into an equivalent q'-state DFA, it is known that $q' = 2^{\Theta(\sqrt{q \log q})}$ is possible but also necessary [4]. So, in a sense, the ETH bound poses the question if there are other algorithms to decide universality for tally NFAs, not using DFA conversion first. Conversely, it might be possible to tighten the upper bound.

We now turn to the equivalence problem for NFAs. From Theorem 6, we obtain:

Corollary 7. *Unless ETH fails, there is no $O^*(2^{o(q^{1/3})})$-time algorithm for deciding equivalence of NFAs A_1 and A_2 on at most q states and input alphabet $\{a\}$.*

We are finally turning towards TALLY-DFA-INTERSECTION and also towards TALLY-NFA-INTERSECTION. CoNP-completeness of this problem, both for DFAs and for NFAs, was indicated in [16], referring to [10,23]. We make this more explicit in the following, in order to also obtain some ETH-based results.

Theorem 8. *Let k DFAs $A_1, \ldots, A_k$ with input alphabet $\{a\}$ be given, each with at most q states. If ETH holds, then there is no algorithm with that decides if $\bigcap_{i=1}^k L(A_i) = \varnothing$ in time $O^*(2^{o(\min(k, \sqrt{q}/\log q))})$.*

Proof. We revisit our previous reduction (from an instance $G = (V, E)$ of 3-COLORING with $|V| = n$ and $|E| = m$ to some NFA instance for UNIVERSALITY), which delivered the union of many simple sets L_i, each of which can be accepted by a DFA A_i whose automaton graph is a simple cycle. These DFAs A_i have $O(n^2)$ states each. The complements $\overline{L_i}$ of the L_i can be also accepted by DFAs $\overline{A_i}$ of the same size. Ignoring constants, originally the union of $O(n + m)$ many such sets was formed. Considering now the intersection of the complements of the mentioned simple sets, we obtain the result. □

Proposition 9. *Let k DFAs $A_1, \ldots, A_k$ with input alphabet $\{a\}$ be given, each with at most q states. There is an algorithm that decides if $\bigcap_{i=1}^k L(A_i) = \varnothing$ in time $O^*(2^{\min(k \log q, 1.5 \cdot q)})$.*

Proof. For the upper bound there are basically two algorithms; the natural approach to solve this intersection problem would be to first build the product automaton, which is of size q^k, and then solve the emptiness problem in linear time on this device. This gives an overall running time of $O^*(q^k) = O^*(2^{k \log q})$; also see [24, Theorem 8.3]. On the other hand, we can test all words up to length $q + 2^{1.5q}$. As each DFA has at most q states in each DFA, processing a word enters a cycle in at most q steps. Also the size of the cycle in each DFA is bounded by q. The least common multiple of all integers bounded by q, i.e., $e^{\psi(q)}$, where ψ is the second Chebyshev function, is bounded by $2^{1.5q}$; see Propositions 3.2 and 5.1. in [6]. This yields an upper bound $O^*(2^{1.5q})$ of the running time. □

Hence in the case where the exponent is dominated by k, the upper and lower bound differ by a factor of $\log q$, and in the other case by a factor of $\sqrt{q} \cdot \log q$.

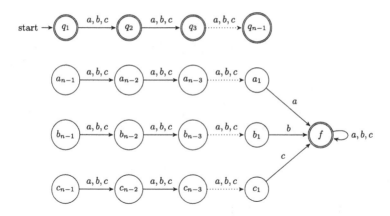

Fig. 1. A sketch of the NFA construction of Theorem 10

3   The Case of Non-unary Inputs

In the classical setting, the automata problems that we study are harder for binary (and larger) input alphabet sizes (PSPACE-complete; see [15]). Also, notice that the best-known algorithms are also slower in this case. This should be reflected in the lower bounds that we can obtain for them (under ETH), too.

Theorem 10. *Assuming ETH, there is no algorithm for solving* UNIVERSALITY *for q-state NFAs with binary input alphabets that runs in time* $O(2^{o(q)})$.

Proof. Let $G = (V, E)$ be an undirected graph, and $V = \{v_1, \ldots, v_n\}$. Let $\Sigma = \{a, b, c\}$ represent three colors. Then there is a natural correspondence of a word in Σ^n to a coloring of the graph, where the i-th letter in the words corresponds to the color of v_i. We construct an automaton with $3(n-1) + 1$ states, as sketched in Fig. 1. Additionally, for each edge (v_i, v_j) with $i < j$ in the graph, we add three types of transitions to the automaton: $q_i \xrightarrow{a} a_{j-i}$, $q_i \xrightarrow{b} b_{j-i}$, $q_i \xrightarrow{c} c_{j-i}$. Inputs of length n encode a coloring of the vertices. First notice that the automaton will accept every word of length not equal to n. Further, our construction enables the check of an improper coloring. We should accept a word $w_1 \ldots w_n$ iff $i < j$ and $(i, j) \in E$ and $w_i = w_j$. Pick such a word and assume, without a loss of generality, that $w_i = a$. Then the automaton will accept w_i, since the additional edge $q_i \xrightarrow{a} a_{j-i}$ allows for an accepting run terminating in the state f. Note that the automaton accepts all words of length at most $n - 1$.

The converse direction is also easily seen. Hence, if there is a valid coloring the automaton does not accept all words.

It is simple to change the construction given above to get away with binary input alphabets (instead of ternary ones). □

We are now turning towards DFA-INTERSECTION and to NFA-INTERSECTION. In the classical perspective, both are PSPACE-complete problems.

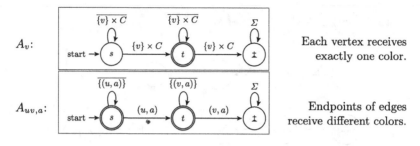

Fig. 2. The DFAs necessary to express a proper coloring

Proposition 11. *Let now $q \geqslant 3$. There is no algorithm that, given k DFAs (or NFAs) $A_1, \ldots, A_k$ with arbitrary input alphabet, each with at most q states, decides if $\bigcap_{i=1}^{k} L(A_i) = \varnothing$ in time $O^*(2^{\log q \cdot o(k)})$ unless ETH fails. The problems can be also solved in time $O^*(2^{\log q \cdot k})$.*

Proof. The hardness is by adaptation of the the 3-COLORING reduction we sketched for UNIVERSALITY. For parameters k and q, we take a graph with $|V| + |E| = k \log_3 q = \Theta(k \log q)$.[3] For the DFAs, choose alphabet $\Sigma = V \times C$, $C = \{1, 2, 3\}$. The states are $s, t, \bot$. For each vertex v, we define the DFA A_v; for each edge uv and each color a, we define the DFA $A_{uv,a}$, as described in Fig. 2. As we can compute the intersection for each block of $\log_3 q$ automata into a single DFA in polynomial time and obtain an automaton with q states, we reduce the number of DFAs to $k = (|E| + |V|)/\log_3 q$. Hence we got k DFAs each with q states. The claimed algorithm is via the well known product automaton construction.   □

We can encode the large alphabet of the previous construction into the binary one, but we get a weaker result. In particular, the DFAs A_v and $A_{uv,a}$ in this revised construction have $O(\log n)$ states, and not constantly many as before.

Proposition 12. *There is no algorithm that, given k DFAs $A_1, \ldots, A_k$ with binary input alphabet, each with at most q states, decides if $\bigcap_{i=1}^{k} L(A_i) = \varnothing$ in time $O^*(2^{o(k)})$ or $O^*(2^{o(2^q)})$ unless ETH fails.*

The following proposition gives a matching upper bound:

Proposition 13. *There is an algorithm that, given k DFAs with binary input alphabet, each with $\leqslant q$ states, decides INTERSECTION in time $O^*(2^{\log q \cdot \min(k, 2^{2q})})$.*

Proof. We will actually give two algorithms that solve this problem. One has a running time of $O^*(2^{k \log q})$ and one a running time of $O^*(2^{2^q \cdot \log q})$. The result then follows.
(a) We can first construct the product automaton of the DFAs $A_1, \ldots, A_k$, which

---

[3] In this proof, we neglect the use of some ceiling functions for the sake of readability.

is a DFA with at most $q^k = 2^{k \log q}$ many states. In this automaton, one can test emptiness in time linear in the number of states.

(b) Notice that there are only finitely many different DFAs with $\leqslant q$ many states. Intersection is easy to compute for DFAs with the same underlying labeled graph. On binary alphabets, each state has exactly two outgoing edges. Thus, there are q^2 possible choices for the outgoing edges of each state. Hence in total there are $(q^2)^q = q^{2q}$ different such DFAs. By merging first all DFAs with the same graph structure we can assume that $k \leqslant q^{2q}$. We now proceed as in (a).   □

4 Related Problems

Our studies so far only touched the peak of an iceberg. Let us mention and briefly discuss at least some related problems for finite automata in this section.

4.1 The Aperiodicity Problem

Recall that a regular language is called star-free (or aperiodic) if it can be expressed, starting from finite sets, with the Boolean operations and with concatenation. (So, Kleene star is disallowed in the set constructions.) We denote the subclass of the regular languages consisting of the star-free languages by SF.

It is known that a language is star-free if and only it its syntactic monoid is aperiodic [21], that is, it does not contain any nontrivial group. Here we will use a purely automata-theoric characterization: A language accepted by some minimum-state DFA A is star-free iff for every input word w, for every integer $r \geqslant 1$ and for every state q, $\delta^*(q, w^r) = q$ implies $\delta^*(q, w) = q$.

This allows a minimal automaton of a star-free language to contain a cycle, but if the word w along a cycle starting at q is a power of another word u, i.e., $w = u^r$ for some r, then u also forms a cycle starting at q.

For this class SF (and in fact for any other subregular language class), one can ask the following decision problem. Given a DFA A, is $L(A) \in$ SF? This problem (called APERIODICITY in the following) was shown to be PSPACE-complete in [2]. Recall the following characterization of aperiodicity: Cho and Huynh present a reduction that first (again) proves that the DFA-INTERSECTION-NONEMPTINESS is PSPACE-complete (by giving a direct simulation of the computations of a polynomial-space bounded TM) and then show how to alter this reduction in order to obtain the desired result. Unfortunately, this type of reductions is not very useful for ETH-based lower-bound proofs. In an earlier paper, Stern [22] proved that APERIODICITY is CoNP-hard. His reduction is from 3SAT (on n variables and m clauses), and it produces a minimum-state DFA with $O(nm)$ many states. Hence, we can conclude a lower bound of $O(2^{o(\sqrt{q})})$ for APERIODICITY on q-state DFAs. This can be improved:

The basic idea of the proof of the next proposition is to reduce the intersection problem (in a restricted version) to aperiodicity. Given language $L_1, L_2, \ldots, L_k$, consider the language $L = (L_1\$L_2\$\ldots\$L_k\$)^*$, and let A be its minimal automaton. One direction is easy: if the intersection of the languages $L_1, \ldots, L_k$ contains a word u, then $(u\$)^k$ forms a cycle in A starting at the initial state, but u does not. This gives the idea to show that if there is a word w in the intersection, then the language L is not aperiodic. The other direction is more involved.

Proposition 14. *Assuming ETH, there is no algorithm for solving* APERIODICITY *for q-state DFAs on arbitrary input alphabets that runs in time $O(2^{o(q)})$.*

If we use the automaton over the binary alphabet from Proposition 13 in the proof of the previous proposition, we get a bound for APERIODICITY over the binary alphabet. Actually the resulting automaton in the reduction will be over a ternary alphabet, but this can be recoded by at most tripling the number of states.

Corollary 15. *Under ETH, there is no algorithm to solve* APERIODICITY *for q-state DFAs on binary input alphabets in time $O(2^{o(q^{1-\epsilon})})$ for any $\epsilon > 0$.*

We are not aware of any published exponential-time APERIODICITY algorithm.

Proposition 16. *There is an algorithm for solving* APERIODICITY *that runs in time $O^*(q^q) = O^*(2^{q \log q})$ on a given q-state DFA with arbitrary input alphabet.*

Another related problem asks whether, given a DFA A, the language $L(A)$ belongs to AC^0. Analyzing the PSPACE-hardness proof from [1], we see that the same lower bound result as stated for APERIODICITY holds for this question.

4.2 Synchronizing Words

A deterministic finite semi-automaton (DFSA) A is given by $A = (Q, \Sigma, \{\mu_a \mid a \in \Sigma\})$, where, for each $a \in \Sigma$, there is a mapping $\mu_a : Q \to Q$. Given a DFSA A and a state set Q_{sync}, a Q_{sync}-*synchronizing word* $w \in \Sigma^*$ satisfies

$$\forall p, p' \in Q_{sync} : \mu_w(p) = \mu_w(p').$$

The Q_{sync}-SYNCHRONIZING WORD (Q_{sync}-SW) problem is the question, given a DFSA A, a set of states Q_{sync} and an integer k, whether there exists a Q_{sync}-synchronizing word w of length at most k for A. Correspondingly, the Q_{sync}-SYNCHRONIZING WORD problem can be stated. Notice that while Q-SW is NP-complete, Q_{sync}-SW is even PSPACE-complete, see [20].

Theorem 17. *There is an algorithm for solving Q_{sync}-SW on bounded input alphabets that runs in time $O^*(2^q)$ for q-state deterministic finite semi-automata. Conversely, assuming ETH, there is no $O^*(2^{o(q)})$-time algorithm for this task.*

Proof. It was already observed in [7] that the algorithm given there for Q-SW transfers to Q_{sync}-SW, as this is only a breadth-first search algorithm on an auxiliary graph (of exponential size, with vertex set 2^Q). The PSPACE-hardness proof contained in [20, Theorem 1.22], based on [19], reduces from DFA-INTERSECTION. Given k automata each with at most s states, with input alphabet Σ, one deterministic finite semi-automaton $A = (Q, \{\mu_a \mid a \in \Sigma\})$ is constructed such that $|Q| \leqslant sk + 2$. Hence, an $O^*(2^{o(|Q|)})$-time algorithm for Q_{sync}-SW would result in an $O^*(2^{o(sk)})$-time algorithm for DFA-INTERSECTION, contradicting Proposition 12. $\square$

5 SETH-Based Bounds: Length-Bounded Problem Variants

Cho and Huynh studied in [3] the complexity of a so-called bounded version of UNIVERSALITY, where in addition to the automaton A with input alphabet Σ, a number k (encoded in unary) is input, and the question is if $\Sigma^{\leqslant k} \subseteq L(A)$. This problem is again CoNP-complete for general alphabets. The proof given in [3] is by reduction from the n-STEP HALTING PROBLEM FOR NTMs. Our reduction from 3-COLORING also shows the mentioned CoNP-completeness result in a more standard way.

In [7], another SETH-based was derived. Namely, it was shown that (under SETH) there is no algorithm that determines, given a DFSA $A = (Q, \Sigma, \{\mu_a \mid a \in \Sigma\})$ and an integer ℓ, whether or not there is a synchronizing word for A in time $O((|\Sigma| - \varepsilon)^\ell)$ for any $\varepsilon > 0$. Here, Σ is part of the input; the statement is also true for fixed binary input alphabets. We will use this result now.

Theorem 18. *There is an algorithm with running time $O^*(|\Sigma|^\ell)$ that, given k DFAs over the input alphabet Σ and an integer ℓ, decides whether or not there is a word $w \in \Sigma^{\leqslant \ell}$ accepted by all these DFAs. Conversely, there is no algorithm that solves this problem in time $O((|\Sigma| - \varepsilon)^\ell)$ for any $\varepsilon > 0$ unless SETH fails.*

Proof. The mentioned algorithm simply tests all words of length up to ℓ. We show how to find a synchronizing word of length at most ℓ for a given DFSA $A = (Q, \Sigma, \{\mu_a \mid a \in \Sigma\})$ and an integer ℓ that runs in time $O((|\Sigma| - \varepsilon)^\ell)$, assuming that there is an algorithm with such a running time for BOUNDED DFA-INTERSECTION. From A, we build $|Q|^2$ many DFAs $A_{s,f}$ (with start state s and with unique final state f, while the transition function of all $A_{s,f}$ is identical, corresponding to $\{\mu_a \mid a \in \Sigma\}$). Let A^ℓ be the automaton that accepts any word of length at most ℓ. Now, we create $|Q|$ many instances of I_f of BOUNDED DFA-INTERSECTION. I_f is given by $\{A_{s,f} \mid s \in Q\} \cup \{A^\ell\}$. A has a synchronizing word of length at most ℓ if and only if for some $f \in Q$, I_f is a YES-instance. □

Clearly, we can use state complementation and a variant of the NFA union construction to show the following result.

Corollary 19. *There is an algorithm with running time $O^*(|\Sigma|^\ell)$ that, given some NFA over the input alphabet Σ and an integer ℓ, decides whether or not there is a word $w \in \Sigma^{\leqslant \ell}$ not accepted by this NFA. Conversely, no algorithm solves this problem in time $O((|\Sigma| - \varepsilon)^\ell)$ for any $\varepsilon > 0$ unless SETH fails.*

Clearly, this implies a similar result for BOUNDED NFA-EQUIVALENCE.

From these reductions, we can borrow quite a lot of other results from [7], dealing with inapproximability and parameterized intractability.

6 Two Further Ways to Interpret Finite Automata

Finite automata cannot be only used to process (contiguous) strings, but they might also jump from one position of the input to another position, or they can process two-dimensional words. We picked these two processing modes for the subsequent analysis, as they were introduced quite recently [8, 18].

6.1   Jumping Finite Automata

A *jumping finite automaton* (JFA) formally looks like a usual string-processing NFA. However, the application of a word looks different: If $p \xrightarrow{a} p'$ is a transition rule, then it can transform the input string u into u' provided that u, u' decompose as $u = u_1 a u_2$ and $u' = u_1 u_2$. This model was introduced in [18] and further studied in [9]. It is relatively easy to see that the languages accepted by JFAs are just the inverses of the Parikh images of the regular languages. In particular, the emptiness problem for JFAs is as simple as for NFAs. On unary input alphabets, JFAs and NFAs just work the same. Hence, UNIVERSALITY is hard for JFAs, as well. Classical complexity considerations on these formalisms are contained in [9,11,14] and the quoted papers; observe that mostly the input is given in the form of Parikh vectors of numbers encoded in binary, while we will consider the input given as words, since JFAs were introduced this way in [18].

Notice however that the uniform word problem for JFAs is NP-hard. Analyzing the proof given in [9, Theorem 54], we can conclude:

Theorem 20. *Under ETH, there is no algorithm that, given a JFA A on q states and a $w \in \Sigma^*$, decides if $w \in L(A)$ in time $O^*(2^{o(q)})$, $O^*(2^{o(|w|)})$ nor $O^*(2^{o(|\Sigma|)})$.*

There is a dynamic programming algorithm solving UNIVERSAL MEMBERSHIP for q-state JFAs (that is not improvable by Theorem 20). A word w allows the transition from state p to state p' iff for some decomposition $w \in w_1 \shuffle w_2$, p can transfer to $\hat{p}$ by reading w_1 and from $\hat{p}$ one can go into p' when reading w_2. For the correct implementation of the shuffle possibilities, we need to store possible translations for all subsets of indices within the input word, yielding a table (and time) complexity of $O^*(2^{|w|})$. We have no other upper bound.

What about the three decidability questions that are central to this paper for these devices? As the behavior of JFA is the same as that of NFA on unary alphabets, we can borrow all results from Sect. 2.

Theorem 21. *Let $k = |\Sigma|$ be fixed. Unless ETH fails, there is no algorithm that solves UNIVERSALITY for q-state JFAs in time $O^*(2^{o(q^{\frac{k}{k+2}})})$.*

Notice that the expression that we claim somehow interpolates between the third root of q (in the exponent of 2), namely, when $k = 1$, and then it also coincides with our earlier findings, and q itself (if k tends to infinity). We can obtain very similar results for EQUIVALENCE for JFAs.

Let us now discuss INTERSECTION. Also the problem of detecting emptiness of the intersection of two JFA languages is NP-hard. This and a related study on ETH-based complexity can be found in [9]. For the intersection of k JFAs, the proof of Proposition 11 actually shows the analogous result also in that case. For bounded alphabets, we use the proof of Theorem 21 to obtain:

Corollary 22. *Let Σ be fixed. Unless ETH fails, there is no algorithm for solving JFA INTERSECTION in time $O^*(2^{o(k)+o(q^{|\Sigma|/2})})$ for k JFAs with $\leqslant q$ states.*

Namely, we can construct to a given 3-COLORING instance with n vertices and m edges a collection of $k = 3m + 1$ many JFAs, each with $n^{|\Sigma|/2}$ many states.

6.2   Boustrophedon Finite Automata

Boustrophedon finite automata (BFAs) have been introduced to describe a simple processing of rectangular-shaped pictures with finite automata that scan these pictures as depicted on the right side.

Without going into formal details, let us mention that we have shown in [8] that the non-emptiness problem for this type of finite automata is NP-hard. We reduced from TALLY-DFA-INTERSECTION. From this (direct) construction, we can immediately deduce:

Proposition 23. *There is no algorithm that, given some BFA A with at most q states, decides if $L(A) = \varnothing$ in time $O^*(2^{o(q^{1/3})})$ unless ETH fails.*

First observe that although this problem is similar to the intersection problem, the only "communication" between the rows is via the state that is communicated and via the length information that is implicitly checked. In particular, we can first convert a given BFA into one with unary input alphabet.

Proposition 24. EMPTINESS *for q-state BFAs can be decided in time $O^*(q^q)$.*

7   Conclusions

So far, there was no systematic study of hard problems for finite automata under ETH. We are only aware of the papers [7,9] on these topics. Returning to the survey of Holzer and Kutrib [12], it becomes clear that there are quite a many hard problems related to finite automata and regular expressions that have not yet been examined with respect to exact algorithms and ETH. This hence gives ample room for future research. Also, there are quite a many modifications of finite automata with hard decision problems. We are currently studying the decidability status of UNIVERSALITY and similar problems for BFAs.

References

1. Beaudry, M., McKenzie, P., Thérien, D.: The membership problem in aperiodic transformation monoids. J. ACM **39**(3), 599–616 (1992)
2. Cho, S., Huynh, D.T.: Finite-automaton aperiodicity is PSPACE-complete. Theoret. Comput. Sci. **88**(1), 99–116 (1991)
3. Cho, S., Huynh, D.T.: The parallel complexity of finite-state automata problems. Inf. Comput. **97**(1), 1–22 (1992)
4. Chrobak, M.: Finite automata and unary languages. Theoret. Comput. Sci. **47**, 149–158 (1986)

5. Cygan, M., Fomin, F., Kowalik, Ł., Lokshtanov, D., Marx, D., Pilipczuk, M., Pilipczuk, M., Saurabh, S.: Parameterized Algorithms. Springer, Heidelberg (2015)
6. Dusart, P.: Estimates of some functions over primes without R.H. Technical report arXiv:1002.0442 [math.NT] (2010)
7. Fernau, H., Heggernes, P., Villanger, Y.: A multi-parameter analysis of hard problems on deterministic finite automata. J. Comput. Syst. Sci. **81**(4), 747–765 (2015)
8. Fernau, H., Paramasivan, M., Schmid, M.L., Thomas, D.G.: Scanning pictures the boustrophedon way. In: Barneva, R.P., Bhattacharya, B.B., Brimkov, V.E. (eds.) IWCIA 2015. LNCS, vol. 9448, pp. 202–216. Springer, Heidelberg (2015). doi:10. 1007/978-3-319-26145-4_15
9. Fernau, H., Paramasivan, M., Schmid, M.L., Vorel, V.: Characterization, complexity results on jumping finite automata. Technical report arXiv:1512.00482 [cs.FL] (2015)
10. Galil, Z.: Hierarchies of complete problems. Acta Informatica **6**, 77–88 (1976)
11. Haase, C., Hofman, P.: Tightening the complexity of equivalence problems for commutative grammars. Technical report arXiv:1506.07774 [cs.FL] (2015)
12. Holzer, M., Kutrib, M.: Descriptional and computational complexity of finite automata - a survey. Inf. Comput. **209**(3), 456–470 (2011)
13. Impagliazzo, R., Paturi, R., Zane, F.: Which problems have strongly exponential complexity? J. Comput. Syst. Sci. **63**(4), 512–530 (2001)
14. Kopczyński, E.: Complexity of problems of commutative grammars. Logical Methods Comput. Sci. **11**(1:9) (2015). http://www.lmcs-online.org/ojs/viewarticle. php?id=1533
15. Kozen, D.: Lower bounds for natural proof systems. In: 18th Annual Symposium on Foundations of Computer Science, FOCS, pp. 254–266. IEEE (1977)
16. Lange, K.-J., Rossmanith, P.: The emptiness problem for intersections of regular languages. In: Havel, I.M., Koubek, V. (eds.) MFCS 1992. LNCS, vol. 629, pp. 346–354. Springer, Heidelberg (1992)
17. Lokshtanov, D., Marx, D., Saurabh, S.: Lower bounds based on the Exponential Time Hypothesis. EATCS Bull. **105**, 41–72 (2011)
18. Meduna, A., Zemek, P.: Jumping finite automata. Int. J. Found. Comput. Sci. **23**(7), 1555–1578 (2012)
19. Rystsov, I.K.: Polynomial complete problems in automata theory. Inf. Process. Lett. **16**(3), 147–151 (1983)
20. Sandberg, S.: 1 Homing and synchronizing sequences. In: Broy, M., Jonsson, B., Katoen, J.-P., Leucker, M., Pretschner, A. (eds.) Model-Based Testing of Reactive Systems. LNCS, vol. 3472, pp. 5–33. Springer, Heidelberg (2005)
21. Schützenberger, M.P.: On finite monoids having only trivial subgroups. Inf. Control **8**, 190–194 (1965). (now Information and Computation)
22. Stern, J.: Complexity of some problems from the theory of automata. Inf. Control **66**(3), 163–176 (1985). (now Information and Computation)
23. Stockmeyer, L.J., Meyer, A.R.: Word problems requiring exponential time: preliminary report. In: Proceedings of the 5th Annual ACM Symposium on Theory of Computing, STOC, pp. 1–9. ACM (1973)
24. Wareham, H.T.: The parameterized complexity of intersection and composition operations on sets of finite-state automata. In: Yu, S., Păun, A. (eds.) CIAA 2000. LNCS, vol. 2088, pp. 302–310. Springer, Heidelberg (2001)

A Practical Algorithm for the Uniform Membership Problem of Labeled Multidigraphs of Tree-Width 2 for Spanning Tree Automata

Akio Fujiyoshi[✉]

Department of Computer and Information Sciences, Ibaraki University,
4-12-1 Nakanarusawa, Hitachi, Ibaraki 316-8511, Japan
akio.fujiyoshi.cs@vc.ibaraki.ac.jp

Abstract. This paper presents a practical algorithm for the uniform membership problem of labeled multidigraphs of tree-width at most 2 for spanning tree automata. Though it has been shown that the membership problem is solvable in linear time for graphs of bounded tree-width, the algorithm obtained in the previous study is unusable in practice because of a big hidden constant.

1 Introduction

In this paper, we study tree automata recognizing labeled multidigraphs. We define that a labeled multidigraph is accepted by a tree automaton if and only if the graph has a spanning tree accepted by the tree automaton. We call this automaton a spanning tree automaton. The membership problem of labeled multidigraphs for a spanning tree automaton has been studied [8]. Though the membership problem is NP-complete because the Hamiltonian path problem can be easily reduced to it, there exists a linear-time algorithm for the membership problem of labeled multidigraphs of bounded tree-width using a theorem of Courcelle [3,4]. However, the algorithm obtained by Courcelle's theorem is unusable in practice because of a big hidden constant. In addition, the tree automaton itself should be a part of the input in practical situations. Thus this paper will present a practical algorithm for the uniform membership problem of labeled multidigraphs of tree-width at most 2 for spanning tree automata.

The motivation of this study is to establish a robust and efficient recognition method for mathematical OCR [2,5,9]. As shown in Fig. 1, a mathematical OCR system constructs a labeled multidigraph representing the adjacency relation of mathematical symbols from a scanned image. The vertex labels represent mathematical symbols, while the edge labels represent types of the adjacency relation of mathematical symbols. From the labeled multidigraph, we want to obtain the spanning tree representing proper connections of mathematical symbols, which should be syntactically reasonable. In order to define the syntax of mathematical formulae and verify candidates of the spanning tree, we make use of spanning tree automata.

© Springer International Publishing Switzerland 2016
Y.-S. Han and K. Salomaa (Eds.): CIAA 2016, LNCS 9705, pp. 101–112, 2016.
DOI: 10.1007/978-3-319-40946-7_9

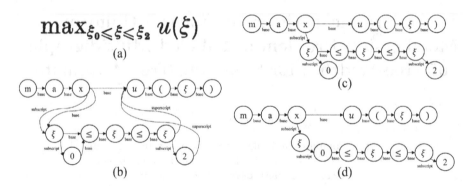

Fig. 1. (a) A scanned image, (b) a labeled multidigraph representing the adjacency relation of symbols, (c) the spanning tree representing proper connections of symbols, and (d) a spanning tree of a misrecognition.

The algorithm presented in this paper can be seen as an extension of the algorithm for directed acyclic graphs (DAGs) [6,7]. Since cycles and multiple sources have to be considered here, the algorithm is more sophisticated.

The time complexity of the algorithm is $O(n \cdot m^5 \cdot 2^w)$ time, where n is the size of a graph, m the number and w the maximum width of transition rules of an automaton. Since the maximum outdegree of spanning trees representing proper connections of mathematical symbols is 7 (meaning the seven directions), w and also 2^w are bounded by a constant. Therefore, the algorithm works in polynomial time. In addition, labeled multidigraphs of tree-width at most 2 are sufficient for the application to mathematical OCR in most cases.

2   Labeled Multidigraphs

2.1   Definitions

A *labeled multidigraph* is a 8-tuple $G = (V, E, tail, head, \Sigma, \Delta, \sigma, \delta)$, where V is a finite set of *vertices*, E is a finite set of *edges*, $tail : E \to V$ is a function assigning to each edge its *tail*, $head : E \to V$ is a function assigning to each edge its *head*, Σ is a finite set of *vertex labels*, Δ is a finite set of *edge labels*, $\sigma : V \to \Sigma$ is a function assigning to each vertex its label, and $\delta : E \to \Delta$ is a function assigning to each edge its label. For a pair of edges $e, e' \in E$, e and e' are *multiple* if $tail(e) = tail(e')$ and $head(e) = head(e')$, and e and e' are *symmetric* if $tail(e) = head(e')$ and $head(e) = tail(e')$. For a vertex $v \in V$, the *incoming edges* of v is the set $in(v) = \{e \in E \mid head(e) = v\}$, the *indegree* of v is $|in(v)|$, the *outgoing edges* of v is the set $out(v) = \{e \in E \mid tail(e) = v\}$, and the *outdegree* of v is $|out(v)|$. A *source* is a vertex of indegree 0, and a *sink* is a vertex of outdegree 0. We define the *size* of a labeled multidigraph G as $|V \cup E|$, the number of vertices plus the number of edges.

Let $E' \subseteq E$ be a subset of edges. The *edge-induced subgraph* of G by E', denoted by $G[E']$, is the labeled multidigraph $(V', E', tail', head', \Sigma, \Delta, \sigma', \delta')$

such that $V' \subseteq V$, $tail' \subseteq tail$, $head' \subseteq head$, $\sigma' \subseteq \sigma$, $\delta' \subseteq \delta$ and every vertex in V' has at least one incoming or outgoing edge in E'. The *spanning subgraph* of G by E', denoted by $G\langle E'\rangle$, is the labeled multidigraph $(V, E', tail', head', \Sigma, \Delta, \sigma, \delta')$ such that $tail' \subseteq tail$, $head' \subseteq head$ and $\delta' \subseteq \delta$.

Let $G = (V, E, tail, head, \Sigma, \Delta, \sigma, \delta)$ be a labeled multidigraph. G is *acyclic* if there is not a subset of edges $E' \subseteq E$ such that $E' \neq \emptyset$ and every vertex of $G[E']$ has indegree 1 and outdegree 1. For a pair of distinct vertices $u, v \in V$, a *simple directed path* of G from u to v is an edge-induced subgraph $G[E']$ for some $E' \subseteq E$ such that $G[E']$ is acyclic and every vertex of $G[E']$ has indegree 1 and outdegree 1 except that u has indegree 0 and outdegree 1 and v has indegree 1 and outdegree 0.

A *labeled rooted tree* is a labeled multidigraph $T = (V, E, tail, head, \Sigma, \Delta, \sigma, \delta)$ such that T is acyclic, T has exactly one source, and there is a unique simple directed path from the source to every other vertex. The source of a tree is also called the *root*, while the sinks are also called *leaves*. An ordered tree can be seen as a special labeled rooted tree such that $\Delta = \{1, 2, \ldots, max_d\}$, max_d is the maximum outdegree of vertices, and the outgoing edges of each vertex are uniquely labeled as $1, 2, 3, \ldots$.

A *labeled tree-pair* is a labeled multidigraph $P = (V, E, tail, head, \Sigma, \Delta, \sigma, \delta)$ obtained as the disjoint union of two labeled rooted trees. When P is the disjoint union of two labeled rooted trees T_1 and T_2, we write $P = T_1 \cup T_2$.

Let $G = (V, E, tail, head, \Sigma, \Delta, \sigma, \delta)$ be a labeled multidigraph. A *spanning tree* of G is a spanning subgraph $G\langle E'\rangle$ for some $E' \subseteq E$ such that $G\langle E'\rangle$ is a labeled rooted tree. A *spanning tree-pair* of G is a spanning subgraph $G\langle E'\rangle$ for some $E' \subseteq E$ such that $G\langle E'\rangle$ is a labeled tree-pair.

2.2   The Base Graph of a Labeled Multidigraph and Reduction Rules for Base Graphs of Tree-Width at Most 2

Let $G = (V, E, tail, head, \Sigma, \Delta, \sigma, \delta)$ be a labeled multidigraph. The *base graph* of G is the undirected simple graph $G' = (V, E')$ such that $E' = \{\{v_1, v_2\} \mid$ there exists an edge $e \in E$ such that $tail(e) = v_1$ and $head(e) = v_2$, or $tail(e) = v_2$ and $head(e) = v_1\}$. We define the tree-width of a labeled multidigraph as the tree-width [3,4] of its base graph.

We assign to each vertex and edge of G' a set of original edges of G, called *representing edges*, as follows:

- For each vertex $v \in V$, the set of representing edges is $rep(v) = \emptyset$.
- For each edge $\{v_1, v_2\} \in E'$, the set of representing edges is $rep(\{v_1, v_2\}) = \{e \in E \mid tail(e) = v_1 \text{ and } head(e) = v_2, \text{ or } tail(e) = v_2 \text{ and } head(e) = v_1\}$.

It is known that any undirected simple graph of tree-width at most 2 can be reduced to a single-vertex graph by the following reduction rules [1]:

1. If the graph has a vertex v_2 incident to exactly 1 edge $\{v_1, v_2\}$, then remove the vertex v_2 and the edge $\{v_1, v_2\}$.
2. If the graph has a vertex v_2 incident to exactly 2 edges $\{v_1, v_2\}, \{v_2, v_3\}$ and there is no edge between v_1 and v_3, then remove the vertex v_2 and the edges $\{v_1, v_2\}, \{v_2, v_3\}$, and connect the vertices v_1 and v_3 by a new edge.

3. If the graph has a vertex v_2 incident to exactly 2 edges $\{v_1, v_2\}, \{v_2, v_3\}$ and there is an edge between v_1 and v_3, then remove the vertex v_2 and the edges $\{v_1, v_2\}, \{v_2, v_3\}$.

During the reduction process using the above reduction rules, the sets of representing edges of a base graph will be changed as follows:

1. When a vertex v_2 and an edge $\{v_1, v_2\}$ are removed, the set of representing edges of v_1 will be $rep(v_1) \cup rep(v_2) \cup rep(\{v_1, v_2\})$.
2. When a vertex v_2 and edges $\{v_1, v_2\}, \{v_2, v_3\}$ are removed and a new edge $\{v_1, v_3\}$ is added, the set of representing edges of $\{v_1, v_3\}$ will be $rep(\{v_1, v_2\}) \cup rep(v_2) \cup rep(\{v_2, v_3\})$.
3. When a vertex v_2 and edges $\{v_1, v_2\}, \{v_2, v_3\}$ are removed and there has already been an edge between v_1 and v_3, the set of representing edges of $\{v_1, v_3\}$ will be $rep(\{v_1, v_2\}) \cup rep(v_2) \cup rep(\{v_2, v_3\}) \cup rep(\{v_1, v_3\})$.

Example 1. A labeled multidigraph, its base graph, a reduction process, and changes of the set of representing edges are illustrated in Fig. 2.

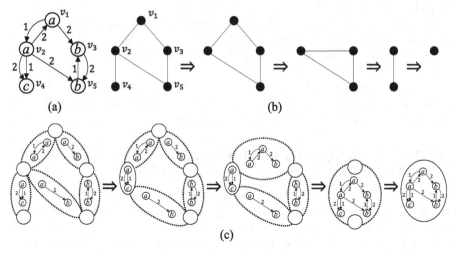

(a) (b) (c)

Fig. 2. (a) A labeled multidigraph, (b) its base graph and a reduction process to a single-vertex graph, and (c) changes of the set of representing edges.

3   Spanning Tree Automaton

Spanning tree automata are similar to well-known nondeterministic top-down tree automata for ordered trees, but the inputs are more general. Edge labels can be arbitrarily specified for the transition rules of a spanning tree automaton.

3.1   Definitions

Let $\mathcal{X} = \{x_1, x_2, \ldots\}$ be a fixed countable set of *variables*.

A *spanning tree automaton* over alphabets Σ and Δ is a quintuple $\mathcal{A} = (Q, \Sigma, \Delta, q_0, R)$ where Q is a finite set of *states*, $q_0 \in Q$ is the *initial state*, and R is a finite set of *transition rules* of the following form:

$$q(f(c_1(x_1), \ldots, c_n(x_n))) \rightarrow f(c_1(q_1(x_1)), \ldots, c_n(q_n(x_n))),$$

where $n \geq 0$, $q, q_1, \ldots, q_n \in Q$, $f \in \Sigma$, $c_1, \ldots, c_n \in \Delta$, and $x_1, \ldots, x_n \in \mathcal{X}$. The number n is called the *width* of a transition rule. When $n = 0$, we write $q(f) \rightarrow f$ instead of $q(f()) \rightarrow f()$. Since we assume widths of transition rules are bounded by a constant in this paper, we define the *size* of a spanning tree automaton $\mathcal{A}$ as $|R|$, the number of transition rules.

Let $r : q(f(c_1(x_1), \ldots, c_n(x_n))) \rightarrow f(c_1(q_1(x_1)), \ldots, c_n(q_n(x_n)))$ be a transition rule. We define $l\text{-}state(r) = q$, $v\text{-}label(r) = f$, $width(r) = n$, $var(r) = \{x_1, \ldots, x_n\}$ and, for each $1 \leq i \leq n$, $r\text{-}state(r, i) = q_i$ and $e\text{-}label(r, i) = c_i$.

Let $T = (V, E, tail, head, \Sigma, \Delta, \sigma, \delta)$ be a labeled rooted tree, and let $v_r \in V$ be the root of T. A *state mapping* on T is a function $\mu : V \rightarrow Q$. A state mapping μ on T is *acceptable* by $\mathcal{A}$, if $\mu(v_r) = q_0$ and, for each $v \in V$, a transition rule $q(f(c_1(x_1), \ldots, c_n(x_n))) \rightarrow f(c_1(q_1(x_1)), \ldots, c_n(q_n(x_n)))$ is in R for some $n \geq 0$, $\mu(v) = q$, $\sigma(v) = f$, and v has exactly n outgoing edges $e_1, \ldots, e_n$ such that $\delta(e_i) = c_i$ and $\mu(head(e_i)) = q_i$ for each $1 \leq i \leq n$. $\mathcal{A}$ *accepts* T if there is an acceptable state mapping on T.

Let $G = (V, E, tail, head, \Sigma, \Delta, \sigma, \delta)$ be a labeled multidigraph. $\mathcal{A}$ *accepts* G if G has a spanning tree T and $\mathcal{A}$ accepts T. A set $\mathcal{S}$ of labeled multidigraphs is *recognizable* if there exists a spanning tree automaton $\mathcal{A}$ such that $\mathcal{S} = \{G \mid G$ is accepted by $\mathcal{A}\}$.

Example 2. The following is an example of a spanning tree automaton, which accepts binary boolean expression trees with true value: $\mathcal{A} = (Q, \Sigma, \Delta, q_T, R)$, where $Q = \{q_T, q_F\}$, $\Sigma = \{\wedge, \vee, \neg, T, F\}$, $\Delta = \{1, 2\}$, and R consists of transition rules:

$$
\begin{aligned}
r_1 &: q_F(\wedge(1(x_1), 2(x_2))) \rightarrow \wedge(q_F(1(x_1)), q_F(2(x_2))), \\
r_2 &: q_F(\wedge(1(x_1), 2(x_2))) \rightarrow \wedge(q_T(1(x_1)), q_F(2(x_2))), \\
r_3 &: q_F(\wedge(1(x_1), 2(x_2))) \rightarrow \wedge(q_F(1(x_1)), q_T(2(x_2))), \\
r_4 &: q_T(\wedge(1(x_1), 2(x_2))) \rightarrow \wedge(q_T(1(x_1)), q_T(2(x_2))), \\
r_5 &: q_F(\vee(1(x_1), 2(x_2))) \rightarrow \vee(q_F(1(x_1)), q_F(2(x_2))), \\
r_6 &: q_T(\vee(1(x_1), 2(x_2))) \rightarrow \vee(q_T(1(x_1)), q_F(2(x_2))), \\
r_7 &: q_T(\vee(1(x_1), 2(x_2))) \rightarrow \vee(q_F(1(x_1)), q_T(2(x_2))), \\
r_8 &: q_T(\vee(1(x_1), 2(x_2))) \rightarrow \vee(q_T(1(x_1)), q_T(2(x_2))), \\
r_9 &: q_F(\neg(1(x_1))) \rightarrow \neg(q_T(1(x_1))), \\
r_{10} &: q_T(\neg(1(x_1))) \rightarrow \neg(q_F(1(x_1))), \\
r_{11} &: q_T(T) \rightarrow T, \text{ and} \\
r_{12} &: q_F(F) \rightarrow F.
\end{aligned}
$$

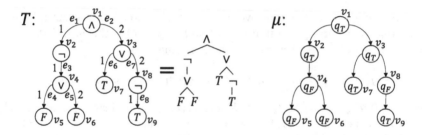

Fig. 3. A labeled rooted tree and an acceptable state mapping on it for $\mathcal{A}$.

Consider a labeled rooted tree T illustrated in Fig. 3. Because there exists an acceptable state mapping μ, the labeled rooted tree T is accepted by $\mathcal{A}$.

3.2 Partial Acceptance

The algorithm solving the uniform membership problem handles spanning trees and spanning tree-pairs of subgraphs of an input graph. For spanning trees and spanning tree-pairs of subgraphs, we need the notion of partial acceptance.

For a labeled rooted tree or a labeled tree-pair $T = (V, E, tail, head, \Sigma, \Delta, \sigma, \delta)$, a pair of vertices $v_1, v_2 \in V$, a pair of transition rules $r_1, r_2 \in R$ and a pair of subsets of variables $\mathcal{X}_1 \subseteq var(r_1)$, $\mathcal{X}_2 \subseteq var(r_2)$, T is $(v_1, r_1, \mathcal{X}_1, v_2, r_2, \mathcal{X}_2)$-*acceptable* by $\mathcal{A}$ if there exists a state mapping $\mu : V \to Q$ such that all of the following four conditions hold:

1. If v is a source of T, $v \neq v_1$ and $v \neq v_2$, then $\mu(v) = q_0$.
2. For each $v \in V$, if $v \neq v_1$ and $v \neq v_2$, then a transition rule $q(f(c_1(x_1), \ldots, c_n(x_n))) \to f(c_1(q_1(x_1)), \ldots, c_n(q_n(x_n)))$ is in R for some $n \geq 0$, $\mu(v) = q$, $\sigma(v) = f$, and v has exactly n outgoing edges $e_1, \ldots, e_n$ such that $\delta(e_i) = c_i$ and $\mu(head(e_i)) = q_i$ for each $1 \leq i \leq n$.
3. If r_1 is of the form $q(f(c_1(x_1), \ldots, c_n(x_n))) \to f(c_1(q_1(x_1)), \ldots, c_n(q_n(x_n)))$ for some $n \geq 0$, and $\mathcal{X}_1$ consists of k distinct variables $\{x_{i_1}, \ldots, x_{i_k}\}$ for some $0 \leq k \leq n$, then $\mu(v_1) = q$, $\sigma(v_1) = f$, and v_1 has exactly k outgoing edges $e_1, \ldots, e_k$ such that $\delta(e_j) = c_{i_j}$ and $\mu(head(e_j)) = q_{i_j}$ for each $1 \leq j \leq k$.
4. If r_2 is of the form $q(f(c_1(x_1), \ldots, c_n(x_n))) \to f(c_1(q_1(x_1)), \ldots, c_n(q_n(x_n)))$ for some $n \geq 0$, and $\mathcal{X}_2$ consists of k distinct variables $\{x_{i_1}, \ldots, x_{i_k}\}$ for some $0 \leq k \leq n$, then $\mu(v_2) = q$, $\sigma(v_2) = f$, and v_2 has exactly k outgoing edges $e_1, \ldots, e_k$ such that $\delta(e_j) = c_{i_j}$ and $\mu(head(e_j)) = q_{i_j}$ for each $1 \leq j \leq k$.

For a labeled rooted tree $T = (V, E, tail, head, \Sigma, \Delta, \sigma, \delta)$, a vertex $v_1 \in V$, a transition rule $r_1 \in R$ and a subset of variables $\mathcal{X}_1 \subseteq var(r_1)$, T is $(v_1, r_1, \mathcal{X}_1)$-*acceptable* by $\mathcal{A}$ if T is $(v_1, r_1, \mathcal{X}_1, v_1, r_1, \mathcal{X}_1)$-*acceptable* by $\mathcal{A}$.

Example 3. Consider the spanning tree automaton $\mathcal{A}$ and the state mapping μ on T for $\mathcal{A}$ in Example 2. The following T_1, T_2, T_3, T_4, P_1 and P_2 are labeled rooted trees and labeled tree-pairs, which are edge-induced subgraphs of T:

$$T_1 = T[\{e_1, e_3, e_4, e_5\}], \qquad T_2 = T[\{e_1, e_2, e_3, e_4, e_5\}],$$
$$T_3 = T[\{e_1, e_2, e_3, e_4, e_6\}], \qquad T_4 = T[\{e_3, e_4\}],$$
$$P_1 = T[\{e_5, e_7, e_8\}], \quad and \quad P_2 = T[\{e_1, e_2, e_5, e_6, e_7, e_8\}].$$

According to the state mapping μ, T_1 is $(v_1, r_4, \{x_1\})$-acceptable, T_2 is $(v_3, r_6, \emptyset)$-acceptable, $(v_3, r_7, \emptyset)$-acceptable and $(v_3, r_8, \emptyset)$-acceptable, T_3 is $(v_3, r_6, \{x_1\}, v_4, r_5, \{x_1\})$-acceptable and $(v_3, r_8, \{x_1\}, v_4, r_5, \{x_1\})$-acceptable, T_4 is $(v_2, r_{11}, \{x_1\}, v_4, r_5, \{x_1\})$-acceptable, P_1 is $(v_3, r_6, \{x_2\}, v_4, r_5, \{x_2\})$-acceptable and $(v_3, r_7, \{x_2\}, v_4, r_5, \{x_2\})$-acceptable, and P_2 is $(v_2, r_{11}, \emptyset, v_4, r_5, \{x_2\})$-acceptable (Fig. 4).

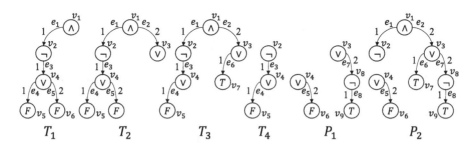

Fig. 4. Partially acceptable labeled rooted trees and labeled tree-pairs.

4   Algorithm for the Uniform Membership Problem

In this section, we present an algorithm for uniform membership problem of labeled multidigraphs of tree-width at most 2 for spanning tree automata. Let $G = (V, E, tail, head, \Sigma, \Delta, \sigma, \delta)$ be a labeled multidigraph, let $G' = (V, E')$ be the base graph of G, and let $\mathcal{A} = (Q, \Sigma, \Delta, q_0, R)$ be a spanning tree automaton.

The main task of the algorithm is to calculate sets: $\alpha[v], \beta[v] \subseteq R \times 2^{\mathcal{X}}$ for each $v \in V$, and $A[v_1, v_2], B[v_1, v_2], C[v_1, v_2], D[v_1, v_2] \subseteq R \times 2^{\mathcal{X}} \times R \times 2^{\mathcal{X}}$ for each $(v_1, v_2) \in V \times V$ provided that $\{v_1, v_2\} \in E'$. They are described as follows:

$(r, \mathcal{X}) \in \alpha[v]$ if and only if a spanning tree of $G[rep(v)]$ exists such that its root is v and it is $(v, r, \mathcal{X})$-acceptable by $\mathcal{A}$.

$(r, \mathcal{X}) \in \beta[v]$ if and only if a spanning tree of $G[rep(v)]$ exists such that its root is not v and it is $(v, r, \mathcal{X})$-acceptable by $\mathcal{A}$.

$(r_1, \mathcal{X}_1, r_2, \mathcal{X}_2) \in A[v_1, v_2]$ if and only if a spanning tree of $G[rep(\{v_1, v_2\})]$ exists such that its root is v_1 and it is $(v_1, r_1, \mathcal{X}_1, v_2, r_2, \mathcal{X}_2)$-acceptable by $\mathcal{A}$.

$(r_1, \mathcal{X}_1, r_2, \mathcal{X}_2) \in B[v_1, v_2]$ if and only if a spanning tree-pair of $G[rep(\{v_1, v_2\})]$ exists such that its sources are v_1 and v_2, and it is $(v_1, r_1, \mathcal{X}_1, v_2, r_2, \mathcal{X}_2)$-acceptable by $\mathcal{A}$.

$(r_1, \mathcal{X}_1, r_2, \mathcal{X}_2) \in C[v_1, v_2]$ if and only if a spanning tree of $G[rep(\{v_1, v_2\})]$ exists such that its root is neither v_1 nor v_2, and it is $(v_1, r_1, \mathcal{X}_1, v_2, r_2, \mathcal{X}_2)$-acceptable by $\mathcal{A}$.

$(r_1, \mathcal{X}_1, r_2, \mathcal{X}_2) \in D[v_1, v_2]$ if and only if a spanning tree-pair of $G[rep(\{v_1, v_2\})]$ exists such that one of its sources is v_2, the other source is not v_1, there is no simple directed path from v_2 to v_1, and it is $(v_1, r_1, \mathcal{X}_1, v_2, r_2, \mathcal{X}_2)$-acceptable by $\mathcal{A}$.

An informal sketch of partially acceptable spanning trees and tree-pairs related to the sets are illustrated in Fig. 5.

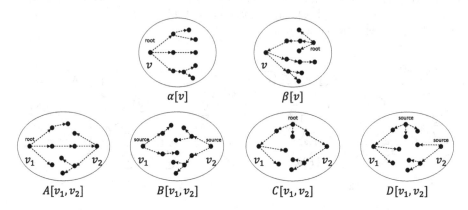

Fig. 5. Partially acceptable spanning trees and tree-pairs related to the sets.

The above sets are maintained to satisfy their requirement during a reduction process of the base graph G'. When the base graph is reduced to a single-vertex graph with the only vertex v, if $(r, var(r)) \in \beta[v]$, or $(r, var(r)) \in \alpha[v]$ and $l\text{-}state(r) = q_0$, then a spanning tree of G accepted by $\mathcal{A}$ exists, or else no such spanning tree exists.

4.1   Initial Setting of the Sets

Considering partially acceptable spanning trees and tree-pairs of $G[rep(v)]$ and $G[rep(\{v_1, v_2\})]$ for each $v \in V$ and $(v_1, v_2) \in V \times V$ provided that $\{v_1, v_2\} \in E'$, the sets are initialized as follows:

$\alpha[v] := \{(r, \emptyset) \mid v\text{-}label(r) = \sigma(v)\}$,
$\beta[v] := \emptyset$,
$A[v_1, v_2] := \{(r_1, \{x_i\}, r_2, \emptyset) \mid \exists e \in E$ such that $tail(e) = v_1$ and $head(e) = v_2$, $v\text{-}label(r_1) = \sigma(v_1)$, $v\text{-}label(r_2) = \sigma(v_2)$, $1 \le i \le width(r_1)$, $r\text{-}state(r_1, i) = l\text{-}state(r_2)$ and $e\text{-}label(r_1, i) = \delta(e)\}$,
$B[v_1, v_2] := \{(r_1, \emptyset, r_2, \emptyset) \mid v\text{-}label(r_1) = \sigma(v_1)$ and $v\text{-}label(r_2) = \sigma(v_2)\}$,
$C[v_1, v_2] := \emptyset$, and
$D[v_1, v_2] := \emptyset$.

4.2 Maintenace of the Sets During a Reduction Process of the Base Graph

When a vertex v_2 and an edge $\{v_1, v_2\}$ are removed, $\alpha[v_1]$ and $\beta[v_1]$ are updated as follows:

$$\alpha[v_1] := \{(r, \mathcal{X}_1 \cup \mathcal{X}_2) \mid (r, \mathcal{X}_1) \in \alpha[v_1], (r, \mathcal{X}_2, r', \mathcal{X}_3) \in A[v_1, v_2], (r', \mathcal{X}_4) \in \alpha[v_2],$$
$$\mathcal{X}_3 \cup \mathcal{X}_4 = var(r') \text{ and } \mathcal{X}_3 \cap \mathcal{X}_4 = \emptyset\}, \text{ and}$$
$$\beta[v_1] := \{(r, \mathcal{X}_1 \cup \mathcal{X}_2) \mid (r, \mathcal{X}_1) \in \beta[v_1], (r, \mathcal{X}_2, r', \mathcal{X}_3) \in A[v_1, v_2], (r', \mathcal{X}_4) \in \alpha[v_2],$$
$$\mathcal{X}_3 \cup \mathcal{X}_4 = var(r') \text{ and } \mathcal{X}_3 \cap \mathcal{X}_4 = \emptyset\}$$
$$\cup \{(r, \mathcal{X}_1 \cup \mathcal{X}_2) \mid (r, \mathcal{X}_1) \in \alpha[v_1], (r', \mathcal{X}_3, r, \mathcal{X}_2) \in A[v_2, v_1], (r', \mathcal{X}_4) \in \beta[v_2],$$
$$\mathcal{X}_3 \cup \mathcal{X}_4 = var(r') \text{ and } \mathcal{X}_3 \cap \mathcal{X}_4 = \emptyset\}$$
$$\cup \{(r, \mathcal{X}_1 \cup \mathcal{X}_2) \mid (r, \mathcal{X}_1) \in \alpha[v_1], (r, \mathcal{X}_2, r', \mathcal{X}_3) \in C[v_1, v_2], (r', \mathcal{X}_4) \in \alpha[v_2],$$
$$\mathcal{X}_3 \cup \mathcal{X}_4 = var(r') \text{ and } \mathcal{X}_3 \cap \mathcal{X}_4 = \emptyset\}$$
$$\cup \{(r, \mathcal{X}_1 \cup \mathcal{X}_2) \mid (r, \mathcal{X}_1) \in \alpha[v_1], (r', \mathcal{X}_3, r, \mathcal{X}_2) \in A[v_2, v_1], (r', \mathcal{X}_4) \in \alpha[v_2],$$
$$\mathcal{X}_3 \cup \mathcal{X}_4 = var(r'), \mathcal{X}_3 \cap \mathcal{X}_4 = \emptyset \text{ and } l\text{-}state(r') = q_0\}.$$

Figure 6 shows all the possible combinations of partially acceptable spanning trees to form new partially acceptable spanning trees for $\alpha[v_1]$ and $\beta[v_1]$.

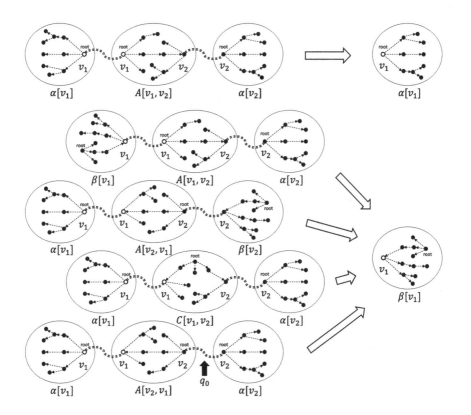

Fig. 6. Partially acceptable spanning trees to update $\alpha[v_1]$ and $\beta[v_1]$.

When a vertex v_2 and edges $\{v_1, v_2\}$, $\{v_2, v_3\}$ are removed, we have to think of the existence of an edge between v_1 and v_3. If there has not been an edge between v_1 and v_3, and a new edge $\{v_1, v_3\}$ is added, then $A[v_1, v_3]$, $A[v_3, v_1]$, $B[v_1, v_3]$, $B[v_3, v_1]$, $C[v_1, v_3]$, $C[v_3, v_1]$, $D[v_1, v_3]$ and $D[v_3, v_1]$ are newly initialized.

For instance, $D[v_1, v_3]$ is newly initialized as follows:

$$
\begin{aligned}
D[v_1, v_3] := &\{(r, \mathcal{X}_1, r', \mathcal{X}_2) \mid (r'', \mathcal{X}_3, r, \mathcal{X}_1) \in A[v_2, v_1], \ (r'', \mathcal{X}_4, r', \mathcal{X}_2) \in B[v_2, v_3], \\
&(r'', \mathcal{X}_5) \in \beta[v_2], \ \mathcal{X}_3 \cup \mathcal{X}_4 \cup \mathcal{X}_5 = \mathrm{var}(r'') \text{ and } \mathcal{X}_3, \mathcal{X}_4, \mathcal{X}_5 \text{ are pairwise disjoint}\} \\
\cup &\{(r, \mathcal{X}_1, r', \mathcal{X}_2) \mid (r, \mathcal{X}_1, r'', \mathcal{X}_3) \in C[v_1, v_2], \ (r'', \mathcal{X}_4, r', \mathcal{X}_2) \in B[v_2, v_3], \\
&(r'', \mathcal{X}_5) \in \alpha[v_2], \ \mathcal{X}_3 \cup \mathcal{X}_4 \cup \mathcal{X}_5 = \mathrm{var}(r'') \text{ and } \mathcal{X}_3, \mathcal{X}_4, \mathcal{X}_5 \text{ are pairwise disjoint}\} \\
\cup &\{(r, \mathcal{X}_1, r', \mathcal{X}_2) \mid (r'', \mathcal{X}_3, r, \mathcal{X}_1) \in A[v_2, v_1], \ (r'', \mathcal{X}_4, r', \mathcal{X}_2) \in D[v_2, v_3], \\
&(r'', \mathcal{X}_5) \in \alpha[v_2], \ \mathcal{X}_3 \cup \mathcal{X}_4 \cup \mathcal{X}_5 = \mathrm{var}(r'') \text{ and } \mathcal{X}_3, \mathcal{X}_4, \mathcal{X}_5 \text{ are pairwise disjoint}\} \\
\cup &\{(r, \mathcal{X}_1, r', \mathcal{X}_2) \mid (r, \mathcal{X}_1, r'', \mathcal{X}_3) \in D[v_1, v_2], \ (r', \mathcal{X}_2, r'', \mathcal{X}_4) \in A[v_3, v_2], \\
&(r'', \mathcal{X}_5) \in \alpha[v_2], \ \mathcal{X}_3 \cup \mathcal{X}_4 \cup \mathcal{X}_5 = \mathrm{var}(r'') \text{ and } \mathcal{X}_3, \mathcal{X}_4, \mathcal{X}_5 \text{ are pairwise disjoint}\} \\
\cup &\{(r, \mathcal{X}_1, r', \mathcal{X}_2) \mid (r'', \mathcal{X}_3, r, \mathcal{X}_1) \in A[v_2, v_1], \ (r'', \mathcal{X}_4, r', \mathcal{X}_2) \in B[v_2, v_3], \\
&(r'', \mathcal{X}_5) \in \alpha[v_2], \ \mathcal{X}_3 \cup \mathcal{X}_4 \cup \mathcal{X}_5 = \mathrm{var}(r''), \ \mathcal{X}_3, \mathcal{X}_4, \mathcal{X}_5 \text{ are pairwise disjoint} \\
&\text{and } l\text{-}state(r'') = q_0\}.
\end{aligned}
$$

Figure 7 shows all the possible combinations of partially acceptable spanning trees and tree-pairs to form new partially acceptable spanning tree-pairs for $D[v_1, v_3]$. The remaining sets can be newly initialized in a similar way.

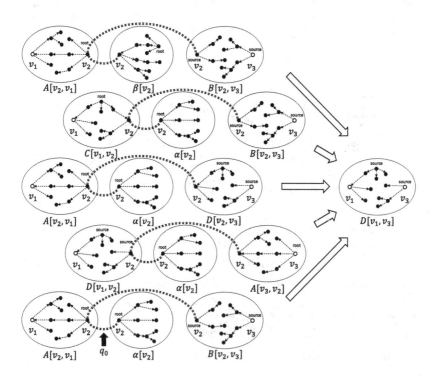

Fig. 7. Partially acceptable spanning trees and tree-pairs to initialize $D[v_1, v_3]$.

On the other hand, if there has already been an edge between v_1 and v_3, then we first calculate $A[v_1, v_3]$, $A[v_3, v_1]$, $B[v_1, v_3]$, $B[v_3, v_1]$, $C[v_1, v_3]$, $C[v_3, v_1]$, $D[v_1, v_3]$ and $D[v_3, v_1]$ of $G[rep(\{v_1, v_2\}) \cup rep(v_2) \cup rep(\{v_2, v_3\})]$. Combining these sets with $A[v_1, v_3]$, $A[v_3, v_1]$, $B[v_1, v_3]$, $B[v_3, v_1]$, $C[v_1, v_3]$, $C[v_3, v_1]$, $D[v_1, v_3]$ and $D[v_3, v_1]$ of $G[rep(\{v_1, v_3\})]$, the sets are updated.

Let A_{v_1, v_3} and B_{v_1, v_3} be the sets $A[v_1, v_3]$ and $B[v_1, v_3]$ of $G[rep(\{v_1, v_2\}) \cup rep(v_2) \cup rep(\{v_2, v_3\})]$. Combining A_{v_1, v_3} and B_{v_1, v_3} with $A[v_1, v_3]$ and $B[v_1, v_3]$ of $G[rep(\{v_1, v_3\})]$, $A[v_1, v_3]$ is updated as follows:

$$A[v_1, v_3] := \{(r, \mathcal{X}_1 \cup \mathcal{X}_2, r', \mathcal{X}_3 \cup \mathcal{X}_4) \mid (r, \mathcal{X}_1, r', \mathcal{X}_3) \in A[v_1, v_3], (r, \mathcal{X}_2, r', \mathcal{X}_4) \in$$
$$B_{v_1, v_3}, \mathcal{X}_1 \cap \mathcal{X}_2 = \emptyset \text{ and } \mathcal{X}_3 \cap \mathcal{X}_4 = \emptyset\}$$
$$\cup \{(r, \mathcal{X}_1 \cup \mathcal{X}_2, r', \mathcal{X}_3 \cup \mathcal{X}_4) \mid (r, \mathcal{X}_1, r', \mathcal{X}_3) \in B[v_1, v_3], (r, \mathcal{X}_2, r', \mathcal{X}_4) \in A_{v_1, v_3},$$
$$\mathcal{X}_1 \cap \mathcal{X}_2 = \emptyset \text{ and } \mathcal{X}_3 \cap \mathcal{X}_4 = \emptyset\}.$$

Figure 8 shows all the possible combinations of partially acceptable spanning trees and tree-pairs to form new partially acceptable spanning trees for $A[v_1, v_3]$. The remaining sets can be updated in a similar way.

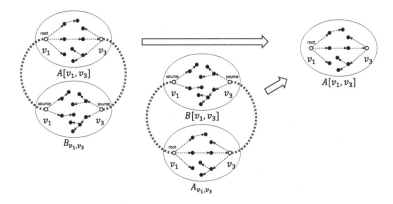

Fig. 8. Partially acceptable spanning tree and tree-pairs to update $A[v_1, v_3]$.

4.3   Correctness of the Algorithm

The correctness of the algorithm can be shown by confirming the following:

- When G' is reduced to a single-vertex v, $G[rep(v)] = G$.
- The sets always satisfy their requirement during a reduction process of G'.
- The combinations of partially acceptable spanning trees and tree-pairs related to the sets α, β, A, B, C and D are sufficient to update the sets.
- In the calculation of updated value of the each set, all the possible combinations partially acceptable spanning trees and tree-pairs are considered.

To confirm the last point, all combinations of partially acceptable spanning trees and tree-pairs related to the sets have been checked.

4.4 Time Complexity of the Algorithm

Because at least one edge is removed from E' for each iteration, the number of iterations is $O(n)$, where n is the size of a graph. The calculations of the sets can be done in $O(m^5 \cdot 2^w)$ time, where m is the number and w the maximum width of transition rules of an automaton. Since we assume widths of transition rules are bounded by a constant in this paper, 2^w is bounded by a constant. Therefore, if inputs are restricted to graphs of tree-width at most 2, the uniform membership problem of labeled multidigraphs for spanning tree automata is solvable in $O(n \cdot m^5)$ time, where n is the size of a graph, m the size of an automaton.

5 Conclusion

A practical algorithm for the uniform membership problem of labeled multidigraphs of tree-width at most 2 for spanning tree automata was presented. Since reduction rules that reduce any undirected simple graph of tree-width at most 3 to a single-vertex graph are also known [1], we think that the algorithm presented in this paper can be extended for graphs of tree-width at most 3.

References

1. Arnborg, S., Proskurowski, A.: Characterization and recognition of partial 3-trees. SIAM J. Algebraic Discrete Methods **7**(2), 305–314 (1986)
2. Chan, K.F., Yeung, D.Y.: Mathematical expression recognition: a survey. Int. J. Doc. Anal. Recogn. **3**(1), 3–15 (2000)
3. Courcelle, B.: The monadic second-order logic of graphs I. Recognizable sets of finite graphs. Inf. Comput. **85**, 12–75 (1990)
4. Courcelle, B., Engelfriet, J.: Graph Structure and Monadic Second-Order Logic - A Language-Theoretic Approach, Encyclopedia of Mathematics and Its Applications, vol. 138. Cambridge University Press, Cambridge (2012)
5. Eto, Y., Suzuki, M.: Mathematical formula recognition using virtual link network. In: Proceedings of the 6th International Conference on Document Analysis and Recognition (ICDAR 2001), pp. 430–437 (2001)
6. Fujiyoshi, A.: Recognition of a spanning tree of directed acyclic graphs by tree automata. In: Maneth, S. (ed.) CIAA 2009. LNCS, vol. 5642, pp. 105–114. Springer, Heidelberg (2009)
7. Fujiyoshi, A.: Recognition of directed acyclic graphs by spanning tree automata. Theor. Comput. Sci. **411**(38–39), 3493–3506 (2010)
8. Fujiyoshi, A.: Recognition of labeled multidigraphs by spanning tree automata. In: Holzer, M., Kutrib, M. (eds.) CIAA 2014. LNCS, vol. 8587, pp. 188–199. Springer, Heidelberg (2014)
9. Fujiyoshi, A., Suzuki, M., Uchida, S.: Verification of mathematical formulae based on a combination of context-free grammar and tree grammar. In: Autexier, S., Campbell, J., Rubio, J., Sorge, V., Suzuki, M., Wiedijk, F. (eds.) AISC/Calculemus/MKM 2008. LNCS (LNAI), vol. 5144, pp. 415–429. Springer, Heidelberg (2008)

A Practical Simulation Result for Two-Way Pushdown Automata

Robert Glück[(✉)]

DIKU, Department of Computer Science, University of Copenhagen,
Copenhagen, Denmark

Abstract. The simulation of *two-way deterministic and nondeterministic pushdown automata* is revisited. A uniform algorithm presented herein decides on a random-access machine in linear time resp. cubic time whether a given pushdown automaton accepts a word, while the actual run of the automaton may take exponential time. The algorithm is practical since it only explores reachable configurations, simulates a class of quasi-deterministic decision problems in linear time even if the pushdown automaton is nondeterministic, and iterates over a simple work list. This is an improvement over previous simulation algorithms.

1 Introduction

This study revisits two classic results of pushdown automata. Cook proved the surprising result [5] that on a random-access machine (RAM) it is possible to decide in linear time whether a *two-way deterministic pushdown automaton* (2DPDA) accepts a word, while the actual run of the automaton may take exponential time. This insight was utilized by Knuth *et al.* [12, p. 339] to find a linear-time solution for the left-to-right string-matching problem, which can easily be expressed as a 2DPDA. This solution has a wide range of applications. The earlier method by Aho *et al.* [1] required $O(n^2)$ steps for simulating a 2DPDA, which was improved to $O(n)$ by Cook. However, whereas Cook's construction applies only to deterministic automata, the Aho *et al.* construction can decide in cubic time whether a *two-way nondeterministic pushdown automaton* (2NPDA) accepts a word (*cf.* [14]).

Unfortunately, both constructions are complicated in that they do not follow the control flow of a pushdown automaton running on a word, but examine all possible flows and thus trace many unreachable computation paths. This makes the constructions not only difficult to follow, but also impractical because of the large number of unreachable configurations given the initial configuration. Jones clarified Cook's construction by using a semantics-based simulator that interprets a 2DPDA in linear time, following the actual control flow and avoiding unreachable branches [10]; a similar solution was proposed by Rytter [13]. Jones' construction simulates the stack of a deterministic pushdown automaton by the call stack in a recursive Algol-like language.

However, to the best of the author's knowledge, no algorithm has been reported to simulate 2NPDA by exploring only the reachable configurations, nor

© Springer International Publishing Switzerland 2016
Y.-S. Han and K. Salomaa (Eds.): CIAA 2016, LNCS 9705, pp. 113–124, 2016.
DOI: 10.1007/978-3-319-40946-7_10

one to simulate both, 2DPDA and 2NPDA, in linear resp. cubic time. The present paper reports such an algorithm. Here we describe a uniform and straightforward algorithm that simulates both types of pushdown automata in linear and cubic time, respectively, by exploring only the space of reachable configurations. The algorithm has several practical advantages over previous constructions:

1. *Constant factor*: as only the set of reachable configurations is explored for a given decision problem, which is usually much smaller than the set of all possible configurations of the pushdown automaton for a given word, the constant factor of the simulation is reduced;
2. *Dynamic adaption*: if a decision problem is quasi-deterministic, a notion derived by inspection of the algorithm in this paper, it can be solved in linear time regardless of whether the pushdown automaton is deterministic or nondeterministic; there is no need for a static analysis of the automaton before the simulation;
3. *Iterative*: the core of the algorithm is a single loop that iterates over a work list; there is no need to rely on the call stack of a recursive implementation language.

The algorithm has been applied to program staging by partial evaluation of recursive flowchart programs and to solve Futamura's challenge of linear-time specialization of a naive string matcher into a linear-time matcher [8].

In Sect. 2, we review the standard move relation and a horizontal relation, the latter being the key to our main technical result, *i.e.* the fast decision algorithm presented in Sect. 3. Termination and correctness of the algorithm are proven in Sect. 4, the complexity is analyzed in Sect. 5, and an application to program staging by partial evaluation is outlined in Sect. 6. Section 7 discusses related work, and Sect. 8 is the conclusion.

2   Pushdown Automata: Configuration Relations

A two-way pushdown automaton has a finite-state control attached to a pushdown stack and a read-only input tape with a two-way head. We review the standard terminology and a relation for horizontal stack layers and terminators.

2.1   Preliminaries

Definition 1 (2NPDA). *A two-way non-deterministic pushdown automaton (2NPDA) is a tuple $M = (Q, \Sigma, \Gamma, \delta, q_0, q_f)$ where Q is a finite set of states, Σ is a finite set of input symbols, Γ is a finite set of stack symbols including the distinguished bottom-of-stack symbol Z, which marks the bottom of the stack, and*

$$\delta : Q \times (\Sigma \cup \{\triangleright, \triangleleft\}) \times \Gamma \longrightarrow \mathcal{P}(Q \times \{-1, 0, 1\} \times \Gamma^*)$$

is a transition function mapping into finite subsets. States $q_0, q_f \in Q$ are the initial and final states, respectively. Integers $d \in \{-1, 0, 1\}$ represent the directions for moving the tape head: left, stay, right. The input of M is a word $w \in \Sigma^*$ that is located on the tape between the left endmarker $\triangleright$ and the right endmarker $\triangleleft$.

Notational conventions: states are denoted by $p, q \in Q$, tape symbols by $a, b \in \Sigma \cup \{\triangleright, \triangleleft\}$, stack symbols by $A, B \in \Gamma$, stacks by $\alpha, \beta \in \Gamma^*$, and tape positions for a word of length n by $i, j \in \mathbb{N}_{n+1} = \{0, 1, \cdots, n+1\}$. Without loss of generality, assume that δ pushes and pops at most one stack symbol. We assume that δ is always defined, that is $\delta(q, a, A) = \emptyset$ if there is no move for (q, a, A). If for no triple (q, a, A) in the domain of δ does $\delta(q, a, A)$ contain more than one element, then M is said to be *deterministic*.

Let $M = (Q, \Sigma, \Gamma, \delta, q_0, q_f)$ be a 2NPDA and $w \in \Gamma^*$ be an input word of length n, fixed through the remainder of this presentation.

Definition 2 (Configuration, ID). *A surface configuration (configuration, for short) is a triple $c = (q, i, A) \in C = (Q \times \mathbb{N}_{n+1} \times \Gamma)$ where $q \in Q$ is a state, $i \in \mathbb{N}_{n+1}$ is a position, $A = \mathsf{top}(c) \in \Gamma$ is the symbol on top of the stack. An* instantaneous description *(ID, for short) is a pair $(c, \alpha) = (C \times \Gamma^+)$ where $c \in C$ is a configuration and $\alpha \in \Gamma^+$ is a stack whose top (left-most symbol) is $\mathsf{top}(c)$.*

Note that an ID (c, α) contains a configuration c and a stack α. The choice of the next move by δ depends only on c, not on α. The number of IDs is infinite, due to the possibility of unbounded stacks, but the number of configurations is finite. There are $O(n)$ possible configurations for a word of length n.

Definition 3 (Computation). *Let word $w = a_1 a_2 \ldots a_n$ and $a_0 = \triangleright$ and $a_{n+1} = \triangleleft$. A move from ID (c, α) to ID (d, β) in a single computation step, written as $(c, \alpha) \vdash (d, \beta)$, is defined for $c = (p, i, A)$ and $d = (q, j, B)$ by*

$$\begin{array}{lll}
(c, \alpha) & \vdash ((q, j, B), B\alpha) & \text{if } (q, j, B) \in \mathsf{ispush}(c), \\
(c, A\alpha) & \vdash ((q, j, B), B\alpha) & \text{if } (q, j, B) \in \mathsf{isop}(c), \\
(c, AB\alpha) & \vdash ((q, j, B), B\alpha) & \text{if } (q, j, \mathsf{Z}) \in \mathsf{ispop}(c),
\end{array}$$

where

$$\begin{array}{l}
\mathsf{ispush}((p, i, A)) = \{(q, i+j, B) \mid (q, j, BA) \in \delta(p, a_i, A)\}, \\
\mathsf{isop}((p, i, A)) = \{(q, i+j, B) \mid (q, j, B) \in \delta(p, a_i, A)\}, \\
\mathsf{ispop}((p, i, A)) = \{(q, i+j, \mathsf{Z}) \mid (q, j, \epsilon) \in \delta(p, a_i, A)\}.
\end{array}$$

As usual, we use the reflexive, transitive closure of $\vdash$ to define a sequence of computation steps, and write $(c, \alpha) \vdash^* (d, \beta)$ if (c, α) leads to (d, β) in zero or more moves. *M halts* if it reaches an ID for which there is no move. We say M *accepts* if it halts in an ID (c_f, Z) where c_f is a final configuration and Z is the bottom-of-stack symbol. A *final configuration* (q_f, j, Z) has no move for any j.

It is convenient to denote the set of all configurations following a pop by

$$\mathsf{follow}(c, d) = \{ e \mid (d, \mathsf{top}(d)\mathsf{top}(c)) \vdash (e, \mathsf{top}(e)) \}.$$

Definition 4 (Acceptance). *M accepts word w if started in ID (c_0, Z) halts in ID (c_f, Z), where $c_0 = (q_0, 0, \mathsf{Z})$ is the* initial configuration *and $c_f = (q_f, j, \mathsf{Z})$ is a final configuration. The language accepted by M is defined by*

$$L(M) = \{ w \mid (c_0, \mathsf{Z}) \vdash^* (c_f, \mathsf{Z}) \text{ on } w \in \Sigma^* \}.$$

Definition 5 (Reachable, Predecessors). *M has for word w the set of* reachable configurations R_w *and the set of* predecessors $P_w(d)$ *of a configuration d:*

$$R_w = \{\, c \mid (c_0, \mathsf{Z}) \vdash^* (c, \alpha) \text{ on } w \,\},$$
$$P_w(d) = \{\, c \in R_w \mid d \in \mathsf{ispush}(c) \vee d \in \mathsf{isop}(c) \text{ on } w \,\}.$$

There are at most $O(n)$ reachable configurations R_w for a word of length n. All configurations in R_w that directly reach d without popping are predecessors of d. This concludes the review of the terminology and assumptions used below.

2.2   Horizontal Layers and Terminators

Another, equivalent form of defining the language of a pushdown automaton can can be given by relating the configurations on the same horizontal layer of the stack and identifying the configurations that end a layer. This structuring of the computation requires only two notions:

1. horizontal configuration relation, written $c \longrightarrow d$, and
2. terminator d of a configuration c, written $\mathsf{term}(c, d)$.

Definition 6 (Horizontal Layer). *Two configurations c and d are on the same horizontal stack layer, written $c \longrightarrow d$, iff there is a sequence of moves for $m \geq 0$,*

$$(c, \mathsf{top}(c)) \vdash (e_1, \alpha_1) \cdots \vdash (e_m, \alpha_m) \vdash (d, \mathsf{top}(d)),$$

where each intermediate ID (e_j, α_j) has a stack height $|\alpha_j| \geq 2$. If $m = 0$, then $(c, \mathsf{top}(c)) \vdash (d, \mathsf{top}(d))$. As usual, $\longrightarrow^*$ is the reflexive, transitive closure of $\longrightarrow$.

Definition 7 (Terminator). *A configuration d is terminal, written $\mathsf{isterm}(d)$, iff $\mathsf{ispop}(d) \neq \emptyset$ or d is final. A terminator of a configuration c is a configuration d (if it exists), written $\mathsf{term}(c, d)$, iff $c \longrightarrow^* d$ and d is terminal.*

The *language defined by the horizontal relation* for M is

$$L'(M) = \{ w \mid c_0 \longrightarrow^* c_f \text{ on } w \in \Sigma^* \}.$$

Theorem 8 (Equivalence). *The languages defined by the standard relation $\vdash$ and the horizontal relation $\longrightarrow$ are identical, $L(M) = L'(M)$.*

Proof. Follows directly from the definition of how a 2NPDA accepts a word. □

Figure 1 illustrates a computation sequence from an initial ID (c_0, Z) to a final ID (c_f, Z). The top symbol of the stack $(A, B, \ldots)$ and the configuration $(a, b, \ldots)$ of each intermediate ID is shown. The reader will detect the growing and shrinking stack and the repetition of a sequence of moves from ID $(b, B\alpha)$ to ID $(e, B\alpha)$. Although the moves on the surface are identical regardless of the stack below, there is no way to shortcut the computation steps $(\vdash)$.

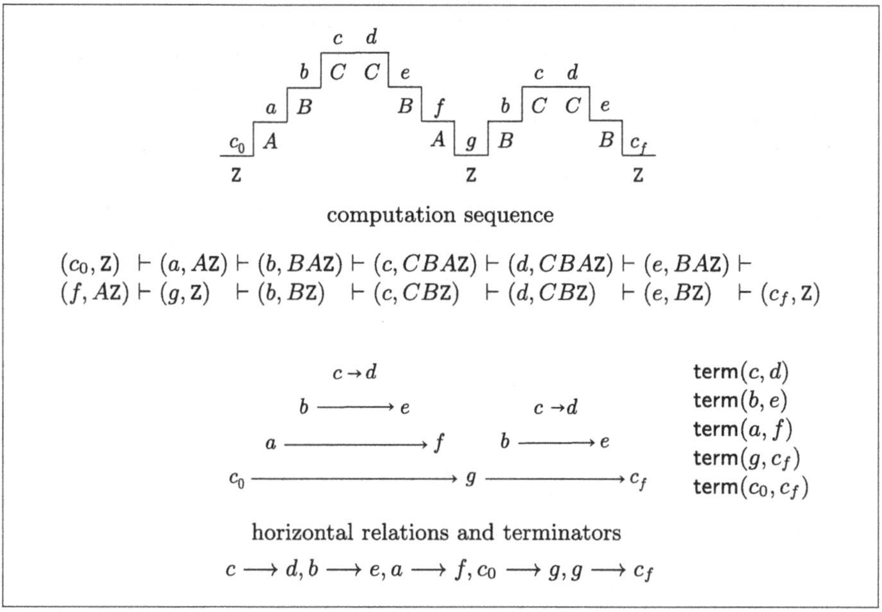
computation sequence
(c0, Z) ⊢ (a, AZ) ⊢ (b, BAZ) ⊢ (c, CBAZ) ⊢ (d, CBAZ) ⊢ (e, BAZ) ⊢
(f, AZ) ⊢ (g, Z) ⊢ (b, BZ) ⊢ (c, CBZ) ⊢ (d, CBZ) ⊢ (e, BZ) ⊢ (cf, Z)
term(c, d)
term(b, e)
term(a, f)
term(g, cf)
term(c0, cf)
horizontal relations and terminators
c ⟶ d, b ⟶ e, a ⟶ f, c0 ⟶ g, g ⟶ cf

Fig. 1. Standard computation sequence vs. horizontal relation of a PDA on some word

The situation is different if we consider the horizontal relation between the configurations. It is immediately clear that the relations $b \longrightarrow e$ and $c \longrightarrow d$ can be reused once they are known. The terminators are also indicated in the figure. For example, c_f is a terminator of c_0 and g, but g is not a terminator of c_0 (it neither pops nor is it a final configuration). The sharing of horizontal relations and terminators is key to the fast simulation algorithm in the following section.

3 An Agenda-Based Decision Algorithm

Using the stack-based computation relation $\vdash$ in Definition 3 may take *exponential time* to reach a final configuration from the initial configuration. The same computation steps may be repeated many times. The computation may be *nonterminating* because the stack may grow forever without ever reaching a final configuration. Both problems may be avoided by taking advantage of the *horizontal transition* relation and *memoizing* all terminators of a configuration because the stack below two configurations c and d that are in a horizontal relation does not matter. More formally, the computation sequence on the surface between c and d is the same regardless of the particular stack below:

Lemma 9. *If* $c \longrightarrow d$ *then* $\forall \alpha \in \Gamma^* . (c, \mathsf{top}(c)\alpha) \vdash^* (d, \mathsf{top}(d)\alpha)$.

For example, if c is reached again in the context of another stack, say β, the intermediate steps between $(c, \mathsf{top}(c)\beta)$ and $(d, \mathsf{top}(d)\beta)$ need not be repeated.

Whenever we reach c, we can go directly to d. Moreover, if d has a terminator t, and we find a horizontal relation $c \longrightarrow d$, then t is also a terminator of c.

Similarly, if t is a terminator of d, $\mathsf{term}(d, t)$, which means t triggers a pop operation, and we find some configuration c that *pushes* into d, then the configuration e directly following t at the next lower stack layer after the pop (the *follow* configuration) will be in horizontal relation $c \longrightarrow e$. We see that it is an advantage to memoize terminators found during the computation and to reuse them whenever possible. The simulation algorithm makes use of the following two lemmas to share known terminators.

Lemma 10. *If $c \longrightarrow d \wedge \mathsf{term}(d, t)$ then $\mathsf{term}(c, t)$.*

Lemma 11. *If $d \in \mathsf{ispush}(c) \wedge \mathsf{term}(d, t) \wedge e \in \mathsf{follow}(c, t)$ then $c \longrightarrow e$.*

The agenda-based simulation algorithm sim (Fig. 2) calculates all reachable configurations and their terminators, and returns the terminator set of the initial configuration c_0. The 2NPDA M and the input word w are global to the algorithm. The *language defined* by the algorithm for M is

$$L''(M) = \{w \mid c_f \in \mathsf{sim}(c_0) \text{ for } M \text{ on } w \in \Sigma^*\}.$$

The algorithm works *concurrently* along all paths between reachable configurations. A *forward step* ($\uparrow$) traverses a forward edge (a transition between two configurations) and a *return step* ($\downarrow$) propagates a terminator backward along an edge. The rules are defined such that the algorithm (i) traverses each forward edge just once, and (ii) returns the same terminator just once to each predecessor of a configuration. For this the algorithm memoizes the *predecessors* (in an array K) and the *terminators* (in an array T) of each reachable configuration. (Relation K can be viewed as a representation of a reverse graph where the edges point from configurations to their predecessors.)

The core of the algorithm is a while loop that iterates over an *agenda* A. A *step* selected from A can be *forward* ($\uparrow$) to a new configuration (as a result of a push or op operation) or *return* ($\downarrow$) a terminator. The forward steps ($\uparrow$) and return steps ($\downarrow$) that can be on the agenda and their intended meaning:

step	implies		step	implies
$\uparrow_c^d$	$c \longrightarrow d$		$\downarrow_c^t$	$\mathsf{term}(c, t)$
$\uparrow_{(c)}^d$	$d \in \mathsf{ispush}(c)$		$\downarrow_{(c)}^t$	$e \in \mathsf{follow}(c, t) \wedge c \longrightarrow e$

In case of a $\uparrow_c^d$-step, the two configurations c and d are in a horizontal relation, and in case of a $\uparrow_{(c)}^d$-step, configuration c pushes into d ("c is a stack layer below d"). As a special case, the initial configuration c_0 in the initial forward step $\uparrow_{()}^{c_0}$ is in horizontal relation with an empty predecessor denoted by ( ). The information about the type of the predecessor of d is important when we return a terminator. In the case of a return step, $\downarrow_c^t$ indicates that t is a terminator of a configuration c and $\downarrow_{(c)}^t$ indicates that t is a terminator returned to a configuration c that is one layer below (and Lemma 11 is used to establish a new horizontal relation).

```
procedure sim(c₀: conf): confset
  A := { ↑^{c₀}_{()} };                           (* initial agenda *)
  K := [∅,...,∅];                                  (* initial predecessor table *)
  while A ≠ ∅ do
    case pick(A) of
      ↑^c_k : if K[c] = ∅ then begin A ∪= steps(c); T[c] := ∅; K[c] ∪= k end else   (Ia)
              if k ∉ K[c] then begin A ∪= { ↓^t_k | t ∈ T[c] };  K[c] ∪= k end;      (Ib)
      ↓^t_c : if t ∉ T[c] then begin A ∪= { ↓^t_k | k ∈ K[c] };  T[c] ∪= t end;      (II)
      ↓^t_{(c)} : A ∪= { ↑^e_c | e ∈ follow(c,t) }                                   (III)
    esac
  end;
  return T[c₀]                                     (* terminators of c₀ *)

procedure steps(c: conf): stepset
  return { ↑^d_{(c)} | d ∈ ispush(c) } ∪ { ↑^d_c | d ∈ isop(c) } ∪ { ↓^c_c | isterm(c) }
```

Fig. 2. Agenda-based decision algorithm

The three forms of predecessors, c, (c), and $()$, are collectively denoted by k in the algorithm. For example, $↑^d_k$ can either be $↑^d_c$ or $↑^d_{(c)}$ or $↑^d_{()}$.

The algorithm starts with an initial agenda A and an empty K. The initial forward step is $↑^{c_0}_{()}$. Each iteration of the loop selects and removes a step from A by pick(A). New steps are added to A if one of the four branches (Ia,Ib,II,III) applies to the step. If no branch applies, the only effect is that the step is removed. Steps can be selected in any order and until A is empty (A can be a list or a queue). We assume that no $↓^t_{()}$-step is added to A.

(Ia) $↑^c_k$: All steps that can be made from c are added to the agenda by $A ∪=$ steps(c) if this was not done before, that is $K[c] = ∅$. The predecessor k of c is recorded by adding k to $K[c]$. Note that each application of (Ia), as well as of (Ib) and (II), disables the condition that enabled it (here, $K[c] ≠ ∅$ afterwards). Shorthand notation like $K[c] ∪= k$ is used for $K[c] := K[c] ∪ \{ k \}$.

(Ib) $↑^c_k$: If c is reached from a new predecessor k, that is $k ∉ K[c]$, all known terminators $t ∈ T[c]$ are returned to k, and k is recorded in $K[c]$. Even if no terminator is available for c, that is $T[c] = ∅$, k is recorded in $K[c]$, so that later, when the first terminator is added to $T[c]$, it is also be returned to k.

(II) $↓^t_c$: A terminator t returned to c is only returned to the predecessors of c in $K[c]$ if t is a new terminator of c, that is $t ∉ T[c]$.

(III) $↓^t_{(c)}$: A terminator t returned to a configuration below means that the next configuration is obtained by simulating a pop (Lemma 11 is used).

The initial role of $K[c]$ is to indicate whether c was already visited. After the first visit, $K[c]$ records all predecessors of c. Thus, K represents the reverse graph of configurations reached during the run of the algorithm. $T[c]$ is initialized only when c is reached. An unreachable configuration has no terminators.

4   Correctness and Termination of the Simulation

In this section we show that the decision algorithm correctly determines whether or not an input word is accepted by the pushdown automaton.

Theorem 12 (Correctness). *Algorithm* sim *in Fig. 2 correctly answers the question "Is $w \in L(M)$?".*

Proof. At the end of the algorithm, $c_f \in T[c_0]$ if and only if $w \in L(M)$. First, we will show that the algorithm terminates, then show its soundness and completeness. We outline the argument due to space constraints.

Termination. Each iteration of the while loop selects and removes an element from A by pick(A). Depending on the form of the element $(\uparrow_k^c, \downarrow_c^t, \downarrow_{(c)}^t)$, one of the cases is selected. Each branch (I-II) is guarded by a condition that is disabled after the branch is run (*e.g.* if $K[c] = \emptyset$ at the entry of (Ia), then $K[c] \neq \emptyset$ afterwards). It follows that no branch (I-II) can be selected more than once for the same element, of which there is only a finite number (the number of configurations is finite). Eventually every element is processed, which also means that (I-II) cannot add more elements to A, including $\downarrow_{(c)}^t$-elements. This limits how often (III) can turn the arrow of an element. So the algorithm always terminates.

Soundness ($\Rightarrow$). It is easy to show simultaneously by induction on the number of iterations of the while loop that the following invariants hold after each iteration. Invariants (1–3) are for the elements in the data structures A, T, and K.

(1) Agenda A: (a) $\downarrow_c^t \in A \Rightarrow$ term(c,t), (b) $\downarrow_{(c)}^t \in A \Rightarrow e \in$ follow$(c,t) \wedge c \longrightarrow e$,
   (c) $\uparrow_c^d \in A \Rightarrow c \longrightarrow d$, (d) $\uparrow_{(c)}^d \in A \Rightarrow d \in$ ispush(c),
(2) Terminators T: $t \in T[c] \wedge K[c] \neq \emptyset \Rightarrow$ term(c,t),
(3) Predecessors K: (a) $c \in K[d] \Rightarrow c \longrightarrow d$, (b) $(c) \in K[d] \Rightarrow c \in$ ispush(d).

Before the first iteration, K is empty, so (2–3) hold. Also, (1a,b,d) are vacuously true. The only element in A is $\uparrow_{()}^{c_0}$ which satisfies (1c) by definition for the initial configuration c_0 and its empty predecessor. For the induction step, we assume as induction hypothesis that the invariants (1–3) are true after n iterations of the loop, and show by case analysis that they are true after one more iteration. For each element $(\uparrow_k^c, \downarrow_c^t, \downarrow_{(c)}^t)$ in A the corresponding branch (I-III) is considered, and Lemmas 10 and 11 are used. We can conclude that (1–3) hold for all $n \geq 0$.

Completeness ($\Leftarrow$): For the converse, we must show that if term(c, d) then the algorithm eventually adds the terminator d to $T[c]$. The proof is by induction on the length of the computation sequence $(c, \text{top}(c)) \vdash^* (d, \text{top}(d))$. We assume as induction hypothesis that $(c, \text{top}(c)) \vdash^n (d, \text{top}(d)) \wedge$ term$(c, d) \wedge \uparrow_k^c \in A$ for some k imply that the algorithm eventually performs $T[c] \cup = d$ and $A \cup = \downarrow_k^d$. For the induction step, let $(c, \text{top}(c)) \vdash^1 (e, \alpha) \vdash^n (d, \text{top}(d))$ and assume that the induction hypothesis holds up to n. Consider for each of the cases $e \in$ isop(c) and $e \in$ ispush(c) the very first time the algorithm selects $\uparrow_k^c$ from A, and suppose the terminator d was already found resp. not yet found. Proceed by analysis of all cases. Note when $e \in$ isop(c), configurations b, f can be found such that $(c, \mathsf{top}(c)) \vdash (e, \mathsf{top}(e)\mathsf{top}(c)) \vdash^i (b, \mathsf{top}(b)\mathsf{top}(c)) \vdash (f, \mathsf{top}(f)) \vdash^j (d, \mathsf{top}(d))$ where $i, j \leq n$, $\mathsf{term}(e, b)$, $\mathsf{term}(f, d)$, and $f \in \mathsf{follow}(c, b)$. Finally, we can conclude that the induction hypothesis holds for all $n \geq 0$.

The algorithm terminates, is sound and complete. It correctly solves the decision problem.   □

5   Complexity of the Simulation

5.1   Cubic Time for 2NPDA

Theorem 13. *The question "Is $w \in L(M)$?" about a 2NPDA M is answered in time $O(|w|^3)$ on a RAM with a uniform cost model.*

Proof. Assume that the simulation algorithm in Fig. 2 uses a data structure for sets in which the union of two sets with cardinalities u and v takes at most time $O(u+v)$, the creation of a set from n elements takes time $O(n)$, and membership testing as well as selecting and removing an element from a set takes constant time. Evaluation of steps and follow, which access M, takes constant time. The correctness of the data collected by the algorithm was shown (Theorem 12).

There are at most $n = O(|w|)$ reachable configurations. K and T have at most $O(n^2)$ elements, $K[c]$ and $T[c]$ have at most $O(n)$ elements, and there are $O(n^2)$ possible steps of the form $\uparrow_k^c$ and $\downarrow_k^t$. Steps can only be added to A if a branch (I-III) applies to a step selected from A. (Ia) can only be applied once to each of the n configurations, and each application adds $O(1)$ steps. (Ib) and (II) can only be applied once to each of the $O(n^2)$ possible steps, and each application adds at most $O(n)$ steps. (I-II) add at most $O(n^3)$ steps in total. (III) can only be applied to a $\downarrow$-step added by (I-II), and each application adds $O(1)$ $\uparrow$-steps. The loop can be repeated at most $O(n^3)$ times until $A = \emptyset$. Thus, the time the algorithm takes is cubic in the length of the input.   □

5.2   Polynomial Time for Multi-head 2NPDA

The number of reachable configurations of a j-head 2NPDA M depends on the number of positions each of the j heads can take. If each head can take $n = |w|+1$ positions, there are at most $O(n^j)$ reachable configurations. The algorithm was shown to be cubic time in the number of reachable configurations for a 1-head 2NPDA M (Theorem 13), and runs in polynomial time for a j-head 2NPDA M. This coincides with the time required to simulate a 2NPDA with j heads on a read-only input tape of length n [1].

5.3   Linear Time for Quasi-deterministic Problems and 2DPDA

We now consider the special case of a *quasi-deterministic* decision problem that can be answered in linear time in the length of word w. It is characterized by a set of reachable configurations R_w in which each reachable configuration has $O(1)$ terminators and $O(1)$ predecessors, except for a constant number of reachable configurations that have either $O(n)$ terminators or $O(n)$ predecessors. Only a linear number of steps is required to collect all terminators of a quasi-deterministic problem. The linear-time performance is a property of the decision problem w, not of the pushdown automaton M in general. For the same M, some decision problems may be solved in linear time and others in cubic time. A trivial example is an automaton containing nondeterministic and deterministic sub-automata whose selection depends on the input word. (In general, the control-flow of an automaton may be more complex than that.)

Definition 14 (Quasi-deterministic). *The set of reachable configurations R_w of a 2NPDA M for a word w of length n is* quasi-deterministic *if it is the disjoint union of three configuration sets, $R_w = X \uplus Y \uplus Z$, whose cardinalities are bounded by $O(1)$, $O(1)$, and $O(n)$, respectively, and each configuration $x \in X$, $y \in Y$, $z \in Z$ have the following number of predecessors $P_w(\cdot)$ and terminators $T_w(\cdot)$.*

	x	y	z
Predecessors	$O(1)$	$O(n)$	$O(1)$
Terminators	$O(n)$	$O(1)$	$O(1)$

Theorem 15. *The question "Is $w \in L(M)$?" about a 2NPDA M is answered in time $O(|w|)$ on a RAM with a uniform cost model if the set of reachable configurations R_w is quasi-deterministic.*

Proof. The time to run the simulation algorithm in Fig. 2 depends on the number of steps added to A. Because the set of reachable configurations R_w is quasi-deterministic, the number of predecessors and the number of terminators in $K[c]$ and $T[c]$ of each $c \in R_w = X \uplus Y \uplus Z$ are bounded according to Definition 14. That is, each of the $O(1)$ configurations $x \in X$ has $O(1)$ predecessors in $K[x]$. In this case (II) can only add $O(1)$ $\downarrow_k^t$-steps, $k \in K[x]$, for each $\downarrow_x^t$-step. By similar case analysis for the other branches and configurations, it is easy to verify that the $O(|w|)$ applications of (I-III) add at most $O(|w|)$ steps to A. It follows that the quasi-deterministic decision problem can be solved in linear time. □

Because a quasi-deterministic decision problem has a bounded form of non-determinism, it can be solved in linear time by the simulation algorithm. Due to the *determinism* of the transition relation of a 2DPDA, *all* decision problems of such an automaton are quasi-deterministic and can be solved in linear time by the simulation algorithm (every configuration has at most one terminator). This coincides with the time required to simulate a 2DPDA by Cook's construction [5].

The characterization of quasi-deterministic problems was motivated by practical concerns. They occur during the partial evaluation of recursive flowchart programs, an application outlined below. For example, the maximally-polyvariant partial evaluation [6] of a pushdown string matcher with a static pattern and a dynamic string corresponds to a quasi-deterministic decision problem [8].

6   Application: Maximally-Polyvariant Partial Evaluation

Recursive flowchart languages [3,6] exhibit characteristics that can be modeled by two-way *deterministic* pushdown automata. Their call and return mechanism correspond to the push and pop mechanism of pushdown automata, except that label-store pairs are pushed and popped, not stack symbols.

Partial evaluation utilizes program transformation techniques [11] that explore the state space of a program with partially known input. Some values of the variables in the store of a program are unknown (dynamic), meaning not all control-flow decisions can be taken deterministically and a partial evaluator must explore all possible control flows. Therefore, the state exploration of recursive programs can be modeled by two-way *nondeterministic* pushdown automata.

A partial evaluator does not solve decision problems, but collects all reachable configurations of a source program to generate a residual program. To collect all reachable configurations fast and precisely, the author has employed the methods in a *maximally-polyvariant partial evaluator*, which solved Futamura's challenge of linear-time specialization of a naive string matcher into a linear-time matcher. It turned out that this corresponds to a quasi-deterministic decision problem [8].

7   Related Work

The classic simulation methods for two-way pushdown automata are not practical because they examine all possible configurations of a decision problem including a large number of unreachable configurations [1,5]; exceptions are the 2DPDA simulations [2,10,13]. Our 2NPDA simulation examines only the configurations that are actually reachable from the initial configuration and handles all pushdown automata. The simulation [7] does not handle left-recursion, *i.e.* requires a loop-free 2NPDA. The transitive closure of the horizontal relation defines the set of Cook's realizable pairs of surface configurations and is related to path systems [4]. Parsing algorithms use sophisticated techniques to achieve linear resp. cubic time performance for deterministic and general context-free grammars (*e.g.*, LR parsers, Earley's algorithm). Generalized LR parsing in particular uses a related graph-structured representation of parsing stacks for sharing derivations [9]. The algorithms usually rely on the one-way nature of parsing, and cannot be adapted to two-way nondeterministic pushdown automata.

8   Conclusion and Further Work

The simulation algorithm presented here is a uniform way to prove two classic simulation results of two-way deterministic and nondeterministic pushdown automata [1,5]. As a special case, the algorithm simulates *one-way* pushdown automata, which represent the class of context-free grammar problems. A closer inspection of the correspondence with known parsing algorithms will be interesting and may reveal unexpected connections. The relation with Generalized LR parsing appears closest due to the use of a graph-structured stack [15]. The class of quasi-deterministic decision problems may correspond to current automata-theoretic notions that could be productive to explore. Another direction of investigation is application to fast partial evaluation and program generation.

Acknowledgments. The author would like to thank Chung-chieh Shan and the anonymous reviewers for their input. It is a great pleasure to thank Akihiko Takano for providing the author with excellent working conditions at the National Institute of Informatics, Tokyo, and Masami Hagiya, Zhenjiang Hu, and Kanae Tsushima for their invaluable support in Japan.

References

1. Aho, A.V., Hopcroft, J.E., Ullman, J.D.: Time and tape complexity of pushdown automaton languages. Inf. Control **13**(3), 186–206 (1968)
2. Amtoft-Hansen, T., Nikolajsen, T., Träff, J.L., Jones, N.D.: Experiments with implementations of two theoretical constructions. In: Meyer, A.R., Taitslin, M.A. (eds.) Logic at Botik 1989. LNCS, vol. 363, pp. 119–133. Springer, Heidelberg (1989)
3. Christensen, N.H., Glück, R.: Offline partial evaluation can be as accurate as online partial evaluation. ACM TOPLAS **26**(1), 191–220 (2004)
4. Cook, S.A.: Characterizations of pushdown machines in terms of time-bounded computers. J. ACM **18**(1), 4–18 (1971)
5. Cook, S.A.: Linear time simulation of deterministic two-way pushdown automata. In: Freiman, C.V., Griffith, J.E., Rosenfeld, J.L. (eds.) Information Processing 71, pp. 75–80. North-Holland, Amsterdam (1972)
6. Glück, R.: A self-applicable online partial evaluator for recursive flowchart languages. Softw. Pract. Experience **42**(6), 649–673 (2012)
7. Glück, R.: Simulation of two-way pushdown automata revisited. Electron. Proc. Theor. Comput. Sci. **129**, 250–258 (2013)
8. Glück, R.: Maximally-polyvariant partial evaluation in polynomial time. In: Mazzara, M., Voronkov, A. (eds.) Perspectives of System Informatics. LNCS, vol. 9609, pp. 149–157. Springer, Heidelberg (2016)
9. Grune, D., Jacobs, C.J.H.: Parsing Techniques: A Practical Guide. Monographs in Computer Science, 2nd edn. Springer, New York (2008)
10. Jones, N.D.: A note on linear time simulation of deterministic two-way pushdown automata. Inf. Process. Lett. **6**(4), 110–112 (1977)
11. Jones, N.D., Gomard, C.K., Sestoft, P.: Partial Evaluation and Automatic Program Generation. Prentice-Hall, Upper Saddle River (1993)
12. Knuth, D.E., Morris, J.H., Pratt, V.R.: Fast pattern matching in strings. SIAM J. Comput. **6**(2), 323–350 (1977)
13. Rytter, W.: A simulation result for two-way pushdown automata. Inf. Process. Lett. **16**(4), 199–202 (1983)
14. Rytter, W.: Fast recognition of pushdown automaton and context-free languages. Inf. Control **67**(1–3), 12–22 (1985)
15. Tomita, M.: Efficient Parsing for Natural Language. Kluwer Academic Publishers, Boston (1986)

Nondeterministic Complexity of Operations on Closed and Ideal Languages

Michal Hospodár, Galina Jirásková$^{(\boxtimes)}$, and Peter Mlynárčik

Mathematical Institute, Slovak Academy of Sciences, Grešaková 6,
040 01 Košice, Slovakia
hosmich@gmail.com, jiraskov@saske.sk, mlynarcik1972@gmail.com

Abstract. We study the nondeterministic state complexity of basic regular operations on the classes of prefix-, suffix-, factor-, and subword-closed regular languages and on the classes of right, left, two-sided, and all-sided ideal regular languages. For the operations of union, intersection, complementation, concatenation, square, star, and reversal, we get the tight upper bounds for all considered classes.

1 Introduction

The nondeterministic state complexity of a regular language L, $\mathrm{nsc}(L)$, is the smallest number of states in any nondeterministic finite automaton (NFA) with a single initial state recognizing the language L. The nondeterministic state complexity of a regular operation is defined as the maximal nondeterministic state complexity of languages resulting from the operation, considered as a function of nondeterministic state complexities of the operands.

The nondeterministic state complexity of basic operations on regular languages has been investigated in [8,9], and on prefix-free and suffix-free languages in [6,7]. In this paper we continue this research and study the nondeterministic complexity of operations on closed and ideal languages. The (deterministic) state complexity of operations on the classes of closed and ideal languages has been studied by Brzozowski et al. in [2,3]. Čevorová in [4] examined the state complexity of the square operation on these classes. The class of prefix-closed languages has been investigated in [5].

In this paper we get the tight upper bounds on the nondeterministic state complexity of operations of union, intersection, complementation, concatenation, square, star, and reversal on the classes of prefix-, suffix-, factor-, and subword-closed languages. We also study the operations on left, right, two-sided and all sided ideals and get tight upper bounds for these classes as well.

To prove tightness, we use a fooling set method [1]. Although the gap between a fooling set for a regular language and the size of a minimal NFA for this language may be exponential [10], here this method is successfully used to get tight upper bounds in all the cases. In most cases we describe witness languages over a binary alphabet.

P. Mlynárčik—Research supported by VEGA grant 2/0084/15.

Y.-S. Han and K. Salomaa (Eds.): CIAA 2016, LNCS 9705, pp. 125–137, 2016.
DOI: 10.1007/978-3-319-40946-7_11

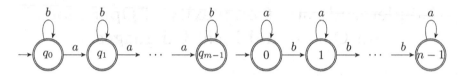

Fig. 1. The DFAs of subword closed languages K and L with $\mathrm{nsc}(K \cup L) = m + n + 1$.

2   Preliminaries

A language L is *prefix (suffix, factor, subword)-closed* iff for every $w \in L$ every prefix (suffix, factor, subword) of w is in L.

Let L be a language over an alphabet Σ. Then we have four classes of ideals. The language L is a *right ideal* iff $L = L\Sigma^*$. The language L is a *left ideal* iff $L = \Sigma^* L$. The language L is *two-sided ideal* iff $L = \Sigma^* L \Sigma^*$. The language L is *all-sided ideal* iff $L = L \sqcup\!\!\sqcup \Sigma^*$, where operation $\sqcup\!\!\sqcup$ is the shuffle operation.

In the paper we investigate the nondeterministic complexity of basic operations on the above mentioned subclasses of regular languages. To prove the minimality of NFAs, we use a fooling set lower-bound technique [1,13].

Definition 1. *A set of pairs of strings* $\{(x_1, y_1), (x_2, y_2), \ldots, (x_n, y_n)\}$ *is called a* fooling set *for a language L if for all i, j in $\{1, 2, \ldots, n\}$,*

    (F1) $x_i y_i \in L$, *and*
    (F2) *if $i \neq j$, then $x_i y_j \notin L$ or $x_j y_i \notin L$.*

Lemma 2 ([1,13]). *Let $\mathcal{F}$ be a fooling set for a language L. Then every NFA (with multiple initial states) for the language L has at least $|\mathcal{F}|$ states.*

Lemma 3 ([11]). *Let $\mathcal{A}$ and $\mathcal{B}$ be sets of pairs of strings and let u and v be two strings such that $\mathcal{A} \cup \mathcal{B}$, $\mathcal{A} \cup \{(\varepsilon, u)\}$, and $\mathcal{B} \cup \{(\varepsilon, v)\}$ are fooling sets for a language L. Then every NFA with a single initial state for L has at least $|\mathcal{A}| + |\mathcal{B}| + 1$ states.*

3   Closed Languages

We start with union and intersection on the class of closed languages.

Theorem 4. *Let $m, n \geq 2$. Let K and L be closed languages with $\mathrm{nsc}(K) = m$ and $\mathrm{nsc}(L) = n$. Then $\mathrm{nsc}(K \cup L) \leq m + n + 1$. The bound is met by binary subword closed languages.*

Proof. The upper bound is the same as for regular languages. To prove tightness, consider the binary subword-closed languages shown in Fig. 1.

Consider the following sets of pairs of strings:
$$\mathcal{A} = \{(b^n a^i, a^{m-1-i} b) \mid 0 \leq i \leq m - 1\}, \ \mathcal{B} = \{(a b^{n-1-j}, b^j a^m) \mid 0 \leq j \leq n - 1\}$$

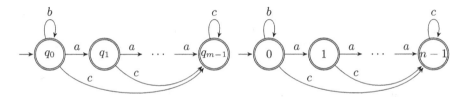

Fig. 2. The subword-closed witnesses K, L for concatenation meeting the bound $m+n$.

Let us show that $\mathcal{A} \cup \mathcal{B}$ is a fooling set. Condition (F1) is satisfied since for each i, j, the strings $b^n a^i \cdot a^{m-1-i} b$ and $ab^{n-1-j} \cdot b^j a^m$ are in $K \cup L$. To prove (F2), we consider three cases:

    (1) if $0 \le i < k \le m-1$, then $b^n a^k \cdot a^{m-1-i} b$ is not in $K \cup L$;

    (2) if $0 \le j < \ell \le n-1$, then $ab^{n-1-j} \cdot b^\ell a^m$ is not in $K \cup L$;

    (3) if $0 \le i \le m-1$ and $0 \le j \le n-1$, then $b^n a^i \cdot b^j a^m$ is not in $K \cup L$.

In addition, $\mathcal{A} \cup \{(\varepsilon, a^m b^{n-1})\}$ and $\mathcal{B} \cup \{(\varepsilon, a^{m-1} b^n)\}$ are fooling sets for $K \cup L$. By Lemma 3, we have that $\mathrm{nsc}(K \cup L) \ge m+n+1$. This holds also for classes of factor-, prefix-, and suffix-closed languages. □

Theorem 5. *Let $m, n \ge 2$. Let K and L be closed languages with $\mathrm{nsc}(K) = m$ and $\mathrm{nsc}(L) = n$. Then $\mathrm{nsc}(K \cap L) \le mn$. The bound is met by binary subword-closed languages.*

Proof. The upper bound is the same as for regular languages. To prove tightness, consider the binary subword-closed languages shown in Fig. 1. Consider the following set of pairs of strings: $\mathcal{F} = \{(a^i b^j, a^{m-1-i} b^{n-1-j}) \mid 0 \le i \le m-1, 0 \le j \le n-1\}$. Let us show that $\mathcal{F}$ is a fooling set for $K \cap L$. Condition (F1) is satisfied since for each i, j, the string $a^i b^j \cdot a^{m-1-i} b^{n-1-j}$ is in $K \cap L$. To prove (F2), let $(i, j) \ne (k, \ell)$. (1) If $i < k$, then $a^k b^\ell \cdot a^{m-1-i} b^{n-1-j}$ is not in $K \cap L$. (2) If $i = k$ and $j < \ell$, then $a^k b^\ell \cdot a^{m-1-i} b^{n-1-j}$ is not in $K \cap L$. Hence $\mathcal{F}$ is a fooling set for $K \cap L$, so $\mathrm{nsc}(K \cap L) \ge mn$. □

Let us continue with concatenation and square.

Theorem 6. *Let K and L be closed languages with $\mathrm{nsc}(K) = m$ and $\mathrm{nsc}(L) = n$. Then $\mathrm{nsc}(KL) \le m + n$. The bound is met by ternary subword-closed languages.*

Proof. The upper bound is the same as for regular languages. To prove tightness, consider the ternary subword-closed languages shown in Fig. 2.

Consider the following set of pairs of strings:
$$\mathcal{F} = \{(a^i, a^{m-1-i} cba^{n-1}) \mid 0 \le i \le m-1\} \cup \{(a^{m-1} cba^j, a^{n-1-j}) \mid 0 \le j \le n-1\}.$$

Let us show that $\mathcal{F}$ is a fooling set for KL. Condition (F1) is satisfied since for each i, j, the strings $a^i \cdot a^{m-1-i} cba^{n-1}$ and $a^{m-1} cba^j \cdot a^{n-1-j}$ are in KL. To prove (F2), notice that KL is a subset of $b^* a^* c^* b^* a^* c^*$ and every string in KL has at most $m-1+n-1$ letters a. We consider three cases.

(1) If $0 \leq i < k \leq m - 1$, then $a^k \cdot a^{m-1-i}cba^{n-1}$ is not in KL, because it has more than $m - 1 + n - 1$ letters a.

(2) If $0 \leq j < \ell \leq n - 1$, then $a^{m-1}cba^{\ell} \cdot a^{n-1-j}$ is not in KL, because it has more than $m - 1 + n - 1$ letters a.

(3) If $0 \leq i \leq m - 1$ and $0 \leq j \leq n - 1$, then $a^{m-1}cba^j \cdot a^{m-1-i}cba^{n-1}$ is not in KL, because this string is not in the form $b^*a^*c^*b^*a^*c^*$.

Hence $\mathcal{F}$ is a fooling set for KL, so $\mathrm{nsc}(KL) \geq m + n$.   □

If $m = n$, then $K = L$ in the proof above, so we get the next result.

Corollary 7. *Let L be a closed language with $\mathrm{nsc}(L) = n$. Then $\mathrm{nsc}(L^2) \leq 2n$. The bound is met by a ternary subword-closed language.*

Theorem 8. *Let L be a closed language over Σ with $\mathrm{nsc}(L) = n$. Then*
   (a) if L is prefix-closed, then $\mathrm{nsc}(L^*) \leq n$, and the bound is tight if $|\Sigma| \geq 2$;
   (b) if L is suffix-closed, then $\mathrm{nsc}(L^*) \leq n$, and the bound is tight if $|\Sigma| \geq 2$;
   (c) if L is factor- or subword-closed, then $\mathrm{nsc}(L^*) = 1$.

Proof. If L is a closed languae, then $\varepsilon \in L$. It follows that $\mathrm{nsc}(L^*) \leq n$. To prove tightness, consider a prefix-closed language shown in Fig. 3 and a suffix-closed language shown in Fig. 4. Lower bound for prefix-closed was proven in [5], lower bound for suffix-closed is n because $L = L^*$. For factor- or subword-closed, let Γ be set of letters present in any string of L. While $L \subseteq \Gamma^*$, every single-letter string from Γ is in L. It follows that $L^* = \Gamma^*$, hence $\mathrm{nsc}(L^*) = 1$.   □

Theorem 9. *Let $n \geq 3$ and L be a closed language with $\mathrm{nsc}(L) = n$. Then $\mathrm{nsc}(L^R) \leq n + 1$. The bound is met by a binary prefix-closed language, by a ternary factor-closed language and by a subword-closed language over an alphabet of size $2n - 2$.*

Proof. The upper bound is the same as for regular languages. To prove tightness, consider the binary prefix-closed language shown in Fig. 3. It was shown in [5] that the reversal of this language requires $n + 1$ states. Now consider the ternary factor-closed language shown in Fig. 5. Consider the following sets of pairs of strings: $\mathcal{A} = \{(b, a^{n-2}c)\}$ and $\mathcal{B} = \{(ba^i, a^{n-2-i}c) \mid 1 \leq i \leq n-2\} \cup \{(ca^{n-1}, \varepsilon)\}$. Let us show that $\mathcal{A} \cup \mathcal{B}$, $\mathcal{A} \cup \{(\varepsilon, a^{n-3}c)\}$, and $\mathcal{B} \cup \{(\varepsilon, a^{n-2}c)\}$ are fooling sets for L^R. Condition (F1) is satisfied since for each i, the string $ba^i \cdot a^{n-2-i}c$ equals $ba^{n-2}c$ that is in L^R since $ca^{n-2}b$ is in L. String ca^{n-1} is also in L^R since $a^{n-1}c$ is in L. To prove (F2), notice that every string of L has at most $n - 2$ continual occurences of a after any c. Thus we consider cases:

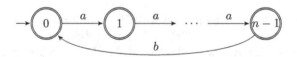

Fig. 3. The prefix-closed witness language L for star and reversal.

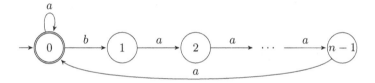

Fig. 4. The suffix-closed witness language L for star meeting the bound n.

(1) If $0 \leq i < j \leq n - 2$, then $ba^j \cdot a^{n-2-i}c$ is not in L^R, because it has more than $n - 2$ continual occurences of a after c. (2) If $0 \leq i \leq n - 2$, then $ca^{n-1} \cdot a^{n-2-i}c$ is not in L^R, because it has more than $n - 2$ continual occurences of a after c.

Sets $\mathcal{A} \cup \{(\varepsilon, a^{n-3}c)\}$ and $\mathcal{B} \cup \{(\varepsilon, a^{n-2}c)\}$ are fooling sets for L^R, because the strings $b \cdot a^{n-3}c$ and $ba^i \cdot a^{n-2}c, i \geq 1$ are not in L^R. Therefore by Lemma 3 $\mathrm{nsc}(L^R) \geq n + 1$. This proof holds also for the class of suffix-closed languages since every factor-closed language is also suffix-closed.

Finally consider the subword-closed language accepted by the DFA shown in Fig. 6. Consider the following sets:
$\mathcal{A} = \{(b_2 b_3 \cdots b_{n-1}, a_1)\}$, $\mathcal{B} = \{(b_1 \cdots b_{i-1}b_{i+1} \cdots b_{n-1}, a_i) \mid 2 \leq i \leq n - 1\} \cup \{(b_1 a_2, \varepsilon)\}$. Let us show that $\mathcal{A} \cup \mathcal{B}$, $\mathcal{A} \cup \{(\varepsilon, a_2)\}$ and $\mathcal{B} \cup \{(\varepsilon, a_1)\}$ are fooling sets for L^R. Condition (F1) for $\mathcal{A} \cup \mathcal{B}$ is satisfied because for every i the string $b_1 \cdots b_{i-1}b_{i+1} \cdots b_{n-1} \cdot a_i$ is in L^R. Next, for every $i \neq j$ the string $b_1 \cdots b_{i-1}b_{i+1} \cdots b_{n-1} \cdot a_j$ is not in L^R, because it has b_j before a_j. Hence (F2) is satisfied. The condition (F1) for $\mathcal{A} \cup \{(\varepsilon, a_2)\}$ and for $\mathcal{B} \cup \{(\varepsilon, a_1)\}$ is satisfied, because the strings a_2 and a_1 are in L^R. The proof of condition (F2) uses the same strings as for $\mathcal{A} \cup \mathcal{B}$. □

We conclude this section with the complementation operation. In [5], a ternary prefix-closed language meeting the upper bound 2^n for complement was described. Now we describe a binary witness language.

Theorem 10. *Let L be a closed language over Σ with $\mathrm{nsc}(L) = n$. Then*
(a) if L is prefix-closed, then $\mathrm{nsc}(L^c) \leq 2^n$, and the bound is tight if $|\Sigma| \geq 2$;
(b) if L is suffix-closed, then $\mathrm{nsc}(L^c) \leq 2^{n-1} + 1$, and the bound is met by a binary factor-closed language;

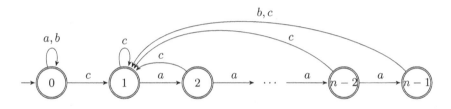

Fig. 5. The factor-closed witness language L for reversal meeting the bound $n + 1$.

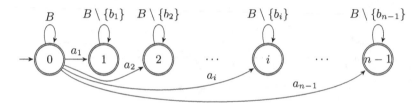

Fig. 6. The DFA of subword-closed language L where $B = \{b_1, \ldots, b_{n-1}\}$.

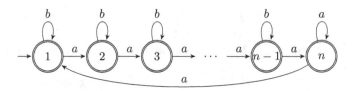

Fig. 7. The NFA of binary witness prefix-closed language L with $\mathrm{nsc}(L^c) = 2^n$.

(c) *if L is subword-closed, then* $\mathrm{nsc}(L^c) \leq 2^{n-1} + 1$, *and the bound is tight if* $|\Sigma| \geq 2^n$.

Proof. (a) The upper bound is the same as for regular languages. To prove tightness, let L be the binary language accepted by the NFA A shown in Fig. 7. First, we prove the reachability of every subset of $\{1, 2, \ldots, n\}$ in the subset automaton of A. Notice that we have $\{1\} \xrightarrow{a^{n-1}} \{n\} \xrightarrow{a^{n-1}} \{1, 2, \ldots, n\}$. Next, we can shift cyclically by one every subset S: we use the string a if $n \notin S$ or if $n \in S$ and $n - 1 \in S$, and we use the string ab otherwise. Finally, we can remove state n from any subset containing n by b. It follows that every subset of $\{1, 2, \ldots, n\}$ is reachable. Thus for every set S, there exists a string u_S such that u_S leads the subset automaton from $\{1\}$ to S.

Now, we define a fooling set for complement of L. For every set S we define a string v_S as follows. First we define $\sigma(i)$, where $i \in \{1, 2, \ldots, n\}$ as

$$\sigma(i) = \begin{cases} ba, & \text{if } i \in S, \\ a, & \text{if } i \notin S. \end{cases}$$

Let $v_S = \sigma(n)\sigma(n-1)\ldots\sigma(2)\sigma(1)$. We show, that such a string is rejected by A from every $i \in S$ and accepted from every $i \notin S$. Let $i \notin S$, then $\sigma(i) = a$, and

$$i \xrightarrow{\sigma(n)} i+1 \xrightarrow{\sigma(n-1)} i+2 \xrightarrow{\sigma(n-2)} \cdots \xrightarrow{\sigma(i+1)} n \xrightarrow{a} 1 \xrightarrow{\sigma(i-1)\ldots\sigma(1)} i,$$

so v_S is accepted since every state is final. If $i \in S$, then $\sigma(i) = ba$, and $i \xrightarrow{\sigma(n)} \{i+1\} \xrightarrow{\sigma(n-1)} \{i+2\} \xrightarrow{\sigma(n-2)} \cdots \xrightarrow{\sigma(i+1)} \{n\}$,
and now A reads the first symbol of $\sigma(i)$ which is b. However, transition on b is not defined in state n, therefore the string v_S is rejected.

Now we show that

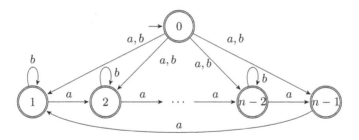

Fig. 8. The factor-closed witness L for complement, with $\mathrm{nsc}(L^c) = 2^{n-1} + 1$.

$\mathcal{F} = \{(u_S, v_S) \mid S \subseteq \{1, 2, \ldots, n\}\}$ is a fooling set for L^c.

(F1) Let $S \subseteq \{1, 2, \ldots, n\}$. The NFA A reaches subset S by u_S, and from every state $q \in S$ the string v_S is rejected. So $u_S v_S$ is rejected by A, so $u_S v_S \in L^c$.

(F2) Let $S, T \subseteq \{1, 2, \ldots, n\}$ and $S \neq T$. Without loss of generality, there exists a state i, such that $i \in S$ and $i \notin T$. So v_T is accepted from i. Hence $u_S v_T$ is accepted by A, and therefore $u_S v_T \notin L^c$. This completes the proof of (a).

(b) We first prove the upper bound. Let $A = (Q, \Sigma, \delta, s, F)$ be a minimal NFA, such that $L(A) = L$. Since A is a minimal NFA, every q in Q is reachable from s and also some final state is reachable from q. Let a state $q \in Q$ be reachable from s by a string u. If a final state is reachable from q by string v, then also uv reaches a final state, so uv is accepted. Since L is suffix-closed, the string v reaches a final state from s. Therefore every subset of Q containing s is equivalent to $\{s\}$ in the subset automaton of NFA A. So subset automaton of A has at most $2^{n-1} + 1$, so $\mathrm{nsc}(L^c) \leq 2^{n-1} + 1$.

To prove tightness, consider the language L accepted by automaton in Fig. 8. If there is an accepting computation from a state q on a string u such that $q \xrightarrow{a(b)} q' \xrightarrow{u'} f$, where $u = au'$ or $u = bu'$ and f is a final state, then there is a computation $s \xrightarrow{a(b)} q' \xrightarrow{u'} f$. It follows that L is suffix-closed. Therefore L is factor-closed. First, we prove the reachability of every subset of $\{1, 2, \ldots, n-1\}$ in the subset automaton of A. Notice that we have $\{0\} \xrightarrow{a} \{1, 2, \ldots, n-1\}$. Next, we can shift cyclically by one every subset S by using the string a. Finally, we can remove state $n-1$ from any subset containing $n-1$ by b. It follows that every subset of $\{1, 2, \ldots, n-1\}$ is reachable. Thus for every set S, there exists a string u_S such that u_S leads the subset automaton from $\{0\}$ to S. Now, we define a fooling set for complement of L. For every set S we define a string v_S as follows. First we define $\sigma(i)$, where $i \in \{1, 2, \ldots, n-1\}$ as $\sigma(i) = ba$ if $i \in S$, and $\sigma(i) = a$ if $i \notin S$. Let $v_S = \sigma(n-1)\sigma(n-2)\cdots\sigma(2)\sigma(1)$. Similarly as in proof in case of prefix-closed in (a) we can show that such a string is rejected by A from every $i \in S$ and accepted from every $i \notin S$. Let $\mathcal{A} = \{(u_S, v_S) \mid S \subseteq \{1, 2, \ldots, n-1\}\}$. We can show that $\mathcal{F} = \mathcal{A} \cup \{(\varepsilon, (ba)^n)\}$ is a fooling set for L^c.

(c) Since subword-closed language is also factor-closed, the upper bound is $2^{n-1} + 1$. To prove tightness consider an NFA A, defined as follows: $A = (Q, \Sigma, \delta, s, F)$, where $Q = \{0, 1, 2, \ldots, n-1\}, s = 0, F = Q$ and $\Sigma =$

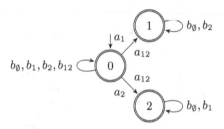

Fig. 9. The subword-closed witness language L with $\mathrm{nsc}(L) = 3$ and $|\Sigma| = 2^n$.

$\{a_S, b_S \mid S \subseteq \{1, 2, \ldots, n-1\}\}$, $\delta(0, a_S) = S$, for $i > 0$ $\delta(i, a_S) = \emptyset$, $\delta(0, b_S) = 0$, for $i > 0$: if $i \notin S$, then $\delta(i, b_S) = \{i\}$ and if $i \in S$, then $\delta(i, b_S) = \emptyset$. Such an NFA is shown in Fig. 9. Consider now the language $L = L(A)$. Let $w \in L$. The string w is accepted in a $i \in S$. Any substring of w is accepted also in the i. Hence L is subword-closed. We can show that $\mathcal{A} = \{(a_S, b_S) \mid S \subseteq \{1, 2, \ldots, n-1\}\} \cup \{(\varepsilon, a_\emptyset\}$ is fooling set for L^c. Therefore $\mathrm{nsc}(L^c) \geq 2^{n-1} + 1$. $\square$

In the end of this section we pay attention to unary closed languages. Consider prefix-closed languages and two cases, finite languages and infinite languages. In the case of finite languages, there is a string with maximum length, so every shorter string also must be in the language. In the case of infinite languages, for arbitrary positive integer i, there is a string w with length at least i and with this string every its prefixes, so such a language is a^*. Moreover suffix-closed, factor-closed and subword-closed coincide.

Theorem 11. *Let K and L be two unary closed languages with $\mathrm{nsc}(K) = m$ and $\mathrm{nsc}(L) = n$. Then $\mathrm{nsc}(K \cup L) \leq \max\{m, n\}$, $\mathrm{nsc}(K \cap L) \leq \min\{m, n\}$, $\mathrm{nsc}(KL) \leq m + n - 1$, $\mathrm{nsc}(L^2) \leq 2n - 1$, $\mathrm{nsc}(L^*) \leq 1$, $\mathrm{nsc}(L^R) \leq n$, and $\mathrm{nsc}(L^c) \leq n + 1$. All these bounds are tight.*

4   Ideal Languages

Let us begin with a useful proposition about some features of automata for left and right ideals.

Proposition 12. *Let L be a regular language. (1) If L is a left ideal, then there exists a minimal NFA A such that $L(A) = L$ and there is a loop on each symbol in the initial state and no transition goes to the initial state from any other state. (2) If L is a right ideal, then there exists a minimal NFA A such that $L(A) = L$ and there is a unique final state in which there is a loop on each symbol and from which no transition goes to any other state.*

Theorem 13. *Let $m, n \geq 1$. Let K and L be ideal languages with $\mathrm{nsc}(K) = m$ and $\mathrm{nsc}(L) = n$. Then $\mathrm{nsc}(K \cap L) \leq mn$. The bound is met by binary all-sided ideals.*

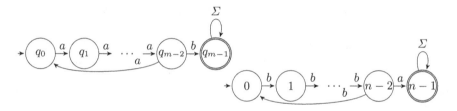

Fig. 10. Witnesses right ideals for union.

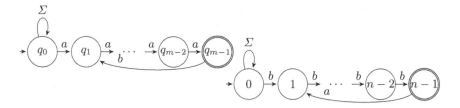

Fig. 11. Witnesses left ideals for union.

Theorem 14. *Let $m, n \geq 3$. Let K and L be ideal languages over an alphabet Σ with $\mathrm{nsc}(K) = m$ and $\mathrm{nsc}(L) = n$. Then*

(a) if K, L are right ideals, then $\mathrm{nsc}(K \cup L) \leq m + n$,

(b) if K, L are left ideals, then $\mathrm{nsc}(K \cup L) \leq m + n - 1$,

(c) if K, L are two-sided or all-sided ideals, then $\mathrm{nsc}(K \cup L) \leq m + n - 2$,

and all the bounds are tight if $|\Sigma| \geq 2$.

Proof. (a) We first prove the upper bound. Let A be a minimal m-state NFA for K and B be a minimal n-state NFA for L. Since K and L are right ideals, A and B have exactly one final state which goes to itself on each symbol. We can get an ε-NFA for $K \cup L$ from NFAs A and B by merging the final states of A and B and by adding a new initial state connnected to the initial states of A and B by ε-transitions. The resulting ε-NFA has $m + n$ states, so the corresponding NFA for $K \cup L$ has also $m + n$ states.

To prove tightness, consider the binary right ideals K and L shown in Fig. 10. Now we show that minimal NFA for $K \cup L$ needs $m + n$ states. To this aim let

$\mathcal{A} = \{(a^{m-1+i}, a^{m-2-i}b) \mid 0 \leq i \leq m - 2\} \cup \{(a^{m-2}b, \varepsilon)\}$, and

$\mathcal{B} = \{(b^{n-1+j}, b^{n-2-j}a) \mid 0 \leq j \leq n - 2\}$.

The sets $\mathcal{A} \cup \mathcal{B}$, $\mathcal{A} \cup \{(\varepsilon, b^{n-2}a)\}$ and $\mathcal{B} \cup \{(\varepsilon, a^{m-2}b)\}$ are fooling sets. By Lemma 3 we have $\mathrm{nsc}(K \cup L) \geq |\mathcal{A}| + |\mathcal{B}| + 1 = m + n$.

(b) We first prove the upper bound. Let A be a minimal m-state NFA for K and B be a minimal n-state NFA for L. Since K and L are left ideals, we may assume by Proposition 12 that A and B have a loop on each symbol in the initial state, and no transition from some other state goes to the initial state.

We can get an NFA for $K \cup L$ from NFAs A and B by merging the initial states. All original transitions from initial states of NFAs A, B go from new merged state to states as before merging. The resulting NFA has $m + n - 1$ states, so $\mathrm{nsc}(K \cup L) \leq m + n - 1$.

To prove tightness, consider two left ideals shown in Fig. 11. Now we show that minimal NFA for $K \cup L$ needs $m + n - 1$ states. To this aim let $\mathcal{A} = \{(a^i, a^{m-1-i}) \mid 0 \leq i \leq m - 1\}$ and $\mathcal{B} = \{(b^j, b^{n-1-j}) \mid 1 \leq j \leq n - 2\} \cup \{(b^{n-1}, ab^{n-2})\}$. The set $\mathcal{A} \cup \mathcal{B}$ is fooling set for $K \cup L$, so $\mathrm{nsc}(K \cup L) \geq m+n-1$, therefore $\mathrm{nsc}(K \cup L) = m + n - 1$.

(c) For upper bound, let A be a minimal m-state NFA for K and B be a minimal n-state NFA for L. Since K and L are left ideals and also right ideals, we may assume by Proposition 12 that A and B have properties claimed there. We can get an NFA for $K \cup L$ from NFAs A and B by merging the initial states, and by merging the final states of A and B. The resulting NFA has $m + n - 2$ states and we leave to the reader to verify the correctness of the construction. To prove tightness, consider languages $K = \{w \in \{a, b\}^* \mid \#_a(w) \geq m - 1\}$ and $L = \{w \in \{a, b\}^* \mid \#_b(w) \geq n - 1\}$, so K and L are all-sided ideals. Notice that each string in $K \cup L$ has at least $m - 1$ symbols a or at least $n - 1$ symbols b. The set $\{(a^i, a^{m-1-i}) \mid 0 \leq i \leq m - 1\} \cup \{(b^j, b^{n-1-j}) \mid 1 \leq j \leq n - 2\}$ is fooling set for $K \cup L$ and contains $m + n - 2$ pairs, so $\mathrm{nsc}(K \cup L) \geq n + m - 2$.   □

In the next theorem we use unary languages to prove tightness.

Theorem 15. *Let $m, n \geq 3$. Let K and L be ideal languages over Σ with $\mathrm{nsc}(K) = m$ and $\mathrm{nsc}(L) = n$. Then $\mathrm{nsc}(KL) \leq m + n - 1$ and the bound is tight if $|\Sigma| \geq 1$. Moreover, $\mathrm{nsc}(L^2) \leq 2n - 1$ and the bound is tight if $|\Sigma| \geq 1$.*

Proof. First, let K, L be left ideals. Let $A = (Q_A, \Sigma, \delta_A, s_A, F_A)$ and $B = (Q_B, \Sigma, \delta_B, s_B, F_B)$ be minimal NFAs for K, L. Since K and L are left ideals, we may assume by Proposition 12 that A and B have a loop on each symbol in the initial state, and no transition from some other state goes to the initial state. We can get an NFA C for KL from NFAs A and B as follows: For every f in F_A add a loop on every symbol and add transitions (f, a, q) when there is a transition (s_B, a, q) in B, where $f \in F_A, a \in \Sigma, q \in Q_B \setminus \{s_B\}$. Set $F_C = F_B, Q_C = Q_A \cup Q_B \setminus \{s_B\}$. The resulting NFA has $m + n - 1$ states, so $\mathrm{nsc}(KL) \leq m + n - 1$.

Now, let K, L be right ideals. Let $A = (Q_A, \Sigma, \delta_A, s_A, \{q_f\})$ be a minimal m-state NFA for K and $B = (Q_B, \Sigma, \delta_B, s_B, \{p_f\})$ be a minimal n-state NFA for L. Since K and L are right ideals, we may assume by Proposition 12 that A and B have a loop on each symbol in a unique final state, and no transition goes from the final state to some other state. We can get an NFA C for KL from NFAs A and B by merging final state of A with initial state of B and excluding of merged state from set of final states as follows: $C = (Q_C, \Sigma, \delta_C, s_A, \{p_f\})$, where $Q_C = (Q_A \setminus \{q_f\}) \cup (Q_B \setminus \{s_B\}) \cup \{n_{AB}\}$ and for every a in Σ we have $\delta_C(n_{AB}, a) = \delta_A(q_f, a) \cup \delta_B(s_B, a)$. The resulting NFA has $m + n - 1$ states, so $\mathrm{nsc}(KL) \leq m + n - 1$.

Two-sided and all-sided ideals are also right ideals, so upper bound is the same as in that cases. To prove tightness, consider all-sided ideal languages $K = \{a^{m-1}a^k \mid k \geq 0\}$ and $L = \{a^{n-1}a^k \mid k \geq 0\}$, with $\mathrm{nsc}(K) = m$ and

$\mathrm{nsc}(L) = n$. The set $\mathcal{F} = \{(a^i, a^{m+n-2-i}) \mid 0 \leq i \leq m+n-2\}$ is fooling set for KL, so $\mathrm{nsc}(KL) \geq |\mathcal{F}| = m+n-1$. It remains to show the case for square. The upper bound follows from general concatenation, when $m = n$. The tightness follows from a coincidence of the forms of witness languages.   □

Theorem 16. *Let $n \geq 2$. Let L be ideal languages over Σ with $\mathrm{nsc}(L) = n$. Then $\mathrm{nsc}(L^*) \leq n+1$ and the bound is met by a binary all-sided ideal.*

Theorem 17. *Let $n \geq 3$. Let L be ideal languages over Σ with $\mathrm{nsc}(L) = n$. (a) If L is right or two-sided or all-sided ideal, then $\mathrm{nsc}(L^R) \leq n$ and the bound is tight if $|\Sigma| \geq 1$. (b) If L is left ideal, then $\mathrm{nsc}(L^R) \leq n+1$ and the bound is tight if $|\Sigma| \geq 3$.*

Theorem 18. *([14]). Let $n \geq 3$. Let L be language over Σ with $\mathrm{nsc}(L) = n$. (a) If L is a right or left ideal, then $\mathrm{nsc}(L^c) \leq 2^{n-1}$. The bound is tight if $|\Sigma| \geq 2$. (b) If L is a two-sided ideal, then $\mathrm{nsc}(L^c) \leq 2^{n-2}$. The bound is tight if $|\Sigma| \geq 2$. (c) If L is an all-sided ideal, then $\mathrm{nsc}(L^c) \leq 2^{n-2}$. The bound is tight if $|\Sigma| \geq 2^{n-2}$.*

In the end we pay attention to unary ideal languages. Let $\Sigma = \{a\}$. If L is a right ideal and a^i is its shortest string, then $L = a^i a^*$. Moreover $L = a^* a^i = a^* a^i a^* = a^* \sqcup a^i$, hence left, right, two-sided and all-sided ideals coincide.

Theorem 19. *Let $m, n \geq 2$. Let K, L be unary ideals with $\mathrm{nsc}(K) = m$, $\mathrm{nsc}(L) = n$. Then $\mathrm{nsc}(K \cap L) = \max\{m, n\}$, $\mathrm{nsc}(K \cup L) = \min\{m, n\}$, $\mathrm{nsc}(KL) = m + n - 1$, $\mathrm{nsc}(L^2) = 2n - 1$, $\mathrm{nsc}(L^*) = n - 1$, $\mathrm{nsc}(L^R) = n$, $\mathrm{nsc}(L^c) = n - 1$.*

5   Conclusions

We investigated the nondeterministic state complexity of basic regular operations on the classes of closed and ideal languages. For each class and for each operation, we obtained the tight upper bounds. To prove tightness we usually used a binary alphabet. In all the cases where we used a larger alphabet for describing witness languages, we do not know whether the obtained upper bounds can be met also by languages defined over smaller alphabets. For both closed and ideal languages, we also considered the unary case. Our results are summarized in the following tables.

Class	$K \cap L, \Sigma$	$K \cup L, \Sigma$	$K \cdot L, \Sigma$
Prefix-closed	$mn, 2$	$m + n + 1, 2$	$m + n, 3$
Suffix-closed	$mn, 2$	$m + n + 1, 2$	$m + n, 3$
Factor-closed	$mn, 2$	$m + n + 1, 2$	$m + n, 3$
Subword-closed	$mn, 2$	$m + n + 1, 2$	$m + n, 3$
Unary closed	$\min\{m, n\}$	$\max\{m, n\}$	$m + n - 1$
Right ideal	$mn, 2$	$m + n, 2$	$m + n - 1, 1$
Left ideal	$mn, 2$	$m + n - 1, 2$	$m + n - 1, 1$
Two-sided ideal	$mn, 2$	$m + n - 2, 2$	$m + n - 1, 1$
All-sided ideal	$mn, 2$	$m + n - 2, 2$	$m + n - 1, 1$
Unary ideal	$\max\{m, n\}$	$\min\{m, n\}$	$m + n - 1$
Regular	$mn, 2$	$m + n + 1, 2$	$m + n, 2$
Unary regular	$mn; \gcd(m, n) = 1$	$m + n; \gcd(m, n) = 1$	$m + n(-1)$

Class	L^2, Σ	L^*, Σ	L^R, Σ	L^c, Σ
Prefix-closed	$2n, 3$	$n, 2$	$n + 1, 2$	$2^n, 2$
Suffix-closed	$2n, 3$	$n, 2$	$n + 1, 3$	$1 + 2^{n-1}, 2$
Factor-closed	$2n, 3$	$1, 1$	$n + 1, 3$	$1 + 2^{n-1}, 2$
Subword-closed	$2n, 3$	$1, 1$	$n + 1, 2n - 2$	$1 + 2^{n-1}, 2^n$
Unary closed	$2n - 1$	1	n	$n + 1$
Right ideal	$2n - 1, 1$	$n + 1, 2$	$n, 1$	$2^{n-1}, 2$
Left ideal	$2n - 1, 1$	$n + 1, 2$	$n + 1, 3$	$2^{n-1}, 2$
Two-sided ideal	$2n - 1, 1$	$n + 1, 2$	$n, 1$	$2^{n-2}, 2$
All-sided ideal	$2n - 1, 1$	$n + 1, 2$	$n, 1$	$2^{n-2}, 2^{n-2}$
Unary ideal	$2n - 1$	$n - 1$	n	$n - 1$
Regular	$2n, 2$	$n + 1, 1$	$n + 1, 2$	$2^n, 2$
Unary regular	$2n(-1)$	$n + 1$	n	$2^{\Theta(\sqrt{n \log n})}$

References

1. Birget, J.C.: Partial orders on words, minimal elements of regular languages, and state complexity. Theoret. Comput. Sci. **119**, 267–291 (1993)
2. Brzozowski, J., Jirásková, G., Li, B.: Quotient complexity of ideal languages. Theoret. Comput. Sci. **470**, 36–52 (2013)
3. Brzozowski, J., Jirásková, G., Zou, C.: Quotient complexity of closed languages. Theor. Comput. Syst. **54**, 277–292 (2014)
4. Čevorová, K.: Square on ideal, closed and free languages. In: Shallit, J., Okhotin, A. (eds.) DCFS 2015. LNCS, vol. 9118, pp. 70–80. Springer, Heidelberg (2015)
5. Čevorová, K., Jirásková, G., Mlynárčik, P., Palmovský, M., Šebej, J.: Operations on automata with all states final. In: Ésik, Z., Fülöp, Z. (eds.) Automata and Formal Languages 2014 (AFL 2014). EPTCS, vol. 151, pp. 201–215 (2014)

6. Han, Y.-S., Salomaa, K.: Nondeterministic state complexity for suffix-free regular languages. In: DCFS 2010, pp. 189–196 (2010)
7. Han, Y.-S., Salomaa, K., Wood, D.: Nondeterministic state complexity of basic operations for prefix-free regular languages. Fundam. Inform. **90**(1–2), 93–106 (2009)
8. Holzer, M., Kutrib, M.: Nondeterministic descriptional complexity of regular languages. Int. J. Found. Comput. Sci. **14**, 1087–1102 (2003)
9. Jirásková, G.: State complexity of some operations on binary regular languages. Theoret. Comput. Sci. **330**, 287–298 (2005)
10. Jirásková, G.: Note on minimal automata and uniform communication protocols. In: Grammars and Automata for String Processing: From Mathematics and Computer Science to Biology, and Back, pp. 163–170. Taylor and Francis (2003)
11. Jirásková, G., Masopust, T.: Complexity in union-free regular languages. Int. J. Found. Comput. Sci. **22**(7), 1639–1653 (2011)
12. Jirásková, G., Mlynárčik, P.: Complement on prefix-free, suffix-free, and non-returning NFA languages. In: Jürgensen, H., Karhumäki, J., Okhotin, A. (eds.) DCFS 2014. LNCS, vol. 8614, pp. 222–233. Springer, Heidelberg (2014)
13. Glaister, I., Shallit, J.: A lower bound technique for the size of nondeterministic finite automata. Inform. Process. Lett. **59**, 75–77 (1996)
14. Mlynárčik, P.: Complement on free and ideal languages. In: Shallit, J., Okhotin, A. (eds.) DCFS 2015. LNCS, vol. 9118, pp. 185–196. Springer, Heidelberg (2015)
15. Sipser, M.: Introduction to the Theory of Computation. PWS Publishing Company, Boston (1997)

On Bounded Semilinear Languages, Counter Machines, and Finite-Index ET0L

Oscar H. Ibarra[1] and Ian McQuillan[2]([✉])

[1] Department of Computer Science, University of California,
Santa Barbara, CA 93106, USA
ibarra@cs.ucsb.edu
[2] Department of Computer Science, University of Saskatchewan,
Saskatoon, SK S7N 5A9, Canada
mcquillan@cs.usask.ca

Abstract. We show that for every trio $\mathcal{L}$ containing only semilinear languages, all bounded languages in $\mathcal{L}$ can be accepted by one-way nondeterministic reversal-bounded multicounter machines (NCM), and in fact, even by the deterministic versions of these machines (DCM). This implies that for every semilinear trio (where these properties are effective), it is possible to decide containment, equivalence, and disjointness concerning its bounded languages. We also provide a relatively simple condition for when the bounded languages in a semilinear trio coincide exactly with those accepted by DCM machines. This is applied to finite-index ET0L systems, where we show that the bounded languages generated by these systems are exactly the bounded languages accepted by DCM. We also define, compare, and characterize several other types of languages that are both bounded and semilinear.

1 Introduction

The notions of bounded languages and semilinear sets and languages are old ones in the area of formal languages (see e.g. [5]), and they have been used and applied extensively. A language $L \subseteq \Sigma^*$ is bounded if there exist words, $w_1, \ldots, w_k \in \Sigma^+$, such that $L \subseteq w_1^* \cdots w_k^*$. The formal definition of semilinear sets $Q \subseteq \mathbb{N}_0^k$ appears in Sect. 2. A language is semilinear if its Parikh map is a semilinear set. Many well-studied language families, such as the context-free languages, and one-way nondeterministic reversal-bounded multicounter languages, only contain languages that are semilinear.

There are various ways of combining these definitions, and in this paper, we consider four. In particular, a language $L \subseteq \Sigma^*$ has been called *bounded semilinear* if there exists a semilinear set $Q \subseteq \mathbb{N}_0^k$ such that $L = \{w \mid w = w_1^{i_1} \cdots w_k^{i_k}, (i_1, \ldots, i_k) \in Q\}$ [11]. In this paper, we refer to these bounded semilinear languages as *bounded Ginsburg semilinear* to disambiguate with other

The research of O.H. Ibarra was supported, in part, by NSF Grant CCF-1117708.
The research of I. McQuillan was supported, in part, by Natural Sciences and Engineering Research Council of Canada Grant 327486-2010.

© Springer International Publishing Switzerland 2016
Y.-S. Han and K. Salomaa (Eds.): CIAA 2016, LNCS 9705, pp. 138–149, 2016.
DOI: 10.1007/978-3-319-40946-7_12

types. Similarly, here we define a language L to be called *bounded Parikh semilinear* if there exists a semilinear set Q such that $L = \{w \mid w = w_1^{i_1} \cdots w_k^{i_k},$ the Parikh map of w is in $Q\}$. A *bounded Ginsburg-Parikh semilinear* language is defined via two semilinear sets as a combination of the bounded Ginsburg and bounded Parikh semilinear concepts. Lastly, we define a language L to be *bounded general semilinear* if L is a bounded language and there is a semilinear set Q such that the Parikh map of L is Q.

It is already known that every bounded Ginsburg semilinear language can be accepted by a one-way nondeterministic reversal-bounded multicounter machine (NCM) [10]. Furthermore, it is known that every NCM machine accepting a bounded language can be converted to a deterministic machine (DCM) accepting the same language [11].

Here, we compare all four notions, and show that bounded Parikh semilinear languages are a strict subset of bounded Ginsburg semilinear languages, which are equal to bounded Ginsburg-Parikh semilinear languages, which are a strict subset of bounded general semilinear languages. In fact, we show that every family whose languages are recursively enumerable cannot contain every bounded general semilinear language (in contrast to the fact that all bounded Ginsburg semilinear languages are in NCM which are all recursive languages). However, we show that for every semilinear trio $\mathcal{L}$ (a family where all languages are semilinear, and is closed under λ-free homomorphism, inverse homomorphism, and intersection with regular languages), every bounded language in $\mathcal{L}$ is bounded Ginsburg semilinear. Hence, all bounded languages in $\mathcal{L}$ are in NCM, and hence also in DCM. This immediately provides several nice algorithmic results for all bounded languages in semilinear trios (where all these properties are effective) such as the ability to test equality, containment, and disjointness, since these are decidable for DCM. Examples of such language families are the context-free languages, finite-index ET0L [12], linear indexed languages [3], multi-push-down languages [2], and many others [7]. We also develop a criterion for when bounded languages within a semilinear trio coincide exactly with those in NCM and DCM; this occurs exactly when the family contains all distinct-letter-bounded Ginsburg semilinear languages (i.e. all bounded Ginsburg semilinear languages over words that are all distinct letters).

We apply our results to L systems to show that the bounded languages in finite-index ET0L coincide exactly with those in NCM and DCM. This is interesting given how different the two types of systems operate. Indeed, NCM is a sequential machine model that operates with multiple independent stores, and the other is a grammar system where rules are applied in parallel. The restriction of *finite-index* on different types of grammar systems enforces that there is an integer k such that, for every word in the language, there is a derivation that uses at most k nonterminals in every sentential form (in particular, a bounded number of active symbols for the case of ET0L). It is known that the family of finite-index ET0L languages coincides with many other families of languages generated by various types of grammar systems with the finite-index condition, such as EDT0L, context-free programmed grammars, ordered grammars, and matrix grammars [13]. Therefore, the bounded languages in these families are equal to those in NCM and DCM as well, and this provides a deterministic machine model accepting exactly the bounded languages in each of these families.

2 Preliminaries

We assume a familiarity with automata and formal languages. We will fix the notation used in this paper. Let Σ be a finite alphabet. Then Σ^* (respectively Σ^+) is the set of all words (non-empty words) over Σ. A word w is any element of Σ^*, while a language is any $L \subseteq \Sigma^*$. The empty word is denoted by λ. Given a language $L \subseteq \Sigma^*$, the complement of L with respect to Σ, $\Sigma^* - L$ is denoted by $\overline{L}$. A language $L \subseteq \Sigma^*$ is *bounded* if there exists (not necessarily distinct) words $w_1, \ldots, w_k$ such that $L \subseteq w_1^* \cdots w_k^*$. L is *letter-bounded* if there exists (not necessarily distinct) letters $a_1, \ldots, a_k$ such that $L \subseteq a_1^* \cdots a_k^*$. If $a_1, \ldots, a_k$ are distinct, then we say L is *distinct-letter-bounded*. Given a language family $\mathcal{L}$, the subset of $\mathcal{L}$ consisting of all bounded languages in $\mathcal{L}$, is $\mathcal{L}^{\text{bd}}$.

Let $\mathbb{N}$ be the set of positive integers, and $\mathbb{N}_0$ the set of non-negative integers. Let $m \in \mathbb{N}_0$. Then, $\pi(m)$ is 1 if $m > 0$ and 0 otherwise. A subset Q of $\mathbb{N}_0^m$ (m-tuples) is a *linear set* if there exist vectors $\boldsymbol{v_0}, \boldsymbol{v_1}, \ldots, \boldsymbol{v_n} \in \mathbb{N}_0^m$ such that $Q = \{\boldsymbol{v_0} + i_1\boldsymbol{v_1} + \cdots + i_n\boldsymbol{v_n} \mid i_1, \ldots, i_n \in \mathbb{N}_0\}$. Then $\boldsymbol{v_0}$ is called the *constant*, and $\boldsymbol{v_1}, \ldots, \boldsymbol{v_n}$ are the *periods*. A finite union of linear sets is called a *semilinear set*.

Let $\Sigma = \{a_1, \ldots, a_m\}$ be an alphabet. The length of a word $w \in \Sigma^*$ is denoted by $|w|$. For $a \in \Sigma$, $|w|_a$ is the number of a's in w, and for any subset X of Σ, $|w|_X = \sum_{a \in X} |w|_a$. The *Parikh map* of w is the vector $\psi(w) = (|w|_{a_1}, \ldots, |w|_{a_m})$, which is extended to languages, by $\psi(L) = \{\psi(w) \mid w \in L\}$. A language is *semilinear* if its Parikh map is a semilinear set. It is known that a language L is semilinear if and only if it has the same Parikh map as some regular language [8]. Furthermore, a language family is called *semilinear* if all the languages in the family are semilinear.

A *one-way k-counter machine* [10] is a tuple $M = (k, Q, \Sigma, \triangleleft, \delta, q_0, F)$, where $Q, \Sigma, \triangleleft, q_0, F$ are respectively the finite set of states, input alphabet, right input end-marker (this is needed for deterministic machines but not nondeterministic machines [9]), initial state (in Q), and accepting states (a subset of Q). The transition function δ is a relation from $Q \times (\Sigma \cup \{\triangleleft, \lambda\}) \times \{0, 1\}^k$ to $Q \times \{-1, 0, +1\}^k$, such that if $\delta(q, a, c_1, \ldots, c_k)$ contains $(p, d_1, \ldots, d_k)$ and $c_i = 0$ for some i, then $d_i \geq 0$ (to prevent negative values in any counter). Then M is deterministic if $|\delta(q, a, i_1, \ldots, i_k) \cup \delta(q, \lambda, i_1, \ldots, i_k)| \leq 1$, for all $q \in Q, a \in \Sigma \cup \{\triangleleft\}, (i_1, \ldots, i_k) \in \{0, 1\}^k$. A configuration of M is a $k + 2$-tuple $(q, w, c_1, \ldots, c_k)$ representing that M is in state q, $w \in \Sigma^* \triangleleft \cup \{\lambda\}$ is still to be read as input, and $c_1, \ldots, c_k \in \mathbb{N}_0$ are the contents of the k counters. The derivation relation $\vdash_M$ is defined between configurations, whereby $(q, aw, c_1, \ldots, c_k) \vdash_M (p, w, c_1 + d_1, \ldots, c_k + d_k)$, if $(p, d_1, \ldots, d_k) \in \delta(q, a, \pi(c_1), \ldots, \pi(c_k))$. Let $\vdash_M^*$ be the reflexive, transitive closure of $\vdash_M$. A word $w \in \Sigma^*$ is accepted by M if $(q_0, w\triangleleft, 0, \ldots, 0) \vdash_M^* (q, \lambda, c_1, \ldots, c_k)$, for some $q \in F, c_1, \ldots, c_k \in \mathbb{N}_0$. Furthermore, M is *l-reversal-bounded* if it operates in such a way that in every accepting computation, the count on each counter alternates between increasing and decreasing at most l times.

For a class of machines $\mathcal{M}$, we use the notation $\mathcal{L}(\mathcal{M})$ to denote the family of languages accepted by machines in $\mathcal{M}$. A language family $\mathcal{L}$ is a *trio* [4] if it is closed under λ-free homomorphism, inverse homomorphism, and intersection with regular languages.

The class of k-counter l-reversal-bounded machines is denoted by $\mathsf{NCM}(k, l)$, and NCM is the class of all reversal-bounded multicounter machines. Similarly, the deterministic variant is denoted by $\mathsf{DCM}(k, l)$ and DCM.

An ET0L system [12] is a tuple $G = (V, \mathcal{P}, S, \Sigma)$, where V is a finite alphabet, $\Sigma \subseteq V$ is the terminal alphabet, $S \in V$ is the axiom, and $\mathcal{P}$ is a finite set of production tables, where each $P \in \mathcal{P}$ is a finite binary relation in $V \times V^*$. It is typically assumed that for all production tables P and each variable $X \in V$, $(X, \alpha) \in P$ for some $\alpha \in V^*$. If $(X, \alpha) \in P$, then we usually write $X \to_P \alpha$. Elements of $V - \Sigma$ are called nonterminals.

Let $x = a_1 a_2 \cdots a_n, a_i \in V, 1 \le i \le n$, and let $y \in V^*$. Then $x \Rightarrow_G y$, if there is a $P \in \mathcal{P}$ such that $y = \alpha_1 \cdots \alpha_n$ where $a_i \to \alpha_i \in P, 1 \le i \le n$. Then $\Rightarrow_G^*$ is the reflexive, transitive closure of $\Rightarrow_G$, and the language generated by G, $L(G) = \{x \in \Sigma^* \mid S \Rightarrow_G^* x\}$. A letter $X \in V$ is *active* if there exists a table $P \in \mathcal{P}$ and a word $\alpha \in V^*$ such that $X \to_P \alpha$ and $\alpha \ne X$. Then A_G are the active symbols of G. Let $k \in \mathbb{N}$. Then G is of index k if, for every word $x \in L(G)$, there exists a derivation $x_0 = S \Rightarrow_G x_1 \Rightarrow_G \cdots \Rightarrow_G x_n = x$ such that, for each $i, 0 \le i \le n, |x_i|_{A_G} \le k$. Then G is of finite index if G is of index k for some $k \ge 1$. If L is an ET0L language, then L is of index k if there exists an ET0L system G of index k such that $L(G) = L$, and L is of finite index if L is of index k for some $k \ge 1$.

The family of languages generated by ET0L systems is denoted by $\mathcal{L}(\mathsf{ET0L})$, and the ET0L languages that are of finite index are denoted by $\mathcal{L}(\mathsf{ET0L_{fin}})$.

3   Bounded Languages and Counter Machines

Next, we define four different types of languages that are both bounded and semilinear.

Definition 1. Let $\Sigma = \{a_1, \ldots, a_n\}$, $w_1, \ldots, w_k \in \Sigma^+$, and $Q_1 \subseteq \mathbb{N}_0^k$ and $Q_2 \subseteq \mathbb{N}_0^n$ be semilinear sets.

1. If $L = \{w \mid w = w_1^{i_1} \cdots w_k^{i_k}, (i_1, \ldots, i_k) \in Q_1\}$, then L is called the *bounded Ginsburg semilinear language induced by* Q_1.
2. If $L = \{w \mid w = w_1^{i_1} \cdots w_k^{i_k}, i_1, \ldots, i_k \in \mathbb{N}_0, (|w|_{a_1}, \ldots, |w|_{a_n}) \in Q_2\}$, then L is called the *bounded Parikh semilinear language induced by* Q_2.
3. If $L = \{w \mid w = w_1^{i_1} \cdots w_k^{i_k}, (i_1, \ldots, i_k) \in Q_1, (|w|_{a_1}, \ldots, |w|_{a_n}) \in Q_2\}$, then L is called the *bounded Ginsburg-Parikh semilinear language* induced by Q_1 and Q_2.
4. If $L \subseteq w_1^* \cdots w_k^*$, and $\psi(L) = Q_2$, then L is called a *bounded general semilinear language* induced by Q_2.

Traditionally, bounded Ginsburg semilinear languages are referred to as simply bounded semilinear languages [11]. However, in this paper, we will use the term bounded Ginsburg semilinear language to disambiguate with other types. Note that a bounded Parikh semilinear language is a special case of bounded general semilinear language.

Example 1. Consider the following languages:

- Let $L_1 = \{w \mid w = (abb)^i(bab)^j(abb)^k, 0 < i < j < k\}$. Here, with the semilinear set $Q_1 = \{(i,j,k) \mid 0 < i < j < k\}$, then $L_1 = \{w \mid w = (abb)^i(bab)^j(abb)^k, (i,j,k) \in Q_1\}$, and therefore L_1 is bounded Ginsburg semilinear.
- Let $L_2 = \{w \mid w = (abb)^i(aba)^j, i, j > 0, 0 < |w|_a = |w_b|\}$. Here, using the semilinear set $Q_2 = \{(n,n) \mid 0 < n\}$, it can be seen that L_2 is bounded Parikh semilinear.
- Let $L_3 = \{w \mid w = (abbb)^i(aab)^j, 0 < i < j, 0 < |w|_a < |w_b|\}$. Using, $Q_1 = \{(i,j) \mid 0 < i < j\}$, and $Q_2 = \{(n,m) \mid 0 < n < m\}$, then L_2 is bounded Ginsburg-Parikh semilinear. Hence, both Q_1 and Q_2 help define L_3. For example, if $i = 2, j = 3$, then $w = (abbb)^2(aab)^3 \in L_3$ since $2 < 3$ and $|w|_a = 8 < |w|_b = 9$. But if $i = 2, j = 4$, then $w = (abbb)^2(aab)^4 \notin L_3$ despite $2 < 4$ since $|w|_a = 10 = |w|_b = 10$.
- Let $L_4 = \{a^{2^i}b \mid i > 0\} \cup \{ba^i \mid i > 0\}$. Then L_4 is bounded as it is a subset of $a^*b^*a^*$, and has the same Parikh map as the regular language $\{ba^i \mid i > 0\}$ and is therefore semilinear, and hence bounded general semilinear. It will become evident from the results in this paper that L_4 is not bounded Ginsburg-Parikh semilinear.

Note that given the semilinear sets and the words $w_1, \ldots, w_k$ in Definition 1, there is only one bounded Ginsburg, bounded Parikh, and bounded Ginsburg-Parikh semilinear language induced by the semilinear sets. But for bounded general semilinear languages, this is not the case, as L_4 in the example above has the same Parikh map as the regular language $\{ba^i \mid i > 0\}$.

We need the following known results:

Proposition 1. *Let* $\Sigma = \{a_1, \ldots, a_n\}$ *and* $w_1, \ldots, w_k \in \Sigma^+$.

1. [10] *If* $L \subseteq w_1^* \cdots w_k^*$ *is in* $\mathcal{L}(\mathsf{NCM})$, *then* $Q_L = \{(i_1, \ldots, i_k) \mid w_1^{i_1} \cdots w_k^{i_k} \in L\}$ *is a semilinear set (every bounded language in* $\mathcal{L}(\mathsf{NCM})$ *is bounded Ginsburg semilinear).*
2. [11] *If* $Q \subseteq \mathbb{N}_0^k$ *is a semilinear set, then* $L_Q = \{w_1^{i_1} \cdots w_k^{i_k} \mid (i_1, \ldots, i_k) \in Q\} \in \mathcal{L}(\mathsf{NCM})$ *(every bounded Ginsburg semilinear language is in* $\mathcal{L}(\mathsf{NCM})$*).*
3. [10] *If* $L \subseteq \Sigma^*$ *is in* $\mathcal{L}(\mathsf{NCM})$, *then* $\psi(L)$ *is a semilinear set.*

Proposition 2. [11] $\mathcal{L}(\mathsf{NCM})^{\mathrm{bd}} = \mathcal{L}(\mathsf{DCM})^{\mathrm{bd}}$.

Corollary 1. *The same as Proposition 1 with* NCM *replaced by a* DCM.

We will also need the following lemma, which is generally known (e.g., it can be derived from the results in [9]). We give a short proof for completeness.

Lemma 1. *Let* $\Sigma = \{a_1, \ldots, a_n\}$. *If* $Q \subseteq \mathbb{N}_0^n$ *is a semilinear set, then* $L_Q = \{w \mid w \in \Sigma^*, \psi(w) \in Q\} \in \mathcal{L}(\text{NCM})$.

Proof. Since $\mathcal{L}(\text{NCM})$ is closed under union, it is sufficient to prove the result for the case when Q is a linear set. Let $Q = \{v \mid v = \boldsymbol{v_0} + i_1\boldsymbol{v_1} + \cdots + i_r\boldsymbol{v_r}$, each $i_j \in \mathbb{N}\}$, where $\boldsymbol{v_0} = (v_{01}, \ldots, v_{0n})$ is the constant and $\boldsymbol{v_j} = (v_{j1}, \ldots, v_{jn})$ $(1 \leq j \leq r)$ are the periods. We construct an NCM M with counters $C_1, \ldots, C_n$ which, when given input $w \in \Sigma^*$, operates as follows:

1. M reads w and stores $|w|_{a_i}$ in counter C_i $(1 \leq i \leq n)$.
2. On λ-moves, M decrements C_i by v_{0i} for each i $(1 \leq i \leq n)$.
3. For $1 \leq j \leq r$, M, on λ-moves, decrements C_i $(1 \leq i \leq n)$ by $k_j v_{ji}$, where k_j is a nondeterministically chosen non-negative integer.
4. M accepts when all counters are zero.

Then, $L(M) = L_Q$. □

Next, we examine the relationship between bounded Ginsburg semilinear languages, bounded Parikh semilinear languages, bounded Ginsburg-Parikh semilinear languages, and bounded general semilinear languages.

To start, we need the following proposition:

Proposition 3. *Let* $\mathcal{L}$ *be any family of languages which is contained in the family of recursively enumerable languages. Then there is a bounded general semilinear language that is not in* $\mathcal{L}$.

Proof. Take any non-recursively enumerable language $L \subseteq a^*$. Let b, c be new symbols, and consider $L' = bLc \cup ca^*b$. Then L' is bounded, since it is a subset of $b^*a^*c^*a^*b^*$. Clearly, L' has the same Parikh map as the regular language ca^*b. Hence, $\psi(L') = \{(n, 1, 1) \mid n \geq 0\}$, which is semilinear. But L' cannot be recursively enumerable, otherwise by intersecting it with the regular language ba^*c, bLc would also be recursively enumerable. But bLc is recursively enumerable if and only if L is recursively enumerable. We get a contradiction, since L is not recursively enumerable. Thus, $L' \notin \mathcal{L}$. □

Therefore, there are bounded general semilinear languages that are not recursively enumerable.

Next, we compare the four types of languages.

Proposition 4. *The family of bounded Parikh semilinear languages is a strict subset of the family of bounded Ginsburg semilinear languages, which is equal to the family of bounded Ginsburg-Parikh semilinear languages, which is a strict subset of the family of bounded general semilinear languages.*

Proof. First note that every bounded Ginsburg semilinear language is a bounded Ginsburg-Parikh semilinear language by setting $Q_2 = \mathbb{N}_0^n$ (in Definition 1). Also, every bounded Parikh semilinear language is a bounded Ginsburg-Parikh semilinear language by setting $Q_1 = \mathbb{N}_0^k$. So, both the families of bounded Ginsburg semilinear languages and bounded Parikh semilinear languages are a subset of the bounded Ginsburg-Parikh semilinear languages.

Next, we will see that every bounded Parikh semilinear language is a bounded Ginsburg semilinear language. Let L be a bounded Parikh semilinear language, induced by semilinear set Q_2. Then $L = \{w \mid w = w_1^{i_1} \cdots w_k^{i_k}, i_1, \ldots, i_k \in \mathbb{N}_0, (|w|_{a_1}, \ldots, |w|_{a_n}) \in Q_2\}$. Let $Q_1 = \{(i_1, \ldots, i_k) \mid w = w_1^{i_1} \cdots w_k^{i_k} \in L\}$. If Q_1 is a semilinear set, then L is bounded Ginsburg semilinear by Proposition 1 Part 2. To see that Q_1 is semilinear: Let $L_1 = \{w \mid w \in \{a_1, \ldots, a_n\}^*, \psi(w) \in Q_2\}$. Clearly, $L = L_1 \cap w_1^* \cdots w_k^*$. Then by Lemma 1, L_1 is in $\mathcal{L}(\text{NCM})$ and since $\mathcal{L}(\text{NCM})$ is closed under intersection with regular sets [10], L is also in $\mathcal{L}(\text{NCM})$. Then, by Proposition 1 Part 1, Q_1 is a semilinear set, and we are done.

Then, notice that the bounded Ginsburg-Parikh semilinear language induced by Q_1, Q_2 is the intersection of the bounded Ginsburg semilinear set induced by Q_1, with the bounded Parikh semilinear language induced by Q_2. From the proof above, every bounded Parikh semilinear language is in fact a bounded Ginsburg semilinear language. Hence, every bounded Ginsburg-Parikh semilinear language is the intersection of two bounded Ginsburg semilinear languages. As every bounded Ginsburg semilinear language is in $\mathcal{L}(\text{NCM})$ by Proposition 1 Part 2, and $\mathcal{L}(\text{NCM})$ is closed under intersection [10], it follows that every bounded Ginsburg-Parikh semilinear language is in $\mathcal{L}(\text{NCM})$. By an application of Proposition 1 Part 1 followed by Part 2, every Ginsburg-Parikh semilinear language must therefore be a bounded Ginsburg semilinear language.

To show that bounded Parikh semilinear languages are properly contained in bounded Ginsburg languages, consider the bounded Ginsburg semilinear language $L = \{a^k b^k a^k \mid k > 0\}$ induced by semilinear set $Q_1 = \{(k, k, k) \mid k > 0\}$. Now the Parikh map of L is the semilinear set $Q_2 = \{(2k, k) \mid k > 0\}$. Thus, if the fixed words are a, b, a (whereby these are the words chosen to define the bounded language), then the bounded Parikh semilinear language induced by Q_2 is $L' = \{a^i b^k a^j \mid i + j = 2k > 0\}$, which is different from L. It is clear that this is true for all fixed words.

It suffices to show that the family of bounded Ginsburg semilinear languages is strictly contained in the family of bounded general semilinear languages. Containment can be seen as follows: Let L be a bounded Ginsburg language. We know that every bounded Ginsburg language is in $\mathcal{L}(\text{NCM})$ by Proposition 1 Part 2, and all $\mathcal{L}(\text{NCM})$ languages are semilinear by Proposition 1 Part 3. Thus L is semilinear, and L is also bounded. Hence, L is bounded general semilinear. Strictness follows from Proposition 3 and the fact that all $\mathcal{L}(\text{NCM})$ languages are recursive [10].   □

In fact, as long as a language family contains a simpler subset of bounded Ginsburg semilinear languages, and is closed under λ-free homomorphism, then it is enough to imply they contain all bounded Ginsburg semilinear languages.

Proposition 5. *Let $\mathcal{L}$ be a language family that contains all distinct-letter-bounded Ginsburg semilinear languages and is closed under λ-free homomorphism. Then $\mathcal{L}$ contains all bounded Ginsburg semilinear languages.*

Proof. Let $w_1, \ldots, w_k \in \Sigma^+$, and let $L \subseteq w_1^* \cdots w_k^*$ be a bounded Ginsburg semilinear language induced by Q_1. Let $b_1, \ldots, b_k$ be new distinct symbols. Consider the bounded Ginsburg semilinear language $L' \subseteq b_1^* \cdots b_k^*$ induced by Q_1. Then $L' \in \mathcal{L}$ by assumption. Finally, apply homomorphism h on L' defined by $h(b_i) = w_i$ for each i. Then $h(L') = L$, which must be in $\mathcal{L}$, since $\mathcal{L}$ is closed under λ-free homomorphism. $\qquad\square$

Furthermore, as long as a language family is a semilinear trio, all bounded languages in the family are bounded Ginsburg semilinear languages.

Proposition 6. *Let* $\Sigma = \{a_1, \ldots, a_n\}$, $w_1, \ldots, w_k \in \Sigma^+$, *let* $\mathcal{L}$ *be a semilinear trio, and let* $L \subseteq w_1^* \cdots w_k^* \in \mathcal{L}$. *Then there is a semilinear set* Q_1 *such that* L *is the bounded Ginsburg semilinear language induced by* Q_1.

Proof. Let $b_1, \ldots, b_k$ be new distinct symbols, and $L_1 = \{b_1^{i_1} \cdots b_k^{i_k} \mid w_1^{i_1} \cdots w_k^{i_k} \in L\}$. Then, since $\mathcal{L}$ is closed under λ-free finite transductions (every trio is closed under λ-free finite transductions [4], Sect. 3.2, Corollary 2), $L_1 \in \mathcal{L}$, as a transducer can read w_1, and output b_1 some number of times (nondeterministically chosen), followed by w_2, etc. Let Q_1 be the Parikh map of L_1, which is semilinear by assumption. It follows that L is the bounded Ginsburg semilinear language induced by Q_1. $\qquad\square$

Hence, all bounded languages in semilinear trios are "well-behaved" in the sense that they are bounded Ginsburg semilinear. For these families, bounded languages, and bounded Ginsburg semilinear languages coincide.

Corollary 2. *Let* $\mathcal{L}$ *be a semilinear trio. Then* $L \in \mathcal{L}$ *is bounded if and only if* L *is bounded Ginsburg semilinear. Hence,* $\mathcal{L}^{\mathrm{bd}} = \{L \mid L \in \mathcal{L} \text{ is bounded Ginsburg semilinear}\}$.

Note that this is not necessarily the case for non-semilinear trios. For example, the language family $\mathcal{L}(\mathsf{ET0L})$ contains the non-semilinear language $\{a^{2^n} \mid n > 0\}$ which is bounded but not semilinear. Hence, $\mathcal{L}(\mathsf{ET0L})$ contains languages that are bounded general semilinear, $\{a^{2^n} \mid n > 0\}b \cup ba^*$, but not bounded Ginsburg semilinear in a similar fashion to Proposition 3. But this cannot happen within semilinear trios.

Then, for an arbitrary semilinear trio $\mathcal{L}$, it is possible to compare all bounded languages in $\mathcal{L}$ to the set of all bounded Ginsburg semilinear languages, which are exactly the bounded languages in $\mathcal{L}(\mathsf{NCM})$.

Proposition 7. *Let* $\mathcal{L}$ *be a semilinear trio. Then* $\mathcal{L}^{\mathrm{bd}} \subseteq \mathcal{L}(\mathsf{NCM})^{\mathrm{bd}} = \mathcal{L}(\mathsf{DCM})^{\mathrm{bd}}$ *and the following conditions are equivalent:*

1. $\mathcal{L}^{\mathrm{bd}} = \mathcal{L}(\mathsf{NCM})^{\mathrm{bd}} = \mathcal{L}(\mathsf{DCM})^{\mathrm{bd}}$,
2. $\mathcal{L}$ *contains all bounded Ginsburg semilinear languages,*
3. $\mathcal{L}$ *contains all bounded Parikh semilinear languages,*
4. $\mathcal{L}$ *contains all distinct-letter-bounded Ginsburg semilinear languages.*

Proof. $\mathcal{L}(\mathsf{NCM})^{\mathrm{bd}} = \mathcal{L}(\mathsf{DCM})^{\mathrm{bd}}$ follows from Proposition 2.

Also, all distinct-letter bounded Ginsburg semilinear languages are bounded Parikh semilinear, and all bounded Parikh semilinear languages are bounded

Ginsburg semilinear languages by Proposition 4, and thus 2 implies 3 and 3 implies 4. The other direction follows from Proposition 5, and thus 4 implies 3 and 3 implies 2. Hence, 2, 3, and 4 are equivalent.

Consider any bounded language $L \subseteq w_1^* \dots w_n^* \in \mathcal{L}$. Then there is a semilinear set Q such that L is the bounded Ginsburg semilinear language induced by Q, by Corollary 2. By Proposition 1 Part 2, $L \in \mathcal{L}(\text{NCM})$.

If $\mathcal{L}$ does not contain all distinct-letter-bounded Ginsburg semilinear languages, then containment is strict, as $\mathcal{L}(\text{NCM})$ does, by Proposition 1 Part 2. Otherwise, if $\mathcal{L}$ does contain all distinct-letter-bounded Ginsburg semilinear languages, then it contains all bounded Ginsburg semilinear languages by Proposition 5, and then by Proposition 1, all bounded languages in $\mathcal{L}(\text{NCM})$ are in $\mathcal{L}$. Hence, 4 is equivalent to 1. □

Lastly, an important note is that for any bounded language L in any semilinear trio $\mathcal{L}$, where the trio properties are effective, and the family is effectively semilinear (i.e. the constant and period vectors of each linear set can be constructed), it is possible to effectively construct a DCM machine accepting L.

Proposition 8. *Let $\mathcal{L}$ be any language family that is effectively closed under the trio operations, and is effectively semilinear. Then, for each bounded language $L \in \mathcal{L}$, it is possible to build a DCM machine accepting L.*

This provides a **deterministic** machine model to accept all bounded languages from these language families defined by nondeterministic machines and grammars. Moreover, DCM machines have many decidable properties, allowing for algorithms to be used on them.

Corollary 3. *Let $\mathcal{L}_1$ and $\mathcal{L}_2$ be two language families effectively closed under the trio operations, and effectively semilinear. Then it is decidable, for $L_1 \in \mathcal{L}_1^{\text{bd}}$, and $L_2 \in \mathcal{L}_2^{\text{bd}}$, whether $L_1 \subseteq L_2$, whether $L_1 = L_2$, and whether $L_1 \cap L_2 \neq \emptyset$.*

Proof. This follows since every bounded language within both language families are in $\mathcal{L}(\text{DCM})$ (effectively) by Proposition 8, and containment, equality, and disjointness are decidable for $\mathcal{L}(\text{DCM})$ [10]. □

4   Application to Finite-Index ETOL

It is known that the family of finite-index ETOL languages is a semilinear trio [12], and therefore all bounded languages in it are DCM languages, by Proposition 8. We will show that the bounded languages in the two families are identical.

Lemma 2. *Let $a_1, \dots, a_k$ be distinct symbols, and $Q \subseteq \mathbb{N}_0^k$ be a semilinear set. Then $L = \{a_1^{i_1} \cdots a_k^{i_k} \mid (i_1, \dots, i_k) \in Q\} \in \mathcal{L}(\text{ETOL}_{\text{fin}})$.*

Proof. Let L be a letter-bounded Ginsburg semilinear language of the form above, and let $\Sigma = \{a_1, \dots, a_k\}$. Then $\psi(L)$ is a finite union of linear sets. Consider each of the linear sets, Q, where there exists $v_0, v_1, \dots, v_n \in \mathbb{N}_0^k$ (v_0 the constant, the rest the periods) with $Q = \{v_0 + i_1 v_1 + \cdots + i_n v_n \mid i_1, \ldots, i_n \in \mathbb{N}_0\}$. Assume that $n \geq 1$, otherwise the set is finite, where the case is obvious.

We create an ET0L system $G_Q = (V, \mathcal{P}, S, \Sigma)$ as follows: $\mathcal{P} = \{P_0, P_1\}$, $V = \Sigma \cup \{Z\} \cup \{X_{i,j} \mid 1 \leq i \leq k, 1 \leq j \leq n\}$, and the productions are:

1. Add $S \to_{P_1} a_1^{v_0(1)} X_{1,1} a_2^{v_0(2)} X_{2,1} \cdots a_k^{v_0(k)} X_{k,1}$ and $S \to_{P_0} Z$.
2. For all $X_{i,j} \in V$, add $X_{i,j} \to_{P_0} a_i^{v_j(i)} X_{i,j}$.
3. For all $X_{i,j} \in V, 1 \leq j < n$, add $X_{i,j} \to_{P_1} X_{i,j+1}$.
4. For all $X_{i,n} \in V$, add $X_{i,n} \to_{P_1} \lambda$.
5. $a \to_P a$ is a production for every $a \in \Sigma \cup \{Z\}$, and $P \in \mathcal{P}$.

Claim. $L(G_Q) = \{a_1^{l_1} \cdots a_k^{l_k} \mid (l_1, \ldots, l_k) \in Q\}$, and G_Q is of index k.

Proof. "$\subseteq$" Let $w \in L(G_Q)$. Thus, there exists $S \Rightarrow_{Q_1} x_1 \Rightarrow_{Q_2} \cdots \Rightarrow_{Q_m} x_m = w \in \Sigma^*$, $Q_l \in \{P_0, P_1\}$, $1 \leq l \leq m$. Then $Q_1 Q_2 \cdots Q_m$ must be of the form

$$P_1 P_0^{i_1} P_1 P_0^{i_2} P_1 \cdots P_0^{i_n} P_1,$$

where $i_j \in \mathbb{N}_0$, by the construction. We will show by induction that, for all $0 \leq j < n$, $x_{i_1 + \cdots + i_j + j + 1}$ (this is the sentential form after the $(j+1)$st application of the production table P_1) is equal to

$$a_1^{v_0(1) + i_1 v_1(1) + \cdots + i_j v_j(1)} X_{1,j+1} a_2^{v_0(2) + i_1 v_1(2) + \cdots + i_j v_j(2)} X_{2,j+1} \cdots \\ a_k^{v_0(k) + i_1 v_1(k) + \cdots + i_j v_j(k)} X_{k,j+1}, \tag{1}$$

and for $j = n$, it is

$$a_1^{v_0(1) + i_1 v_1(1) + \cdots + i_n v_n(1)} a_2^{v_0(2) + i_1 v_1(2) + \cdots + i_n v_n(2)} \cdots a_k^{v_0(k) + i_1 v_1(k) + \cdots + i_n v_n(k)}.$$

The base case, $j = 0$, follows since $x_1 = a_1^{v_0(1)} X_{1,1} a_2^{v_0(2)} X_{2,1} \cdots a_k^{v_0(k)} X_{k,1}$ using the production of type 1.

Let $0 \leq j < n$ and assume that $x_{i_1 + \cdots + i_j + j + 1}$ is equal to the string in Eq. (1). Then, productions created in step 2 must get applied i_{j+1} times, followed by one application created in step 3 if $j + 1 < n$, or one application created in step 4 if $j + 1 = n$. Then it is clear that the statement holds for $j + 1$ as well.

It is also immediate that every sentential form in G_Q has at most k active symbols, and therefore it is of index k.

"$\supseteq$" Let $w = a_1^{l_1} \cdots a_k^{l_k}$, with $(l_1, \ldots, l_k) \in Q$. Then $(l_1, \ldots, l_k) = v_0 + i_1 v_1 + \cdots + i_n v_n$, for some $i_1, \ldots, i_n \in \mathbb{N}_0$. Then, by applying a production table sequence of the form $P_1 P_0^{i_1} P_1 \cdots P_0^{i_k} P_1$, this changes the derivation as follows:

$$S \Rightarrow a_1^{v_0(1)} X_{1,1} a_2^{v_0(2)} X_{2,1} \cdots a_k^{v_0(k)} X_{k,1}$$
$$\Rightarrow^* a_1^{v_0(1) + i_1 v_1(1)} X_{1,2} a_2^{v_0(2) + i_1 v_1(2)} X_{2,2} \cdots a_k^{v_0(k) + i_1 v_1(k)} X_{k,2}$$
$$\Rightarrow^* a_1^{v_0(1) + i_1 v_1(1) + \cdots + i_n v_n(1)} a_2^{v_0(2) + i_1 v_1(2) + \cdots + i_n v_n(2)} \cdots a_k^{v_0(k) + i_1 v_1(k) + \cdots + i_n v_n(k)}$$
$$= a_1^{l_1} \cdots a_k^{l_k}.$$

Hence G_Q can generate all strings in $\{a_1^{l_1} \cdots a_k^{l_k} \mid (l_1, \ldots, l_k) \in Q\}$. Since L is semilinear, then it is the finite union of linear sets. Then L can be generated in this manner since k-index ETOL is closed under union [12].   □

Next, finite-index ETOL languages coincides with languages accepted by other types of finite-index grammars, such as EDTOL, context-free programmed grammars (denoted by CFP), ordered grammars (denoted by O), and matrix grammars (denoted by M) (with the 'fin' subscript used for each family) [13].

Proposition 9. *The bounded languages in the following families coincide,*

- $\mathcal{L}(NCM)$,
- $\mathcal{L}(DCM)$,
- $\mathcal{L}(ETOL_{fin}) = \mathcal{L}(EDTOL_{fin}) = \mathcal{L}(CFP_{fin}) = \mathcal{L}(O_{fin}) = \mathcal{L}(M_{fin})$,
- *the family of bounded Ginsburg semilinear languages.*

Proof. ETOL$_{fin}$ coincides with languages generated by all the other grammar systems of finite-index [13], and so it follows that the bounded languages within each coincide as well. The rest follows from Proposition 7 and Lemma 2.   □

From Proposition 9, we know the bounded languages within NCM and ETOL$_{fin}$ coincide (which are strictly included in the bounded languages within ETOL as the non-semilinear language $\{a^{2^n} \mid n \geq 0\}$ is in ETOL). Next, we will address the relationship between NCM and ETOL$_{fin}$ (over non-bounded languages).

We observe that there are $\mathcal{L}(ETOL_{fin})$ languages that are not in $\mathcal{L}(NCM)$.

Lemma 3. *There exists a language* $L \in \mathcal{L}(EDTOL_{fin}) - \mathcal{L}(NCM)$.

Proof. Consider $L = \{x\#x \mid x \in \{a, b\}^+\}$. It is easy to construct an ETOL system of finite index to generate L. We will show that L cannot be accepted by any NCM.

It was shown in [1] that for any NCM M, there is a constant c (which depends only on M) such that if w is accepted by M, then w is accepted by M within cn steps, where $n = |w|$. So suppose L is accepted by M. Consider a string $x\#x$, where $n = |x| \geq 1$. Then M's input head will reach $\#$ within cn steps. If M has k counters, the number of configurations (state and counter values) when M reaches $\#$ is $O(s(cn)^k)$, where s is the number of states (as each counter can grow to at most cn in cn moves). Since there are 2^n strings of the form $x\#x$, where $x \in \{a, b\}^+$ and $|x| = n$, it would follow that for large enough n, there are distinct strings x and y of length n such that $x\#y$ would be accepted by M. This is a contradiction. Hence L cannot be accepted by any NCM.   □

We leave as open problem whether there are languages in $\mathcal{L}(NCM)$ that are not in EDTOL$_{fin}$. We conjecture that over the alphabet $\Sigma_k = \{a_1, \ldots, a_k\}$, the language $\{w \mid |w|_{a_1} = \cdots = |w|_{a_k}\}$ is not in $\mathcal{L}(EDTOL_{fin})$. A candidate witness language that we initially thought of is the one-sided Dyck language on one letter which is not in ETOL$_{fin}$ [14]. However, this language cannot be accepted by any blind counter machine, which is equivalent to an NCM [6].

5   Conclusions and Open Problems

Here, different restrictions of languages that are bounded and semilinear are defined, and their capacity is studied. In particular, bounded Ginsburg semilinear languages are particularly important as they are exactly the bounded languages accepted by both NCM and DCM, and are also a superset of the bounded languages within every semilinear trio (such as the context-free languages). Thus, they can all be accepted by DCM machines. This provides several decidability properties via DCM machines. We also provide a property that can be used to show that the bounded languages in semilinear trios are identical to those accepted by DCM, and show that this holds for finite-index ET0L.

There are several open questions, such as whether there is an NCM language that cannot be accepted by finite-index ET0L. Also, an investigation of bounded languages in known semilinear trios would be of interest. In addition, there is no known characterization such as Proposition 7 for non-semilinear families.

References

1. Baker, B.S., Book, R.V.: Reversal-bounded multipushdown machines. J. Comput. Syst. Sci. **8**(3), 315–332 (1974)
2. Breveglieri, L., Cherubini, A., Citrini, C., Reghizzi, S.: Multi-push-down languages and grammars. Int. J. Found. Comput. Sci. **7**(3), 253–291 (1996)
3. Duske, J., Parchmann, R.: Linear indexed languages. Theoret. Comput. Sci. **32**(1–2), 47–60 (1984)
4. Ginsburg, S.: Algebraic and Automata-Theoretic Properties of Formal Languages. North-Holland Publishing Company, Amsterdam (1975)
5. Ginsburg, S.: The Mathematical Theory of Context-Free Languages. McGraw-Hill Inc., New York (1966)
6. Greibach, S.: Remarks on blind and partially blind one-way multicounter machines. Theoret. Comput. Sci. **7**, 311–324 (1978)
7. Harju, T., Ibarra, O., Karhumäki, J., Salomaa, A.: Some decision problems concerning semilinearity and commutation. J. Comput. Syst. Sci. **65**(2), 278–294 (2002)
8. Harrison, M.: Introduction to Formal Language Theory. Addison-Wesley Series in Computer Science. Addison-Wesley Pub. Co., Boston (1978)
9. Ibarra, O., McQuillan, I.: The effect of end-markers on counter machines and commutativity. Theoret. Comput. Sci. **627**, 71–81 (2016)
10. Ibarra, O.H.: Reversal-bounded multicounter machines and their decision problems. J. ACM **25**(1), 116–133 (1978)
11. Ibarra, O.H., Seki, S.: Characterizations of bounded semilinear languages by one-way and two-way deterministic machines. Int. J. Found. Comput. Sci. **23**(6), 1291–1306 (2012)
12. Rozenberg, G., Vermeir, D.: On ET0L systems of finite index. Inf. Control **38**, 103–133 (1978)
13. Rozenberg, G., Vermeir, D.: On the effect of the finite index restriction on several families of grammars. Inf. Control **39**, 284–302 (1978)
14. Rozoy, B.: The Dyck language $D_1^{\prime *}$ is not generated by any matrix grammar of finite index. Inf. Comput. **74**(1), 64–89 (1987)

Kuratowski Algebras Generated by Prefix-Free Languages

Jozef Jirásek Jr. and Juraj Šebej[(⊠)]

Institute of Computer Science, Faculty of Science,
P. J. Šafárik University, Jesenná 5, 040 01 Košice, Slovakia
jirasekjozef@gmail.com, juraj.sebej@gmail.com

Abstract. We study Kuratowski algebras generated by prefix-free languages under the operations of star and complement. Our results are as follows. Five of 12 possible algebras cannot be generated by any prefix-free language. Two algebras are generated only by trivial prefix-free languages, the empty set and the language $\{\varepsilon\}$. Each of the remaining five algebras can be generated, for every $n \geq 4$, by a regular prefix-free language of state complexity n, which meets the upper bounds on the state complexities of all the languages in the resulting algebra.

1 Introduction

A language is prefix-free if it does not contain two distinct strings such that one is a prefix of the other. Motivated by prefix codes, the class of prefix-free regular languages has been recently investigated [3,5,6,12,14].

It is known that a minimal deterministic finite automaton (DFA) recognizes a prefix-free language if it has exactly one final state, from which only the empty string is accepted. Using this characterization, tight upper bounds on the deterministic and nondeterministic state complexity of basic regular operations on prefix-free languages have been obtained in [5,6,14].

In particular, if a prefix-free language is accepted by an n-state DFA, then the star of this language is accepted by a DFA of at most n states. In the general case of regular languages the tight upper bound for the star operation is $3/4 \cdot 2^n$. The simplicity of the star operation on prefix-free languages has been used several times in the literature to get tight upper bounds for such operations as cyclic shift [8], boundary [9], complement-star [3] and star-complement-star [15].

In this paper we continue this research and study Kuratowski algebras generated by prefix-free languages. The famous Kuratowski's 14-theorem states that, in a topological space, at most 14 sets can be produced by applying the operations of closure and complement to a given set [4,13]. Kuratowski's theorem in the setting of formal languages has been studied by Brzozowski, Grant, and Shallit in [2]. It has been shown that at most 14 languages can be produced by repeatedly applying the operations of Kleene closure and complement to a given

J. Šebej was supported by the Slovak Grant Agency for Science under contracts VEGA 1/0142/15 and VEGA 2/0084/15.

Y.-S. Han and K. Salomaa (Eds.): CIAA 2016, LNCS 9705, pp. 150–162, 2016.
DOI: 10.1007/978-3-319-40946-7_13

language. All formal languages have been classified according to the structure of the algebras they generate under star and complement. It has been proved that there are precisely 12 such algebras, and each of them is generated by a regular language.

We inspect these 12 algebras in detail, and ask which of them can be generated by a prefix-free language (whether regular or not). If an algebra is generated by a prefix-free language, then our next question is whether it can be generated by a regular prefix-free language of an arbitrarily large state complexity. Then we ask what are the upper bounds on state complexities of languages in a particular algebra, and whether or not this algebra can be generated by a language accepted by an n-state DFA meeting these upper bounds for all the languages in the resulting algebra.

Our results are as follows. Five algebras, those in cases (2a), (3a), (3b), (4), and (5) in Table 2 [2, p.312], cannot be generated by any prefix-free language. Two algebras, those in cases (1a) and (1b), are generated only by trivial prefix-free languages $\{\varepsilon\}$ and the empty language, respectively. Any other algebra can be generated by an n-state DFA prefix-free language that meets the upper bounds on the state complexities of all the languages in the generated algebra. To get these results, we use known results concerning the state complexity of the operations of star, complement-star, and star-complement-star [3,5,15], and a careful inspection of automata for these operations.

2   Preliminaries

We assume that the reader is familiar with basic concepts of regular languages and finite automata. For details, we refer to [7,16,17].

If Σ is a finite alphabet, then Σ^* is the set of strings over Σ, including the empty string ε. A language is any subset of Σ^*. The complement of a language L is the language $L^c = \Sigma^* \setminus L$. The concatenation of languages K and L is the language $KL = \{uv \mid u \in K \text{ and } v \in L\}$. The Kleene closure, or star, of L is defined as $L^* = \cup_{i \geq 0} L^i$, while the positive closure of L is $L^+ = \cup_{i \geq 1} L^i$, where $L^0 = \{\varepsilon\}$ and $L^{i+1} = L^i L$. To simplify the explanation, we use an exponent notation, so for example, L^{c*} and L^{*c*} stand for $(L^c)^*$ and $((L^*)^c)^*$, respectively.

A language is positive-closed if it is closed under positive closure. It is positive-open, if its complement is positive-closed, and it is clopen if it is both closed and open. The terms Kleene-closed and Kleene-open are defined analogously. The positive interior of a language L is $L^\oplus = L^{c+c}$; The Kleene interior is $L^\circledast = L^{c*c}$. Notice that L is positive-open if $L = L^\oplus$. Next, L^+ is closed and $L^\oplus$ is open. We will use the following observation several times.

Lemma 1. *The language $L^\oplus$ contains those strings of L that cannot be expressed as a concatenation of strings of L^c.*

Proof. Recall that $L^\oplus = L^{c+c}$. We have $L^c \subseteq L^{c+}$, and therefore $L^{c+c} \subseteq L$. The language L^{c+} contains all strings w such that there exists a partition $w = w_1 w_2 \cdots w_k$ with $w_i \in L^c$. Hence in L^{c+c} we have all the strings in L that cannot be partitioned as $w = w_1 w_2 \cdots w_k$ with $w_i \in L^c$.   □

Let $B(L)$ be the family of all languages generated from L by positive closure and positive interior; see [2, Subsect. 4.1]. Let $D(L)$ be the family of all languages generated from L by complementation and Kleene closure. Let $E(L)$ be the family of all languages generated from L by Kleene closure and Kleene interior. It is shown in [2, Lemma 20] that $D(L) = E(L) \cup \{M \mid M^c \in E(L)\}$. Moreover, if L is neither open nor closed, then $E(L) = \{L\} \cup \{M \cup \{\varepsilon\} \mid M \in B(L)$ and M is closed$\} \cup \{M \setminus \{\varepsilon\} \mid M \in B(L)$ and $M$is open$\}$ [2, Lemma 22].

For each language L, the family $D(L)$ has at most 14 distinct languages, and Table 2 in [2, p. 312] describes 12 possible algebras, each of which is generated by a regular language. Notice that there is an oversight in cases (2a) and (2b): In case (2a) we should have $\varepsilon \notin L$, $|E(L)| = 3|$, $|D(L)| = 6|$, and it is generated by $\{a\}$. In case (2b) we should have $\varepsilon \in L$, $|E(L)| = 4|$, $|D(L)| = 8|$, and it is generated by $\{\varepsilon, a\}$ [1]. The corrected table is shown below.

Table 1. Classification of formal languages by the structure of $(E(L),^*,^{\circledast})$.

Case	Necessary and Sufficient Conditions	$|E(L)|$	$|D(L)|$	Example
(1a)	L is clopen; $\varepsilon \in L$	2	4	a^*
(1b)	L is clopen; $\varepsilon \notin L$	2	4	a^+
(2a)	L is open but not clopen; $\varepsilon \notin L$	3	6	a
(2b)	L is open but not clopen; $\varepsilon \in L$	4	8	$a \cup \varepsilon$
(3a)	L is closed but not clopen; $\varepsilon \notin L$	3	6	aaa^*
(3b)	L is closed but not clopen; $\varepsilon \in L$	4	8	$aaa^* \cup \varepsilon$
(4)	L is neither open nor closed; L^+ is clopen and $L^{\oplus+} = L^+$	4	8	$a \cup aaa$
(5)	L is neither open nor closed; $L^{\oplus}$ is clopen and $L^{+\oplus} = L^{\oplus}$	4	8	aa
(6)	L is neither open nor closed; L^+ is open but $L^{\oplus}$ is not closed; $L^{\oplus+} \neq L^+$	6	12	$G := a \cup abaa$
(7)	L is neither open nor closed; $L^{\oplus}$ is closed but L^+ is not open; $L^{+\oplus} \neq L^{\oplus}$	6	12	$(a \cup b)^+ \setminus G$
(8)	L is neither open nor closed; $L^{\oplus}$ is not closed and L^+ is not open; $L^{+\oplus} = L^{\oplus+}$	5	10	$a \cup bb$
(9)	L is neither open nor closed; $L^{\oplus}$ is not closed and L^+ is not open; $L^{+\oplus} \neq L^{\oplus+}$	7	14	$a \cup ab \cup bb$

A language is prefix-free if it does not contain two distinct strings, one of which is a prefix of the other. It is known that a minimal DFA A accepts a prefix-free language if and only if A has exactly one final state, from which all the transitions go to the dead state, that is, to a non-final state from which no string is accepted. In what follows we will assume that a regular prefix-free language L with state complexity n is accepted by a minimal DFA $A = (\{s, 1, 2, \ldots, n-3, q_f, q_d\}, \Sigma, \cdot, s, \{q_f\})$ in which q_d is the dead state. Recall that the state complexity of a regular language L, $sc(L)$, is the number of states in the minimal DFA for L. It is well known that a DFA is minimal if all its states are reachable and pairwise distinguishable; recall that a state q is reachable if there exists a string w in Σ^* such that $q = s \cdot w$, and two states p and q are distinguishable if there exists a string w such that exactly one of the states $p \cdot w$ and $q \cdot w$ is final. Throughout the paper, we will use the following observations.

Proposition 2 [6,14]. *Let A be a minimal DFA accepting a prefix-free language L. Then L^+ is accepted by the DFA A^+ obtained from A by replacing each transition (q_f, a, q_d) by the transition $(q_f, a, s \cdot a)$. Moreover, if we make the state q_f to be a unique initial state of A^*, then the resulting DFA A^* accepts L^*.*   □

Proposition 3 [11]. *Let L be a prefix-free language accepted by a minimal n-state DFA A. Then $n - 1 \le sc(L^+) \le n$ and $n - 2 \le sc(L^*) \le n$. Next, if the dead state can be reached from a state in $\{s, 1, 2, \ldots, n - 3\}$, then $sc(L^+) = n$. If, moreover, the initial state s has an in-transition, then $sc(L^*) = n$.*   □

Lemma 4 [15, **Lemma 3,4**]. *Let L be a prefix-free language with $sc(L) = n$. Then $sc(L^{*c*}) \le 2^{n-3}+2$, and this bound is met by the binary prefix-free language accepted by the DFA shown in Fig. 1.*   □

3   Constructions

Let $A = (\{s, 1, 2, \ldots, n - 3, q_f, q_d\}, \Sigma, \cdot, s, \{q_f\})$ be a minimal DFA for a prefix-free language L, in which q_d is the dead state. We will use the following automata constructed from the DFA A as described below.

NFA N_1 for L^{+c+}:
(a) Replace each transition (q_f, a, q_d) in A with the transition $(q_f, a, s \cdot a)$.
(b) Interchange the final and non-final states in the DFA A.
(c) Add ε-transitions from each state in $\{1, 2, \ldots, n - 3\}$ to the initial state s.

DFA D_1 for L^{+c+}:
Apply the subset construction to NFA N_1 to get a DFA D_1 for L^{+c+}.

DFA D_1' for L^{+c+c}:
Interchange the final and non-final states in the DFA D_1.

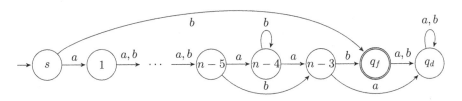

Fig. 1. The binary prefix-free witness language L with $sc(L^{*c*}) = 2^{n-3} + 2$.

NFA N_2 for L^{c+}:
(a) Interchange the final and non-final states of A.
(b) Add ε-transitions from each state in $\{1, 2, \ldots, n-3\}$ to the initial state s.

DFA D_2 for L^{c+}:
Apply the subset construction to the NFA N_2 to get a DFA D_2 for L^{c+}.

DFA D_2' for L^{c+c+}:
(a) Interchange the final and non-final states in the DFA D_2.
(b) Replace each transition $(\{q_f\}, a, \{q_d\})$ in D_2 with transition $(\{q_f\}, a, \{s\} \cdot a)$.
  Now we examine the individual algebras given in Table 1.

3.1 Cases (1a) and (1b)

We start with cases (1a) and (1b). In case (1a), a language L should be clopen with $\varepsilon \in L$; see the column Necessary and Sufficient Conditions in Table 1. The resulting algebra $B(L)$ in case (1a) is shown in Fig. 2 next to the conditions. The algebra $E(L)$ is also in this figure. In what follows we always give the necessary and sufficient conditions for a particular case in the left part of a figure, and we display the algebra $(B(L), ^+, ^\oplus)$ in its right part. Below the conditions, we give the algebra $(E(L), ^*, ^\circledast)$. To get $E(L)$, we use [2, Lemmas 21, 22].

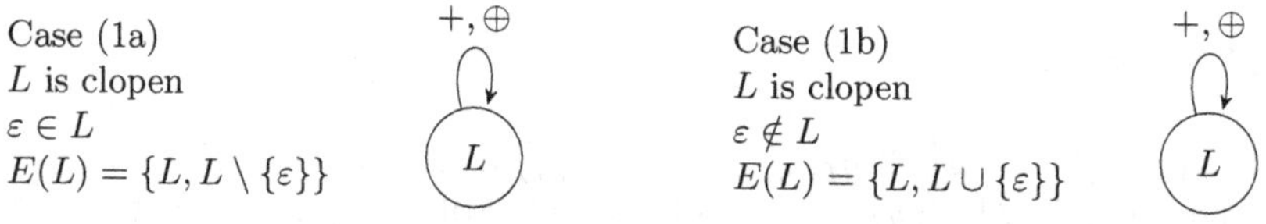

Case (1a) $+, \oplus$
L is clopen
$\varepsilon \in L$
$E(L) = \{L, L \setminus \{\varepsilon\}\}$ L

Case (1b) $+, \oplus$
L is clopen
$\varepsilon \notin L$
$E(L) = \{L, L \cup \{\varepsilon\}\}$ L

Fig. 2. The conditions and algebras $B(L)$ and $E(L)$ in cases (1a) (left) and (1b) (right).

In cases (1a) and (1b), we need clopen prefix-free languages. The next observation shows that only two prefix-free languages are closed.

Lemma 5. *Let L be a closed prefix-free language. Then either $L = \emptyset$ or $L = \{\varepsilon\}$. Both of these languages are also open.* □

  Hence since we have only two clopen prefix-free languages — the empty language and $\{\varepsilon\}$ — we get the following results.

Theorem 6 (Case (1a)). *The 4-element Kuratowski algebra in case (1a) with $E(L) = \{L, L \setminus \{\varepsilon\}\}$ is generated by the prefix-free language $\{\varepsilon\}$. No other prefix-free language generates this algebra. Hence we have $E(\{\varepsilon\}) = \{\{\varepsilon\}, \emptyset\}$, and the state complexities of languages in $E(\{\varepsilon\})$ are (2,1).* □

Theorem 7 (Case (1b)). *The 4-element Kuratowski algebra in case (1b) with $E(L) = \{L, L \cup \{\varepsilon\}\}$ is generated by the prefix-free language $\emptyset$. No other prefix-free language generates this algebra. Hence we have $E(\emptyset) = \{\emptyset, \{\varepsilon\}\}$, and the state complexities of languages in $E(\emptyset)$ are (1,2).* □

Case (2a)	Case (2b)
L is open	L is open
L is not closed	L is not closed
$\varepsilon \notin L$	$\varepsilon \in L$
$E(L) = \{L, L^*, L^+\}$	$E(L) = \{L, L \setminus \{\varepsilon\}, L^*, L^* \setminus \{\varepsilon\}\}$

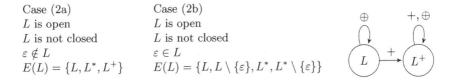

Fig. 3. The conditions and algebras $B(L)$ and $E(L)$ in cases (2a) and (2b).

3.2 Cases (2a) and (2b)

Now we consider cases (2a) and (2b) with conditions and algebras $B(L)$ and $E(L)$ as shown in Fig. 3. In the first theorem we present a prefix-free language that generates the Kuratowski algebra in case (2a). Then we continue with case (2b).

Theorem 8 (Case (2a)). *Let $n \geq 4$. There exists a ternary prefix-free language L with $\mathrm{sc}(L) = n$ which generates the 6-element Kuratowski algebra in case (2a). In addition, all the languages in $E(L) = \{L, L^*, L^+\}$ meet the upper bounds (n, n, n) on the state complexity of corresponding languages in this case.*

Proof Let L be the ternary prefix-free language accepted by the DFA shown in Fig. 5. We can show that L is open. Since the initial state s has an in-trasition on b, and the dead state is reached from $n - 3$ by a, we have $\mathrm{sc}(L^*) = n$ and $\mathrm{sc}(L^+) = n$ by Proposition 3.   □

Theorem 9 (Case (2b)). *The 8-element Kuratowski algebra in case (2b) cannot be generated by any prefix-free language.*   □

Proof. The only prefix-free language containing the empty string is $\{\varepsilon\}$ which is clopen by Lemma 5.   □

3.3 Cases (3a) and (3b)

Consider the cases (3a) and (3b) with conditions as shown in Fig. 4. Since by Lemma 5, there are no prefix-free languages which are closed but not open, we have the following result.

Theorem 10. *The Kuratowski algebras in cases (3a) and (3b) cannot be generated by prefix-free languages.*   □

Case (3a)	Case (3b)
L is closed	L is closed
L is not open	L is not open
$\varepsilon \notin L$	$\varepsilon \in L$
$E(L) = \{L, L^\circledast, L^\circledast \cup \{\varepsilon\}\}$	$E(L) = \{L, L \cup \{\varepsilon\}, L^\circledast, L^\circledast \cup \{\varepsilon\}\}$

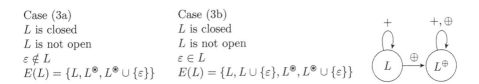

Fig. 4. The conditions and algebras $B(L)$ and $E(L)$ in cases (3a) and (3b).

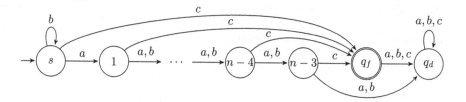

Fig. 5. The ternary witness language L for case (2a) with $\mathrm{sc}(L^*) = n$.

L is neither open nor closed
L^+ is clopen
$L^{\oplus+} = L^+$
$E(L) = \{L, L^+ \cup \{\varepsilon\}, L^+ \setminus \{\varepsilon\}, L^\oplus \setminus \{\varepsilon\}\}$

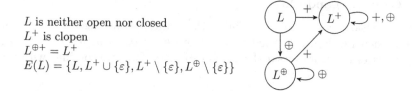

Fig. 6. The conditions and algebras $B(L)$ and $E(L)$ in case (4).

3.4 Case (4)

Our next aim is to show that the Kuratowski algebra in case (4) with conditions as shown in Fig. 6 cannot be generated by any prefix-free language. We start with the following lemma; we omit its proof due to space constraints.

Lemma 11. *Let K and L be prefix-free languages. If $K^+ = L^+$, then $K = L$.*

Theorem 12. *The Kuratowski algebra in case (4) cannot be generated by any prefix-free language.*

Proof. We show that there is no prefix-free language L such that L is not open and $L^{\oplus+} = L^+$. Let L be a prefix-free language. By Lemma 1, the language $L^\oplus$ is a subset of L, therefore $L^\oplus$ is also prefix-free. Since $L^{\oplus+} = L^+$ and both $L^\oplus$ and L are prefix-free, we have $L^\oplus = L$ by Lemma 11. It follows that L is open. This concludes the proof. ☐

L is neither open nor closed
$L^\oplus$ is clopen
$L^{+\oplus} = L^\oplus$
$E(L) = \{L, L^*, L^\oplus \cup \{\varepsilon\}, L^\oplus \setminus \{\varepsilon\}\}$

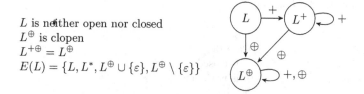

Fig. 7. The conditions and algebras $B(L)$ and $E(L)$ in case (5).

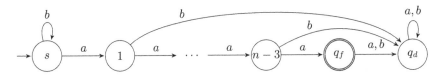

Fig. 8. The binary witness language L with $sc(L^{c*}) = 2^{n-3} + 2$.

3.5   Case (5)

Now consider the Kuratowski algebra in case (5) with conditions as shown in Fig. 7. Let us start with the following observations.

Proposition 13. *Let L be a prefix-free language over Σ such that $L^\oplus$ is closed. Then for each symbol a in Σ, we have $a \notin L$.*

Proof. Suppose for a contradiction that there exists a symbol a in Σ such that $a \in L$. Since a cannot be partitioned into strings in L^c, we have $a \in L^\oplus$, and therefore $aa \in L^{\oplus+}$. Since L is prefix-free, we have $aa \notin L$, and therefore $aa \notin L^\oplus$. Hence $L^\oplus$ is not closed, a contradiction. □

Proposition 14. *Let L be a language over Σ such that $\varepsilon \notin L$ and $a \notin L$ for each symbol a in Σ. Then $L^\oplus = \emptyset$ and $L^{+\oplus} = \emptyset$.*

Proof. Since $\varepsilon \in L^c$ and $a \in L^c$ for each symbol a in Σ, each string in L can be partitioned into one-symbol strings in L^c. Hence $L^\oplus = \emptyset$. The same argument applies to L^+ since $\varepsilon \notin L$ and $a \notin L$ imply $\varepsilon \notin L^+$ and $a \notin L^+$. □

Theorem 15 (Case (5)). *Let $n \geq 4$. There exists a binary prefix-free language L with $sc(L) = n$ which generates the 8-element Kuratowski algebra in case (5). In addition, all the languages in $E(L) = \{L, L^*, L^\oplus \cup \{\varepsilon\}, L^\oplus \setminus \{\varepsilon\}\}$ meet the upper bounds $(n, n, 2, 1)$ on the state complexity of corresponding languages in this case.*

Proof. Let $\Sigma = \{a, b\}$ and consider the language $L = b^*a^{n-2}$ over Σ accepted by the DFA shown in Fig. 8. We can show that L satisfies all the conditions. Now let us consider the complexities of languages in $E(L)$. The upper bounds are $(n, n, 2, 1)$ since if a prefix-free language K generates case (5), then $\varepsilon \notin K$ and $K^\oplus$ is closed. Then, by Propositions 13 and 14, we must have $K^\oplus = \emptyset$. Since in the DFA shown in Fig. 7 the initial state s has an in-transition, and the dead state is reached from $n - 3$ by b, we have $sc(L^*) = n$ by Proposition 3. Next $sc(L^\oplus \cup \{\varepsilon\}) = 2$ and $sc(L^\oplus \setminus \{\varepsilon\}) = 1$. □

3.6   Case (6)

Next we consider the Kuratowski algebra in case (6) with conditions as shown in Fig. 9. We start with three observations we will use for this case, and then describe an n-state DFA prefix-free language generating this algebra, and producing languages in $E(L)$ with maximal possible complexities.

L is neither open nor closed
L^+ is open
$L^\oplus$ is not closed
$L^{\oplus+} \neq L^+$

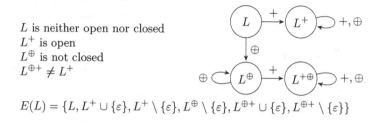

$$E(L) = \{L, L^+ \cup \{\varepsilon\}, L^+ \setminus \{\varepsilon\}, L^\oplus \setminus \{\varepsilon\}, L^{\oplus+} \cup \{\varepsilon\}, L^{\oplus+} \setminus \{\varepsilon\}\}$$

Fig. 9. The conditions and algebras $B(L)$ and $E(L)$ in case (6).

Lemma 16. *Let L be a prefix-free language over Σ such that L^+ is open. Let $a \in \Sigma$ and $u \in \Sigma^*$. If $ua \in L$ then $a \in L$.*

Proof. Let $ua \in L$. Since L is prefix-free, we have $u \notin L$. Moreover $u \in L^{+c}$ because otherwise we would have $ua = u_1 \cdots u_k a \in L^+$ with $k \geq 2$ and $u_1 \in L$, a contradiction with prefix-freeness of L. Next if $a \notin L$, then $a \in L^{+c}$. Hence $u \in L^{+c}$ and $a \in L^{+c}$, but $ua \notin L^{+c}$, so L^{+c} is not closed. This is a contradiction with L^+ open. □

Lemma 17. *Let L be a prefix-free language over Σ such that L^+ is open. Let $\Sigma_f = \{a \in \Sigma \mid ua \in L \text{ for some } u \in \Sigma^*\}$. Then $L^{c+} = L^{+c+} \cup \Sigma^* \Sigma_f \Sigma^+$.*

Proof. ($\subseteq$) Let $w \in L^{c+}$ then $w = w_1 \cdots w_k$ with $w_i \in L^c$.
(a) If for each i, we have $w_i \in L^{+c}$, then $w \in L^{+c+}$.
(b) Let there exist an i such that $w_i \notin L^{+c}$, thus $w_i \in L^+$. Since $w_i \in L^c$, we have $w_i = u_{i1} \cdots u_{i\ell}$, where $\ell \geq 2$, $u_{ij} \in L$, $u_{ij} \neq \varepsilon$. Then we have $w = w_1 \cdots w_{i-1} u'_{i1} \cdot a u_{i2} \cdots u_{i\ell} w_{i+1} \cdots w_k$ for an a in Σ_f. Hence $w \in \Sigma^* \Sigma_f \Sigma^+$.

($\supseteq$) Since $L \subseteq L^+$, we have $L^{c+} \supseteq L^{+c+}$. Let $w \in \Sigma^* \Sigma_f \Sigma^+$. Then $w = uav$ where $u \in \Sigma^*$, $a \in \Sigma_f$, and $v \in \Sigma^+$. Since $a \in \Sigma_f$, by Lemma 16 we have $a \in L$, thus $av \in L^c$. If $u \in L^c$, then $w = u \cdot av \in L^{c+}$. If $u \in L$ and L is prefix-free, then $w = uav \in L^c \subseteq L^{c+}$. In both cases $w \in L^{c+}$, which completes the proof. □

Lemma 18. *Let L be a prefix-free language satisfying all the conditions in case (6). Then $\mathrm{sc}(L^{c+}) \leq n$.*

Proof. Since L^+ is open we have $L^{+c+} = L^{+c}$. By Lemma 17, we get

$$L^{c+} = L^{+c} \cup \Sigma^* \Sigma_f \Sigma^+.$$

Let $A = (\{s, 1, \ldots, n-3, q_f, q_d\}, \Sigma, \cdot, s, \{q_f\})$ be a minimal n-state DFA for L. Let $B = (\{q_0, q_1, q_2\}, \Sigma, \cdot, q_0, \{q_2\})$ be a minimal 3-state DFA for $\Sigma^* \Sigma_f \Sigma^+$, where for each $\sigma \in \Sigma$, we have $q_0 \cdot \sigma = q_0$ if $\sigma \in \Sigma \setminus \Sigma_f$, $q_0 \cdot \sigma = q_1$ if $\sigma \in \Sigma_f$, $q_1 \cdot \sigma = q_2$, $q_2 \cdot \sigma = q_2$. Construct a DFA A' for L^{+c} from A as follows:

(a) replace each transition (q_f, a, q_d) with the transition $(q_f, a, s \cdot a)$, and
(b) interchange the final and non-final states.

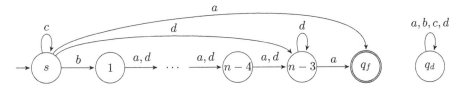

Fig. 10. The DFA A for case (6); all the undefined transitions go to the dead state q_d.

Construct an NFA N for $L^{c+} = L(A') \cup L(B)$ from A' and B by making states s and q_0 initial. Since A' and B are deterministic, each reachable state in the subset automaton of N is a two-element subset $\{p, q_i\}$, where p is a state of A' and q_i is state of B. Notice that each set $\{q_d, q_i\}$ is equivalent to $\{q_d, q_2\}$. Next, each set $\{p, q_2\}$ is equivalent to $\{q_d, q_2\}$ as well. Moreover, each set $\{p, q_1\}$ with $p \neq q_f$ is equivalent to $\{q_d, q_2\}$. Next, the set $\{q_f, q_0\}$ is unreachable, and the set $\{q_f, q_1\}$ is the only reachable non-final set. So the subset automaton of N, which accepts L^{c+}, has at most n reachable and pairwise distinguishable subsets: $\{p, q_0\}$ with $p \in \{s, 1, \dots, n-3\}$, $\{q_f, q_1\}$, and $\{q_d, q_2\}$.   □

Theorem 19. *Let $n \geq 5$. There exists a prefix-free language L over $\{a, b, c, d\}$ with $\mathrm{sc}(L) = n$ which generates the 12-element Kuratowski algebra in case (6). In addition, all the languages in $E(L) = \{L, L^+ \cup \{\varepsilon\}, L^+ \setminus \{\varepsilon\}, L^{\oplus} \setminus \{\varepsilon\}, L^{\oplus+} \cup \{\varepsilon\}, L^{\oplus+} \setminus \{\varepsilon\}\}$ meet the upper the upper bounds (n, n, n, n, n, n) on the state complexities of the corresponding languages in this case.*

Proof. Let L be the language accepted by the DFA A shown in Fig. 10. We can show that L satisfies all the conditions in case (6). Moreover all the languages in $E(L)$ have state complexities n.   □

3.7   Case (7)

Now we consider the Kuratowski algebra in case (7) with conditions as shown in Fig. 11.

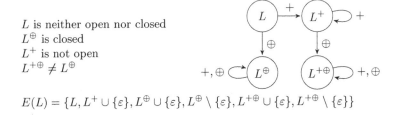

L is neither open nor closed
$L^{\oplus}$ is closed
L^+ is not open
$L^{+\oplus} \neq L^{\oplus}$

$E(L) = \{L, L^+ \cup \{\varepsilon\}, L^{\oplus} \cup \{\varepsilon\}, L^{\oplus} \setminus \{\varepsilon\}, L^{+\oplus} \cup \{\varepsilon\}, L^{+\oplus} \setminus \{\varepsilon\}\}$

Fig. 11. The conditions and algebras $B(L)$ and $E(L)$ in case (7).

L is neither open nor closed
$L^\oplus$ is not closed
L^+ is not open
$L^{+\oplus} = L^{\oplus+}$

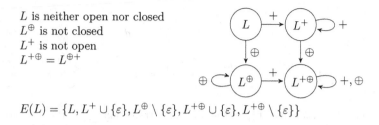

$$E(L) = \{L, L^+ \cup \{\varepsilon\}, L^\oplus \setminus \{\varepsilon\}, L^{+\oplus} \cup \{\varepsilon\}, L^{+\oplus} \setminus \{\varepsilon\}\}$$

Fig. 12. The conditions and algebras $B(L)$ and $E(L)$ in case (8).

Theorem 20. *The 12-element Kuratowski algebra in case (7) cannot be generated by any prefix-free language.*

Proof. Let L be a prefix-free language different from $\emptyset$ and $\{\varepsilon\}$. Then $\varepsilon \notin L$. If there exists a symbol a in Σ such that $a \in L$, then $L^\oplus$ is not closed by Proposition 13. If $a \notin L$ for each a in Σ, then $L^{+\oplus} = L^\oplus = \emptyset$ by Proposition 14. Hence no prefix-free language satisfies the conditions in case (7).   □

3.8   Case (8)

Let us continue with the Kuratowski algebra in case (8) with conditions as shown in Fig. 12. We get the following result, the proof of which is omitted.

Theorem 21 (Case (8)). *Let $n \geq 4$. There exists a prefix-free language L with $\mathrm{sc}(L) = n$ which generates the 10-element Kuratowski algebra in case (8). In addition, all the languages in $E(L) = \{L, L^+ \cup \{\varepsilon\}, L^\oplus \setminus \{\varepsilon\}, L^{+\oplus} \cup \{\varepsilon\}, L^{+\oplus} \setminus \{\varepsilon\}\}$ meet the upper bounds $(n, n, 2^{n-3}+2, 2^{n-3}+2, 2^{n-3}+2)$ on the state complexity on the corresponding languages in this case.*   □

3.9   Case (9)

Finally, let us consider the Kuratowski algebra in case (9) with conditions as shown in Fig. 13. We start with a simple language over $\{a, b\}$ satisfying all the conditions.

L is neither open nor closed
$L^\oplus$ is not closed
L^+ is not open
$L^{+\oplus} \neq L^{\oplus+}$
$E(L) = \{L, L^+ \cup \{\varepsilon\}, L^\oplus \setminus \{\varepsilon\},$
$\qquad\quad L^{+\oplus} \cup \{\varepsilon\}, L^{+\oplus} \setminus \{\varepsilon\},$
$\qquad\quad L^{\oplus+} \cup \{\varepsilon\}, L^{\oplus+} \setminus \{\varepsilon\}\}$

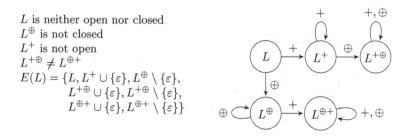

Fig. 13. The conditions and algebras $B(L)$ and $E(L)$ in case (9).

Proposition 22. *The language $\{a, ba, bb\}$ satisfies all the conditions in case (9).*

Now we are going to describe a prefix-free languages over $\{a, b, c, d, e, f, g\}$ such that its intersection with $\{a, b\}^*$ will be the language $\{a, ba, bb\}$ from the lemma above. Since all the conditions in case (9) are given by some inequalities, the new language will satisfy all of them. We can prove the following result.

Theorem 23 (Case (9)). *Let $n \geq 6$. There exists a prefix-free regular language L over the alphabet $\{a, b, c, d, e, f, g\}$ with $\mathrm{sc}(L) = n$ which generates the 14-element Kuratowski algebra in case (9). In addition, all the languages in $E(L) = \{L, L^+ \cup \{\varepsilon\}, L^\oplus \setminus \{\varepsilon\}, L^{+\oplus} \cup \{\varepsilon\}, L^{+\oplus} \setminus \{\varepsilon\}, L^{\oplus +} \cup \{\varepsilon\}, L^{\oplus +} \setminus \{\varepsilon\}\}$ meet the upper bounds $(n, n, 2^{n-3} + 2, 2^{n-3} + 2, 2^{n-3} + 2, 2^{n-3} + 2, 2^{n-3} + 2)$ on the state complexity of the corresponding languages in this case.* □

4 Conclusions

We investigated Kuratowski algebras generated by prefix-free regular languages under the operations of star and complement. We showed that five of 12 possible algebras described in [2, Cases(2a), (3a), (3b), (4), (5)] cannot be generated by any prefix-free language (whether regular or not). Two algebras [2, Cases(1a), (1b)] are generated only by trivial prefix-free languages, the language $\{\varepsilon\}$ and the empty language, respectively. Finally, we proved that each of the remaining five algebras can be generated, for every $n \geq 4$, by a regular prefix-free language of state complexity n, which meets the upper bounds on the state complexities of all the languages in the resulting algebra.

References

1. Brzozowski, J.: Kuratowski algebras generated from L by applying the operators of Kleene closure and complement. Personal communication (2016)
2. Brzozowski, J., Grant, E., Shallit, J.: Closures in formal languages and Kuratowski's theorem. Int. J. Found. Comput. Sci. **22**, 301–321 (2011)
3. Eom, H.-S., Han, Y.-S.: State complexity of boundary of prefix-free regular languages. Int. J. Found. Comput. Sci. **26**, 697–708 (2015)
4. Fife, J.H.: The Kuratowski closure-complement problem. Math. Mag. **64**, 180–182 (1991)
5. Han, Y.-S., Salomaa, K., Wood, D.: Nondeterministic state complexity of basic operations for prefix-free regular languages. Fundam. Inform. **90**, 93–106 (2009)
6. Han, Y.-S., Salomaa, K., Wood, D.: Operational state complexity of prefix-free regular languages. In: Ésik, Z., Fülöp, Z. (eds.) AFL 2009. Institute of Informatics, pp. 99–115. University of Szeged, Hungary (2009)
7. Hopcroft, J.E., Ullman, J.D.: Introduction to Automata Theory, Languages, and Computation, 1st edn. Addison-Wesley, Reading (1979)
8. Jirásek, J., Jirásková, G.: Cyclic shift on prefix-free languages. In: Bulatov, A.A., Shur, A.M. (eds.) CSR 2013. LNCS, vol. 7913, pp. 246–257. Springer, Heidelberg (2013)

9. Jirásek, J., Jirásková, G.: The boundary of prefix-free languages. In: Potapov, I. (ed.) DLT 2015. LNCS, vol. 9168, pp. 300–312. Springer, Heidelberg (2015)
10. Jirásek, J., Jirásková, G.: On the boundary of regular languages. Theoret. Comput. Sci. **578**, 42–57 (2015)
11. Jirásková, G., Palmovský, M., Šebej, J.: Kleene closure on regular and prefix-free languages. In: Holzer, M., Kutrib, M. (eds.) CIAA 2014. LNCS, vol. 8587, pp. 226–237. Springer, Heidelberg (2014)
12. Jirásek, J., Jirásková, G., Krausová, M., Mlynárčik, P., Šebej, J.: Prefix-free languages: left and right quotient and reversal. Theoret. Comput. Sci. **610**, 78–90 (2016)
13. Kuratowski, C.: Sur l'opration Ā de l'analysis situs. Fund. Math. **3**, 182–199 (1922)
14. Krausová, M.: Prefix-free regular languages: closure properties, difference, and left quotient. In: Kotásek, Z., Bouda, J., Černá, I., Sekanina, L., Vojnar, T., Antoš, D. (eds.) MEMICS 2011. LNCS, vol. 7119, pp. 114–122. Springer, Heidelberg (2012)
15. Palmovský, M., Šebej, J.: Star-complement-star on prefix-free languages. In: Shallit, J., Okhotin, A. (eds.) DCFS 2015. LNCS, vol. 9118, pp. 231–242. Springer, Heidelberg (2015)
16. Sipser, M.: Introduction to the Theory of Computation. PWS Publishing Company, Boston (1997)
17. Yu, S.: Regular languages. In: Rozenberg, G., Salomaa, A. (eds.) Handbook of Formal Languages, vol. 1, pp. 41–110. Springer, Heidelberg (1997)

A Logical Characterization of Small 2NFAs

Christos A. Kapoutsis[(✉)] and Lamana Mulaffer

Carnegie Mellon University in Qatar, Doha, Qatar
cak@cmu.edu, lamanamulaffer@gmail.com

Abstract. Let 2N be the class of families of problems solvable by families of *two-way nondeterministic finite automata* of polynomial size. We characterize 2N in terms of families of formulas of *transitive-closure logic*. These formulas apply the transitive-closure operator on a quantifier-free disjunctive normal form of *first-order logic with successor and constants*, where (i) apart from two special variables, all others are equated to constants in every clause, and (ii) no clause simultaneously relates these two special variables and refers to fixed input cells. We prove that automata with polynomially many states are as powerful as formulas with polynomially many clauses and polynomially large constants. This can be seen as a refinement of Immerman's theorem that nondeterministic logarithmic space matches positive transitive-closure logic ($\mathsf{NL} = \mathsf{FO} + \mathsf{pos}\,\mathsf{TC}$).

1 Introduction

A formal machine M and a logical formula φ are equivalent if they determine the same language: a string w is accepted by M iff it satisfies φ. Such comparisons between machines and formulas are the topic of Descriptive Complexity Theory [4]. Its inaugural result was Fagin's Theorem, which says that polynomial-time nondeterministic Turing machines (NTMs) are equivalent to formulas of *existential second-order logic* ($\mathsf{NP} = \exists \mathsf{SO}$) [2]. An analogous result for space complexity is Immerman's theorem that logarithmic-space NTMs are equivalent to formulas of *positive transitive-closure logic* ($\mathsf{NL} = \mathsf{FO} + \mathsf{pos}\,\mathsf{TC}$) [3]. Today we know many such 'logical characterizations' of various computational complexity classes [4].

When it comes to finite automata (on finite strings), an old result of this kind is Büchi's Theorem, that *one-way nondeterministic finite automata* (1NDFAs) are equivalent to formulas of *monadic second-order logic with successor* (MSO[S]) [1] — and thus so are, too, all automata recognizing the regular languages, including the deterministic and/or two-way variants (1DFAs, 2DFAs, 2NFAs). But this is a 'computability result', in the sense that the equivalence involves no restriction on the automata resources — as opposed to Fagin's and Immerman's 'complexity results', where the NTMs are restricted to use only polynomial time or logarithmic space, respectively. What if we focus on automata where the main resource, the number of states, is restricted to be polynomial (in a given parameter)?

L. Mulaffer—Supported by the CMUQ Student-Initiated Undergraduate Research Program 2013.

Y.-S. Han and K. Salomaa (Eds.): CIAA 2016, LNCS 9705, pp. 163–175, 2016.
DOI: 10.1007/978-3-319-40946-7_14

We first asked this in [6], in the context of building a size-complexity theory of two-way finite automata, or 'Minicomplexity Theory' [5]. Specifically, we asked: What is an analog of Fagin's Theorem when we replace NTMs and time with 2NFAs and size? Unfortunately, however, we failed to answer in full generality. Instead, we proved such analogs only for the one-way, rotating, and sweeping restrictions of 2NFAs (where the input head can, respectively, only move forward; or only move forward and jump to the start; or turn only on the end-markers).

The present paper contains the full answer to that question of [6]. In what can be seen as a refinement of Immerman's theorem from above, we prove that polynomial-size 2NFAs are equivalent to a certain class of formulas of FO + pos TC. Specifically, we focus on formulas consisting of a single, positive application of the transitive-closure operator on a quantifier-free disjunctive normal form of first-order logic with successor and constants, where (i) each of the conjunctive clauses equates every variable, except for two special ones, to some constant, and (ii) none of these clauses can *both* relate the two special variables *and* refer to a fixed input cell. We call such formulas *weak one-dimensional graph-accessibility disjunctive-normal-forms* (weak GA/DNF$_1$s) and prove that they are equivalent to polynomial-size 2NFAs, if their clauses are only polynomially many and their constants are only polynomially large. We thus complete our first step, started in [6], into what one could call 'Descriptive Minicomplexity Theory'.

2 Preparation

Let $\mathbb{Z}$ be all integers and $\mathbb{Z}^{\pm} := \mathbb{Z}-\{0\}$. If $n \geq 0$, then we let $[n] := \{0,\dots,n-1\}$, $\mathbb{Z}_n^+ := \{1,\dots,n\}$, and $\mathbb{Z}_n^- := \{-n,\dots,-1\}$. If $w \in \Sigma^*$ is a finite string over some alphabet Σ, then $|w|$ and w_x are its length and x-th symbol (if $1 \leq x \leq |w|$).

2.1 Finite Automata

A *two-way nondeterministic finite automaton* is a tuple $N = (S, \Sigma, \delta, q_s, q_a)$ of a set of *states* S, an *alphabet* Σ, a *start state* $q_s \in S$, an *accept state* $q_a \in S$, and a set of *transitions* $\delta \subseteq S \times (\Sigma \cup \{\vdash, \dashv\}) \times S \times \{L,R\}$, where $\vdash, \dashv \notin \Sigma$ are the two *end-markers* and L,R are the two directions of motion for the input head.

A word $w \in \Sigma^*$ is presented to N between the end-markers, as $\vdash w \dashv$. The computation starts at q_s on $\vdash$. At each step, the next state and head motion may be any of those derived from δ and the current state and symbol. End-markers are never violated, except for $\dashv$ if the next state is q_a. So, each branch of N's computation can *hang* inside the input; or *loop*; or fall off $\dashv$ into q_a, in which case we call it *accepting*. If at least one branch is accepting, we say N *accepts* w.

Let $n = |w|$. A *configuration* of N on w is a pair $(p, x) \in S \times [n+3]$; it means N is at state p reading w_x, if $x \leq n+1$ (we let $w_0 := \vdash$ and $w_{n+1} := \dashv$); or has fallen off $\dashv$ into p, if $x = n+2$. The *configuration graph* $G_{N,w}$ of N on w (Fig. 1a)

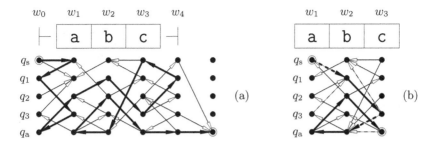

Fig. 1. (a) The configuration graph of a 5-state 2NFA N on a word $w = \text{abc}$. Bold arrows show an accepting branch. (b) The *inner* configuration graph of N on w. Dashed arrows are caused by computations on $\vdash w_1$ or $w_3 \dashv$. E.g., $(q_3, 1) \to (q_1, 2)$ is caused by the path $(q_3, 1) \to (q_a, 0) \to (q_2, 1) \to (q_1, 2)$ in (a).

is the directed graph where vertices are configurations of N on w and an edge $(p, x) \to (q, y)$ exists iff N can switch from (p, x) to (q, y) in a single step, i.e., iff

$$y = x{+}1 \quad \& \quad (p, w_x, q, \text{R}) \in \delta \qquad \text{or} \qquad y = x{-}1 \quad \& \quad (p, w_x, q, \text{L}) \in \delta. \qquad (1)$$

Clearly, N *accepts* w *iff* $G_{N,w}$ has a path $(q_s, 0) \rightsquigarrow (q_a, n{+}2)$.

When $n \geq 2$, a denser representation is the *inner configuration graph* $G'_{N,w}$ (Fig. 1b), where now the vertices are only the *inner configurations* $S \times \mathbb{Z}_n^+$ and an edge, or *inner step*, $(p, x) \to (q, y)$ exists iff any of the following holds:

- N can switch from (p, x) to (q, y) in a single step, as in (1);
- $x = 1$, $y = 2$, and the switch can happen by a U-turn computation on $\vdash w_1$;
- $x = n$, $y = n{-}1$, and the switch can happen by a U-turn on $w_n \dashv$.

We will need to say that N accepts w iff $G'_{N,w}$ has a path $(q_s, 1) \rightsquigarrow (q_a, n)$. But, in general, this is false; it becomes true, if N is in the form of Definition 1 (Fact 1ii). Conveniently, with two more states, every 2NFA can be put in this form (Fact 1i).

Definition 1. *A* 2NFA $N = (\,.\,, \Sigma, \delta, q_s, q_a)$ *is in* inner normal form (INF) *if*

   i. δ *contains* $(q_s, \vdash, q_s, \text{R})$, *but no other tuple* $(q_s, \vdash, \,.\,, \,.\,)$; *and*
  ii. δ *contains every* (q_a, a, q_a, R) *for* $a \in \Sigma \cup \{\dashv\}$, *but no other tuple* $(\,.\,, \,.\,, q_a, \text{R})$.

Fact 1.

   i. *Every s-state* 2NFA *is equivalent to a $(s{+}2)$-state* 2NFA *in* INF.
  ii. *If N is in* INF, *then N accepts w iff $G'_{N,w}$ has a path* $(q_s, 1) \rightsquigarrow (q_a, n)$.

2.2 Logical Formulas

In *quantifier-free first-order logic with successor and constants* over alphabet Σ ($\text{Q·FO}_\Sigma^+[\text{S},\mathbb{Z}^\pm]$), formulas are built out of first-order *variables* $x_0, x_1, \ldots$, *constants* $\pm 1, \pm 2, \ldots \in \mathbb{Z}^\pm$, one *cell* predicate $\alpha(\,.\,)$ for each $\alpha \subseteq \Sigma$, the *equality* predicate $.\,=\,.$, the *successor* predicate $\text{S}(\,.\,,\,.\,)$, and the *connectives* $\neg, \wedge, \vee$.

⊥	a
x_1	1
x_2	0	(a)

	1	2	3	4	5
⊥	a	a	b	a	b
x_1	1	0	0	0	0
x_2	0	0	1	0	0	(b)

	1	2	3	4	5
⊥	a	a	b	a	b
x_1	1	0	0	0	0
x_2	0	0	1	0	0
x_5	0	1	0	0	0	(c)

Fig. 2. (a) A column from $\Sigma|V$, for $\Sigma = \{\mathsf{a},\mathsf{b}\}$, $V = \{x_1, x_2\}$. (b) A well-formed $\hat{w}$ over $\Sigma|V$; here, $\hat{w}(\bot) = \mathsf{aabab}$, $\hat{w}(x_2) = 3$, $\hat{w}(+1) = 1$, $\hat{w}(-1) = 5$. (c) The word $\hat{w}[x_5/2]$.

A formula φ is either an *atom*, of the form $\alpha(t)$, $t = t'$, or $\mathsf{S}(t,t')$, where each of the *terms* t, t' is either a variable or a constant; or *compound*, of the form $\neg\phi$, $\phi \wedge \psi$, or $\phi \vee \psi$, where ϕ, ψ are simpler formulas. An atom is either *local*, of the form $\alpha(t)$; or *relational*, of the form $t = t'$ or $\mathsf{S}(t,t')$. An atom or negation of atom is a *literal*. A conjunction (resp., disjunction) of literals is an $\wedge$-*clause* (resp., an $\vee$-*clause*); a disjunction (resp., conjunction) of $\leq m$ such clauses is an m-clause *disjunctive normal form*, or m-DNF (resp., an m-clause *conjunctive normal form*, or m-CNF). A formula is *non-trivial* if it is not identically true or identically false.

The *length* $|\varphi|$ of a formula φ is the number of occurences of symbols in it, ignoring punctuation and counting each variable, constant, and cell predicate as a single symbol. More carefully, we define $|\varphi|$ by structural induction on φ:

- for all α, t, t': $|\alpha(t)| = 2$ and $|t = t'| = |\mathsf{S}(t,t')| = 3$;
- for all ϕ, ψ: $|\neg\phi| = 1 + |\phi|$ and $|\phi \wedge \psi| = |\phi \vee \psi| = |\phi| + 1 + |\psi|$.

The *margin* of φ is the maximum absolute value of a constant in it; or 0, if φ has no constants. We write $\varphi(x_2, x_5, \dots)$ to indicate that all variables appearing in φ are among $x_2, x_5, \dots$ (note that all variables are free, as there are no quantifiers).

Semantics. For a set of variables V, let $\Sigma|V$ be the alphabet of all functions $u : \{\bot\} \cup V \to \Sigma \cup \{\mathsf{0},\mathsf{1}\}$ which map $\bot$ into Σ and variables into $\{\mathsf{0},\mathsf{1}\}$ (namely, $u(\bot) \in \Sigma$ and $u(x_i) \in \{\mathsf{0},\mathsf{1}\}$ for all $x_i \in V$). Intuitively, every such u is a column of $1+|V|$ cells, labelled by the elements of $\{\bot\} \cup V$ and filled by the respective values of u (Fig. 2a). Likewise, each word $\hat{w} = \hat{w}_1 \cdots \hat{w}_n \in (\Sigma|V)^*$ is a table of n columns and $1+|V|$ rows: one row is labelled by $\bot$ and hosts an n-long word over Σ; the rest are labelled by variables and host n-long bitstrings (Fig. 2b).

We say $\hat{w}$ is *well-formed* if $n \geq 2$ and the row of each variable hosts exactly one 1 (Fig. 2b). Then, $\hat{w}(\bot)$ is the word $\hat{w}_1(\bot) \cdots \hat{w}_n(\bot) \in \Sigma^*$ hosted in the $\bot$-row; $\hat{w}(x_i)$ is the index x of the one column $\hat{w}_x$ which has a 1 in the x_i-row; and, for $c \in \mathbb{Z}^\pm$, $\hat{w}(c)$ is the index c of the c-th leftmost column, if $c > 0$, or the index $n - |c| + 1$ of the $|c|$-th rightmost column, if $c < 0$. Moreover, for any $x_i \notin V$ and index $x \in \mathbb{Z}_n^+$, $\hat{w}[x_i/x]$ is the well-formed word over $\Sigma|(V \cup \{x_i\})$ derived from $\hat{w}$ by adding a row labelled x_i with its x-th bit 1 and all others 0 (Fig. 2c).

Now, given a n-long well-formed $\hat{w}$ over $\Sigma|V$ and a formula φ whose variables are all in V, we say $\hat{w}$ *satisfies* φ and write $\hat{w} \models \varphi$, if what φ 'says' about $\hat{w}(\bot)$ is true when each variable x_i is interpreted as in the x_i-row, namely iff:

for $\varphi \equiv \alpha(t)$: $\hat{w}(\bot)_{\hat{w}(t)} \in \alpha$ for $\varphi \equiv \neg\phi$: $\hat{w} \not\models \phi$

for $\varphi \equiv t = t'$: $\hat{w}(t) = \hat{w}(t')$ for $\varphi \equiv \phi \wedge \psi$: $\hat{w} \models \phi$ and $\hat{w} \models \psi$

for $\varphi \equiv S(t,t')$: $\hat{w}(t) + 1 = \hat{w}(t')$ for $\varphi \equiv \phi \vee \psi$: $\hat{w} \models \phi$ or $\hat{w} \models \psi$.

Transitive Closure. Let $\varphi(\overline{x}, \overline{y})$ be a $\text{Q·FO}_\Sigma^+[\text{S},\mathbb{Z}^\pm]$ formula over $2k + 2$ variables $\overline{x} = x_0, \ldots, x_k$ and $\overline{y} = y_0, \ldots, y_k$. Given an n-long $w \in \Sigma^*$, this defines a binary relation R_φ on $k+1$-tuples of indices in $\mathbb{Z}_n^+$. As usual, the *transitive closure* of R_φ is the binary relation R_φ^* which contains a pair $(\overline{u}, \overline{v})$ iff there is a sequence of tuples $\overline{r}_0, \overline{r}_1, \cdots, \overline{r}_\ell$ such that $\overline{u} = \overline{r}_0$; every $(\overline{r}_i, \overline{r}_{i+1})$ is in R_φ; and $\overline{r}_\ell = \overline{v}$.

We augment our logic with the *transitive closure* operator 'TC', which checks if two tuples of indices are in the relation R_φ^* defined by some $\varphi(\overline{x}, \overline{y})$: given φ and two tuples of terms $\overline{t}, \overline{t}'$, the formula $\text{TC}_\varphi(\overline{t}, \overline{t}')$ (or, more legibly, $\text{TC}[\varphi(\overline{x}, \overline{y})](\overline{t}, \overline{t}')$) has length $1 + |\varphi| + 2k + 2$ and the following semantics, for all well-formed $\hat{w}$:

$$\hat{w} \models \text{TC}_\varphi(\overline{t}, \overline{t}') \qquad \text{iff} \qquad \big((\hat{w}(t_0), \ldots, \hat{w}(t_k)), (\hat{w}(t_0'), \ldots, \hat{w}(t_k')) \big) \in R_\varphi^* .$$

Intuitively, let $G_{\varphi, \hat{w}}$ be the directed graph with vertex set $(\mathbb{Z}_n^+)^{k+1}$ and all edges $(\overline{u}, \overline{v})$ such that $\hat{w}[\overline{x}/\overline{u}, \overline{y}/\overline{v}] \models \varphi(\overline{x}, \overline{y})$; then $\hat{w} \models \text{TC}_\varphi(\overline{t}, \overline{t}')$ iff $G_{\varphi, \hat{w}}$ has a path $(\hat{w}(t_0), \ldots, \hat{w}(t_k)) \rightsquigarrow (\hat{w}(t_0'), \ldots, \hat{w}(t_k'))$. We call this new logic $\text{Q·FO}_\Sigma^+[\text{S},\mathbb{Z}^\pm]+\text{TC}$.

2.3 Finite Automata Versus Logical Formulas

A (*promise*) *problem* over alphabet Σ is any pair $\mathfrak{L} = (L, \tilde{L})$ of disjoint subsets of Σ^*. An automaton N *solves* $\mathfrak{L}$ if it accepts all $w \in L$ but no $w \in \tilde{L}$. A formula φ *solves* $\mathfrak{L}$ if it is satisfied by all $w \in L$ but no $w \in \tilde{L}$.

A family of automata $\mathcal{N} = (N_h)_{h \geq 1}$ (resp., of formulas $\mathcal{F} = (\varphi_h)_{h \geq 1}$) *solves* a family of problems $(\mathfrak{L}_h)_{h \geq 1}$ if every N_h (resp., φ_h) solves the respective $\mathfrak{L}_h$. The automata of $\mathcal{N}$ (resp., the formulas of $\mathcal{F}$) are *small* if every N_h has $\leq p(h)$ states (resp., every φ_h has length $\leq p(h)$), for some polynomial p. Therefore, the set

$$\text{2N} := \left\{ (\mathfrak{L}_h)_{h \geq 1} \;\middle|\; \begin{array}{l} \text{there exist 2NFAs } (N_h)_{h \geq 1} \text{ and a polynomial } p \\ \text{such that every } N_h \text{ solves } \mathfrak{L}_h \text{ with } \leq p(h) \text{ states} \end{array} \right\}$$

is the class of problem families which are solvable by families of small 2NFAs.

A formula $\varphi(\overline{x})$ of $\text{Q·FO}_\Sigma^+[\text{S},\mathbb{Z}^\pm]+\text{TC}$ is *equivalent* to a 2NFA N over $\Sigma|\overline{x}$ if for all well-formed $\hat{w} \in (\Sigma|\overline{x})^*$: $\hat{w}$ satisfies φ iff N accepts $\hat{w}$ (note that $|\hat{w}| \geq 2$).

3   Graph-Accessibility Sentences and Our Theorem

A formula of $\text{Q·FO}_\Sigma^+[\text{S},\mathbb{Z}^\pm]$ is *local*, if all its atoms are local (i.e., it talks only about the contents of certain cells); *quasi-local*, if every relational atom in it uses at least one constant (i.e., it talks only about certain cells' contents and distance from the end-markers); and *relational*, if all its atoms are relational (i.e., it talks only about the order of certain cells). Orthogonally, the formula is *floating*, if all its terms are variables; *quasi-floating*, if every atom uses at least one variable; and *anchored*, if all its terms are constants. Finally, inside an $\wedge$-clause, a variable x is *anchored* if it appears in at least one literal of the form $x = c$ or $c = x$ (without negation), for some constant c; otherwise, it is *floating*.

Given a $Q \cdot FO_\Sigma^+[S, \mathbb{Z}^\pm]$ formula $\varphi(\overline{x}, \overline{y})$ with $\overline{x} = x_0, \ldots, x_k$ and $\overline{y} = y_0, \ldots, y_k$, a *graph-accessibility sentence* (GAS) with *core* φ and *arity* $k+1$ is any formula

$$\mathsf{TC}\big[\,\varphi(\overline{x}, \overline{y})\,\big](\overline{s}, \overline{t}) \tag{2}$$

where $\overline{s} = s_0, \ldots, s_k$ and $\overline{t} = t_0, \ldots, t_k$ are constants. If φ is a DNF, namely

$$\varphi(\overline{x}, \overline{y}) \equiv \bigvee_{i=1}^{m} \varphi_i(\overline{x}, \overline{y})$$

where each φ_i is an $\wedge$-clause and the *degree* m is ≥ 1, then we say (2) is a GA/DNF. If $x_1, \ldots, x_k$ and $y_1, \ldots, y_k$ are all anchored in every φ_i (so that only x_0, y_0 may be floating), then we say (2) is *one-dimensional* (GA/DNF$_1$). Finally, we say (2) is *weak* if no φ_i contains both anchored local atoms and floating relational ones.

Our theorem states that 2NFAs of polynomial size are as powerful as weak GA/DNF$_1$s of polynomial degree and margin; and that this holds already when the margin is 1 and we also require polynomial length and logarithmic arity.

Theorem 1. *The following are equivalent, for every family of problems $\mathcal{L}$:*

1. *$\mathcal{L}$ has small 2NFAs.*
2. *$\mathcal{L}$ has small weak GA/DNF$_1$s of small degree, margin 1, and logarithmic arity.*
3. *$\mathcal{L}$ has weak GA/DNF$_1$s of small degree and small margin.*

Proof. [(1) $\Rightarrow$ (2)] By Lemma 1. [(2) $\Rightarrow$ (3)] Trivial. [(3) $\Rightarrow$ (1)] By Lemma 5. □

4   From Automata to Formulas

The simpler conversion, from automata to formulas, is treated in the next lemma.

Lemma 1. *Every s-state 2NFA is equivalent to a weak GA/DNF$_1$ of degree $O(s^2)$, margin 1, arity $O(\log s)$ and length $O(s^2 \log s)$.*

Proof. Pick any s-state 2NFA N. We first switch to an equivalent 2NFA $\tilde{N}$ which is in INF (Definition 1) and has $\tilde{s} = 2^r$ states, for some r; easily, $\tilde{s} \leq 2s+2$. Without loss of generality, assume $\tilde{N} = ([\tilde{s}], \Sigma, \tilde{\delta}, 0, \tilde{s}-1)$.

We need a weak GA/DNF$_1$ $\mathsf{TC}[\varphi(\overline{x}, \overline{y})](\overline{s}, \overline{t})$ such that, for all w of length $n \geq 2$:

$$\tilde{N} \text{ accepts } w \quad \Longleftrightarrow \quad w \models \mathsf{TC}[\,\varphi(\overline{x}, \overline{y})\,](\overline{s}, \overline{t}). \tag{3}$$

By definition, the right-hand side holds iff the graph $G_{\varphi,w}$ induced by φ (cf. p. 5) has a path $(s_0, s_1, \ldots, s_k) \rightsquigarrow (t_0, t_1, \ldots, t_k)$, where $k+1$ the arity of φ. By Fact 1ii, the left-hand side holds iff the inner configuration graph $G'_{\tilde{N},w}$ induced by $\tilde{N}$ has a path $(0, 1) \rightsquigarrow (\tilde{s}-1, n)$. So, we simply need to pick φ so that $G_{\varphi,w}$ is actually $G'_{\tilde{N},w}$, and then pick $\bar{s}, \bar{t}$ so that they are actually $(0, 1)$ and $(\tilde{s}-1, n)$.

First, we must represent each vertex of $G'_{\tilde{N},w}$, namely each inner configuration $(p, x) \in [\tilde{s}] \times \mathbb{Z}_n^+$, as a vertex of $G_{\varphi,w}$, namely a tuple $\bar{u} = (u_0, u_1, \ldots, u_k)$ of indices from $\mathbb{Z}_n^+$. Of course, x can be represented by any component of $\bar{u}$, say u_0. As for p, we represent it in 'binary' using the other components $u_1, \ldots, u_k$: we pick $k := r = \lg \tilde{s}$ (to ensure we have enough 'bits') and use indices 1 and n (which are distinct, as $n \geq 2$) as 0 and 1, respectively. E.g., if $\tilde{s} = 16$ (so, $k = 4$) and $n = 50$, then the configuration $(p, x) = (2, 22)$ maps to $\bar{u} = (22, 1, 1, 50, 1)$. Note that $(0, 1)$ and $(\tilde{s}-1, n)$ map to $(1, 1, \ldots, 1)$ and $(n, n, \ldots, n)$, i.e., to the interpretations of the tuples of constants $(+1, +1, \ldots, +1)$ and $(-1, -1, \ldots, -1)$.

Given this representation, we now need a $\varphi(\bar{x}, \bar{y})$ which states that the edge $(\bar{x}, \bar{y})$ exists in $G'_{\tilde{N},w}$, namely that $\tilde{N}$ can switch in a single inner step from the inner configuration $(x_0, x_1, \ldots, x_k)$ to the inner configuration $(y_0, y_1, \ldots, y_k)$.

As a start, for every state $p \in [\tilde{s}]$ we need a formula $\xi_p(\bar{u})$ which says that the state of the inner configuration $(u_0, u_1, \ldots, u_k)$ is p. E.g., if $n = 50$ and $p = 2$ as above, then $w \models \xi_p(\bar{u})$ should hold iff $\bar{u}$ is of the form $(., 1, 1, 50, 1)$, and thus $\xi_p(\bar{u})$ should be $u_1 = +1 \wedge u_2 = +1 \wedge u_3 = -1 \wedge u_4 = +1$. In general, we let

$$\xi_p(\bar{u}) := \bigwedge_{i=1}^{k} (u_i = p_i), \tag{4}$$

where each p_i is either $+1$ or -1 depending on whether the i-th most significant bit in the k-bit binary representation of p is 0 or 1.

Additionally, for every two states $p, q \in [\tilde{s}]$ and each direction of head motion, we need the set of symbols which allow the corresponding transition:

$$\alpha_{p,q}^{\mathrm{L}} := \{a \in \Sigma \mid (p, a, q, \mathrm{L}) \in \tilde{\delta}\}, \qquad \alpha_{p,q}^{\mathrm{R}} := \{a \in \Sigma \mid (p, a, q, \mathrm{R}) \in \tilde{\delta}\}. \tag{5}$$

Similarly, for every two states $p, q \in [\tilde{s}]$ and each end-marker, we need the set of symbols which, together with the end-marker, allow the corresponding U-turn:

$$\alpha_{p,q}^{\vdash} := \{a \in \Sigma \mid \text{computing on } \vdash a \text{ from } p \text{ on } a, \tilde{N} \text{ can exit right into } q\},$$
$$\alpha_{p,q}^{\dashv} := \{a \in \Sigma \mid \text{computing on } a\dashv \text{ from } p \text{ on } a, \tilde{N} \text{ can exit left into } q\}. \tag{6}$$

Using the $\wedge$-clauses of (4) and the cell predicates for the sets of (5) and (6), we now build a formula $\varphi(\bar{x}, \bar{y})$ which says that, in a single inner step, $\tilde{N}$ can switch from cell x_0 and 'state' $(x_1, \ldots, x_k)$ to cell y_0 and 'state' $(y_1, \ldots, y_k)$:

$$\varphi(\bar{x}, \bar{y}) := \bigvee_{p,q \in [\tilde{s}]} \Big\{ \big[\xi_p(\bar{x}) \wedge \xi_q(\bar{y}) \wedge \mathsf{S}(x_0, y_0) \wedge \alpha_{p,q}^{\mathrm{R}}(x_0) \big] \tag{7}$$
$$\vee \big[\xi_p(\bar{x}) \wedge \xi_q(\bar{y}) \wedge \mathsf{S}(y_0, x_0) \wedge \alpha_{p,q}^{\mathrm{L}}(x_0) \big] \tag{8}$$
$$\vee \big[\xi_p(\bar{x}) \wedge \xi_q(\bar{y}) \wedge x_0 = +1 \wedge \mathsf{S}(x_0, y_0) \wedge \alpha_{p,q}^{\vdash}(x_0) \big] \tag{9}$$
$$\vee \big[\xi_p(\bar{x}) \wedge \xi_q(\bar{y}) \wedge x_0 = -1 \wedge \mathsf{S}(y_0, x_0) \wedge \alpha_{p,q}^{\dashv}(x_0) \big] \Big\} \tag{10}$$

Intuitively, φ says that there exist states p, q such that the state of inner configuration $\overline{x}$ is p, the state of inner configuration $\overline{y}$ is q, and: $\overline{y}$ is exactly to the right of $\overline{x}$, and the symbol read in $\overline{x}$ allows a right-moving transition $p \rightarrow q$ (line (7)); or $\overline{y}$ is exactly to the left of $\overline{x}$, and the symbol read in $\overline{x}$ allows a left-moving transition $p \rightarrow q$ (line (8)); or $\overline{x}, \overline{y}$ are on cells $1, 2$ and the symbol read in $\overline{x}$ together with $\vdash$ allows a left U-turn from p to q (line (9)); or $\overline{y}, \overline{x}$ are on cells $n-1, n$ and the symbol read in $\overline{x}$ together with $\dashv$ allows a right U-turn from p to q (line (10)).

Overall, our GAS is that of (3) with φ as in (7)–(10) and $\overline{s} = (+1, \ldots, +1)$ and $\overline{t} = (-1, \ldots, -1)$. As promised, the *margin* is 1 (all constants are ± 1) and the *arity* is $k+1 = O(\log s)$. Also, each bracket in (7)–(10) is an $\wedge$-clause of length $O(\log s)$, as the conjunction of two $\wedge$-clauses of length $O(k) = O(\log s)$ and two or three atoms of length $O(1)$; hence, φ is a disjunction of $4\overline{s}^2 = O(s^2)$ $\wedge$-clauses, of total length $O(s^2 \log s)$; and thus our GAS in (3) is a GA/DNF of *degree* $O(s^2)$ and *length* $O(s^2 \log s)$, too. Finally, each bracket in (7)–(10) anchors each one of $x_1, \ldots, x_k$ and $y_1, \ldots, y_k$ (inside ξ_p and ξ_q) and contains no anchored local atoms, making our GAS in (3) both *one-dimensional* and *weak*, as promised. $\square$

5   From Formulas to Automata

We now show how to convert a weak GA/DNF$_1$ to a 2NFA. Facts 2–5 analyze the structure of the given sentence and its sub-formulas; their proofs are straightforward and mostly syntactic. Lemmas 2–4 build two-way automata which simulate those sub-formulas. The final 2NFA for the given sentence is built in Lemma 5.

Fact 2. *Let* $\varphi(\overline{x}, \overline{y}) = \bigvee_{i=1}^{m} \varphi_i(\overline{x}, \overline{y})$ *be the core of a* GA/DNF$_1$ *of arity* $k+1$. *Then every* $\wedge$-*clause* $\varphi_i(\overline{x}, \overline{y})$ *is equivalent to an* $\wedge$-*clause of the form*

$$(x_1 = c_1) \wedge \cdots \wedge (x_k = c_k) \wedge (y_1 = d_1) \wedge \cdots \wedge (y_k = d_k) \wedge \hat{\varphi}(x_0, y_0),$$

for some constants $c_1, \ldots, c_k, d_1, \ldots, d_k$ *and some* $\wedge$-*clause* $\hat{\varphi}(x_0, y_0)$.

Proof. Pick any φ_i. By one-dimensionality, x_1 is anchored in φ_i, so at least one literal is of the form $x_1 = c_1$ or $c_1 = x_1$, for some constant c_1. Consider the following modifications: (1) if the literal is $c_1 = x_1$, change it to $x_1 = c_1$; (2) bring the literal upfront; (3) replace any other occurence of x_1 with c_1. Easily, this brings φ_i into the equivalent form $(x_1 = c_1) \wedge \vartheta_1(x_0, x_2, x_3, \ldots, x_k, \overline{y})$. Similarly, x_2 is also anchored in φ_i, so by repeating modifications (1)–(3) for it, we bring φ_i to the equivalent form $(x_1 = c_1) \wedge (x_2 = c_2) \wedge \vartheta_2(x_0, x_3, x_4, \ldots, x_k, \overline{y})$. Continuing like this for all anchored variables, we eventually get the desired equivalent form $(x_1 = c_1) \wedge \cdots \wedge (x_k = c_k) \wedge (y_1 = d_1) \wedge \cdots \wedge (y_k = d_k) \wedge \vartheta_{2k}(x_0, y_0)$. $\square$

Fact 3. *Every non-trivial* $\wedge$-*clause* $\varphi(x, y)$ *is equivalent to a formula of the form* $\phi \wedge \chi(x) \wedge \psi(y) \wedge \omega(x, y)$, *where each of* ϕ, χ, ψ, ω *is an* $\wedge$-*clause;* ϕ *is anchored local;* χ, ψ *are quasi-floating quasi-local; and* ω *is floating relational.*

Lemma 2. *Suppose φ is an anchored local $\wedge$-clause of margin τ. Then there exists a $O(\tau)$-state 2DFA which, whenever run on a string w from the cell of $\vdash$, returns on that same cell and accepts iff $w \models \varphi$.*

Proof. Formula φ is a conjunction of literals of the form $\alpha(c)$ and $\neg\alpha(c)$, where $\alpha \subseteq \Sigma$ and $c \in \mathbb{Z}_\tau^+ \cup \mathbb{Z}_\tau^-$. We may assume that every such c appears in exactly one literal of the form $\alpha(c)$: Indeed, if it appears in none, then we add the true literal $\Sigma(c)$; if it appears in exactly one, but of the form $\neg\gamma(c)$, then we replace this with the equivalent $\overline{\gamma}(c)$; if it appears in more than one, then we replace the conjunction $\beta_1(c) \wedge \cdots \wedge \beta_r(c) \wedge \neg\gamma_1(c) \wedge \cdots \wedge \neg\gamma_s(c)$ of these literals with the equivalent single literal $\alpha(c)$ where $\alpha := \beta_1 \cap \cdots \cap \beta_r \cap \overline{\gamma}_1 \cap \cdots \cap \overline{\gamma}_s$.

So, φ is essentially a list of 2τ conditions, one for each of the τ leftmost and the τ rightmost cells of w, and $w \models \varphi$ iff all are true. To test this, a 2DFA M starting from $\vdash$ scans the leftmost cells, counting up to τ and confirming all respective conditions; then sweeps to $\dashv$; then scans the rightmost cells backwards, again counting up to τ and confirming all respective conditions; then sweeps to $\vdash$ and accepts —if any condition fails or any cell does not exist (because w is too short), then M rejects. Easily, this can be implemented with $O(\tau)$ states.   □

Fact 4. *Every quasi-local formula is equivalent to a formula in which every atom is of the form $\alpha(\,.\,)$ or $x = c$, where $\alpha \subseteq \Sigma$, x is a variable, and c is a constant.*

Lemma 3. *Suppose $\varphi(x)$ is a quasi-floating quasi-local $\wedge$-clause of margin τ. Then there exists a $O(\tau)$-state 2DFA which, whenever run on a string w from a cell $1 \leq x^* \leq |w|$, returns on that same cell and accepts iff $w[x/x^*] \models \varphi(x)$.*

Proof. By Fact 4, by the margin τ, and since φ is quasi-floating with x as the only variable, we may assume that every atom is of the form $\alpha(x)$ or $x = c$, where $\alpha \subseteq \Sigma$ and $c \in \mathbb{Z}_\tau^+ \cup \mathbb{Z}_\tau^-$.

So, each literal has the form $\alpha(x)$, $\neg\alpha(x)$, $x = c$, or $\neg(x = c)$, for some α and c. As in the proof of Lemma 2, we may assume the first two forms contribute exactly one literal: the literal $\alpha(x)$, for α the intersection of Σ, of all β from occuring literals $\beta(x)$, and of all $\overline{\gamma}$ from occuring literals $\neg\gamma(x)$. We may also assume that the third form contributes at most one literal for collectively all $c > 0$ and at most one literal for collectively all $c < 0$: if there are two literals $x = c_1$, $x = c_2$ for distinct $c_1, c_2 > 0$, then φ is always false, and thus the 2DFA is just the trivial one which simply halts and rejects —similarly for $c_1, c_2 < 0$.

Overall, without loss of generality, we may assume that $\varphi(x)$ consists of: exactly one $\alpha(x)$ for $\alpha \subseteq \Sigma$; an optional $x = c$ for $c \in \mathbb{Z}_\tau^+$; an optional $x = c$ for $c \in \mathbb{Z}_\tau^-$; and zero or more $\neg(x = c)$ for $c \in \mathbb{Z}_\tau^+ \cup \mathbb{Z}_\tau^-$.

To test $w[x/x^*] \models \varphi$, a 2DFA run on w from cell x^* first verifies $\alpha(x)$ by testing that $w_{x^*} \in \alpha$. It then scans left counting down from τ, until its counter is 0 or it sees $\vdash$ (whichever happens first), and then returns to cell x^*; during this trip, it tests the optional $x = c$ and the zero or more $\neg(x = c)$ for $c > 0$. It then performs a symmetric trip of $\leq \tau$ steps to the right of cell x^* and back, during which it tests the optional $x = c$ and the zero or more $\neg(x = c)$ for $c < 0$. Finally, it accepts if all tests succeeded. Easily, $O(\tau)$ states are enough.   □

Fact 5. *Every not-identically-false floating relational $\wedge$-clause $\varphi(x, y)$ is equivalent to $S(x, y)$, $x = y$, $S(y, x)$, or a conjunction of $\neg S(x, y)$, $\neg(x = y)$, $\neg S(y, x)$.*

Lemma 4. *Suppose $\varphi(x, y)$ is an $\wedge$-clause of margin τ which does not contain both anchored local and floating relational atoms. Then there exists a $O(\tau)$-state 2NFA which, whenever run on a string w from a cell $1 \leq x^* \leq |w|$, computes so that, for all $1 \leq y^* \leq |w|$:*

$$\begin{array}{c} \text{a computation path which} \\ \text{halts \& accepts on cell } y^* \text{ exists} \end{array} \iff w[x/x^*, y/y^*] \models \varphi(x, y). \quad (11)$$

Proof. If φ is trivial, then the 2NFA is also trivial. So, assume φ is non-trivial. Let $\phi, \chi(x), \psi(y), \omega(x, y)$ be the $\wedge$-clauses given by Fact 3. Since anchored local and floating relational atoms cannot co-exist, at least one of ϕ and ω is empty.

Case 1. Suppose ω is empty. Then, when run on w from cell x^*, our 2NFA N must create nondeterministic branches which collectively accept on every cell y^* such that $\phi \wedge \chi(x) \wedge \psi(y)$ holds if $x = x^*$ and $y = y^*$. For this, N first checks χ on cell x^*; then resets its head (forgetting x^*) and reads the ends of w to check ϕ; then sweeps w and, on every cell y^*, guesses and verifies that ψ is true on y^*.

Specifically, let Φ, X, Ψ be the $O(\tau)$-state 2DFAs given by Lemma 2 for ϕ and by Lemma 3 for χ and ψ, respectively. Starting on cell x^*, N first simulates X. This brings it back to x^* having checked χ on x^*. Then N goes to $\vdash$ and starts simulating Φ. This brings it back to $\vdash$ having checked ϕ. Then N scans w and, on every cell y^*, spawns a new branch which simulates Ψ, eventually returning to y^* having checked ψ on y^*. Finally, N accepts (in that branch) iff all checks succeeded. Easily, N satisfies (11) and has size $O(|\Phi| + |X| + |\Psi|) = O(\tau)$.

Case 2. Suppose ϕ is empty. Then φ is equivalent to $\chi(x) \wedge \psi(y) \wedge \omega(x, y)$, where ω is not identically false (since φ is non-trivial), and thus is equivalent to one of $S(x, y)$, $x = y$, $S(y, x)$, or to a conjunction of their negations (Fact 5).

2a. If ω is equivalent to $S(x, y)$: Then the branches of N must collectively accept on every cell y^* such that $\chi(x) \wedge \psi(y) \wedge S(x, y)$ holds when $x = x^*$ and $y = y^*$. Because of $S(x, y)$, the only possible y^* of this kind is $x^* + 1$. So, N should just accept on cell $x^* + 1$ iff $\chi(x) \wedge \psi(y)$ holds when $x = x^*$ and $y = x^* + 1$. Hence, N starts on x^* by simulating X. This brings it back to x^* having checked χ on x^*. Then it moves one cell to the right, checks that it is not $\dashv$, and starts simulating Ψ, eventually returning to the cell, having checked ψ on $x^* + 1$. In the end, N accepts iff all checks succeeded. Note that N is, in fact, deterministic.

2b. If ω is equivalent to $x = y$: Then φ is equivalent to $\chi(x) \wedge \psi(y) \wedge x = y$, so the only possible y^* is x^*. Hence, N works as in Case 2a, but without the one step to the right between the simulations of X and of Ψ.

2c. If ω is equivalent to $S(y, x)$: Then φ is equivalent to $\chi(x) \wedge \psi(y) \wedge S(y, x)$, so the only possible y^* is $x^* - 1$. So, N works as in Case 2a, except that, between the simulations of X and of Ψ, it moves left and checks that it does not read $\vdash$.

2d. If ω is equivalent to a conjunction of $\neg S(x, y), \neg(x = y), \neg S(y, x)$: Then ω excludes a certain set of cells $Y_\omega \subseteq \{x^* - 1, x^*, x^* + 1\}$ from being accepted. So, N must accept on cell y^* iff $\chi(x) \wedge \psi(y)$ holds for $x = x^*, y = y^*$ and $y^* \notin Y_\omega$. As above, N starts on x^* by simulating X, and returns on it after checking χ on x^*.

Then it spawns five branches, one for each of the five cases as to where cell y^* is with respect to cell x^*: before x^*-1, on x^*-1, on x^*, on x^*+1, or after x^*+1.

- In the first branch: N moves left by two cells, checking that neither is $\vdash$. It then sweeps up to $\vdash$ and, on each cell y^*, spawns a branch which simulates Ψ and eventually returns on y^* having checked ψ on it.
- In the second branch: If $x^*-1 \in Y_\omega$ (i.e., if ω contains $\neg S(y, x)$), then N just rejects. Otherwise, it moves left once, checks that it is not on $\vdash$, then simulates Ψ. This brings it back to the same cell x^*-1, having checked ψ on it.
- In the third and fourth branches: N works similarly to the second one. It just rejects, if $x^* \in Y_\omega$ (i.e., if ω contains $\neg(x=y)$) or if $x^*+1 \in Y_\omega$ (i.e., if ω contains $\neg S(x, y)$), respectively. Otherwise, it simulates Ψ after, respectively, not moving at all or moving once to the right.
- In the last branch: N works symmetrically to the first one. It moves right by two cells checking against $\dashv$, and then simulates Ψ on each cell before $\dashv$.

In all cases, N accepts in a given branch iff all checks along it have succeeded.

Easily, in all four cases, N satisfies (11) and contains one copy of each of X and Ψ, plus $O(1)$ more states, for a total size of $O(|X|+|Y|) = O(\tau)$. □

Lemma 5. *Every weak* GA/DNF$_1$ *of degree m and margin τ is equivalent to a* 2NFA *with $O(m\tau)$ states.*

Proof. Let $\psi = \mathsf{TC}[\varphi(\overline{x}, \overline{y})](\overline{s}, \overline{t})$ be as in the statement. Let the arity be $k+1$. Then $s_0, \ldots, s_k, t_0, \ldots, t_k \in \mathbb{Z}_\tau^+ \cup \mathbb{Z}_\tau^-$ and the core φ has the form $\bigvee_{i=1}^m \varphi_i(\overline{x}, \overline{y})$, where (Fact 2) each φ_i is equivalent to:

$$(x_1 = c_1^i) \wedge \cdots \wedge (x_k = c_k^i) \ \wedge \ (y_1 = d_1^i) \wedge \cdots \wedge (y_k = d_k^i) \ \wedge \ \hat{\varphi}_i(x_0, y_0), \quad (12)$$

for some constants $c_1^i, \ldots, c_k^i, d_1^i, \ldots, d_k^i \in \mathbb{Z}_\tau^+ \cup \mathbb{Z}_\tau^-$ and an $\wedge$-clause $\hat{\varphi}_i$ of margin τ where anchored local and floating relational atoms do not co-exist (as ψ is weak).

We build a 2NFA N which accepts an input $w \in \Sigma^*$ of length $n \geq 2$ iff $w \models \psi$, i.e., iff the graph $G_{\varphi,w}$ (see p. 5) has a path from vertex $\overline{s}$ to vertex $\overline{t}$. To check this, N nondeterministically guesses such a path *in stages*, in the standard way: starting each stage, it remembers only the last vertex $\overline{u}$ of the path guessed so far (originally, $\overline{u} := \overline{s}$); then it checks whether $\overline{u} = \overline{t}$ and, if so, accepts; otherwise, it nondeterministically selects a neighbor $\overline{v}$ of $\overline{u}$ and updates its memory to $\overline{u} := \overline{v}$, completing the stage. Below, we describe how N implements this algorithm.

Central in this implementation is how N remembers $\overline{u}$. Clearly, $\overline{u}$ will always be a vertex reachable from $\overline{s}$, so the following fact becomes important:

Claim. If $\overline{u}$ is reachable from $\overline{s}$, then $(u_1, \ldots, u_k) = (s_1, \ldots, s_k)$ or there is $i = 1, \ldots, m$ such that $(u_1, \ldots, u_k) = (d_1^i, \ldots, d_k^i)$; either way, $u_1, \ldots, u_k \in \mathbb{Z}_\tau^+ \cup \mathbb{Z}_\tau^-$.

Proof. If $\overline{u} = \overline{s}$, the claim is trivial. Suppose $\overline{u} \neq \overline{s}$. Then the path $\overline{s} \rightsquigarrow \overline{u}$ has ≥ 1 step. Let $\overline{v} \to \overline{u}$ be the last one. Then $(\overline{v}, \overline{u})$ is an edge in $G_{\varphi,w}$, so $w[\overline{x}/\overline{v}, \overline{y}/\overline{u}]$ satisfies $\varphi(\overline{x}, \overline{y})$; hence, it satisfies some $\varphi_i(\overline{x}, \overline{y})$; so, it satisfies the corresponding $(y_1 = d_1^i) \wedge \cdots \wedge (y_k = d_k^i)$; which implies that $(u_1, \ldots, u_k) = (d_1^i, \ldots, d_k^i)$. ⊡

So, N separates $\overline{u}$ into (1) its 'bounded components' $u_1, \ldots, u_k \in \mathbb{Z}_\tau^+ \cup \mathbb{Z}_\tau^-$; and (2) its 'unbounded component' $u_0 \in \mathbb{Z}_n^+$. To remember (1), it keeps in its state an index $0 \leq i \leq m$ such that $(u_1, \ldots, u_k) = (d_1^i, \ldots, d_k^i)$ —for convenience, let $(d_1^0, \ldots, d_k^0) := (s_1, \ldots, s_k)$. To remember (2), it places its head on cell u_0 of w.

Overall, each state of N is of the form (i, σ), where i identifies (as described) the list $u_1, \ldots, u_k$ and σ shows the status of the current stage. As a special case, $\sigma = \mathsf{B}$ means the stage has just begun. So, if N is in state (i, B) on cell u^*, then it has reached vertex $\overline{u} = (u^*, d_1^i, \ldots, d_k^i)$ and is now beginning the next stage.

With this representation, the search for a path $\overline{s} \leadsto \overline{t}$ takes N through configurations $((i_0, \mathsf{B}), u_0^*), ((i_1, \mathsf{B}), u_1^*), \ldots, ((i_l, \mathsf{B}), u_l^*)$, where $u_0^* = s_0, i_0 = 0$; and the search succeeds iff $u_l^* = t_0$ and $(d_1^{i_l}, \ldots, d_k^{i_l}) = (t_1, \ldots, t_k)$. To complete the description of N, we must explain how N navigates through these configurations.

In a special first stage, N alters its configuration from $(q_\mathsf{s}, 0)$ to $((i_0, \mathsf{B}), u_0^*) = ((0, \mathsf{B}), s_0)$. For this, it moves its head to cell s_0 (by counting s_0 steps from $\vdash$, if $s_0 > 0$; or by moving to $\dashv$ and counting s_0 steps backwards, if $s_0 < 0$) and switches to state $(0, \mathsf{B})$. Easily, this can be done with $O(s_0) = O(\tau)$ states.

From then on, whenever at a configuration $((i, \mathsf{B}), u^*)$, our N works as follows.

First, it checks if $\overline{u} = \overline{t}$, i.e., if (1) $u^* = t_0$ and (2) $(d_1^i, \ldots, d_k^i) = (t_1, \ldots, t_k)$. Check 2 is hardwired, so it needs no extra states. Check 1 involves a trip to the left (if $t_0 > 0$) or right (if $t_0 < 0$) for t_0 steps or up to the end-marker, and back to cell u^*. There, if both checks succeeded, N accepts; otherwise, it switches to a special state (i, C). Overall, this uses $O(t_0) = O(\tau)$ states of the form $(i, .)$.

State (i, C) means that N is about to choose the next vertex $\overline{v}$ among the out-neighbors of $\overline{u}$, so as to switch to the appropriate next configuration $((., \mathsf{B}), .)$. Note that $\overline{v}$ is an out-neighbor of $\overline{u}$ iff $(\overline{u}, \overline{v})$ is an edge of $G_{\varphi, w}$; i.e., iff $w[\overline{x}/\overline{u}, \overline{y}/\overline{v}]$ satisfies some $\wedge$-clause $\varphi_j(\overline{x}, \overline{y})$ as in (12); i.e., iff there exists j such that

- the bounded components $v_1, \ldots, v_k$ of $\overline{v}$ are equal to the second tuple of constants $d_1^j, \ldots, d_k^j$ in one of the φ_j whose first tuple of constants $c_1^j, \ldots, c_k^j$ are the bounded components $d_1^i, \ldots, d_k^i$ of $\overline{u}$, namely:

$$(d_1^i = c_1^j) \wedge \cdots \wedge (d_k^i = c_k^j) \ \wedge \ (v_1 = d_1^j) \wedge \cdots \wedge (v_k = d_k^j) ; \quad \text{and}$$

- the unbounded component v_0 of $\overline{v}$ together with the unbounded component u^* of $\overline{u}$ satisfy the respective $\hat{\varphi}$: $w[x_0/u^*, y_0/v_0] \models \hat{\varphi}_j(x_0, y_0)$.

So, to nondeterministically choose such a $\overline{v}$, our N works in two sub-stages:

- First, it chooses $v_1, \ldots, v_k$, by simply choosing the index j of some $\wedge$-clause (if any) whose first tuple of constants is exactly $d_1^i, \ldots, d_k^i$. This selection is hardwired and takes N to a special state (j, D) still on cell u^*.
- Then, it chooses v_0, by simulating the $O(\tau)$-state 2NFA given by Lemma 4 for $\hat{\varphi}_j$, from cell u^* up to every cell v^* such that $w[x_0/u^*, y_0/v^*] \models \hat{\varphi}_j(x_0, y_0)$. This needs $O(\tau)$ states of the form $(j, .)$ and ends at a state (j, B).

Overall, the result is a nondeterministic computation whose accepting branches take N to all configurations $((j, \mathsf{B}), v^*)$ such that $\overline{v} = (v^*, d_1^j, \ldots, d_k^j)$ is an out-neighbor of $\overline{u} = (u^*, d_1^i, \ldots, d_k^i)$. This concludes our description of a full stage.

In total, N uses $O(\tau)$ states for the special first stage and, for each i, another $O(\tau) + O(1) + O(\tau) = O(\tau)$ states for every stage that starts after state (i, B). So, the total number of states is $O(\tau) + (1+m) \cdot O(\tau) = O(m\tau)$, as promised. $\square$

6 Conclusion

Completing [6], we descriptively characterized 2N. We can show that Theorem 1 is tight, in that its sentences can solve non-regular problems, if two-dimensional; or problems outside 2N, if the core is in CNF. It would be nice to see descriptive characterizations for other minicomplexity classes, too.

References

1. Büchi, R.J.: Weak second-order arithmetic and finite automata. Zeitschrift für mathematische Logik und Grundlagen der Mathematik **6**(1–6), 66–92 (1960)
2. Fagin, R.: Generalized first-order spectra and polynomial-time recognizable sets. In: Karp, R.M. (ed.) Complexity of Computation. AMS-SIAM Symposia in Applied Mathematics, vol. VII, pp. 43–73 (1974)
3. Immerman, N.: Nondeterministic space is closed under complementation. SIAM J. Comput. **17**(5), 935–938 (1988)
4. Immerman, N.: Descriptive Complexity. Springer, New York (1998)
5. Kapoutsis, C.: Minicomplexity. J. Automata Lang. Comb. **17**(2–4), 205–224 (2012)
6. Kapoutsis, C.A., Lefebvre, N.: Analogs of Fagin's Theorem for small nondeterministic finite automata. In: Yen, H.-C., Ibarra, O.H. (eds.) DLT 2012. LNCS, vol. 7410, pp. 202–213. Springer, Heidelberg (2012)

Experiments with Synchronizing Automata

Andrzej Kisielewicz, Jakub Kowalski, and Marek Szykuła[✉]

Department of Mathematics and Computer Science,
University of Wrocław, Wrocław, Poland
andrzej.kisielewicz@math.uni.wroc.pl, {jko,msz}@cs.uni.wroc.pl

Abstract. We have improved an algorithm generating synchronizing automata with a large length of the shortest reset words. This has been done by refining some known results concerning bounds on the reset length. Our improvements make possible to consider a number of conjectures and open questions concerning synchronizing automata, checking them for automata with a small number of states and discussing the results. In particular, we have verified the Černý conjecture for all binary automata with at most 12 states, and all ternary automata with at most 8 states.

1 Introduction

A deterministic finite automaton $\mathcal{A}$ is $\langle Q, \Sigma, \delta \rangle$, where Q is the set of the states, Σ is the input alphabet, and $\delta \colon Q \times \Sigma \to Q$ is the (complete) transition function. Throughout the paper, by n we denote the number of states $|Q|$. If $|\Sigma| = k$ then $\mathcal{A}$ is called k-ary. The transition function δ is naturally extended to a function $2^Q \times \Sigma^* \to 2^Q$. The *image* of $S \subseteq Q$ under the action of a word $w \in \Sigma^*$ is $Sw = \{\delta(q, w) \mid q \in S\}$. The *rank* of a word $w \in \Sigma^*$ is $|Qw|$, and the *rank* of $\mathcal{A}$ is the minimal rank of a word over $\mathcal{A}$. For a non-empty subset $\Sigma' \subseteq \Sigma$, we may define the automaton $\mathcal{A}' = \langle Q, \Sigma', \delta' \rangle$, where δ' is the natural restriction of δ to Σ'. In such a case $\mathcal{A}$ is called an *extension* of $\mathcal{A}'$. The automata of rank 1 are called *synchronizing*, and each word w with $|Qw| = 1$ is called a *synchronizing* (or *reset*) word for $\mathcal{A}$. An automaton is *irreducibly synchronizing* if it is not an extension of a synchronizing automaton over a smaller alphabet.

We are interested in the length of a shortest reset word for $\mathcal{A}$ (there may be more than one word of the same shortest length). We call it the *reset length* of $\mathcal{A}$. The famous Černý conjecture states that every synchronizing automaton $\mathcal{A}$ with n states has a reset word of length $\leq (n-1)^2$ [9]. This conjecture was formulated by Černý in 1969 and is considered the longest-standing open problem in combinatorial theory of finite automata. So far, the conjecture has

---

A. Kisielewicz—Supported in part by the National Science Centre, Poland under project number 2012/07/B/ST1/03318.

J. Kowalski—Supported in part by the National Science Centre, Poland under project number 2015/17/B/ST6/01893.

M. Szykuła—Supported in part by the National Science Centre, Poland under project number 2013/09/N/ST6/01194.

Y.-S. Han and K. Salomaa (Eds.): CIAA 2016, LNCS 9705, pp. 176–188, 2016.
DOI: 10.1007/978-3-319-40946-7_15

been proved only for a few special classes of automata, and a cubic upper bound $(n^3 - n)/6 - 1$ [19] has been established, which was not improved for over 30 years (see [15, 24] for excellent surveys). The bound $(n - 1)^2$ is met for every n by the Černý automata [9], which is the only known infinite series of automata meeting this bound (besides that, there are 8 known particular examples with $n \leq 6$ states [22] also meeting the bound).

There were several efforts to check computationally the conjecture for all automata with a small number of states. In particular, Ananichev et al. [3, 4] have checked all binary automata with at most $n = 9$ states, and the checking for all automata with at most $n = 10$ states was reported in [22]. In [16], using a dedicated algorithm, we have verified the conjecture for all binary automata with $n \leq 11$ states.

In this paper, first we describe improvements to our algorithm from [16], which are aimed at making possible verifying the conjectures for larger automata. While these are results of a rather technical nature, and may be not very interesting from theoretical point of view, they make possible to restrict the computation process to much smaller class of relevant automata, and thus to consider also automata with a larger number of states.

We extend verification of the Černý conjecture up to 12 states and present an extensive experimental study on important problems and conjectures closely related to upper bounds on reset lengths. We consider known conjectures, and restate or state new ones basing on our experiments. Most of them imply an improvement for the general cubic bound, and hence are very hard but stand as possible ways to attack the main problem. All of the conjectures are experimentally confirmed for automata with a small number of states and/or letters.

2   Reset Lengths of Extensions

In this section we describe two theoretical results we apply in the improved algorithm. We are interested mainly in estimating the reset length of synchronizing automata that arise as extensions of non-synchronizing automata by one letter. In some cases, we are able to provide better upper bounds than the general bound $(n^3 - n)/6 - 1$ [19].

In particular, we search for synchronizing automata with relatively large reset length. We improve the algorithm from [16] which takes a set of $(k - 1)$-ary automata with n states and generates all their nonisomorphic one-letter extensions. To perform an exhaustive search over the k-ary automata with n states with some property, we need to progressively run the algorithm $k - 1$ times starting from the complete set of non-isomorphic unary automata. However, in each run, if we know that any extension of an automaton $\mathcal{A}$ cannot have the desired property, we can safely drop $\mathcal{A}$ from further computations. Since the number of generated automata grows rapidly, suitable knowledge saves a lot of computational time and extends the class of the automata investigated. The technical details of the algorithm and proofs can be found in [17].

A subset $M \subseteq Q$ of the states is called *compressible*, if there is a word w such that $|Mw| < |M|$. Let $\mathcal{A} = \langle Q, \Sigma, \delta \rangle$ be a finite automaton. We say that a sequence (M_i, x_i, y_i), $(1 \le i \le \ell)$ of m-subsets (subsets of size m) M_i of Q and pairs of states $x_i, y_i \in Q$ is an *m-subset Frankl-Pin sequence* if the following conditions are satisfied

1. $x_i, y_i \in M_i$ for $1 \le i \le \ell$;
2. either x_i or y_i is not in M_j for all $1 \le j < i \le \ell$.

If all the pairs $\{x_i, y_i\}$ belong to a set P of pairs, we will say that this sequence is *over* P. Given a set P of *compressible* pairs, by the *synchronizing height* $h(P)$ of P we mean the minimal h such that for each pair $\{x, y\} \in P$ there exists a word w of length h such that $xw = yw$.

It is known that a shortest word compressing M cannot be longer than the length of the Frankl-Pin sequence starting from M [12] (this, in fact, is used to obtain the bound $(n^3 - n)/6$ mentioned above). Our first technical improvement is that if the synchronizing height is smaller than the maximal length of a Frankl-Pin sequence over P, then we have

Theorem 1. *Let P be a set of compressible pairs in $\mathcal{A}$, $h(P)$ the synchronizing height of P, and $p(P)$ the maximal length of a Frankl-Pin sequence over P. Then, for every compressible m-subset M of Q $(2 \le m \le n)$, there is a word compressing M whose length does not exceed*

$$\binom{n - m + 2}{2} - p(P) + h(P).$$

This result improves the estimation in [12] by the negative summand $(p(P) - h(P))$. It is to be combined and compared with the result by Pin [18] saying that if w is a word of rank r and there exists a word of rank $\le r - 1$, then there is such a word of length $\le 2|w| + n - r + 1$. There are other results of this kind that can be used for providing bounds for extensions, as that in [6]. Unfortunately, for small values of n that are within our considerations, this does not overcome the bound from Theorem 1.

Recall that an automaton $\mathcal{A} = \langle Q, \Sigma, \delta \rangle$ is *one-cluster*, if it has a letter $a \in \Sigma$ such that for every pair $q, s \in Q$ there are $i, j \ge 1$ such that $qa^i = sa^j$. This means that the graph of the transformation induced by a is connected. In particular, it has a unique cycle $C \subseteq Q$ with the property $Ca^i = C$ for every $i \ge 0$, and there is $\ell \ge 0$ such that $Qa^\ell = C$. The least such ℓ is called the *level* of $\mathcal{A}$. Steinberg [21] proved that if the length m of the cycle is prime, then the one-cluster automaton $\mathcal{A}$ has a reset word of length at most

$$n - m + 1 + 2\ell + (m - 2)(n + \ell). \tag{1}$$

We generalize this result to arbitrary lengths and get an additional negative summand. We refine the proof of Steinberg [21] and the summand is expressed in algebraic terms of the proof. Therefore, to present the result we have to recall basic notations from [21].

Given a one cluster automaton with the notation as above, we consider the matrix representation $\pi : \Sigma^* \to M_n(\mathbb{Q})$ defined by $\pi(w)_{q,r} = 1$ if $qw = r$, and

0, otherwise. Given $S \subseteq Q$ we define $[S]$ to be the characteristic row vector of S in $\mathbb{Q}^n$, $[S]^T$ its transpose, and $\gamma_S = [S]^T - (|S|/|C|)[Q]^T$. By $w\gamma_S$ we denote the product of corresponding matrices; in particular, the word w represents the matrix $\pi(w)$, and the product is a vector in the space $\mathbb{Q}^n$. We consider the subspace $W_S = \text{Span}\{a^{\ell+j}\gamma_S \in \mathbb{Q}^n \mid 0 \le j \le m - 1\}$ (cf. [21]), and the *cyclic period* q_S of S, understood as the least number q such that $Sa^q = S$. Now, we define $D^*(m, k)$ to be the minimal value of $m - q_S + \dim W_S$ taken over all vectors S with $|S| = k$. Then we prove the following:

Theorem 2. *Let $\mathcal{A} = \langle Q, \Sigma, \delta \rangle$ be a synchronizing automaton with n states, such that there exists a word w of length s inducing a one-cluster transformation with level ℓ and cycle C of length $m > 1$. Then $\mathcal{A}$ has a reset word of length at most*

$$s(\ell + m - 2)(m - 1) + (n + 1)(m - 1) + s\ell - \sum_{k=1}^{m-1} D^*(m, k).$$

One can demonstrate that this results generalizes and improves earlier bounds in [8, 21], and a careful estimation of the summand $D^*(m, k)$ yields the currently best general bound for reset lengths of one-cluster automata:

Corollary 3. *A synchronizing one-cluster automaton $\mathcal{A}$ with n states and the cycle of length m has a reset word of length at most*

$$2nm - 4m \ln \frac{m + 3}{2} + 2m - n + 1 \tag{2}$$

Nevertheless, for small values of m we can compute the exact values of $D^*(m, k)$, and this yields considerably better bounds than the general estimation above.

3   Experiments and Conjectures

In this section we discuss the results of our experiments with the improved algorithm concerning various conjectures and open problems in the area.

The Černý Conjecture. We have verified the Černý conjecture for several cases. In particular, we confirmed it for all binary automata with $n \le 12$ states, and for all ternary automata with $n \le 8$.

Verifying the Černý conjecture for binary automata with $n = 12$ states was the most difficult computation that we have performed here. The total time of a single processor core spent for this computation was about 100 years. We performed this on a grid in parallel using mostly about 200 cores of Quad-Core AMD Opteron(tm) Processor 8350, 2.0 GHz. The total number of automata generated by our algorithm in this case was about 10^{15}.

For ternary automata with $n = 8$ states the computation took 1.25 years of a single processor core, and we had to generate and check about 2.1×10^{10} automata. One may compare these numbers with the numbers of non-isomorphic initially connected automata that one would need to generate applying the technique described in [3]. The corresponding numbers are: about 2.2×10^{17} for binary automata with $n = 12$ states, and 5.7×10^{17} for ternary automata with $n = 8$ states.

Within the range we have considered, the only automata meeting the bound $(n-1)^2$ other than the Černý series are known examples with $n \leq 6$ states that were presented in [22].

Slowly Synchronizing Automata. For the case of binary automata $n = 12$ states, we have obtained also the complete list of strongly connected synchronizing automata with reset length ≥ 94.

Table 1. The numbers of all non-isomorphic strongly connected synchronizing binary automata with 12 states with reset length ≥ 94.

Reset length	94	95–98	99	100	101	102	103–109	110	111	112	113–120	121
Number of automata	3	0	3	21	9	2	0	2	1	1	0	1
Series						$\mathscr{H}_n,\mathscr{\ddot{H}}_n$		$\mathscr{E}_n,\mathscr{D}''_n$	$\mathscr{W}_n$	$\mathscr{D}'_n$		$\mathscr{C}_n$

Table 1 shows the exact numbers of automata in this range, and the corresponding series according to naming from [3,4,16]. Here, all automata with reset length ≥ 99 has a similar structure of one long cycle and a small gadget (cf. [4]), and they can be generalized to series of length $n^2 - O(n)$ as well. We confirm, for $n \leq 12$, [4, Conjecture 1], which is a generalization of the Černý conjecture, describing all binary synchronizing automata with reset length $\geq n^2 - 4n + 8$ (104 for $n = 12$) and stating that up to isomorphism this list is complete.

As observed in [4,16,22], there are gaps in the set of possible reset lengths near the Černý bound $(n-1)^2$. We confirm for binary automata that for $n = 6, 7, 8$ there is one gap, for $n = 9, 10$ there are two gaps, and for $n = 11, 12$ there are three gaps.

There is no binary strongly connected automaton with 12 states and reset length 95, but we have constructed such an automaton over a ternary alphabet (Fig. 1). Similarly, we know an automaton for $n = 9$ with reset length 53 (second gap), and for $n = 11$ with reset length 79 (third gap). This shows that the gaps, except the first one, are not necessarily preserved over larger alphabets.

Extending Words in One-Cluster Automata. One-cluster automata are an important class of synchronizing automata for which a quadratic bound on reset length has been found [5,21].

Despite several attempts [5,8,10,20,21] at improving the bounds, so far, the Černý conjecture has been proved only for one-cluster automata with a cycle of length n (circular automata) or with a prime-length cycle. In [21] an algebraic argument making use of ascending chain of linear subspaces and averaging trick has been applied. The proof is based on the claim that any subset $S \subset C$ on the

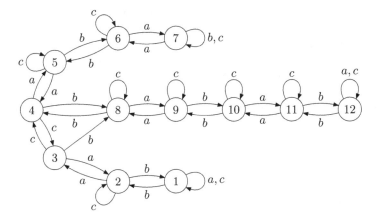

Fig. 1. An irreducibly synchronizing strongly connected ternary automaton with 12 states and reset length 95.

cycle C can be extended on this cycle by a word of length at most $\ell + n$ (we apply here the notation of Sect. 2). It is demonstrated that this holds in the case of prime length of C. Proving it for non-prime lengths would provide the proof of the Černý conjecture for the whole class of one-cluster automata.

We have exhaustively searched for small examples of one-cluster automata with a non-prime cycle length such that the length $\ell + n$ of extending words is exceeded for some subset S, but found out that $\ell + n$ is sufficient in all tested cases, instead of the value $n + \ell + |C| - D^*(|C|, |S|)$ used to prove the bound from Theorem 2. Also, we found out that we can always use an extending word of the form wa^ℓ with $|w| \leq n$, which is the form used in the proof for prime $|C|$.

Conjecture 1. Let $\mathcal{A}$ be a one-cluster synchronizing automaton with a one-cluster letter a with the cycle C and level ℓ. For any non-empty proper subset $S \subset C$ there is a word w such that $|S(wa^\ell)^{-1} \cap C| > |S|$ and $|w| \leq n$.

In all the cases tested, for any ℓ, non-prime $|C| < n$, and $|S|$ with $1 \leq |S| < |C|$, we found an automaton for which we needed a word w of length exactly n. So, it seems that the bound $|w| \leq n$ is tight.

Worst Cases for the Greedy Compressing Algorithm. The *greedy compressing algorithm* is a well known approach for finding a reset word [11,19,24]. It starts from $S = Q$, and iteratively finds a shortest word w such that $|Sw| < |S|$ and uses Sw for next iteration, until $|S| = 1$. The concatenated words w form the found reset word. The length of the resulted reset word can vary, since there is ambiguity in selection of shortest words w. By bounding the length of the found reset word we also obtain an upper bound for the reset length, and in fact, the upper bound $(n^3 - n)/6$ for the reset length is obtained by bounding the lengths of words w for $|S| = 2, \ldots, n$ and summing these bounds [19]. It is known that this algorithm finds a word of length $\Omega(n^2 \log n)$ for the Černý automaton [15], but it was not clear whether it is the worst case example.

We experimentally tested the greedy algorithm for the worst cases. Here, we restricted the studied class to irreducibly synchronizing automata, as otherwise we would get a lot of trivial examples derived from automata over a smaller alphabet. By the *worst case length* we mean the maximum length of the found word by the algorithm over all selections of shortest compressing words that can be taken by the algorithm. For example, for automaton $\mathcal{G}_1$ from Fig. 2, the worst case length is 19 and a sequence of subsets considered by the greedy algorithm in the worst case can be the following:

$$Q = \{1, 2, 3, 4, 5\} \xrightarrow{b} \{1, 2, 4, 5\} \xrightarrow{aca} \{3, 4, 5\} \xrightarrow{bcbacb} \{1, 4\} \xrightarrow{acbbcbaca} \{3\}.$$

While this requires potentially very expensive computation, the worst case length can be computed by a kind of dynamic algorithm and $n-1$ iterations of breadth-first search in the power automaton.

It may be surprising that the Černý automata generally do not exhibit the worst case length. We have observed that for some values $n \geq 10$ the slowly synchronizing series $\mathscr{W}_n$, $\mathscr{D}_n''$, and $\mathscr{G}_n$ (see [3,4]) exceed the worst case length of the Černý automaton with the same number of states. In addition, we have found out four particular ternary examples shown in Fig. 2 exceeding the worst case length of the Černý automaton with the same number of states, which do not seem generalizable to series. Up to isomorphism, there are no more such examples within the range we have considered (Table 2).

The results we have collected do not allow to state a reasonable conjecture. So far, $\mathscr{W}_n$ is the best candidate for the largest worst case lengths for $n \geq 10$, and the Černý automata for $n \leq 9$, except $\mathcal{G}_1$ and $\mathcal{G}_2$ from Fig. 2 for $n = 5, 6$.

Problem 2. What are the largest worst case lengths of the greedy compressing algorithm of automata with n states?

It is noticeable that the dual *greedy extending algorithm*, which starts from a singleton and uses shortest extending words rather than compressing ones, seem to have generally larger worst case lengths. For example, for the case of binary $n = 7$ in the worst case it can find a reset word of length 48 for some strongly connected automaton, whereas the greedy compressing algorithm finds a word of length at most 43.

Aperiodic Synchronizing Automata. Recall that an automaton is *aperiodic* if there is no word inducing a transformation with a cycle of length ≥ 2 (the transition semigroup has only trivial subgroups). In [24] Volkov mentioned that although a quadratic upper bound for the reset length of aperiodic synchronizing automata has been proved, the largest reset length for known aperiodic automata does not exceed $n + \lfloor n/2 \rfloor - 2$. This length is reached by a series of binary automata constructed by Ananichev [2]. In this connection, it may be interesting

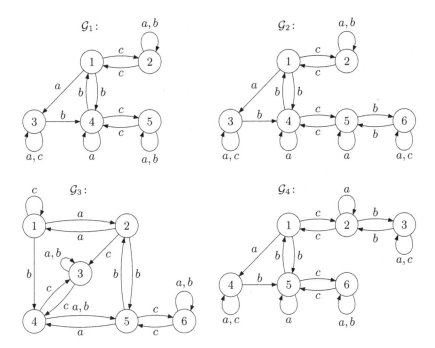

Fig. 2. Automata $\mathcal{G}_1$, $\mathcal{G}_2$, $\mathcal{G}_3$, and $\mathcal{G}_4$, with the worst case length 19, 30, 28, and 28, and reset lengths 15, 22, 20, and 20, respectively.

to know that the same bound is also reached for every $n > 1$ by a series of irreducibly ternary aperiodic automata. It has a quite simple definition and an easy proof for the reset length (comparing with [2]). Let $\mathcal{A}_n = \langle Q, \{a, b, c\}, \delta \rangle$, where $Q = \{v_1, \ldots, v_n\}$, $\delta(v_i, a) = v_{i+1}$ for $1 \le i \le n - 2$, $\delta(v_i, b) = v_{i-1}$ for $2 \le i \le n - 1$, $\delta(v_{\lfloor n/2 \rfloor}, c) = v_n$, and $\delta(v_i, x) = v_i$, otherwise $(x \in \Sigma)$ (shown in Fig. 3).

Volkov[1] has also pointed out that $n - 1$ may be an upper bound for the reset length in the class of *strongly connected* synchronizing aperiodic automata, but

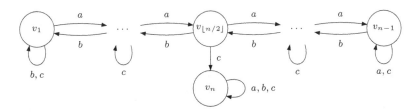

Fig. 3. A ternary irreducibly synchronizing n-state aperiodic automaton with the reset length $n + \lfloor n/2 \rfloor - 2$.

---

[1] Personal communication.

there was not enough evidence. The bound can be met trivially if the underlying digraph of the automaton is a *bidirectional path*: $Q = 1, \ldots, n$, for every $1 \le i \le n - 1$ there are the directed edges $(i, i + 1)$ and $(i + 1, i)$, and every edge that is not a loop is of that form.

Since our verifications involve a huge number of aperiodic automata, we experimentally support the following conjectures:

Conjecture 3 (cf. [24]). Every synchronizing aperiodic automaton with $n > 1$ states has a reset word of length at most $\le n + \lceil n/2 \rceil - 2$.

Conjecture 4 (Volkov). Every strongly connected synchronizing automaton has a reset word of length at most $n - 1$. Moreover, if this bound is met, then the underlying digraph of the automaton is a bidirectional path.

Avoiding States. In a recent short note [13] the authors state the following problem related to the recent unsuccessful attempt of improving the general upper bound on reset length [23]: Given a strongly connected synchronizing automaton, what is the minimal length ℓ such that for any $q \in Q$ there is a word w of length $\le \ell$ and such that $q \notin Qw$. If $\ell \in O(n)$, then we would obtain a better upper bound than $(n^3 - n)/6$.

Experimentally, we have found out what the value of ℓ for a given n might be, and provided support for the following conjecture:

Conjecture 5. In a synchronizing strongly connected automaton, for any $q \in Q$ there is a word w of length $\le 2n - 2$ and such that $q \notin Qw$. This bound is tight for $n \ge 4$ over a ternary alphabet.

Recently, Vojtěch Vorel[2] discovered an infinite series of binary automata whose minimal length in question is $2n - 4$, which is currently the best theoretical lower bound for the problem.

New Rank Conjecture. Pin [19] proposed the following generalization of the Černý conjecture: For every $0 < d, n$, if there is a word of rank $\le n - d$, then there is such a word of length $\le d^2$. Pin proved this for $d \le 3$. However, Kari [14] found a celebrated counterexample to this conjecture for $d = 4$, which is a binary automaton $\mathcal{K}$ with 6 states (Fig. 4). As a consequence, a modification of this generalized conjecture was proposed restricting it to d being the rank of the considered automaton (see for example [1]). However this seems to be a quite radical restriction.

In our computations, we have found no other counterexample to Pin's conjecture except for trivial extensions and modifications. This may suggest that Kari construction works due to the number of involved states small enough, and is, in fact, an exception. By a *trivial extension* of an automaton over alphabet Σ we mean one obtained by adding letters to Σ that acts either as the identity transformation or as any letter in Σ. So a trivial extension has the same number

---

[2] Personal communication, unpublished.

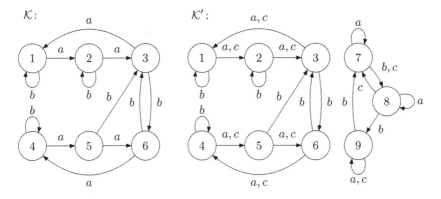

Fig. 4. The Kari automaton $\mathcal{K}$ [14], and a Kari-like automaton $\mathcal{K}'$.

of the states and the transition semigroup, and trivial extensions of the Kari automaton $\mathcal{K}$ are counterexamples to the Pin's conjecture, for $d = 4$, as well. By a *disjoint union* of two automata $\mathcal{A} = \langle Q, \Sigma, \delta \rangle$ and $\mathcal{A}' = \langle Q', \Sigma, \delta' \rangle$ we mean the construction where the automata have the same alphabet Σ, and disjoint sets of states Q, Q', and the union is simply $\mathcal{A} = \langle Q \cup Q', \Sigma, \delta \cup \delta' \rangle$. If we take a disjoint union of $\mathcal{K}$ with any *permutation automaton* (one whose letters act like permutations, or in other words, one of rank equal to its size), then again we get a counterexample to the Pin's conjecture, for $d = 4$. Yet, in all these automata the failure is caused by the same Kari construction on the set of the 6 states. In our experiments, we have discovered no other counterexample. This may be treated as an evidence for the conjecture we state below.

Consider the smallest class of automata containing $\mathcal{K}$ and closed on taking trivial extension and disjoint union with permutation automata. Let us call automata in this class *Kari-like* automata (see Fig. 4). Then we have

Conjecture 6. For every d, if an automaton $\mathcal{A}$ has a word of rank at most $n - d$, then there is such a word of length at most d^2, unless $\mathcal{A}$ is a Kari-like automaton and $d = 4$ (in which case there is a word of rank $n - 4$ of length $d^2 + 1 = 17$).

Subset Synchronization. The last conjecture was posed by Ângela Cardoso:

Conjecture 7 (Cardoso [7]). In a synchronizing automaton, for any subset S of states there is a word w with $|Sw| = 1$ of length at most

$$(n - 1)^2 - \left\lceil \frac{n - |S|}{|S|} \right\rceil \left(2n - |S| \left\lceil \frac{n}{|S|} \right\rceil - 1 \right).$$

This is another generalization of the Černý conjecture, and it can be viewed as a counterpart for the rank conjecture, where we bound the length of words compressing Q to a subset of the given size, rather than a subset to a singleton.

Conjecture 7 has been proved for several special classes of automata, and the formula is tight for any subset size in the Černý series. Besides confirmation for small automata, we identified 18 particular examples of irreducibly synchronizing automata with $n \in \{3, 4, 5, 6\}$ states meeting the bound for some subset S that are not isomorphic to the Černý automata. Note that the conjecture is not true in general for non-synchronizing automata, as Vorel [25] has constructed a series of non-synchronizing strongly connected binary automata with subsets whose shortest synchronizing words are of exponential length.

3.1   Summary

Table 2 summarizes the ranges for which the discussed conjectures have been confirmed or the problems checked. The ranges vary due to different numbers of automata that have to be checked, computational complexity of verification for a single automaton, and computation time devoted for each of the problems.

Table 2. Experimental verification of conjectures. The numbers denote the size of the alphabet up to which the given conjecture has been checked. The symbol ∞ denotes that the problem has been verified for all automata with the given number of states and any number of letters.

Problem	Number of states n								
	≤ 4	5	6	7	8	9	10	11	12
Černý conjecture and [4, Conjecture 1]	∞	∞	6	4	3	2	2	2	2
Conjecture 1 (one-cluster)	∞	5	4	3	2	2	2		
Problem 2 (greedy algorithm)	∞	6	4	3	2	2	2		
Conjecture 3 (aperiodic)	∞	5	3	3	2	2	2		
Conjecture 4 (strongly connected aperiodic)	∞	8	5	3	2	2	2	2	
Conjecture 5 (avoiding states)	∞	8	4	3	2	2	2		
Conjecture 6 (new rank conjecture)	∞	∞	5	3	3	2	2		
Conjecture 7 (subset synchronization)	∞	∞	5	4	3	2	2		

Acknowledgments. We thank Mikhail Volkov for suggesting Conjecture 4, and Mikhail Berlinkov for observing that the bound for one-cluster automata can be improved for periodic subsets on the cycle, which leaded to an improvement of our algorithm. We thank also Vojtěch Vorel for discussing the problem of avoiding states and sharing the series. The main part of the computations was performed on a grid that belongs to Institute of Computer Science of Jagiellonian University (thanks to Adam Roman).

References

1. Almeida, J., Steinberg, B.: Matrix mortality and the Černý-Pin conjecture. In: Diekert, V., Nowotka, D. (eds.) DLT 2009. LNCS, vol. 5583, pp. 67–80. Springer, Heidelberg (2009)
2. Ananichev, D.S.: The mortality threshold for partially monotonic automata. In: De Felice, C., Restivo, A. (eds.) DLT 2005. LNCS, vol. 3572, pp. 112–121. Springer, Heidelberg (2005)
3. Ananichev, D., Gusev, V., Volkov, M.: Slowly synchronizing automata and digraphs. In: Hliněný, P., Kučera, A. (eds.) MFCS 2010. LNCS, vol. 6281, pp. 55–65. Springer, Heidelberg (2010)
4. Ananichev, D.S., Volkov, M.V., Gusev, V.V.: Primitive digraphs with large exponents and slowly synchronizing automata. J. Math. Sci. **192**(3), 263–278 (2013)
5. Béal, M.-P., Berlinkov, M.V., Perrin, D.: A quadratic upper bound on the size of a synchronizing word in one-cluster automata. Int. J. Found. Comput. Sci. **22**(2), 277–288 (2011)
6. Berlinkov, M., Szykuła, M.: Algebraic synchronization criterion and computing reset words. In: Italiano, G.F., Pighizzini, G., Sannella, D.T. (eds.) MFCS 2015. LNCS, vol. 9234, pp. 103–115. Springer, Heidelberg (2015)
7. Cardoso, Â.: The Černý Conjecture and Other Synchronization Problems. Ph.D. thesis, University of Porto, Portugal (2014)
8. Carpi, A., D'Alessandro, F.: Independent sets of words and the synchronization problem. Adv. Appl. Math. **50**(3), 339–355 (2013)
9. Černý, J.: Poznámka k homogénnym experimentom s konečnými automatami. Matematicko-fyzikálny Časopis Slovenskej Akadémie Vied **14**(3), 208–216 (1964)
10. Dubuc, L.: Sur les automates circulaires et la conjecture de Černý. Informatique théorique et applications **32**, 21–34 (1998)
11. Eppstein, D.: Reset sequences for monotonic automata. SIAM J. Comput. **19**, 500–510 (1990)
12. Frankl, P.: An extremal problem for two families of sets. Eur. J. Comb. **3**, 125–127 (1982)
13. Gonze, F., Jungers, R.M., Trahtman, A.N.: A note on a recent attempt to improve the Pin-Frankl bound. Discrete Math. Theor. Comput. Sci. **17**(1), 307–308 (2015)
14. Kari, J.: A counter example to a conjecture concerning synchronizing word in finite. EATCS Bull. **73**, 146–147 (2001)
15. Kari, J., Volkov, M.V.: Černý's conjecture and the road coloring problem. In: Handbook of Automata. European Science Foundation (2013)
16. Kisielewicz, A., Szykuła, M.: Generating small automata and the Černý conjecture. In: Konstantinidis, S. (ed.) CIAA 2013. LNCS, vol. 7982, pp. 340–348. Springer, Heidelberg (2013)
17. Kisielewicz, A., Szykuła, M.: Generating Synchronizing Automata with Large Reset Lengths (2016). http://arxiv.org/abs/1404.3311
18. Pin, J.-E.: Utilisation de l'algèbre linéaire en théorie des automates. In: Actes du 1er Colloque AFCET-SMF de Mathématiques Appliquées II, pp. 85–92 (1978)
19. Pin, J.-E.: On two combinatorial problems arising from automata theory. In: Proceedings of the International Colloquium on Graph Theory and Combinatorics. North-Holland Mathematics Studies, vol. 75, pp. 535–548 (1983)
20. Steinberg, B.: The averaging trick and the Černý conjecture. Int. J. Found. Comput. Sci. **22**(7), 1697–1706 (2011)

21. Steinberg, B.: The Černý conjecture for one-cluster automata with prime length cycle. Theoret. Comput. Sci. **412**(39), 5487–5491 (2011)
22. Trahtman, A.N.: An efficient algorithm finds noticeable trends and examples concerning the Černy conjecture. In: Královič, R., Urzyczyn, P. (eds.) MFCS 2006. LNCS, vol. 4162, pp. 789–800. Springer, Heidelberg (2006)
23. Trahtman, A.N.: Modifying the upper bound on the length of minimal synchronizing word. In: Owe, O., Steffen, M., Telle, J.A. (eds.) FCT 2011. LNCS, vol. 6914, pp. 173–180. Springer, Heidelberg (2011)
24. Volkov, M.V.: Synchronizing automata and the Černý conjecture. In: Martín-Vide, C., Otto, F., Fernau, H. (eds.) LATA 2008. LNCS, vol. 5196, pp. 11–27. Springer, Heidelberg (2008)
25. Vorel, V.: Subset synchronization of transitive automata. In: AFL, pp. 370–381 (2014)

Implementation of Code Properties
via Transducers

Stavros Konstantinidis[1]([⊠]), Casey Meijer[1], Nelma Moreira[2], and Rogério Reis[2]

[1] Saint Mary's University, Halifax, NS, Canada
`s.konstantinidis@smu.ca, dylanyoungmeijer@gmail.com`
[2] CMUP & DCC, Faculdade de Ciências da Universidade do Porto,
Rua do Campo Alegre, 4169-007 Porto, Portugal
`{nam,rvr}@dcc.fc.up.pt`

Abstract. The FAdo system is a symbolic manipulator of formal language objects, implemented in Python. In this work, we extend its capabilities by implementing methods to manipulate transducers and we go one level higher than existing formal language systems and implement methods to manipulate objects representing classes of independent languages (widely known as code properties). Our methods allow users to define their own code properties and combine them between themselves or with fixed properties such as prefix codes, suffix codes, error detecting codes, etc. The satisfaction and maximality decision questions are solvable for any of the definable properties. The new online system LaSer allows one to query about a code property and obtain the answer in a batch mode. Our work is founded on independence theory as well as the theory of rational relations and transducers, and contributes with improved algorithms on these objects.

Keywords: Automata · Codes · FAdo · Implementation · Language properties · Regular languages · Symbolic computation · Transducers · Program generation

1 Introduction

Several programming platforms are nowadays available, providing methods to transform and manipulate various formal language objects: Grail/Grail+ [10,24], Vaucanson 1 [5], Vaucanson-R [30], FAdo [1,9], JFLAP and OpenFST [22]. Some of these systems allow one to manipulate such objects within simple script

---

Due to the page limit we chose to omit algorithmic details and proofs of correctness, and focus on providing a somewhat comprehensive presentation on implementation aspects and the new capabilities of FAdo. Details can be found in [17].

N. Moreira and R. Reis are partially supported by CMUP (UID/MAT/00144/2013), which is funded by FCT with national and European structural funds through the programs FEDER, under the partnership agreement PT2020. S. Konstantinidis and C. Meijer are supported by NSERC, Canada.

Y.-S. Han and K. Salomaa (Eds.): CIAA 2016, LNCS 9705, pp. 189–201, 2016.
DOI: 10.1007/978-3-319-40946-7_16

environments. Grail for example, one of the oldest systems, provides a set of filters manipulating automata and regular expressions on a UNIX command shell. Similarly, FAdo provides a set of methods manipulating such objects on a Python shell. Software environments for symbolic manipulation of formal languages are widely recognized as important tools for theoretical and practical research. They allow easy prototyping of new algorithms, testing algorithm performance with large datasets, corroborate or disprove descriptional complexity bounds for manipulations of formal system representations, etc. Due to the combinatorial nature of formal language representations, their calculations are almost impossible without computational aid.

In this work, we extend the capabilities of FAdo and LaSer [8,19] by implementing transducer methods and by going to the higher level of implementing objects representing classes of independent formal languages, also known as code properties. More specifically, the contributions of the present paper are as follows. **(a)** Implementation of transducer objects and several transducer methods (various product constructions, rational operations, transducer functionality test) *(Sect. 3)*. **(b)** Definitions of objects representing code properties and methods for their manipulation, which to our knowledge is a new development in software related to formal language objects. In addition to some fixed code properties (such as prefix code, infix code, hypercode), these methods can be used to construct new code properties and combine existing properties, including various error-detecting properties *(Sect. 4)*. **(c)** Enhancement and implementation of decision algorithms for code properties of regular languages. In particular, such algorithms have been implemented and enhanced so as to provide witnesses in case of a negative answer *(Sect. 5)*. To our knowledge such implementations are not openly available. **(d)** A mathematical definition of what it means to simulate (and hence implement) a hierarchy of properties and the proof that there is no complete simulation of the set of error-detecting properties *(Sect. 4)*. **(e)** Generation of executable Python code based on the requested question about a given code property. This is mostly of use in the online LaSer, which receives client requests and attempts to compute answers *(Sect. 6)*. **(f)** All the above classes and methods are open source (GPL). Our work is founded on independence theory [15,29] as well as the theory of rational relations and transducers [3,26].

2   Terminology and Background

Sets, Alphabets, Words, Languages. If S is a set, then $|S|$ denotes the cardinality of S, and 2^S denotes the set of all subsets of S. An *alphabet* is a finite nonempty set of symbols. In this paper, we write Σ, Δ for any arbitrary alphabets. The set of all words, or strings, over an alphabet Σ is written as Σ^*, which includes the *empty word* ε. A *language* (over Σ) is any set of words. We use standard operations and notation on words and languages [13,20,25].

Codes, Properties, Independent Languages. A *code property*, or *independence*, [15], is a set $\mathcal{P}$ of languages for which there is $n \in \mathbb{N} \cup \{\aleph_0\}$ such that $L \in \mathcal{P}$, if and only if $L' \in \mathcal{P}$, for all $L' \subseteq L$ with $0 < |L'| < n$. If L is in $\mathcal{P}$

then we say that L *satisfies* $\mathcal{P}$. Thus, L satisfies $\mathcal{P}$ exactly when all nonempty subsets of L with less than n elements satisfy $\mathcal{P}$. A language $L \in \mathcal{P}$ is called $\mathcal{P}$-*maximal*, or a maximal $\mathcal{P}$ code, if $L \cup \{w\} \notin \mathcal{P}$ for any word $w \notin L$. We note that every L satisfying $\mathcal{P}$ is included in a maximal $\mathcal{P}$ code [15]. As far as we know, all code related properties in the literature [4,6,8,11,15,23,28] are code properties as defined here. The focus of this work is on 3-independences that can also be viewed as independences with respect to a binary relation in the sense of [29].

Automata [26,32]. A nondeterministic finite automaton with ε-transitions, for short *automaton* or ε-*NFA*, is a quintuple $\mathbf{a} = (Q, \Sigma, T, I, F)$ such that Q is the finite set of states, Σ is an alphabet, $I, F \subseteq Q$ are the sets of start (or initial) states and final states, respectively, and $T \subseteq Q \times (\Sigma \cup \varepsilon) \times Q$ is the finite set of *transitions*. The ε-NFA $\mathbf{a}$ is called *trim*, if every state appears in some accepting path of $\mathbf{a}$. The automaton $\mathbf{a}$ is called an *NFA*, if no transition label is ε, that is, $T \subseteq Q \times \Sigma \times Q$.

Transducers and (Word) Relations [3,26,32]. A (word) *relation* over Σ and Δ is a subset of $\Sigma^* \times \Delta^*$, that is, a set of pairs (x, y) of words over the two alphabets (respectively). The *inverse* of a relation ρ, denoted by ρ^{-1}, is the relation $\{(y, x) \mid (x, y) \in \rho\}$. A (finite) *transducer* is a sextuple $\mathbf{t} = (Q, \Sigma, \Delta, T, I, F)$ such that Q, I, F are exactly the same as those in ε-NFAs, Σ is now called the *input* alphabet, Δ is the *output* alphabet, and $T \subseteq Q \times \Sigma^* \times \Delta^* \times Q$ is the finite set of transitions. We write $(p, x/y, q)$ for a transition – the *label* here is (x/y), with x being the input and y being the output label. The *size* of $(p, x/y, q)$ is the number $1 + |x| + |y|$. The size $|\mathbf{t}|$ of $\mathbf{t}$ is the sum of $|Q|$ and the sizes of all transitions. The relation $\mathrm{R}(\mathbf{t})$ *realized by* the transducer $\mathbf{t}$ is the set of labels in all the accepting paths of $\mathbf{t}$. We write $\mathbf{t}(x)$ for the set of *possible outputs of* $\mathbf{t}$ on input x, that is, $y \in \mathbf{t}(x)$ iff $(x, y) \in \mathrm{R}(\mathbf{t})$. The *domain* of $\mathbf{t}$ is the set of all words w such that $\mathbf{t}(w) \neq \emptyset$. The *inverse* of a transducer $\mathbf{t}$, denoted as $\mathbf{t}^{-1}$, realizes the inverse of the relation realized by $\mathbf{t}$. The transducer $\mathbf{t}$ is said to be in *standard form*, if each transition $(p, x/y, q)$ is such that $x \in (\Sigma \cup \varepsilon)$ and $y \in (\Delta \cup \varepsilon)$. If $\mathbf{s}$ and $\mathbf{t}$ are transducers, then there is a transducer $\mathbf{s} \vee \mathbf{t}$ realizing $\mathrm{R}(\mathbf{s}) \cup \mathrm{R}(\mathbf{t})$.

3   Transducer Object Classes and Methods

Here we discuss some aspects of the implementation of transducer objects and related methods. These are contained in the module `transducers.py` and can be imported as follows:

```
from FAdo.transducers import *
```

The FAdo class `GFT`, for General Form Transducer, is a subclass of `NFA`, which is the FAdo class for ε-NFAs. A transducer $\mathbf{t} = (Q, \Sigma, \Delta, T, I, F)$ is implemented as an object `t` with six instance variables `States`, `Sigma`, `Output`, `delta`, `Initial`, `Final` corresponding to the six components of $\mathbf{t}$. Specifically, `States` is a list of unique state names, meaning that each state name has an index which is the position of the state name in the list, with 0 being the first index value. The variables `Sigma`, `Output`, `Initial` and `Final` are sets, where the latter two are sets of state indexes. For efficiency reasons, the set of transitions T is implemented as a Python dictionary

$$\texttt{delta: } \{0,\dots,n-1\} \to (\texttt{Sigma} \to 2^{\texttt{Output} \times \{0,\dots,n-1\}}),$$

where n is the number of states. Thus, for any $p \in \{0,\dots,n-1\}$, `delta[p]` is a dictionary, and for any input label x, `delta[p][x]` is a set of pairs (y,q) corresponding to all transitions $\{(p, x/y, q) \in T \mid y \in \texttt{Output}, q \text{ is a state index}\}$.

Standard form transducers are objects of the FAdo class `SFT`, which is a subclass of `GFT`. The class `SFT` is very important from an algorithmic point of view, as most product constructions require a transducer to be in standard form. The conversion from GFT to SFT is done using the method `toSFT()`.

Example 1. The following code defines a string `s` containing a transducer description, and then constructs an SFT transducer `t` from `s` via a method of the module `fio`, which contains input/output methods for formal language objects. On input x, `t` returns the set of all proper suffixes of x—see also Fig. 1.[1] It has an initial state 0 and a final state 1, and deletes at least one of the input symbols.

```
s = '@Transducer 1 * 0\n'\
    '0 a @epsilon 0\n0 b @epsilon 0\n'\
    '0 a @epsilon 1\n0 b @epsilon 1\n'\
    '1 a a 1\n1 b b 1\n'
t = fio.readOneFromString(s)
```

As usual, \n denotes the *end of line character*, so the string `s` consists of 7 lines: the first indicates the type of object followed by the final states (in this case 1) and the start states after * (in this case 0); the second line contains the transition $(0, a/\varepsilon, 0)$; the last line contains the transition $(1, b/b, 1)$. Here `t.Sigma={a,b}`.

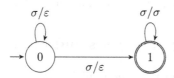

Fig. 1. On input x, the above transducer outputs any proper suffix of x.

Recall, for a transducer **t** and word w, $\mathbf{t}(w)$ is the set of possible outputs of **t** on input w. Note that this set can be empty, finite, or even infinite. In any case, it

---

[1] Note: In transducer figures, the input and output alphabets are equal. An arrow with label σ/σ represents a set of transitions with labels σ/σ, for each alphabet symbol σ; and similarly for an arrow with label σ/ε. An arrow with label σ/σ' represents a set of transitions with labels σ/σ' for all distinct alphabet symbols σ, σ'.

is always a regular language. The FAdo method `t.runOnWord(w)` assumes that t is an SFT object and returns an automaton accepting the language `t(w)`.

Example 2. The following code is a continuation of Example 1. It prints the set of all proper suffixes of the word ababb, which are all of length ≤ 4.

```
a = t.runOnWord('ababb')
n = len('ababb')
print a.enumNFA(n)
```

Assuming t is an SFT object, the following methods are available: "`t.inverse()`" returns the inverse of t; "`t.evalWordP((x,y))`" returns `True` or `False`, depending whether the pair `(x,y)` belongs to the relation realized by t; "`t.nonEmptyW()`" returns some word pair `(x, y)` which belongs to the relation realized by t, if nonempty; otherwise, it returns the pair (`None, None`).

Product Constructions [3,16,32]. The next methods are adaptations of the standard product construction [13] between two NFAs which produces an NFA accepting the intersection of the corresponding languages. Assume that t and s are SFT objects and a is an NFA object: "`t.inIntersection(a)`" returns a transducer realizing all word pairs (x, y) such that x is accepted by a and (x, y) is realized by t; "`t.outIntersection(a)`" as above except that y is accepted by a; "`t.runOnNFA(a)`" returns the NFA accepting the language $\bigcup_{w \in L(a)} t(w)$; "`t.composition(s)`" returns a transducer realizing the composition $R(t) \circ R(s)$.

Rational Operations [3]. A relation ρ is a *rational relation*, if it is equal to $\emptyset$, or $\{(x, y)\}$ for some words x and y, or can be obtained from other ones by using a finite number of times any of the three (rational) operators: union, concatenation, Kleene star. A classic result on transducers says that a relation is rational if and only if it can be realized by a transducer. The following methods are now available in FAdo, where we assume that s and t are SFT transducers: `t.union(s)`; `t.concat(s)`; `t.star()`. The implementation of these methods mimics the implementation of the corresponding methods on automata.

Witness of Transducer **Non-*functionality*.** A transducer t is called *functional* if $|t(w)| \leq 1$, for every word w. Transducer functionality can be tested in polynomial time [2]. A triple of words (w, z, z') is called a *witness of* t*'s nonfunctionality*, if $z \neq z'$ and $z, z' \in t(w)$. We have implemented the SFT method `t.nonFunctionalW()`, which returns a witness of t's non-functionality, or the triple (`None,None,None`) if t is functional. Our method is based on the decision test and uses extra bookkeeping for producing the desired witness.

Theorem 1. *The FAdo method* `t.nonFunctionalW()` *computes a size* $O(|t|^2)$ *witness of* t*'s non-functionality, if and only if one exists.*

The proof of correctness can be found in [17]. There is a sequence (t_n) of transducers such that $|t_n| \to \infty$ and the minimal witness of each t_n is of size $\Theta(|t_n|^2)$.

4   Object Classes Representing Code Properties

In this section we discuss our implementation of objects representing code properties. We are interested in methods that allow one to formally describe code properties. Three such formal methods are the implicational conditions of [14], where a property is described by a first order formula of a certain type, the regular trajectories of [6], where a property is described by a regular expression over $\{0, 1\}$, and the transducers of [8], where a property is described by a transducer. These formal methods can describe most properties of practical interest. The formal methods of regular trajectories and transducers are implemented here, as the transducer formal method follows naturally our implementation of transducers, and every regular expression of the regular trajectory formal method can be converted efficiently to a transducer object of the transducer formal method.

Input-Altering Transducer Properties [8]. A transducer t is *input-altering* if, for all words w, $w \notin t(w)$. In this formal method such a transducer t describes the code property $\mathcal{P}_t^{al}$ consisting of all languages L such that

$$t(L) \cap L = \emptyset. \tag{1}$$

With this formal method we can define the *suffix code* property: L is a suffix code if no L-word is a proper suffix of an L-word. The transducer defined in Example 1 is input-altering and describes the suffix code property over the alphabet $\{a, b\}$.

Error-Detecting Properties via Input-Preserving Transducers [8,16]. A transducer t is *input-preserving* if, for all words w in the domain of $R(t)$, $w \in t(w)$. Such a transducer t is also called a *channel transducer*, in the sense that an input message w can be transmitted via t and the output can always be w (no transmission error), or a word other than w (error). In this formal method the transducer t describes the *error-detecting for t* property $\mathcal{P}_t^{ed}$ consisting of all languages L over the input alphabet of t such that

$$t(w) \cap (L - w) = \emptyset, \quad \text{for all words } w \in L. \tag{2}$$

Every input-altering transducer property is an error-detecting property [8].

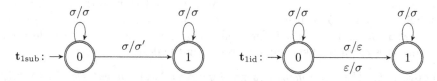

Fig. 2. On input x, t_{1sub} outputs either x, or any word that results by substituting exactly one symbol in x. On input x, t_{1id} outputs either x, or any word that results by deleting, or inserting, exactly one symbol in x.

Example 3. Consider the property *1-substitution error-detecting code* over {a, b}, where error means the substitution of one symbol by another symbol. The following channel transducer defines this property—see also Fig. 2. The transducer will substitute at most one symbol of the input word with another symbol.

```
s1  = '@Transducer 0 1 * 0\n0 a a 0\n0 b b 0\n'\
         '0 b a 1\n0 a b 1\n1 a a 1\n1 b b 1\n'
t1 = fio.readOneFromString(s1)
```

We note that the transducer approach to defining error-detecting code properties is very powerful, as it allows one to model insertion and deletion errors, in addition to substitution errors—see Fig. 2. Codes for such errors are actively under investigation—see [23], for instance.

4.1 Implementation in FAdo

We have defined the Python classes `TrajProp`, `IATProp` and `ErrDetectProp` corresponding to the types of properties discussed above. These property types are described, respectively, by regular trajectory expressions, input-altering transducers, and input-preserving transducers. In all cases, given a transducer object, an object of the class is created. An object `p` of the class `IATProp`, say, is defined via some transducer `t` and represents a particular code property, that is, the class of languages satisfying Eq. (1). The class `ErrDetectProp` is a super-class of the others. These classes and all related methods and functions are in the module `codes.py` and can be imported as follows.

```
import FAdo.codes as codes
```

Although each of the above four classes requires a transducer to create an object of the class, we have defined a set of what we call *build functions* as a user interface for creating code property objects. These build functions are shown next in use with specific arguments from previous examples.

Example 4. Consider again Examples 1, 3 in which the strings s and s1 are defined containing, respectively, the proper suffixes transducer and the transducer permitting up to 1 substitution error. The following object definitions are possible with the FAdo package

```
icp = codes.buildTrajPropS('1*0*1*', {'a', 'b'})
scp = codes.buildIATPropS(s)
s1dp = codes.buildErrorDetectPropS(s1)
pcp = codes.buildPrefixProperty({'a', 'b'})
icp2 = codes.buildInfixProperty({'a', 'b'})
```

In the first statement, `icp` represents the infix code property over the alphabet {a, b} and is defined via the trajectory expression 1*0*1*. In the next two statements, `scp, s1dp` represent, respectively, the suffix code property and the 1-substitution error-detecting property. The last two statements are explained below—`pcp` and `icp2` represent the prefix code and infix code properties, respectively.

Fixed Properties. We have created specific classes for the well-known properties prefix, suffix, infix, outfix, and hypercodes. As before, users need only to know about the `build`-interfaces for creating objects of these classes. For example, `buildPrefixProperty(Sigma)` returns an object of the class `PrefixProp` that represents all prefix codes over the alphabet `Sigma`.

4.2   Combining Code Properties

In many cases it is desirable to talk about languages satisfying more than one property. For example, most of the practical 1-substitution error-detecting codes are infix codes (in fact *block codes*, that is, those whose words are of the same length). We have defined the operation `&` between any two error-detecting properties independently of how they were created. This operation returns an object representing the class of all languages satisfying both properties. This object is constructed via the transducer that results by taking the union of the two transducers describing the two properties—see Rational Operations in Sect. 3.

Example 5. Using the properties `icp`, `s1dp` created above in Example 4, we can create the conjunction `p1` of these properties, and using the properties `pcp`, `scp` we can create their conjunction `bcp` which is known as the *bifix code property*.

```
p1 = icp & s1dp
bcp = pcp & scp
```

The object `p1` is of type `ErrDetectProp`. If, however, the two properties involved are input-altering then our implementation makes sure that the object returned is also of type input-altering—this is the case for `bcp`.

Our top Python superclass is `ErrDetectProp`. When viewed as a set of (potential) objects, this class implements the set of properties

$$\mathcal{P}^{ed} = \{\mathcal{P}^{ed}_t \mid t \text{ is an input-preserving transducer}\}. \tag{3}$$

In fact, we have also implemented the methods '&' and '≤' in a way that the triple (`ErrDetectProp`, &, ≤) constitutes a syntactic hierarchy (see further below). This means that '&' simulates intersection between properties and '≤' simulates subset relationship between two properties such that the following desirable statements hold true, for any `ErrDetectProp` objects `p`, `q`

p & p returns p;   p ≤ q if and only if p & q   returns p

Our implementation associates to each `ErrDetectProp` object p a nonempty set p.ID of names. If p is a fixed property object, p.ID has one hardcoded name. If p is built from a transducer t, p.ID has one name, the name of t—this name is based on a string description of t. If p = q&r, then p.ID is the union of q.ID and r.ID minus any fixed property name N for which another fixed property name M exists in the union such that the M-property is contained in the N-property—see [17] for details.

Next we define what it means to simulate a set of code properties $\mathcal{Q} = \{\mathcal{Q}_j \mid j \in J\}$ via a syntactic hierarchy $(G, \&, \leq)$, which can ultimately be implemented (as is the case here) in a programming language. The idea is that each $g \in G$ represents a property $[g] = \mathcal{Q}_j$, for some index j, and G is the set of generators of the semigroup $(\langle G \rangle, \&)$ whose operation '$\&$' simulates the process of combining properties in $\mathcal{Q}$, that is $[x\&y] = [x] \cap [y]$, and the partial order '$\leq$' simulates subset relation between properties, that is $x \leq y$ implies $[x] \subseteq [y]$, for all $x, y \in \langle G \rangle$. We show that there is an efficient simulation of the set of properties $\mathcal{P}^{ed}$ in (3) and that there can be no *complete* simulation of that set of properties.

Definition 1. *A syntactic hierarchy is a triple $(G, \&, \leq)$ where G is a non-empty set and (a) $(\langle G \rangle, \&)$ is the commutative semigroup generated by G with computable operation '$\&$'. (b) $(\langle G \rangle, \leq)$ is a decidable partial order (reflexive, transitive, antisymmetric). (c) For all $x, y \in \langle G \rangle$, we have that $x \leq y$ implies $x\&y = x$, and that $x\&y \leq x$.*

Definition 2. *Let $\mathcal{Q} = \{\mathcal{Q}_j \mid j \in J\}$ be a set of properties, for some index set J. A (syntactic) simulation of $\mathcal{Q}$ is a quintuple $(G, \&, \leq, [\,], \varphi)$ such that $(G, \&, \leq)$ is a syntactic hierarchy; $[\,] : \langle G \rangle \to \mathcal{Q}$ is a surjective mapping; $\varphi : J \to \langle G \rangle$ with $[\varphi(j)] = \mathcal{Q}_j$; for all $x, y \in \langle G \rangle$, $x \leq y$ implies $[x] \subseteq [y]$; and for all $x, y \in \langle G \rangle$, $[x\&y] = [x] \cap [y]$. The simulation is called complete if, for all x, y, $[x] \subseteq [y]$ implies $x \leq y$. The simulation is called linear if J has a size function $|\cdot|$ and $\langle G \rangle$ has a size function $\|\cdot\|$ such that $\|\varphi(j)\| = O(|j|)$, for all $j \in J$, and for all x, y, $\|x\&y\| = O(\|x\| + \|y\|)$.*

By a size function on a set X, we mean any function f of X into $\mathbb{N}_0$.

Theorem 2. *There is a linear simulation of the set of properties $\mathcal{P}^{ed}$.*

Theorem 3. *There is no complete simulation of the set of properties $\mathcal{P}^{ed}$.*

The above result implies that for any FAdo `ErrDetectProp` objects p, q defined via transducers t and s with $\mathcal{P}^{ed}_t \subseteq \mathcal{P}^{ed}_s$ it does not always hold that p $\leq$ q. On the other hand, our implementation of the set of the five fixed properties constitutes a complete simulation of these properties, when the same alphabet is used. Using the notation of Example 4, this implies that

```
pcp & icp2   returns   icp2
```

5   Methods of Code Property Objects

In the context of the research on code properties, we consider the following three algorithmic problems as fundamental. *Satisfaction problem:* Given the description of a code property and the description of a language, decide whether the language satisfies the property. In the *witness version* of this problem, a negative answer is also accompanied by an appropriate set of words showing how the property is violated. *Maximality problem:* Given the description of a code property and the description of a language L, decide whether the language is maximal with respect to the property. In the *witness version* of this problem, a negative answer is also accompanied by a word w that can be added to the language L. *Construction problem:* Given the description of a code property and two positive integers n and ℓ, construct a language that satisfies the property and contains n words of length ℓ (if possible). It is assumed that the code property can be implemented as p via a transducer t and, in the first two problems, the language is given via an NFA a. Next we discuss the implementation of methods for the satisfaction problem, all of which work in polynomial time. Due to the page limit we omit details on the maximality problems. Aspects of the construction problem are discussed in [18].

Methods p.satisfiesP(a). Equation (1) implies that, if the property p is described by an input-altering transducer t, the method p.satisfiesP(a) can be implemented as follows, where & is NFA intersection

```
c = t.runOnNFA(a)
return (a & c).emptyP()
```

If p is an error-detecting property, the transducer t is input-preserving and Eq. (2) is tested via transducer functionality. In FAdo this test can be done as follows, where functionalP() returns whether a transducer is functional.

```
s = t.inIntersection(a)
return s.outIntersection(a).functionalP()
```

Methods with Witnesses: p.notSatisfiesW(a). For input-altering transducer and error-detecting properties, the witness version of p.satisfiesP(a) returns either a pair of *different* words $u, v \in L(a)$ violating the property, that is, $v \in t(u)$ or $u \in t(v)$, or they return the pair (None, None). In the former case, the pair (u, v) is called a *witness of the non-satisfaction of p by* the language $L(a)$. We accomplish this by changing appropriately the implementations of p.satisfiesP(a) shown before—see [17] for details.

Example 6. The following Python interaction shows that a^*b is a prefix and 1-error-detecting code. The strings st, s1 contain the descriptions of an NFA accepting a^*b, and a transducer allowing up to 1 substitution error on the input word.

```
>>> a = fio.readOneFromString(st)
>>> pcp = codes.buildPrefixProperty({'a','b'})
>>> s1dp = codes.buildErrDetectPropS(s1)
>>> p2 = pcp & s1dp
>>> p2.notSatisfiesW(a)
(None, None)
```

Uniquely Decipherable Codes. The property of unique decipherability, *UD code property* for short, is probably the first historically property of interest in coding theory from the points of view of both information theory [27] as well as formal languages [21]. This property is not defined via a transducer and is treated differently. In particular the witness version of the satisfaction problem is solved based on the decision algorithm of [12]—see [17] for details. Next we only show an example of how one can use the satisfaction method.

Example 7. The following Python interaction produces a witness of the non-satisfaction of the UD code property by the finite language $L = \{ab, abba, bab\}$.

```
>>> a = L.toNFA()
>>> p = codes.buildUDCodeProp(a.Sigma)
>>> p.notSatisfiesW(a)
(['ab', 'bab', 'abba', 'bab'], ['abba', 'bab', 'bab', 'ab'])
```

The two word lists are different, but their concatenations form equal words.

6   LaSer and Program Generation

The first version of LaSer [8] was a limited and self-contained set of C++ automaton and transducer methods with a web interface having the following functionality: a user uploads a file containing an automaton and a file containing either a trajectory automaton, or an input altering-transducer, and LaSer would respond with an answer to the witness version of the satisfaction problem for input-altering transducer properties. The new version discussed here is based on the FAdo set of automaton and transducer methods and allows clients to request a response about the witness versions of the satisfaction and maximality problems for input-altering transducer, error-detecting and error-correcting properties. We call the above type of functionality, where LaSer computes and returns the answer, the *online service* of LaSer. A feature of the new version of LaSer, which we believe to be original in the community of software on automata and formal languages, is the *program generation service*. This is the capability to generate a self-contained Python program that can be downloaded on the client's machine and executed on that machine returning thus the desired answer. This feature is useful as the execution of certain algorithms, even of polynomial time complexity, can be quite time consuming for a server software.

7   Concluding Remarks

There are a few directions for future research. First, the existing implementation of transducers is not always efficient when it comes to describing code properties. For example, the transducer defined in Example 3 consists of 6 transitions. In general, if the alphabet has size s, then that transducer would require $s + s(s - 1) + s = s^2 + s$ transitions. However, a symbolic notation for transitions would be more compact and can possibly be used by modifying the appropriate transducer methods—certain symbolic transducers are investigated in [31]. Formal methods for defining code properties need to be evolved further with the aim of ultimately implementing these properties and answering efficiently the satisfaction problem. These methods should be capable of allowing to express properties that cannot be expressed in the transducer methods. In particular, as all transducer properties in this work are 3-independences, they do not include properties like comma-free code property. The formal method of [14] is quite expressive, using a certain type of first order formulae to describe properties. We also note that if the defining method is too expressive then even the satisfaction problem could become undecidable—see for example the method of [7].

References

1. Almeida, A., Almeida, M., Alves, J., Moreira, N., Reis, R.: FAdo and GUItar: tools for automata manipulation and visualization. In: Maneth, S. (ed.) CIAA 2009. LNCS, vol. 5642, pp. 65–74. Springer, Heidelberg (2009)
2. Béal, M.P., Carton, O., Prieur, C., Sakarovitch, J.: Squaring transducers: an efficient procedure for deciding functionality and sequentiality. Theoret. Comput. Sci. **292**(1), 45–63 (2003)
3. Berstel, J.: Transductions and Context-Free Languages. B.G. Teubner, Stuttgart (1979)
4. Berstel, J., Perrin, D., Reutenauer, C.: Codes and Automata. Cambridge University Press, New York (2009)
5. Claveirole, T., Lombardy, S., O'Connor, S., Pouchet, L.-N., Sakarovitch, J.: Inside vaucanson. In: Farré, J., Litovsky, I., Schmitz, S. (eds.) CIAA 2005. LNCS, vol. 3845, pp. 116–128. Springer, Heidelberg (2006)
6. Domaratzki, M.: Trajectory-based codes. Acta Informatica **40**, 491–527 (2004)
7. Domaratzki, M., Salomaa, K.: Codes defined by multiple sets of trajectories. Theoret. Comput. Sci. **366**, 182–193 (2006)
8. Dudzinski, K., Konstantinidis, S.: Formal descriptions of code properties: decidability, complexity, implementation. IJFCS **23**(1), 67–85 (2012)
9. FAdo: Tools for formal languages manipulation. http://fado.dcc.fc.up.pt/
10. Grail: Grail+. http://www.csit.upei.ca/~ccampeanu/Grail/
11. Hamming, R.W.: Error detecting and error correcting codes. Bell Syst. Tech. J. **26**(2), 147–160 (1950)
12. Head, T., Weber, A.: Deciding code related properties by means of finite transducers. In: Capocelli, R., De Santis, A., Vaccaro, U. (eds.) Sequences II, Methods in Communication, Security, and Computer Science, pp. 260–272. Springer, New York (1993)
13. Hopcroft, J.E., Ullman, J.D.: Introduction to Automata Theory, Languages, and Computation. Addison-Wesley, Reading (1979)
14. Jürgensen, H.: Syntactic monoids of codes. Acta Cybernetica **14**, 117–133 (1999)
15. Jürgensen, H., Konstantinidis, S.: Codes. In: Rozenberg and Salomaa [25], pp. 511–607
16. Konstantinidis, S.: Transducers and the properties of error-detection, error-correction and finite-delay decodability. JUCS **8**, 278–291 (2002)
17. Konstantinidis, S., Meijer, C., Moreira, N., Reis, R.: Symbolic manipulation of code properties. Computing Research Repository (2015). arXiv:1504.04715v1
18. Konstantinidis, S., Moreira, N., Reis, R.: Channels with synchronization/substitution errors and computation of error control codes. Computing Research Repository (2016). arXiv:1601.06312v1
19. LaSer: Independent LAnguage SERver. http://laser.cs.smu.ca/independence/

20. Mateescu, A., Salomaa, A.: Formal languages: an introduction and a synopsis. In: Rozenberg and Salomaa [25], pp. 1–39
21. Nivat, M.: Elements de la théorie générale des codés. In: Automata Theory, pp. 278–294 (1966)
22. OpenFst: OpenFst Library. http://www.openfst.org/
23. Paluncic, F., Abdel-Ghaffar, K., Ferreira, H.: Insertion/deletion detecting codes and the boundary problem. IEEE Trans. Info. Theory **59**(9), 5935–5943 (2013)
24. Raymond, D., Wood, D.: Grail: a C++ library for automata and expressions. J. Symbolic Comput. **17**(4), 341–350 (1994)
25. Rozenberg, G., Salomaa, A. (eds.): Handbook of Formal Languages, vol. I. Springer-Verlag, Berlin (1997)
26. Sakarovitch, J.: Elements of Automata Theory. Cambridge University Press, New York (2009)
27. Sardinas, A.A., Patterson, G.W.: A necessary and sufficient condition for the unique decomposition of coded messages. IRE Int. Conven. Rec. **8**, 104–108 (1953)
28. Shyr, H.J.: Free Monoids and Languages, 2nd edn. Hon Min Book Company, Taichung (1991)
29. Shyr, H.J., Thierrin, G.: Codes and binary relations. In: Malliavin, M.P. (ed.) Séminaire d'Algèbre Paul Dubreil Paris 1975–1976 (29ème Année). Lecture Notes in Mathematics, vol. 586, pp. 180–188. Springer, Heidelberg (1977)
30. Vaucanson: The Vaucanson Project. http://vaucanson-project.org/
31. Veanes, M.: Applications of symbolic finite automata. In: Konstantinidis, S. (ed.) CIAA 2013. LNCS, vol. 7982, pp. 16–23. Springer, Heidelberg (2013)
32. Yu, S.: Regular languages. In: Rozenberg and Salomaa [25], pp. 41–110

On Synchronizing Automata
and Uniform Distribution

Emil Lerner[✉]

Faculty of Computational Mathematics and Cybernetics,
Lomonosov Moscow State University, Moscow, Russia
neex.emil@gmail.com

Abstract. Let $m > 1, [m] = \{0, 1, \ldots, m - 1\}$, $[m]^\infty$ be a set of all one-side infinite sequences with elements from $[m]$. Consider a function $g : [m]^\infty \to [m]^\infty$ which is a bijection defined by a deterministic finite transducer (DFT) whose input/output alphabets are $[m]$. Denote the prefix of length n of an infinite word w by $w \bmod m^n$. A function $f : [m]^\infty \to [m]^\infty$ is said to be compatible if from $w_1 \bmod m^n = w_2 \bmod m^n$ it follows $f(w_1) \bmod m^n = f(w_2) \bmod m^n$. It is known that all functions defined by DFT are compatible. A function f is said to be a uniformly distributed function over $[m]^\infty$ if the set $\left\{ \frac{\overline{f(z) \bmod m^n}}{m^n} : z \in [m]^n \right\}$ is uniformly distributed as $n \to \infty$ (here $\overline{f(z) \bmod m^n}$ stands for the number whose base-m expansion is first n symbols of $f(z)$). We prove a necessary and sufficient condition for composite function $f \odot g$ to be uniformly distributed for any uniformly distributed compatible function $f : [m]^\infty \to [m]^\infty$. The condition is based on a generalization of the notion of synchronizing automaton.

Keywords: Deterministic finite transducer · Uniform distribution · Synchronizing automata · m-adic number

1 Introduction

Synchronizing automata are known to have tight relationship with many areas of mathematics and applications (for survey, see [9]). Present paper reveals a connection between the notion of synchronizing automaton and the theory of uniform distribution of sequences.

Let $m > 1, [m] = \{0, 1, \ldots, m - 1\}$. Denote the set of words of length n with elements from $[m]$ by $[m]^n$, denote the set of finite words of any length by $[m]^*$, and denote the set of one-side infinite sequences by $[m]^\infty$.

The set $[m]^n$ can be naturally associated to $\{0, 1, \ldots, m^n - 1\}$ by using little-endian (i.e. least significant digits go first) base-m representations of numbers. Let $w \in [m]^n$. Denote by $\overline{w}$ the integer from $\{0, 1, \ldots, m^n\}$, whose little-endian base-m expansion is w (for example, if $m = 10$ and w is 0123456789, $\overline{w}$ is the number 9876543210). Similarly, the set $[m]^\infty$ can naturally be considered as the set of m-adic integers $\mathbb{Z}_m$, see [8].

© Springer International Publishing Switzerland 2016
Y.-S. Han and K. Salomaa (Eds.): CIAA 2016, LNCS 9705, pp. 202–212, 2016.
DOI: 10.1007/978-3-319-40946-7_17

Suppose $u \in [m]^\infty, n \in \{1, 2, 3, \ldots\}$. Denote by u mod m^n initial segment (prefix) of u of length n. Note that mod m^n is the reduction modulo m^n in terms of m-adic numbers.

Denote a deterministic finite transducer (DFT) G by a four-tuple (S, T, O, s_0), where S is a finite set of states, $T : S \times [m] \to S$ is a transition function, $O : S \times [m] \to [m]$ is an output function and s_0 is initial state. The input and output alphabet of the transducer are $[m]$. A directed graph whose set of vertices is S and an edge goes from s_1 to s_2 if and only if there exists $c\colon T(s_1, c) = s_2$ is called *underlying graph* of a DFT.

A DFT defines a mapping $g : [m]^\infty \to [m]^\infty$. It is known that the mapping g is *compatible*; that is, for any $x_1, x_2 \in [m]^\infty, n \in \mathbb{N}$ from x_1 mod $m^n = x_2$ mod m^n it follows that $g(x_1)$ mod $m^n = g(x_2)$ mod m^n. Vice versa, given a compatible mapping g there exists a deterministic transducer (though not necessarily finite, i.e., whose set of states is not necessarily finite) which defines the mapping g, see e.g. [1].

Let $f\colon [m]^\infty \to [m]^\infty$ be a compatible function (but not necessary defined by a DFT). Note that by compatibility the mapping $f\colon [m]^* \to [m]^*$ is well defined (as mapping of finite initial parts of infinite strings). Suppose $w \in [m]^*$. Denote

$$
P_n(f; w) = \frac{1}{m^n} \cdot \# \left\{ x \in [m]^n : \frac{\overline{w}}{m^{|w|}} \leq \frac{\overline{f(x) \text{ mod } m^n}}{m^n} < \frac{\overline{w} + 1}{m^{|w|}} \right\},
$$

where $|w|$ stands for the length of the word w. Loosely speaking, $P_n(f; w)$ is the probability of event "w is a suffix of $f(x)$ mod m^n" when x is chosen uniformly from $[m]^n$. Note that suffix is specified by the most significant digits in base-m representation of $\overline{f(x)}$ mod m^n, not by the least.

Definition 1. *A compatible function f is called uniformly distributed if for any word $w \in [m]^*$ it holds $\lim_{n \to \infty} P_n(f; w) = m^{-|w|}$.*

For more detailed information on the theory of uniformly distributed sequences we refer to [4].

Denote a deterministic finite automaton (DFA) A by a tuple (S, T, s_0), where S is a set of states, $T : S \times [m] \to S$ is a transition function and s_0 is initial state. The input alphabet of the automaton is $[m]$.

Definition 2. *DFA A is called synchronizing if there exists a word $w \in [m]^*$ such that the state A reaches after being fed by word w does not depend on s_0, i.e. A reaches the same state after being fed by word w when the feeding starts from state s for any $s \in S$.*

The notion of synchronizing DFA was introduced in [3].

Let $G = (S, T, O, s_0)$ be a DFT. By $s(w), s \in S, w \in [m]^*$ we denote the state the transducer reaches after being fed by the word w and the feeding starts when G is at the state s. We say that the pair of states s_1, s_2 is *simultaneously reachable* if there exist $w_1, w_2 \in [m]^*, |w_1| = |w_2|$ such that $s_0(w_1) = s_1, s_0(w_2) = s_2$.

Definition 3. *A DFA is called cc-synchronizing if for any simultaneously reachable pair of states $s_1, s_2 \in S$ there exists a word $w_0 \in [m]^*$ such that $s_1(w_0) = s_2(w_0)$.*

Here, "cc" is a reminder for "cyclic class", the notion from the theory of Markov chains [5, p. 405].

Recall that directed graph of Markov chain with finite number of states determines possibility of transition between states. In case when the graph has only one strongly connected component all states are split into c classes $S_0, S_1, \ldots, S_{c-1}$. Markov chain with one initial state will fall in one of states of S_i after n steps if and only if $n \bmod c = i$. In case of DFT with the same underlying graph we have same partition for the set of states. The notion of synchronizing DFA covers the case of $c = 1$ (if $c > 1$ the automaton is not synchronizing). The notion of cc-synchronizing DFT covers the case of arbitrary c. We will later define used notions of the theory of Markov Chains in more details.

The property of a DFT being cc-synchronizing is more general than the property of a corresponding DFA (i.e. obtained from the DFT by omitting output function) being syncrhonizing. That means, if the DFA obtained by omitting output function from a DFT is synchronizing, then this DFT is cc-synchronizing. The converse does not hold.

Consider a transducer function g; that is, g is a mapping $g \colon [m]^\infty \to [m]^\infty$ defined by a DFT. Let G be a minimal (in terms of number of states) DFT whose function is g. The function g is called cc-synchronizing if G is cc-synchronizing.

The goal of this paper is to prove that being cc-synchronizing is a necessary and sufficient condition for a bijective transducer function $g \colon [m]^\infty \to [m]^\infty$ to preserve the uniform distribution, i.e. for any uniformly distributed compatible $f \colon [m]^\infty \to [m]^\infty$ the function $f \odot g$ is uniformly distributed as well.

The problem is motivated by theory of pseudorandom number generators: a transducer $h \colon [m]^\infty \to [m]^\infty$ can be used to generate pseudorandom numbers by iterating h and dividing these values by m^n: so $x_0/m^n, h(x_0) \bmod m^n/m^n, h(h(x_0)) \bmod m^n/m^n, \ldots$ is considered as a sequence of pseudorandom numbers from $[0, 1)$.

Such sequences can be used for applications (e.g., for numerical methods like Monte Carlo, or for cryptography) when the function h is uniformly distributed (see more on transducers which generate uniformly distributed sequences in e.g. [2]). Therefore our main result can be used to construct a variety of pseudorandom number generators out of a given one.

2   Markov Chain Facts

Recall some notions and facts from the theory of Markov chains.

Let $\Gamma = (V, E)$ be directed graph, V is the set of vertexes and E is the set of edges. Let Γ be such that exactly m edges come out from every vertex. $C \subset V$ is called *strongly connected component* (SCC) if following conditions hold:

1. for every $v_1, v_2 \in C$ both v_1 is reachable from v_2 and v_2 is reachable from v_1;

2. for every $v_1 \in C, v_2 \notin C$ either v_1 is not reachable from v_2 or v_2 is not reachable from v_1.

A strongly connected component C is called *final* if no other SCCs are reachable from any vertex of C.

Theorem 1. *Let v be a vertex of final SCC of directed graph Γ. Then there exists $c, n_0 \in \mathbb{N}, \varepsilon > 0$ such that a random walk which starts from vertex v and chooses next edge with equal probability $1/m$ will return at vertex v at steps $nc, n \geq n_0$ with probability at least ε.*

Vertexes of a final SCC are called *recurrent*. From Theorem 1 it follows that if v is recurrent then the number of ways (cycles) from v to v of length nc is at least εm^{nc}.

Theorem 2. *Let Γ consist of exactly one SCC (which is obviously final). Let c be greatest common divisor of lengths of all cycles of G. Then all vertexes are split in c pairwise disjoint sets V_i: $V = V_0 \cup V_1 \cup \ldots \cup V_{c-1}$ such that all edges from vertexes of V_i go to vertexes of class $V_{(i+1) \bmod c}$.*

Sets V_i are called *cyclic classes*.
For proofs of Theorems 1 and 2, see [5].

3 Sufficient Condition for Uniform Distribution

Lemma 1. *Let $g : [m]^\infty \to [m]^\infty$ be a bijective transducer function such that corresponding DFT $G = (S, T, O, s_0)$ is cc-synchronizing. Then a composite function $f \odot g$ is uniformly distributed for any uniformly distributed compatible $f : [m]^\infty \to [m]^\infty$.*

Proof. It follows from the Definition 3 that the underlying graph of G will have only one final SCC.

First, we consider the case when the underlying graph of G has only one cyclic class. In this case all pairs of recurrent states are simultaneously reachable. By A_k denote the set of words $a, |a| = k$, such that for every simultaneously reachable pair of states s_1, s_2 it holds $s_1(a) = s_2(a)$. As G is cc-synchronizing A_{k_0} is not empty for some $k_0 \in \mathbb{N}$. Obviously, for any $a \in A_k$ and any $c \in [m]$ words $c||a, a||c$ are contained in A_{k+1} (here $w_1||w_2$ stands for concatenation of words w_1 and w_2). So, if $a \in A_k$ we have $w_1||a||w_2 \in A_{k+|w_1|+|w_2|}$ for every finite words w_1, w_2. Thus, we have $\lim_{k \to \infty} m^{-k}|A_k| = 1$.

Denote by $S(a)$ the state G reaches after being fed by input a, where $a \in A_k$ and the feeding starts from any state. It follows from the definition of set A_k and from the fact that underlying graph of G has only one cyclic class that $S(a)$ is well defined (does not depend on the state from which the feeding starts).

Since g is a bijection, then for any state s the function defined by transducer (S, T, O, s) (i.e. the transducer obtained from G by choosing s as initial state) is a bijection as well. That means that for any state s and word w there exists a word w_s^{-1} such that w is output of G on input w_s^{-1} starting from state s.

For all $w \in [m]^*, k, n \in \mathbb{N}, n \geq k + |w|$,

$$P_n(f \odot g; w) \geq \sum_{a \in A_k} P_n(f; a||w_{S(a)}^{-1})$$

Here transducer G is fed first by $n - k - |w|$ unknown symbols (which is output of f), then by k symbols of a (and goes to the state $S(a)$) and then by $w_{S(a)}^{-1}$ (and its output is w). Not all alternatives for probability $P_n(f \odot g; w)$ are present in the sum in the right part of inequality (that is why we get an inequality rather than equality).

Since k is chosen arbitrarily, it follows that

$$\liminf_{n \to \infty} P_n(f \odot g; w) \geq \lim_{k \to \infty} |A_k| \min_{a \in A_k} \lim_{n \to \infty} P_n(f; a||w_{S(a)}^{-1}) = (*)$$

As f is uniformly distributed, we have $\lim_{n \to \infty} P_n(f; a||w_{S(a)}^{-1}) = m^{-|w|-k}$, so

$$(*) = \lim_{k \to \infty} |A_k| m^{-|w|-k} = m^{-|w|}$$

Since $\sum_{w \in [m]^k} P_n(f \odot g; w) = 1$ for any $k, n \in \mathbb{N}, n \geq k$ and $|[m]^k| = m^k$ the lower bound for limit inferior is actually the value of $\lim_{n \to \infty} P_n(f \odot g; w)$. That follows from the fact that if we have two sequences a_n, b_n such that $\liminf_{n \to \infty} a_n \geq a, \liminf_{n \to \infty} b_n \geq b$ and $a_n + b_n = a + b$ for every n then $\lim_{n \to \infty} a_n = a$ and $\lim_{n \to \infty} b_n = b$ (we use this fact for m^k sequences $P_n(f \odot g; w), w \in [m]^k$).

So we have $\lim_{n \to \infty} P_n(f \odot g; w) = m^{-|w|}$ in case when the underlying graph of G has only one cyclic class.

Now, let it have $c, c > 1$ cyclic classes.

Given a word w, we can prefix w with $k = c - |w| \bmod c$ arbitrary symbols, so that total length is divisible by c. Summing over all such prefixes for $n \geq k + |w|$ we obtain

$$P_n(f \odot g; w) = \sum_{w' \in [m]^k} P_n(f \odot g; w'||w)$$

Thus, without loss of generality we can consider only w such that $|w| \bmod c = 0$.

Let $y \in [m]^*$. Denote

$$P_{n,y}(f; w) = \frac{1}{m^n} \cdot \# \left\{ x \in [m]^n : \frac{\overline{w}}{m^{|w|}} \leq \frac{\overline{f(x) \bmod m^n}}{m^n} < \frac{\overline{w} + 1}{m^{|w|}}, x \bmod m^{|y|} = y \right\}$$

Let $r \in \mathbb{N}$. We have

$$P_n(f \odot g, w) = \sum_{y \in [m]^r} P_{n,y}(f \odot g, w)$$

Let $y \in [m]^r$. Denote by G'_y the DFT whose input and output alphabet are words from $[m]^c$, the set of states is S, initial state is $s_0(y)$ and a transition goes from s_1 to s_2 inputting $w_I \in [m]^c$ and outputting w_O if $s_1(w_I) = s_2$ in original transducer G, and w_O is the concatenation of all outputs of transitions of path from s_1 to s_2 labeled by w_I.

The assumptions of previous paragraph hold for the reduced transducer G'_y, so we have (for original transducer G)

$$\lim_{n \to \infty} P_{nc+r,y}(f \odot g; w) = m^{-|w|-r}$$

Summing over all possible y and considering all $r \in [c]$, we obtain $\lim_{n \to \infty} P_n(f \odot g; w) = m^{-|w|}$. Thus, $f \odot g$ is uniformly distributed.

4 Necessity of the Condition and the Main Theorem

Let us consider an example in Fig. 1. The transducer is not cc-synchronizing. Let $x \in [2]^\infty$, $x_0 x_1 \ldots x_i \ldots$ be the corresponding sequence of elements of $[2]$, and $f(x) = y_0 y_1 \ldots y_i \ldots$ be a uniformly distributed function, constructed as follows:

- $y_i = x_i$ if i is not a power of 2;
- $y_i = \sum_{j=0}^{i-1} y_j \bmod 2$ if i is a power of 2 (all $y_j, j < i$ are already defined).

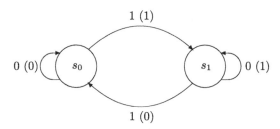

1 (1)

0 (0) s_0 s_1 0 (1)

1 (0)

Fig. 1. "Bitwise sum" transducer. Label $a(b)$ means the transition is performed on input symbol a, outputting b. The output of the transducer after being fed by word $w = w_0 w_1 \ldots w_n \ldots$ is the word $w' = w'_0 w'_1 \ldots w'_n \ldots$ such that $w'_i = \sum_{j=0}^{i} w_j \bmod 2$.

Let us prove that f is uniformly distributed. Let $w \in [2]^*$ and $n > 2|w|$. There exist no more than one $t \in \mathbb{N} : 2^t \in [n - |w| + 1, n]$. If no such t exist $P_n(f; w) = 2^{-|w|}$ obviously holds.

Otherwise $2^{t-1} < n - |w|$. So, as $\sum_{i=0}^{2^t-1} y_i \bmod 2 = 0$, we have $y_{2^t} = \sum_{i=2^{t-1}+1}^{2^t-1} x_i \bmod 2$. If w is a suffix of $x \bmod 2^n$ then y_{2^t} is unambiguously defined by $x_{2^{t-1}+1}, \ldots, x_{n-|w|}$, namely $y_{2^t} = (\sum_{i=2^{t-1}+1}^{n-|w|} x_i + \delta) \bmod 2$, where δ is sum of all elements of word $(w \bmod 2^{|w|+2^t-n})$. Note that the number of possible values for $x \bmod m^n$ is the same if $\sum_{i=2^{t-1}+1}^{n-|w|} x_i \bmod 2$ equal to 0 and 1 (this number is $2^{n-|w|-2^{t-1}-1}$). So we have $P_n(f; w) = 2^{-|w|}$ in this case too. As n is arbitrary number greater than $2|w|$ that means that f is uniformly distributed.

However, $f \odot g$ is not uniformly distributed. Let $g(f(x)) = z$, where $z = z_0 z_1 \ldots z_i \ldots$. For every $t \in \mathbb{N}$, z_{2^t} is zero for every $x \in [2]^\infty$. That means that sequence $P_n(f \odot g; 0)$ contains infinite number of ones and $\lim_{n \to \infty} P_n(f \odot g; 0) = 1/2$ does not hold.

Our proof generalizes this idea.

First of all, we introduce a concept of labeled directed graph of pairs, denoted by G^2 (see [3]). The vertexes of the graph are:

– all ordered pairs of different simultaneously reachable states (s_1, s_2);
– one special vertex T.

The set of edges of G^2 is constructed as follows

– an edge labeled by $c \in [m]$ goes from (s_1, s_2) to (s_3, s_4) if $s_1(c) = s_3$ and $s_2(c) = s_4$ ($s_3 \neq s_4$ as the set of vertexes contains only different pairs of states);
– an edge labeled by $c \in [m]$ goes from (s_1, s_2) to T if $s_1(c) = s_2(c)$;
– an edge labeled by c goes from T to T for all $c \in [m]$.

An example of graph of pairs is shown in Fig. 2.

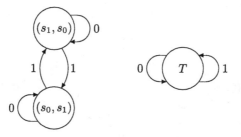

Fig. 2. The graph of pairs for "bitwise sum" transducer (see Fig. 1). The graph contains two SCCs. Note that if G^2 contains multiple SCCs then G is not cc-synchronizing, and vice versa.

Now, let us prove the remaining part of the main result.

Lemma 2. *Let $g : [m]^\infty \to [m]^\infty$ be a bijective transducer function such that minimal (in terms of number of states) corresponding DFT G is not cc-synchronizing. Then there exists a uniformly distributed compatible $f: [m]^\infty \to [m]^\infty$ such that the composite function $f \odot g$ is not uniformly distributed.*

Proof. Since G is not cc-synchronizing, it follows that there exists a pair of simultaneously reachable states s_1, s_2 such that T is not reachable from (s_1, s_2) in G^2. We can choose s_1, s_2 such that (s_1, s_2) is recurrent in G^2 as a recurrent vertex is reachable from every vertex of the directed graph G^2. Furthermore as G

is minimal transducer for g we can choose s_1, s_2 such that output functions are not equal in this pair of states (that is, $O(s_1, x) \neq O(s_2, x)$ for some $x \in [m]$). As g is a bijection, that means that there exist different symbols $c_1, c_2 \in [m]$ such that the output of G is the same when G is fed by c_1 and feeding starts from state s_1 and G is fed by c_2 and feeding starts from s_2. Denote that output symbol by c_O (i.e. $O(s_1, c_1) = O(s_2, c_2) = c_O$ and $c_1 \neq c_2$).

Let $k_i, i \in \mathbb{N}$ be an increasing sequence of numbers such that for some $\varepsilon > 0$ and for every $i \in \mathbb{N}$ it holds

$$\frac{1}{m^{k_i}} \cdot \# \left\{ w \colon w \in [m]^{k_i}, s_1(w) = s_1, s_2(w) = s_2 \right\} \geq \varepsilon.$$

The fact that such sequence exists follows from that (s_1, s_2) is recurrent in G^2 and every w of length k_i that labels a loop from (s_1, s_2) to (s_1, s_2) satisfies the condition. Denote the set $\{w \colon w \in [m]^{k_i}, s_1(w) = s_1, s_2(w) = s_2\}$ by W_i.

Let l_0 be such that there exist a pair of words $u_1, u_2 \in [m]^{l_0} \colon s_0(u_1) = s_1, s_0(u_2) = s_2$. Such l_0 exists as s_1, s_2 are simultaneously reachable. Let $U_0^1 = \{u_1\}, U_0^2 = \{u_2\}$.

As (s_1, s_2) is a recurrent state in G^2, both s_1 and s_2 are recurrent in the underlying graph of G. That means we can choose $\varepsilon' > 0$ such that there exist $c, n_0 \in \mathbb{N}$ such that the number of ways from s_0 to s_t of length $n_0 + nc$ is at least $\varepsilon' m^{nc}$. Without loose of generality let $\varepsilon' < m^{-l_0}$.

Let $l_j, j > 0$ be sequence of natural numbers and $U_j^1, U_j^2 \colon U_i^t \subset [m]^{l_j}$ two sequences of sets which will be defined later.

Let $x \in [m]^\infty, x_0 x_1 \ldots x_i \ldots$ be the corresponding sequence of elements of $[m]$, and let f, $f(x) = y_0 y_1 \ldots y_i \ldots$ be the function $[m]^\infty \to [m]^\infty$ constructed as follows:

1. $y_i = c_t, t \in \{1, 2\}$ if $i = l_j + k_j$ for some $j \in \mathbb{N}$, U_j^t contains the word $y_0 y_1 \ldots y_{l_j-1}$ and W_j contains the word $y_{l_j} y_{l_j+1} \ldots y_{l_j+k_j-1}$ and $x_{l_j+k_j} \in \{c_1, c_2\}$;
2. $y_i = x_i$ otherwise.

Let's choose l_j and sets $U_j^t, t \in \{1, 2\}$ such that following properties hold:

1. $l_j > l_{j-1} + k_j$;
2. for $t \in \{1, 2\}$ and every $w \in U_j^t$ it holds $s_0(w) = s_t$;
3. for $t \in \{1, 2\} \colon m^{-l_j} \#\{x \in [m]^{l_j} \colon f(x) \bmod l_j \in U_j^t\} > \varepsilon'$;
4. $\#\{x \in [m]^{l_j} \colon f(x) \bmod l_j \in U_j^1\} = \#\{x \in [m]^{l_j} \colon f(x) \bmod l_j \in U_j^2\}$.

So, U_j^t and l_j is constructed iteratively: we choose l_j and U_j^t after l_{j-1} and U_{j-1}^t are choosen. When using f in construction of sets U_j^t, we assume that all $l_{j'}, j' \geq j$ are big enough such that $f(x) \bmod m^{l_j}$ is defined.

The fact that such l_j and U_j^t exist (in particular, property 2 for U_j^t holds) on every step follows from Theorem 1, from the fact that s_1, s_2 are recurrent in underlying graph of G and from the definition of ε'.

Now we need to prove that f is uniformly distributed. Let $w \in [m]^*$. Let i be such that $k_i > |w|$ (such i exists as $k_i \to \infty$). For $n > l_i$ we have $P_n(f; w) =$

$m^{-|w|}$: it is obvious if none of $l_j + k_j$ fall into $[n - |w| + 1, n]$ and follows from the property 3 of U_j^t and the fact that no more than one such j exists otherwise.

Consider $f \odot g$. Let $g(f(x))$ be a sequence $z_0 z_1 \ldots z_i \ldots$ of elements of $[m]$.

Let $j \in \mathbb{N}$. We have $z_{l_j + k_j} = c_O$ either if assumptions of case 4 in definition of f hold or if it is case 4 and $x_{l_j + k_j} = c_O$. Because of property 4 of U_j^t and $m^{-k_j} |W_j| > \varepsilon$, we obtain $P_{l_j + k_j}(f \odot g; c_O) > 1/m + \varepsilon' \varepsilon$ for every $j \in \mathbb{N}$. That contradicts the uniform distribution of $f \odot g$.

Thus we have the compatible uniformly distributed function $f \colon [m]^\infty \to [m]^\infty$ such that $f \odot g$ is not uniformly distributed.

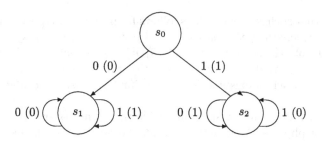

Fig. 3. An example of DFT whose graph contains two SCCs. The output of the transducer after being fed by word $w = w_0 w_1 \ldots w_n \ldots$ is the word $w' = w_0' w_1' \ldots w_n' \ldots$ such that $w_0' = w_0$ and $w_n' = (w_0 + w_n) \bmod 2$ for $n > 0$. The graph of pairs for this DFT contains three separate vertexes: (s_1, s_2), (s_2, s_1) and T each having two loops labeled by 0 and 1.

Let us trace the proof of Lemma 2 on another example shown in Fig. 3. This is an example of DFT G whose underlying graph has two SCCs. That means it is not cc-synchronizing. First, note that pair s_1, s_2 satisfies all requirements imposed by the proof with $c_1 = 0, c_2 = 1$ and $c_O = 0$.

Let $k_i = i$. Every word in $[2]^i$ labels a cycle in G^2 as it contains only loops. Thus $\varepsilon = 0.99, k_i = i$ and $W_i = [2]^i$ satisfy appropriate conditions.

Let $l_0 = 1$, u_1 be the word "0", u_2 be the word "1". For every $w \in [2]^*$ it holds $s_0(0||w) = s_1$ and $s_0(1||w) = s_2$. So let $\varepsilon' = 0.499$, $l_i = l_{i-1} + k_i + 1$ for $i > 0$ and $U_i^1 = \{0||w, w \in [2]^{l_i - 1}\}, U_i^2 = \{1||w, w \in [2]^{l_i - 1}\}$.

Thus we have $l_i = 1 + i(i + 3)/2$ and following function f (as above, $f(x_0 x_1 \ldots) = y_0 y_1 \ldots$):

- $y_i = x_0$ if $i = l_j + k_j$ for some $j \in \mathbb{N}$;
- $y_i = x_i$ otherwise.

It is uniformly distributed, as for $n > 2|w|^2$ each of $y_i, i \in [n - |w| + 1, n]$ is equal some $x_{i'}$, and i' is different for different i (actually, $i' = i$ for all i except no more than one, and $i' = 0$ it the latter case). However $f \odot g$ is not uniformly distributed: $z_{l_i + k_i}$ doesn't depend on x and is always zero for every i (as above, $f(g(x)) = z_0 z_1 \ldots$).

From Lemmas 1 and 2 the main result of this paper follows.

Theorem 3. *Let $g : [m]^\infty \to [m]^\infty$ be a bijective transducer function, and let $G = (S, T, O, s_0)$ be a minimal (in terms of number of states) corresponding DFT. A composite function $f \odot g$ is uniformly distributed for any uniformly distributed compatible $f : [m]^\infty \to [m]^\infty$ if and only if G is cc-synchronizing.*

5   Discussion

From Theorem 3 it follows that cc-synchronizing bijective DFT preserves uniform distribution when composed with arbitrary compatible uniformly distributed function f; that is, a composition of a cc-synchronizing DFT (whose function is bijective) with arbitrary uniformly distributed deterministic transducer produces a uniformly distributed sequence. This fact is used to prove uniform distribution of sequences generated by polynomials.

Consider a polynomial f with integer coefficients. For given prime p and $s \in \mathbb{N}$ consider the following set

$$\Omega(f; s) = \left\{ \left(\frac{x \bmod p^n}{p^n}, \frac{f(x) \bmod p^n}{p^n}, \ldots, \frac{f^{(s)}(x) \bmod p^n}{p^n} \right), n \in \mathbb{N}, x \in [p^n] \right\}$$

Here, $f^{(i)}$ is the i-th iteration of f, and $x \bmod p^n$ is the least nonnegative residue of x modulo p^n. The coordinates of points of $\Omega(f; s)$ are s successive values of a pseudorandom generator, whose transition function is f.

If f is linear then $\Omega(f; s)$ is known to be distributed non-uniformly, see [6]. Lemma 1 used in proof of the fact that $\Omega(f; s)$ is uniformly distributed when $\deg f \geq 2$, see [7].

Given a DFT G, if G^2 introduced in the proof of Theorem 3 contains only one SCC then G is cc-synchronizing, and vice versa. The proof of this fact is similar to that of the criterion that DFA is synchronizing, which could be found at [3] (and follows from that single vertex T is always an SCC in G^2).

Note that in case when $m = p^k$ we can consider input as k different inputs in alphabet $[p]$ (same for output). In terms of p-adic functions, that means that we can superpose transducer function $\mathbb{Z}_p^k \to \mathbb{Z}_p^k$ with a function $f : \mathbb{Z}_p^k \to \mathbb{Z}_p^k$ such that f is uniformly distributed in the Euclidean hypercube $[0, 1)^k$. This gives a method to construct new generators of uniformly distributed sequences out of a given such generator.

Further work includes generalization of Theorem 3 for injective mappings g and defining classes of generators of uniformly distributed sequences that can be constructed out of known ones using this method.

Acknowledgment. The author is grateful to prof. Vladimir Anashin for constant attention to this work and for useful discussions. This research was partially supported by Russian Foundation for Basic Research, research project No. 16-01-00470.

References

1. Anashin, V.: The Non-Archimedean theory of discrete systems. Math. Comput. Sci. **6**(4), 375–393 (2012)
2. Anashin, V., Khrennikov, A.: Applied Algebraic Dynamics. Walter de Gruyter, Berlin (2009)
3. Černý, J.: Poznamka k homogennym eksperimentom s konecnymi automatami. Matematicko-fyzikalny Casopis Slovensk. Akad. Vied **14**(3), 208–216 (1964). (in Slovak)
4. Drmota, M., Tichy, R.F.: Sequences, Discrepancies and Applications. LNM, vol. 1651. Springer-Verlag, Heidelberg (1997)
5. Kemeny, J., Snell, J.: Finite Markov Chains. D. van Nostrand Co. Inc., Princeton (1960)
6. Knuth, D.: The Art of Computer Programming, vol. 2, 3rd edn. Addison-Wesley Longman Publishing Co. Inc., Boston (1997)
7. Lerner, E.E.: Uniform distribution of sequences generated by iterated polynomials. Doklady Math. **92**(3), 704–706 (2015)
8. Mahler, K.: p-Adic Numbers and their Functions. Cambridge University Press, Cambridge (1981)
9. Volkov, M.V.: Synchronizing automata and the Černý conjecture. In: Martín-Vide, C., Otto, F., Fernau, H. (eds.) LATA 2008. LNCS, vol. 5196, pp. 11–27. Springer, Heidelberg (2008)

Looking for Pairs that Hard to Separate: A Quantum Approach

Aleksandrs Belovs[1], J. Andres Montoya[2(✉)], and Abuzer Yakaryılmaz[3]

[1] CWI, Amsterdam, The Netherlands
stiboh@gmail.com
[2] Universidad Nacional de Colombia, Bogotá, Colombia
jamontoyaa@unal.edu.co
[3] National Laboratory for Scientific Computing, Petrópolis, RJ, Brazil
abuzer@lncc.br

Abstract. Determining the minimum number of states required by a deterministic finite automaton to separate a given pair of different words (to accept one word and to reject the other) is an important challenge. In this paper, we ask the same question for quantum finite automata (QFAs). We classify such pairs as easy and hard ones. We show that 2-state QFAs with real amplitudes can separate any easy pair with zero-error but cannot separate some hard pairs even in nondeterministic acceptance mode. When using complex amplitudes, 2-state QFAs can separate any pair in nondeterministic acceptance mode, and here we conjecture that they can separate any pair also with zero-error. Then, we focus on (a more general problem) separating a pair of two disjoint finite set of words. We show that QFAs can separate them efficiently in nondeterministic acceptance mode, i.e., the number of states is two to the power of the size of the small set.

Keywords: Quantum finite automaton · Zero-error · Nondeterminism · Succinctness · Promise problems

1 Introduction

Determining the minimum number of states required by a deterministic finite automaton (DFA) to separate any given pair of words is one of the famous open problems in automata theory [5]. We can generalize this question in a straightforward way by considering different computational models (e.g. see [16]).

We focus on quantum finite automata (QFAs). We classify such pairs as easy and hard ones. We show that 2-state QFAs with real amplitudes can separate any easy pair with zero-error but cannot separate some hard pairs even in nondeterministic acceptance mode. When using complex amplitudes, 2-state QFAs can separate any pair in nondeterministic acceptance mode and here we conjecture that they can separate any pair also with zero-error. Then, we focus on (a more general problem) separating a pair of two disjoint finite set of words.

© Springer International Publishing Switzerland 2016
Y.-S. Han and K. Salomaa (Eds.): CIAA 2016, LNCS 9705, pp. 213–223, 2016.
DOI: 10.1007/978-3-319-40946-7_18

We show that QFAs can separate them efficiently in nondeterministic acceptance mode, i.e., the number of states is two to the power of the size of the small set.

In the next section, we provide the necessary background. The results on separating pairs are given in Sect. 3. The results on separating two finite sets are presented in Sect. 4.

2   Background

We refer the reader to [13] for a pedagogical introduction to quantum finite automata (QFAs), to [2] for a comprehensive survey on QFAs, and to [11] for a complete reference on quantum computation.

We denote the alphabet by Σ, and we suppose that it does not contain the right end-marker \$. For any given word $x \in \Sigma$, $|x|$ represents the length of x, $|x|_\sigma$ represents the number of occurrences of symbol σ in x, and x_j represents the j-th symbol of x, where $\sigma \in \Sigma$ and $1 \leq j \leq |x|$. As a special case, if $|\Sigma| = 1$, then the automaton and languages can be called unary.

2.1   Easy and Hard Pairs

Throughout the paper, a pair of words (x, y) refers to two different words defined on the same alphabet. A pair of words (x, y) is called *easy* if x and y have different numbers of occurrences of a symbol, i.e., $\exists \sigma \in \Sigma\, (|x|_\sigma \neq |y|_\sigma)$. Otherwise, the pair is called *hard*. Remark that any pair with different lengths (and so any unary pair) is easy.

Any hard pair defined on an alphabet with at least three elements can be mapped to a binary hard pair as follows. Let (x, y) be a hard pair defined on $\{\sigma_1, \ldots, \sigma_k\}$ for some $k > 2$. Since the pair is hard, we have

$$|x|_{\sigma_i} = |y|_{\sigma_i}$$

for each $1 \leq i \leq k$. Then, there should be an index j ($1 \leq j \leq |x| = |y|$) such that $x_j = \sigma_i \neq y_j = \sigma_{i'}$ for $i \neq i'$. If we delete all the other symbols and keep only σ_is and $\sigma_{i'}$s in x and y, we obtain two new words: x' and y', respectively. It is clear that (x', y') is a hard pair. So, instead of separating the hard pair (x, y), we can try to separate (x', y'). Algorithmically, we apply the identity operators on the symbols other than σ_i and $\sigma_{i'}$. Hence, unless otherwise specified, we focus on unary and binary words throughout the paper.

2.2   A Motivating Problem: Looking for Pairs that are Truly Hard to separate

Let w, v be two words of length n. What is the size of a minimal DFA separating those two words? The best upper bound is $O\left(n^{\frac{2}{5}} \log^{\frac{3}{5}}(n)\right)$ (see [12]), but we do not know of a set of pairs requiring such a large number of states.

Recall that DFAs can perform modular counting, and modular counting can be used to separate easy pairs using logarithmic number of states. Unfortunately the best lower bound is also $\Omega(\log(n))$ (see [8]), which was given by using the following set of pairs

$$S = \left\{ \left(0^{n-1}, 0^{n-1+lcm(1,2,\dots,n)} \right) : n \geq 1 \right\}.$$

Thus, the hardest set of pairs registered in the literature is a set of easy pairs. We call those pairs as GK pairs (the initials of Goralcik and Koubek [8]). There are many reasons to believe that the set of GK pairs cannot be the hardest set of pairs. We can provide some evidence concerning this issue by considering some different models of automata for which it can be proved that the GK pairs are not the hardest pairs. Perhaps, more interesting, we can get some clues that could be used in the construction of a harder set of pairs, an infinite set of pairs requiring a superlogarithmic number of states.

Previous to this work we studied alternating finite state automata. We proved that easy pairs can be separated by those automata using $O(\log(\log(n)))$ states. And, on the other hand, it was proved that there exists an infinite set of pairs requiring $\Omega\left(\sqrt{\log(n)}\right)$ states. The proof of the lower bound is nonconstructive, and we do not know which are the pairs that require $\Omega\left(\sqrt{\log(n)}\right)$ states.

We have begun this work classifying pairs into two classes: easy and hard pairs. It is a rough classification which should be refined. Remark that hard pairs are not always hard to separate. Consider for instance a pair (x, y) such that $x_1 \neq y_1$. This pair can be separated by using a DFA with three states. We believe that studying some other models of automata can help us to establish a very much finer and pertinent classification.

In this work we consider quantum finite automata. We prove that easy pairs can be separated by QFAs with real amplitudes using two states. On the other hand, we prove that there are hard pairs that cannot be separated using only two states. We also consider QFAs with complex amplitudes. We conjecture that any pair can be separated by those automata using two states, and we prove that such a conjecture (and our motivating problem) has unexpected relations with some problems in the theory of Lie groups.

2.3 QFAs

Quantum finite automata (QFAs) are a non-trivial generalization of probabilistic finite automata [9,18]. Here we give the definition of the known simplest QFA model, called Moore-Crutchfield QFAs (MCQFAs) [10] since we can present our results (and our conjecture) based on this model.

An n-state MCQFA M, which operates on n-dimensional Hilbert space ($\mathcal{H}_n$, i.e., $\mathbb{C}^n$ with the inner product) is a 5-tuple

$$M = (Q, \Sigma, \{U_\sigma \mid \sigma \in \Sigma\}, |u_0\rangle, Q_a),$$

where $Q = \{q_1, \ldots, q_n\}$ is the set of states, $U_\sigma \in \mathbb{C}^{n \times n}$ is a unitary transition matrix whose (i,j)th entry represent the transition amplitude from the state q_j to the state q_i when reading symbol $\sigma \in \Sigma$ $(1 \leq i, j \leq n)$, $|u_0\rangle \in \mathbb{C}^n$ is the column vector representing the initial quantum state, and $Q_a \subseteq Q$ is the set of accepting states. The basis of $\mathcal{H}_n$ is formed by $\{|q_j\rangle \mid 1 \leq j \leq n\}$ where $|q_j\rangle$ has 1 at the j-th entry and 0s in the remaining entries. At the beginning of the computation, M is in $|u_0\rangle$, either one of the basis states or a superposition (a linear combination) of basis states. Let $x \in \Sigma^*$ be a given input word. During reading the input x from left to right symbol by symbol, the quantum state of M is changed as follows:

$$|u_j\rangle = U_{x_j} |u_{j-1}\rangle,$$

where $1 \leq j \leq |x|$. After reading the whole word, the quantum state is measured to determine whether M is in an accepting state or not (a measurement on computational basis). Let the final quantum state, represented as $|u_f^x\rangle$ or $|u_f\rangle$, have the following amplitudes

$$|u_f^x\rangle = |u_f\rangle = |u_{|w|}\rangle = \begin{pmatrix} \alpha_1 \\ \alpha_2 \\ \vdots \\ \alpha_n \end{pmatrix}.$$

Since the probability of observing jth state is $|\alpha_j|^2$, the input is accepted with probability $\sum_{q_j \in Q_a} |\alpha_j|^2$.

2.4   Promise Problems

The disjoint languages $X \subseteq \Sigma^*$ and $Y \subseteq \Sigma^*$ are said to be separated by M exactly or zero-error if any $x \in X$ is accepted by M with probability 1 and any $y \in Y$ is accepted by M with probability 0, or vice versa. If $|X| = |Y| = 1$, then it is said that the corresponding pair is separated by M exactly. In case of one-sided bounded error, any $x \in X$ is accepted with probability 1 and any $y \in Y$ is accepted with probability at most $p < 1$, or vice versa. If $|X| = |Y| = 1$, then it is said that the pair is separated by M with one-sided bounded-error.

Nondeterministic QFA is a theoretical model and it is defined as a special acceptance mode of a QFA, also known as recognition with cutpoint 0 [17]. The disjoint languages $X \subseteq \Sigma^*$ and $Y \subseteq \Sigma^*$ are said to be separated by a nondeterministic MCQFA M if any $x \in X$ is accepted by M with some nonzero probability and any $y \in Y$ is accepted by M with probability 0, or vice versa. If $|X| = |Y| = 1$, then it is said that the pair is separated by nondeterministic M.

3   Separating Pairs with 2 States

In this section, we present our results on separating pairs.

3.1   MCQFAs with Real Amplitudes

First at all we prove that any easy pair can be separated by a 2-state MCQFA with real amplitudes (all components of the initial states and the transition matrices are real numbers).

Theorem 1. *Any given pair of unary words (a^d, a^{d+t}) $(d \geq 0$ and $t > 0)$ can be exactly separated by a MCQFA, say $R_{d,t}$.*

Proof. We define a unary 2-state MCQFA denoted with the symbol $R_{d,t}$. Let $\{q_1, q_2\}$ be the set of states. Note that any possible quantum state of such automaton is a point on the unit circle, where $|q_1\rangle$ is $(1, 0)$ and $|q_2\rangle$ is $(0, 1)$. Automaton $R_{d,t}$ is defined by the following specifications (remark that R stands for rotation).

- The initial state is $\cos(\frac{d\pi}{2t}) |q_1\rangle - \sin(\frac{d\pi}{2t}) |q_2\rangle$, the point on the unit circle obtained by making a clockwise rotation with angle $\frac{d\pi}{2t}$ (d times $\frac{\pi}{2t}$) when starting at the point $|q_1\rangle$.
- The single unitary operator is a counter-clockwise rotation with angle $\frac{\pi}{2t}$.
- The single accepting state is q_1.

After reading a^d, the automaton is in $|q_1\rangle$ and so it is accepted with probability 1, and, after reading a^{d+t}, the automaton is in $|q_2\rangle$ and so it is accepted with probability 0.   □

Corollary 1. *Any easy pair of words can be separated exactly by a 2-state MCQFA with real amplitudes.*

There exist hard pairs of words that can be exactly separated by a 2-state MCQFA with real amplitudes, for instance, the pair (ab, ba): Let $\left(\frac{1}{\sqrt{2}} \quad \frac{1}{\sqrt{2}} \right)^T$ be the initial state, and we apply U_a and U_b when reading symbols a and b, respectively, where

$$U_a = \begin{pmatrix} \frac{1}{\sqrt{2}} & \frac{1}{\sqrt{2}} \\ \frac{1}{\sqrt{2}} & \frac{-1}{\sqrt{2}} \end{pmatrix} \text{ and } U_b = \begin{pmatrix} 1 & 0 \\ 0 & -1 \end{pmatrix}.$$

Then, after reading the words ab and ba, we obtain the following final states:

$$|u_f^{ab}\rangle = \begin{pmatrix} 1 & 0 \\ 0 & -1 \end{pmatrix} \begin{pmatrix} \frac{1}{\sqrt{2}} & \frac{1}{\sqrt{2}} \\ \frac{1}{\sqrt{2}} & \frac{-1}{\sqrt{2}} \end{pmatrix} \begin{pmatrix} \frac{1}{\sqrt{2}} \\ \frac{1}{\sqrt{2}} \end{pmatrix} = \begin{pmatrix} 1 \\ 0 \end{pmatrix}$$

and

$$|u_f^{ba}\rangle = \begin{pmatrix} \frac{1}{\sqrt{2}} & \frac{1}{\sqrt{2}} \\ \frac{1}{\sqrt{2}} & \frac{-1}{\sqrt{2}} \end{pmatrix} \begin{pmatrix} 1 & 0 \\ 0 & -1 \end{pmatrix} \begin{pmatrix} \frac{1}{\sqrt{2}} \\ \frac{1}{\sqrt{2}} \end{pmatrix} = \begin{pmatrix} 0 \\ 1 \end{pmatrix}.$$

Therefore, the pair (ab, ba) can be exactly separated by 2-state MCQFAs with real amplitudes.

However, such automata cannot distinguish all pairs of words, as exemplified by the following simple result.

Theorem 2. *No 2-state non-deterministic MCQFA with real entries can separate two words $x, y \in \{a^2, b^2\}^*$ provided that $|x|_a = |y|_a$ and $|x|_b = |y|_b$.*

Proof. Consider any such MCQFA, and let U_a and U_b be the transition matrices corresponding to a and b, respectively. The operators U_a^2 and U_b^2 are rotations in $\mathbb{R}^2$, hence, they commute. Thus,

$$\left|u_f^x\right\rangle = U_b^{|x|_b} U_a^{|x|_a} \left|u_0\right\rangle = U_b^{|y|_b} U_a^{|y|_a} \left|u_0\right\rangle = \left|u_f^y\right\rangle,$$

and no final measurement can distinguish these two identical final states.   □

Remark 1. It follows from the above results that GK pairs can be separated by using 2 states. On the other hand, it was constructively proved that there exist pairs requiring at least three states.

3.2 MCQFAs with Complex Amplitudes

In the previous section, we show that 2-state MCQFAs with real entries cannot separate all pairs of words. We conjecture that 2-state MCQFAs with complex entries (some components of the initial states and the transition matrices can be complex numbers) can exactly separate any pair of words. This conjecture is related to some problems in the theory of Lie groups (see below).

Theorem 3. *Any pair of words can be separated by a 2-state MCQFA with complex amplitudes in nondeterministic acceptance mode.*

Proof. Now, we describe an explicit 2-state nondeterministic MCQFA that can separate any given pair. For our purpose, we use an already known QFA algorithm given in [1]. For a given binary word $x \in \{a, b\}^*$, M_x is a 3-state ($\{q_1, q_2, q_3\}$) MCQFA. The initial state is $|q_1\rangle$ and the accepting states are q_2 and q_3. The unitary operators for symbols a and b are given below:

$$U_a = \frac{1}{5}\begin{pmatrix} 4 & 3 & 0 \\ -3 & 4 & 0 \\ 0 & 0 & 5 \end{pmatrix} \text{ and } U_b = \frac{1}{5}\begin{pmatrix} 4 & 0 & 3 \\ 0 & 5 & 0 \\ -3 & 0 & 4 \end{pmatrix}$$

We define the initial state as follows:

$$|u_0\rangle = U_{x_1}^{-1} U_{x_2}^{-1} \cdots U_{x_{|x|}}^{-1} |q_1\rangle$$

Then, the final quantum state for x is

$$\left|u_f^x\right\rangle = U_{x_{|x|}} \cdots U_{x_2} U_{x_1} U_{x_1}^{-1} U_{x_2}^{-1} \cdots U_{x_{|x|}}^{-1} |q_1\rangle = |q_1\rangle$$

So, the accepting probability of M on x is zero. i.e., $f_M(x) = 0$. On the other hand, for any given word $y \neq x$, the final quantum state for y is different from $|q_1\rangle$:

$$\left|u_f^y\right\rangle = U_{y_{|y|}} \cdots U_{y_2} U_{y_1} U_{x_1}^{-1} U_{x_2}^{-1} \cdots U_{x_{|x|}}^{-1} |q_1\rangle = \alpha_1 |q_1\rangle + \alpha_2 |q_2\rangle + \alpha_3 |q_3\rangle, \quad (1)$$

where $|\alpha_1|$ is always less than 1 when $x \neq y$ [1]. Then, the accepting probability of M on y is nonzero. i.e., $f_M(y) > 0$.

Therefore, we can say that nondeterminsitic MCQFA M_x separates x from any other word. Based on a conversion technique given in [1], we can convert M_x into a 2-state ($\{p_1, p_2\}$) MCQFA, say N_x, defined on $\mathbb{C}^2$ such that, after reading the same word, the probability of observing q_1 is 1 if and only if the probability of observing p_1 is 1.   □

We conjecture that any pair can be separated by a 2-state MCQFA with complex amplitudes and zero-error. Let us discuss some facts concerning the conjecture.

Let $w \in \{a, b, a^{-1}, b^{-1}\}^n$, and let $SU(2)$ be the group of 2×2 unitary matrices whose determinant is equal to 1. Suppose that $w = w_1 \cdots w_n$, and let $f_w : SU(2) \times SU(2) \to SU(2)$ be the word map defined by

$$f_w(M, N) = \prod_{i \leq n} A_i$$

where given $i \leq n$, the matrix $A_i \in SU(2)$ is defined as

$$A_i = \begin{cases} M & \text{if } w_i = a \\ N & \text{if } w_i = b \\ M^{-1} & \text{if } w_i = a^{-1} \\ N^{-1} & \text{if } w_i = b^{-1} \end{cases}.$$

Remark 2. The notion of word map can be extended in a straightforward way to any group different of $SU(2)$. We are interested in the word maps that are defined over the special unitary groups.

Given $x, y \in \{a, b\}^*$, if $y = y_1 \cdots y_n$ and $x = x_1 \cdots x_n$, we set

$$yx^{-1} = y_n \cdots y_1 x_1^{-1} \cdots x_n^{-1}$$

Notice that if the matrix $R_{\frac{\pi}{2}} = \begin{pmatrix} i & 0 \\ 0 & -i \end{pmatrix}$ belongs to the image of $f_{yx^{-1}}$ (i.e., if the image of $f_{yx^{-1}}$ contains a rotation by $\frac{\pi}{2}$), then one can choose $(M, N) \in f_{yx^{-1}}^{-1}(R_{\frac{\pi}{2}})$, and use the pair (M, N) to built a 2-state MCQFA separating the pair (x, y) with zero-error. Thus, the problem of separating any pair using two quantum states is closely related to the problem of surjectivity of word maps in the special unitary group $SU(2)$.

The word map f_w is a continuous map defined over a topological space (the Lie group $SU(2)$) that is compact and connected. Moreover, it satisfies the following condition:

For all $M, N, U \in SU(2)$, the equality

$$f_w(U^\dagger MU, U^\dagger NU) = U^\dagger f_w(M, N) U$$

holds.

The above facts imply that the image of f_w is of the form

$$\{V \in SU\,(2) : U \text{ has eigenvalues } e^{\pm\theta}\ 0 \le \theta \le \alpha\}$$

for some real $\alpha = \alpha\,(w)$. Remark that if $\alpha\,(yx^{-1}) \ge \frac{\pi}{2}$, then the pair (x, y) can be separated with zero-error.

A famous result of Borel [4] implies that the image of f_w is dense in the Zariski topology. However, it does not imply that the image is dense in the ordinary topology. Actually, it can be very far from that. As shown by Thom [14], given $\varepsilon > 0$, there exists a word w_ε such that $\alpha\,(w_\varepsilon) < \varepsilon$. The results of Thom do not imply the existence of a pair (x, y) such that $\alpha\,(yx^{-1}) < \frac{\pi}{2}$. Actually, there are some additional results in the theory of word maps suggesting that such a bad pair cannot exist.

Let F_2 be the free group with two generators, it consists of finite words over $\{a, b, a^{-1}, b^{-1}\}^*$ with the concatenation operation, modulo the relations $aa^{-1} = a^{-1}a = bb^{-1} = b^{-1}b = \epsilon$, where ϵ is the empty word. Notice that if w and u are equal, as elements of F_2, then $f_w = f_u$. Given $x, y \in F_2$, the commutator of x and y is the element $xyx^{-1}y^{-1}$, which we denote with the symbol $[x, y]$. The derived subgroup of F_2, denoted with the symbol $F_2^{(1)}$, is the subgroup generated by all the commutators. The second derived subgroup, denoted with the symbol $F_2^{(2)}$, is the derived subgroup of $F_2^{(1)}$. Elkasapy and Thom [7] showed that if $w \notin F_2^{(2)}$, then the corresponding word map f_w : $SU\,(n) \times SU\,(n) \to SU\,(n)$ is surjective for infinitely many n. We prove, below, that for all pair (x, y) the word $yx^{-1} \notin F_2^{(2)}$. Notice that if for all $w \notin F_2^{(2)}$ the word map $f_w : SU\,(2) \times SU\,(2) \to SU\,(2)$ is surjective, then any pair can be separated by using two qubits and zero-error. This last fact provides additional motivation to study this type of word maps.

Theorem 4. *For any two different words $x, y \in \{a, b\}^*$, the element xy^{-1} lies outside of the second derived subgroup $F_2^{(2)}$.*

Proof. An element $w \in F_2$ lies in the first derived subgroup of F_2, if and only if, the total degree of both a and b in w is equal to zero. That is, $xy^{-1} \notin F_2^{(1)}$ if and only if (x, y) is an easy pair.

Now assume that (x, y) is a hard pair. It is well known that $F_2^{(1)}$ is a free group. A set of generators for $F_2^{(1)}$ is the set

$$T = \{[a^k, b^l] : k, l > 0\}$$

Notice that $[a^k, b^l]^{-1} = [b^l, a^k]$. Again, $w \in F_2^{(1)}$ lies in $F_2^{(2)}$, if and only if, the unique decomposition of w into the elements of T contains each $[a^k, b^l]$ with total degree 0.

Given $x \in \{a, b\}^*$, we have a decomposition of the form

$$x = \prod_i [a^{k_i}, b^{l_i}]^{\varepsilon_i} \cdot a^{|x|_a} b^{|x|_b}$$

where, for all i, we have $k_i + l_i > k_{i-1} + l_{i-1}$ and $\varepsilon_i = \pm 1$. Now, it is not hard to see that $xy^{-1} \in F_2^{(2)}$, if and only if, $x = y$.   □

Remark 3. We say that (x, y) is a bad pair if $\alpha\left(yx^{-1}\right) < \frac{\pi}{2}$. If there exist such bad pairs, then it would be interesting to find the minimum number of states that are necessary to separate such bad pairs by using a DFA. Recall that one of our motivating problems is the construction of a set of pairs requiring a superlogarithmic number of states to be separated. It would also be interesting if this problem is related to the theory of word maps and Lie groups.

4   Separating Two Finite Sets

In this section, we focus on a more general problem: Separating two finite languages. Let $X = \{x_1, \ldots, x_m\}$ and $Y = \{y_1, \ldots, y_n\}$ be two disjoint set of binary words by assuming that $m \leq n$ (the sets are exchanged, otherwise). We consider the case of nondeterministic MCQFAs.

4.1   Nondeterministic MCQFAs

We use the MCQFA algorithm given at the end of the proof of Theorem 3. Let $N(X) = \{N_{x_1}, N_{x_2}, \ldots, N_{x_m}\}$ be the set of 2-state MCQFAs mentioned there. We can obtain a MCQFA, say N_X, by tensoring all MCQFAs in $N(X)$,

$$N_X = N_{x_1} \otimes N_{x_2} \otimes \cdots \otimes N_{x_m},$$

i.e., executing all of them in parallel. The tensor product is obtained in a straightforward way. The set of states of N_X is $\{p_1, p_2\}^m$. If $|u_{j,0}\rangle$ is the initial state of N_{x_j} and $U_{j,a}$ ($U_{j,b}$) is the unitary operator for symbol a (b), then the initial state of N_X is

$$|u_{1,0}\rangle \otimes |u_{2,0}\rangle \otimes \cdots \otimes |u_{m,0}\rangle$$

and the unitary operator for symbol a (b) is

$$U_{1,a} \otimes U_{2,a} \otimes \cdots \otimes U_{m,a} \quad (U_{1,b} \otimes U_{2,b} \otimes \cdots \otimes U_{m,b}),$$

where $1 \leq j \leq m$. Similarly, if $\left|u_{j,f}^y\right\rangle$ is the final state of N_{x_j} and β_j is the amplitude of the state $|p_2\rangle$ after reading binary word y, then the final state of N_X on y will be

$$\left|u_{1,f}^y\right\rangle \otimes \left|u_{2,f}^y\right\rangle \otimes \cdots \otimes \left|u_{m,f}^y\right\rangle$$

and so the amplitude of $|(p_2, p_2, \ldots, p_2)\rangle$ will be

$$\beta = \beta_1 \beta_2 \cdots \beta_m.$$

Therefore, it is clear that, if $x_j = y$, then β will be zero since β_j is zero. More generally, $\beta = 0$ if and only if $y \in X$. Thus, by picking $(p_2, p_2, \ldots, p_2)$ as the single accepting state of M_X, we can obtain the machine that separates any given word from a word in X. Remark that the number of states of N_X is 2^m.

Theorem 5. *The disjoint binary finite languages X and Y $(1 \leq |X| \leq |Y|)$ can be separated by nondeterministic MCQFAs with $2^{|X|}$ states.*

5   Concluding Remarks

The motivating problem of our research is the problem of quantifying the number of states that are required to separate a given pair of words using DFAs. This problem has its roots in machine learning [16], and it has been intensively studied, but in despite of all the efforts, so few is known about it. We believe that we can shed some light on this elusive problem, by considering the same kind of questions for different models of automata. In previous research we studied alternating finite state automata. In this work we studied QFAs, and in the extended version of this paper [3] we consider the novel model of affine automata [6,15]. We think that these questions are interesting in their own rigth, and that they deserve further investigation.

Acknowledgement. We thank Andreas Thom for the discussions on our conjecture and anonymous reviewers for their helpful comments. The first author acknowledges the support provided by FP7 FET Proactive project QALGO. The second author acknowledges the support provided by Universidad Nacional de Colombia project Hermes 32083. The third author acknowledges the support provided by CAPES, grant 88881.030338/2013-01. Moreover, some parts of the work were done while the third author was visiting Bogotá, Colombia in December 2014.

References

1. Ambainis, A., Watrous, J.: Two-way finite automata with quantum and classical states. Theor. Comput. Sci. **287**(1), 299–311 (2002)
2. Ambainis, A., Yakaryılmaz, A.: Automata: from mathematics to applications. In: Automata and Quantum Computing (to appear). arXiv:1507.01988
3. Belovs, A., Montoya, J.A., Yakaryılmaz, A.: Can one quantum bit separate any pair of words with zero-error? Technical report (2016). arXiv:1602.07967
4. Borel, A.: On free subgroups of semisimple groups. L'Enseignement Mathématique **29**, 151–164 (1983)
5. Demaine, E.D., Eisenstat, S., Shallit, J., Wilson, D.A.: Remarks on separating words. In: Holzer, M. (ed.) DCFS 2011. LNCS, vol. 6808, pp. 147–157. Springer, Heidelberg (2011)
6. Díaz-Caro, A., Yakaryılmaz, A.: Affine computation and affine automaton. In: Computer Science - Theory and Applications. LNCS, vol. 9691, pp. 1–15. Springer (2016). arXiv:1602.04732
7. Elkasapy, A., Thom, A.: About Gotô's method showing surjectivity of word maps. Indiana Univ. Math. J. **63**(5), 1553–1565 (2014). arXiv:1207.5596
8. Goralčík, P., Koubek, V.: On discerning words by automata. In: Kott, L. (ed.) Automata, Languages and Programming. LNCS, vol. 226. Springer, Heidelberg (1986)
9. Hirvensalo, M.: Quantum automata with open time evolution. Int. J. Nat. Comput. **1**(1), 70–85 (2010)
10. Moore, C., Crutchfield, J.P.: Quantum automata and quantum grammars. Theor. Comput. Sci. **237**(1–2), 275–306 (2000)
11. Nielsen, M., Chuang, I.: Quantum Computation and Quantum Information, 10th edn. Cambridge University Press, Cambridge (2010)

12. Robson, J.M.: Separating strings with small automata. Inf. Process. Lett. **30**(4), 209–214 (1989)
13. Say, A.C.C., Yakaryılmaz, A.: Quantum finite automata: a modern introduction. In: Calude, C.S., Freivalds, R., Kazuo, I. (eds.) Gruska Festschrift. LNCS, vol. 8808, pp. 208–222. Springer, Heidelberg (2014)
14. Thom, A.: Convergent sequences in discrete groups. Can. Math. Bull. **56**(2), 424–433 (2013). arXiv:1003.4093
15. Villagra, M., Yakaryılmaz, A.: Language recognition power and succintness of affine automata. In: Calude, C.S., Dinneen, M.J. (eds.) UCNC 2015. LNCS, vol. 9252. Springer, Heidelberg (2015)
16. Yakaryılmaz, A., Montoya, J.A.: On discerning strings with finite automata. In: 2015 Latin American Computing Conference, pp. 1–5. IEEE (2015)
17. Yakaryılmaz, A., Say, A.C.C.: Languages recognized by nondeterministic quantum finite automata. Quantum Inf. Comput. **10**(9&10), 747–770 (2010)
18. Yakaryılmaz, A., Say, A.C.C.: Unbounded-error quantum computation with small space bounds. Inf. Comput. **279**(6), 873–892 (2011)

Prefix Distance Between Regular Languages

Timothy Ng[✉]

School of Computing, Queen's University, Kingston, ON K7L 3N6, Canada
ng@cs.queensu.ca

Abstract. The prefix distance between two words x and y is defined as the number of symbol occurrences in the words that do not belong to the longest common prefix of x and y. We show how to model the prefix distance using weighted transducers. We use the weighted transducers to compute the prefix distance between two regular languages by a transducer-based approach originally used by Mohri for an algorithm to compute the edit distance. We also give an algorithm to compute the inner prefix distance of a regular language.

1 Introduction

Distance measures are used in a variety of applications to measure the similarity of data. For instance, the Hamming distance counts the number of positions in which two words of equal length differ. Another common measure is the Levenshtein distance, also called the edit distance, which counts the number of insertion, deletion, and substitution operations that are needed to transform one word to another. However, counting the number of edit operations to transform one word into another is not the only relevant way to measure the similarity between words. The prefix distance is defined in terms of the longest common prefix of two words. For the words x and y, their prefix distance is the number of symbols that do not belong to their longest common prefix. We can define the suffix and subword distances in a similar way in terms of the longest common suffix or subword of two words.

These distance measures can be extended in various ways to distances between sets of words, or languages. A common extension of a distance function for languages L_1 and L_2 takes the minimum distance between a word u in L_1 and a word v in L_2. An alternative extension is called the relative distance [4]. The relative distance from a language L_1 to a language L_2 is the supremum over all words w in L_1 of the smallest distance between w and L_2. Another notion of distance on languages is the inner distance of a language [11]. For a language L, the inner distance is the smallest distance between two words u and v in L.

Much of the work on computing distances on languages has been focused on the edit distance and its variants. Pighizzini [15] studied the hardness of computing the edit distance between a word and a language. Mohri [14] showed how to compute the edit distance and its variants between two regular languages in polynomial time. Benedikt et al. [1,2] showed how to compute the

© Springer International Publishing Switzerland 2016
Y.-S. Han and K. Salomaa (Eds.): CIAA 2016, LNCS 9705, pp. 224–235, 2016.
DOI: 10.1007/978-3-319-40946-7_19

relative edit distance between regular languages. Han et al. [8] gave a polyno-mial time algorithm for computing the edit distance between a regular language and context-free language. Konstantinidis [11] gave an algorithm for computing the inner edit distance of a regular language in quadratic time. Kari et al. [10] gave a quadratic time algorithm for computing the inner Hamming distance of a regular language. Konstantinidis and Silva [12] showed how to compute the inner distance for variants of the edit distance.

Naturally, the same extensions to languages can be applied to the prefix, suffix, and subword distances and some of these extensions have already been studied. Bruschi and Pighizzini [3] studied the hardness of computing the prefix distance between a word to a language in the context of intrusion detection. Choffrut and Pighizzini [4] showed that the relative prefix distance between two regular languages is computable. Kutrib et al. [13] considered a parameter-ized prefix distance between languages to measure fault tolerance of finite-state devices.

In this paper, we show how to compute the prefix distance between two regular languages. We show how to model prefix distance using edit systems and construct transducers which realize these models. We use these transducers to compute distances using a similar approach to Mohri's edit distance algorithm for weighted automata from [14]. We also show how to use the weighted transducer approach to compute the inner prefix distance of a given regular language. We also give polynomial time algorithms based on the transducer-based approach to compute the suffix distance and the subword distance between two regular languages.

2   Preliminaries

Here we briefly recall some definitions and notation used in the paper. For all unexplained notions on finite automata and regular languages the reader may consult the textbook by Shallit [16] or the survey by Yu [17]. More on weighted automata and transducers can be found in the textbook by Droste et al. [7]. A survey of distances is given by Deza and Deza [6].

In the following, Σ is always a finite alphabet, the set of all words over Σ is denoted Σ^*, and ε denotes the empty word. The reversal of a word $w \in \Sigma^*$ is denoted by w^R. The length of a word w is denoted by $|w|$. The cardinality of a finite set S is denoted $|S|$ and the power set of S is 2^S. A word $w \in \Sigma^*$ is a *subword* or *factor* of x if and only if there exist words $u, v \in \Sigma^*$ such that $x = uwv$. If $u = \varepsilon$, then w is a *prefix* of x. If $v = \varepsilon$, then w is a *suffix* of x.

A *nondeterministic finite automaton* (NFA) is a tuple $A = (Q, \Sigma, \delta, Q_0, F)$ where Q is a finite set of states, Σ is an alphabet, δ is a multi-valued transition function $\delta : Q \times \Sigma \to 2^Q$, $Q_0 \subseteq Q$ is a set of initial states, and $F \subseteq Q$ is a set of final states. We extend the transition function δ to $Q \times \Sigma^* \to 2^Q$ in the usual way. A word $w \in \Sigma^*$ is *accepted* by A if for some $q_0 \in Q_0$, $\delta(q_0, w) \cap F \neq \emptyset$ and the language recognized by A consists of all words accepted by A. An ε-NFA is the extension of an NFA where transitions can be labeled by the empty word ε.

It is known that every ε-NFA has an equivalent NFA without ε-transitions with the same number of states. An NFA is a *deterministic finite automaton* (DFA) if $|Q_0| = 1$ and for all $q \in Q$ and $a \in \Sigma$, $\delta(q, a)$ either consists of one state or is undefined. The size of A, denoted $|A|$, is defined as the sum of the number of states and transitions of A, $|Q| + |\delta|$.

A *weighted finite-state transducer* [7] with weights in the $(\min, +)$-semiring $\mathbb{K}$ is a 6-tuple $T = (Q, \Sigma, \Delta, I, F, E)$ where Q is a finite set of states, Σ is the input alphabet, Δ is the output alphabet, $I \subseteq Q$ is the set of initial states, $F \subseteq Q$ is the set of final states, $E \subseteq Q \times (\Sigma \cup \{\varepsilon\}) \times (\Delta \cup \{\varepsilon\}) \times \mathbb{K} \times Q$ is a finite set of transitions with weights in $\mathbb{K}$. The size of T, denoted $|T|$, is defined as the sum of the number of states and transitions of T, $|Q| + |E|$.

A *path* or *computation* of T is a word π over the alphabet of transitions E

$$\pi = (p_1, u_1, v_1, w_1, q_1) \cdots (p_n, u_n, v_n, w_n, q_n)$$

with $q_i = p_{i+1}$ for $1 \leq i < n$. A path π from p to q is accepted if $p \in I$ and $q \in F$. Let $\omega : E^* \to \mathbb{K}$ be a weight function for paths defined by $\omega(\pi) = \sum_{i=1}^{n} w_i$, the sum of the weights of each transition in π. The label of a path π, denoted $\ell(\pi)$ is the pair of words (x, y) with $x = u_1 \cdots u_n$ and $y = v_1 \cdots v_n$. Let $w : \Sigma^* \times \Sigma^* \to \mathbb{K}$ be a weight function for labels (x, y) defined as the weight of the minimum weight accepted path labeled by (x, y),

$$w(x, y) = \min_{\pi \in E^*} \{\omega(\pi) \mid \ell(\pi) = (x, y)\}.$$

A function $d : \Sigma^* \times \Sigma^* \to \mathbb{N} \cup \{0\}$ is a *distance* if it satisfies for all $x, y \in \Sigma^*$

1. $d(x, y) = 0$ if and only if $x = y$,
2. $d(x, y) = d(y, x)$,
3. $d(x, z) \leq d(x, y) + d(y, z)$ for $z \in \Sigma^*$.

A distance between words can be extended to a distance between a word $w \in \Sigma^*$ and a language $L \subseteq \Sigma^*$ by

$$d(w, L) = \min\{d(w, w') \mid w' \in L\}.$$

We generalize this to a distance between two languages L_1 and L_2,

$$d(L_1, L_2) = \min\{d(w_1, w_2) \mid w_1 \in L_1, w_2 \in L_2\}.$$

The *inner distance* of a language L (also called the *self distance*) is the minimal distance between any two distinct words that both belong to L.

$$d(L) = \min\{d(w_1, w_2) \mid w_1, w_2 \in L, w_1 \neq w_2\}.$$

The *prefix distance* of x and y counts the number of symbols which do not belong to the longest common prefix of x and y. It is defined by

$$d_p(x, y) = |x| + |y| - 2 \cdot \max_{z \in \Sigma^*} \{|z| \mid x, y \in z\Sigma^*\}.$$

Similarly, the *suffix distance* of x and y counts the number of symbols which do not belong to the longest common suffix of x and y and is defined

$$d_s(x, y) = |x| + |y| - 2 \cdot \max_{z \in \Sigma^*} \{|z| \mid x, y \in \Sigma^* z\}.$$

The *subword distance* of x and y counts the number of symbols which do not belong to the longest common subword of x and y and is defined

$$d_f(x, y) = |x| + |y| - 2 \cdot \max_{z \in \Sigma^*} \{|z| \mid x, y \in \Sigma^* z \Sigma^*\}.$$

3   Edit Strings and Edit Systems

Edit systems, also called error systems, were first studied extensively by Kari and Konstantinidis in [9] as a formalization for errors in terms of formal languages. Informally, an edit system is a formal language over the alphabet of edit operations and are used to model different types of errors. We present some basic definitions for edit systems and model the prefix, suffix, and subword distances using edit systems.

For an alphabet Σ, let $\mathcal{E}_\Sigma$ be the alphabet of *edit operations* over Σ,

$$\mathcal{E}_\Sigma = \{(a/b) \mid a, b \in \Sigma \cup \{\varepsilon\}, ab \neq \varepsilon\}.$$

We use $\mathcal{E}$ whenever Σ is obvious from the context. An *edit string* or *alignment* of two words is an element of $\mathcal{E}^*$. For an edit string $e = (a_1/b_1)(a_2/b_2) \cdots (a_n/b_n)$, we call $a_1 a_2 \cdots a_n$ the *input part* of e and $b_1 b_2 \cdots b_n$ the *output part*. We define the edit morphism to be the morphism $h : \mathcal{E}^* \to \Sigma^* \times \Sigma^*$ by $h(e) = (a_1 \cdots a_n, b_1 \cdots b_n)$.

We can define subsets of the alphabet of edit operations which correspond to the classical edit operations of substitution, insertion, and deletion and the identity operation by

- the set of substitution operations $\mathcal{S} = \{(a/b) \mid a \neq b, a, b \in \Sigma\}$,
- the set of insertion operations $\mathcal{I} = \{(\varepsilon/a) \mid a \in \Sigma\}$,
- the set of deletion operations $\mathcal{D} = \{(a/\varepsilon) \mid a \in \Sigma\}$,
- the set of identity operations $\mathcal{E}_0 = \{(a/a) \mid a \in \Sigma\}$.

We define a cost function $c : \mathcal{E} \to \mathbb{N}$ which assigns a cost to each element of the edit alphabet. Note that the standard definition of edit distance assigns the cost of every non-identity symbol $(a/b) \in \mathcal{E} \setminus \mathcal{E}_0$ to be 1. However, the prefix, suffix, and subword distances count each additional symbol in both words, so our cost function c is defined

- $c((a/a)) = 0$, for all $(a/a) \in \mathcal{E}_0$,
- $c((\varepsilon/a)) = 1$, for all $(\varepsilon/a) \in \mathcal{I}$,
- $c((a/\varepsilon)) = 1$, for all $(a/\varepsilon) \in \mathcal{D}$,
- $c((a, b)) = 2$, for all $(a/b) \in \mathcal{S}$.

The cost of an edit string $e = e_1 e_2 \cdots e_n$ is then the sum of the cost of each symbol

$$c(e) = \sum_{i=1}^{n} c(e_i).$$

A language defined over $\mathcal{E}$ is called an *edit system*. A regular edit system can be modeled by a finite automaton defined over $\mathcal{E}$. Such an edit system may also be realized as a finite-state transducer, where for each symbol $(a/b) \in \mathcal{E}$, the transitions on (a/b) are considered transition labels with a as the input part and b as the output part. Thus, a computation path of a finite state transducer over $\mathcal{E}$ corresponds with an edit string.

We now define the language of edit strings for the prefix distance L_p by

$$L_p = \mathcal{E}_0^*(\mathcal{E} \setminus \mathcal{E}_0)^*.$$

Informally, this is the set of edit strings with a prefix of identity operations followed by non-identity edit operations. We define the function $d_p' : \Sigma^* \times \Sigma^* \to \mathbb{N}$ on $x, y \in \Sigma^*$ by

$$d_p'(x, y) = \min_{e \in L_p} \{c(e) \mid h(e) = (x, y)\}.$$

In the following proposition, we show that $d_p'(x, y)$ is exactly the prefix distance of x and y.

Proposition 1. *Let $x, y \in \Sigma^*$ be two words. Then $d_p'(x, y) = d_p(x, y)$.*

Proof. Consider two words x and y with $x = px'$ and $y = py'$, where p is the longest common prefix of x and y. By definition, we have $d_p(x, y) = |x| + |y| - 2|p| = |x'| + |y'|$. Now consider an edit string $e \in L_p$ with $h(e) = (x, y)$. Since $e \in L_p$, we split e into two parts $e = e_0 e_1$, where $e_0 \in \mathcal{E}_0^*$ and $e_1 \in (\mathcal{E} \setminus \mathcal{E}_0)^*$. To minimize the cost $c(e)$, we require e_0 to be as long as possible and minimize the length of e_1, since $c(e_0) = 0$. Thus, e_0 corresponds to a string of identity operations for the longest common prefix p of x and y. This means that e_1 is the edit string such that $h(e_1) = (x', y')$. Thus, $c(e) = c(e_1)$ and since $e' \in (\mathcal{E} \setminus \mathcal{E}_0)^*$, we have $c(e') = |x'| + |y'|$. □

We can similarly define the same notions for suffix and infix distances. Let $L_s = (\mathcal{E} \setminus \mathcal{E}_0)^* \mathcal{E}_0^*$ be the language of edit strings for suffix distance and let $L_f = (\mathcal{E} \setminus \mathcal{E}_0)^* \mathcal{E}_0^* (\mathcal{E} \setminus \mathcal{E}_0)^*$ be the language of edit strings for infix distance. We define the functions

$$d_s'(x, y) = \min_{e \in L_s} \{c(e) \mid h(e) = (x, y)\},$$

$$d_f'(x, y) = \min_{e \in L_f} \{c(e) \mid h(e) = (x, y)\}.$$

The following result is proved analogously as Proposition 1

Proposition 2. *Let $x, y \in \Sigma^*$ be two words. Then*

1. $d_s(x, y) = d_s'(x, y)$, *and*
2. $d_f(x, y) = d_f'(x, y)$.

4   Computing the Prefix Distance Between Regular Languages

We give a polynomial time algorithm to compute the prefix distance between two languages given by nondeterministic finite automata. Mohri [14] gave an algorithm for computing the edit distance between two regular languages by using weighted transducers. We use this approach by defining a weighted transducer with paths which correspond to edit strings in L_p.

We define the transducer $T_p = (Q, \Sigma, \Delta, I, F, E)$, by setting $Q = \{0, 1\}$, $\Delta = \Sigma$, $I = \{0\}$, $F = \{0, 1\}$, and the transition set E is given by

- $(0, a, a, 0, 0)$ for all $a \in \Sigma$,
- $(0, a, \varepsilon, 1, 1)$ for all $a \in \Sigma$,
- $(0, \varepsilon, a, 1, 1)$ for all $a \in \Sigma$,
- $(0, a, b, 2, 1)$, with $a \neq b$ for all $a, b \in \Sigma$,
- $(1, a, \varepsilon, 1, 1)$ for all $a \in \Sigma$,
- $(1, \varepsilon, a, 1, 1)$ for all $a \in \Sigma$,
- $(1, a, b, 2, 1)$, with $a \neq b$ for all $a, b \in \Sigma$.

The transducer is shown with $\Sigma = \{a, b\}$ in Fig. 1. We claim that the transducer T_p takes as input some word w and outputs a word x such that any accepting computation path of T_p on w corresponds to an edit string in L_p which transforms w into x and that the weight of this path is the cost of the corresponding edit string. We prove this in the following lemma.

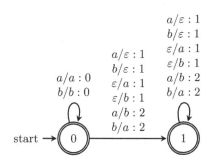

Fig. 1. The transducer T_p over the alphabet $\Sigma = \{a, b\}$.

Lemma 1. *The set of accepting paths of the transducer T_p over Σ corresponds to exactly the set of edit strings over Σ belonging to L_p. If π is an accepting path of T_p and e_π is the corresponding edit string, then the weight of π is $c(e_\pi)$.*

Proof. Let φ be a morphism $\varphi : E^* \to \mathcal{E}^*$ that maps a computation path of T_p to an edit string over $\mathcal{E}$ defined by $\varphi((p, a, b, i, q)) = (a/b)$. Consider an accepting path $\pi = \pi_1 \cdots \pi_n$ of T_p. Since both states 0 and 1 are final states, an accepting path may end in either state. If π ends in state 0, then π never leaves state 0 and π is of the form

$$\pi = (0, a_1, a_1, 0, 0) \cdots (0, a_n, a_n, 0, 0),$$

where $a_i \in \Sigma$ for all $1 \leq i \leq n$. Then $\varphi(\pi) = (a_1/a_1) \cdots (a_n/a_n) \in \mathcal{E}_0^*$. Note that every transition going from state 0 to itself has weight 0 and π therefore has weight 0. The cost of $\varphi(\pi)$ is also 0, as it consists only of identity operations, which have a cost of 0.

Now consider when π ends in state 1. Then π can be decomposed into $\pi = \pi_0\pi_1$ where for some $k < n$, we have

$$\pi_0 = (0, a_1, a_1, 0, 0) \cdots (0, a_{k-1}, a_{k-1}, 0, 0)$$
$$\pi_1 = (0, a_k, a_k', i_k, 1)(1, a_{k+1}, a_{k+1}', i_{k+1}, 1) \cdots (1, a_n, a_n', i_n, 1)$$

where $a_i \in \Sigma$ for $1 \leq i < k$ and $a_j, a_j' \in \Sigma \cup \{\varepsilon\}$ with $a_j \neq a_j'$ and $i_j \in \{1, 2\}$ for $k \leq j \leq n$. As in above, π_0 is a path which ends in state 0 and thus $\varphi(\pi_0)$ maps to a word over $\mathcal{E}_0$ with cost 0. The first transition in π_1 takes the machine to state 1. Since there are no transitions of the form $(1, a, a, 0, 1)$, the word $\varphi(\pi_1)$ contains no symbols from $\mathcal{E}_0$. In other words, $\varphi(\pi_1)$ is a word over the alphabet $\mathcal{E} \setminus \mathcal{E}_0$.

Now consider an edit string $e \in L_p$. We can decompose e into two parts $e = e_0 e_1$ with $e_0 \in \mathcal{E}_0^*$ and $e_1 \in (\mathcal{E} \setminus \mathcal{E}_0)^*$. Then e_0 corresponds to a computation path that ends in state 0 and e_1 corresponds to a path which begins with a transition from state 0 to state 1 and ends on state 1. Thus any edit string in L_p corresponds to an accepting path in T_p.

It remains to be shown that the cost of $\varphi(\pi_1)$ is the same as the weight of the path π_1. By definition of T_p, each transition with a label a/ε or ε/a has weight 1 for all $a \in \Sigma$ and every transition with a label a/b with $a \neq b$ has weight 2. This corresponds to the costs assigned by the cost function c and thus the weight of π_1 is exactly the cost of $\varphi(\pi_1)$.

Thus, we have $\varphi(\pi) = \varphi(\pi_0)\varphi(\pi_1) \in \mathcal{E}_0^*(\mathcal{E} \setminus \mathcal{E}_0)^* = L_p$ and $w(\pi) = w(\pi_0) + w(\pi_1) = c(\varphi(\pi_0)) + c(\varphi(\pi_1)) = c(\varphi(\pi))$. $\qquad\square$

Observe that if π is a minimum weight accepting path of T_p transforming a word w into a word x, then the weight of π is $d_p(w, x)$. This leads to the following result.

Proposition 3. *Let $x, y \in \Sigma^*$. Then the weight $w(x, y)$ of x and y in T_p is exactly $d_p(x, y)$.*

Proof. Recall that, by definition, the weight of a pair of words (x, y) in T_p is the minimum weight of all accepting paths of T_p with label (x, y). By Lemma 1, each path π in T_p corresponds to an edit string e_π in L_p and has weight equivalent to $c(e_\pi)$. Thus, we have

$$w(x, y) = \min_{e \in L_p} \{c(e) \mid h(e) = (x, y)\},$$

which is exactly $d_p(x, y)$ by Proposition 1. $\qquad\square$

Now we move to the main result. We wish to compute the prefix distance of two given regular languages L_1 and L_2. To do this, we compute a transducer for which pairs of words (x, y) with $x \in L_1$ and $y \in L_2$ have weight equal to $d_p(x, y)$. Let A_1 and A_2 be finite automata recognizing regular languages L_1 and L_2, respectively. Recall that an unweighted finite automaton over Σ may be viewed as a weighted transducer with input and output alphabets Σ and in which each transition labeled by $a \in \Sigma$ is labeled by a/a and has weight 0.

The composition $T_1 \otimes T_2 = (Q, \Sigma, \Gamma, I, F, E)$ of two weighted transducers $T_1 = (Q_1, \Sigma, \Delta, I_1, F_1, E_1)$ and $T_2 = (Q_2, \Delta, \Gamma, I_2, F_2, E_2)$ is defined by $Q = Q_1 \times Q_2$, $I = I_1 \times I_2$, $F = Q \cap (F_1 \times F_2)$, and the transition set E consists of transitions of the form $((q_1, q_1'), a, c, w_1 + w_2, (q_2, q_2'))$ for each transition $(q_1, a, b, w_1, q_2) \in E_1$ and $(q_1', b, c, w_2, q_2') \in E_2$. The composition $T_1 \otimes T_2$ can be computed in $O(|T_1||T_2|)$ time.

Now consider the weighted transducer $T = A_1 \otimes T_p \otimes A_2$. We show in the following lemma that for $x \in L_1$ and $y \in L_2$, the weight of (x, y) in T is $d_p(x, y)$.

Theorem 1. *Let L_1 and L_2 be regular languages recognized by NFAs A_1 and A_2, respectively. If $x \in L_1$ and $y \in L_2$, then (x, y) is the label of an accepting path of $T = A_1 \otimes T_p \otimes A_2$ and the weight of (x, y) in T is $d_p(x, y)$.*

Proof. Consider two words $x \in L_1$ and $y \in L_2$. By definition of composition, for any accepting path of T, the input part must be recognized by A_1, the output part must be recognized by A_2, and the path must correspond to an edit string in the language L_p. Thus, there is an accepting path π of T with label $\ell(\pi) = (x, y)$ which corresponds to an edit string $e_\pi \in L_p$ with $h(e_\pi) = (x, y)$. By Proposition 3, the weight $w(x, y)$ of T must be $d_p(x, y)$.   □

This result implies that the weight of the minimal weight path of $A_1 \otimes T_p \otimes A_2$ is the prefix distance between $L(A_1)$ and $L(A_2)$. This leads us to an efficient algorithm to compute the prefix distance between two regular languages.

Theorem 2. *For given NFAs A_1 and A_2 recognizing the languages L_1 and L_2, respectively, the value $d_p(L_1, L_2)$ can be computed in polynomial time.*

Proof. Recall that the prefix distance between L_1 and L_2 is defined

$$d_p(L_1, L_2) = \min\{d_p(x, y) \mid x \in L_1, y \in L_2\}.$$

By Theorem 1, for two words $x \in L_1$ and $y \in L_2$, the weight of (x, y) in the weighted transducer $T = A_1 \otimes T_p \otimes A_2$ is $d_p(x, y)$. By definition, this is the weight of the minimal weight path with label (x, y) accepted by T. Then the weight of a minimal weight accepting path in T from the initial state to a final state must be $d_p(L_1, L_2)$ by definition.

With T_p fixed, in the worst case, the composition of the weighted transducer $T = A_1 \otimes T_p \otimes A_2$ can be computed in time $O(|A_1||A_2|)$ and the size of T is $O(|A_1||A_2|)$ [7]. To compute $d_p(L_1, L_2)$, we compute T and find the shortest path from the initial state of T to a final state of T. Since there are no negative cycles, we use Dijkstra's single-source shortest path algorithm, which has running time

$O(|E| + |Q| \log |Q|)$, where E is the transition set of T and Q is the state set of T [5]. Thus, $d_p(L_1, L_2)$ can be computed in polynomial time.   □

In Proposition 2, we have characterized the suffix distance and the subword distance, respectively, in terms of the edit systems L_s and L_f. By using a weighted transducer based construction analogous to the one used for the prefix distance in Theorem 2, we can get a polynomial time algorithm for computing the suffix distance and subword distance between regular languages.

Theorem 3. *For given NFAs A_1 and A_2 recognizing the languages L_1 and L_2, respectively,*

1. *$d_s(L_1, L_2)$ can be computed in polynomial time, and*
2. *$d_f(L_1, L_2)$ can be computed in polynomial time.*

5   Computing the Inner Prefix Distance of a Regular Language

Kari et al. [10] give an algorithm for computing the inner Hamming distance of a regular language using a similar approach with NFAs over the edit alphabet. In the development of the algorithm, a crucial observation was the necessity of excluding all edit strings with cost 0, since $d(x, y) = 0$ if and only if $x = y$. Thus, for our algorithm, we need to modify the language L_p to exclude all edit strings with cost 0 and define a corresponding weighted transducer.

We define the language of edit strings for the prefix distance excluding all edit strings which result in identity,

$$L_p^{(1)} = \mathcal{E}_0^* (\mathcal{E} \setminus \mathcal{E}_0)^+.$$

The language $L_p^{(1)}$ is almost exactly the same as the language L_p with the exception that no edit strings $e \in \mathcal{E}_0^*$ are in $L_p^{(1)}$. That is, every edit string in $L_p^{(1)}$ must contain at least one symbol with nonzero cost.

Now, we define the transducer $T_p^{(1)} = (Q, \Sigma, \Delta, I, F, E)$ by choosing $Q = \{0, 1\}$, $\Delta = \Sigma$, $I = \{0\}$, $F = \{1\}$, and the transition set E is as in the definition of T_p. The transducer $T_p^{(1)}$ is the transducer T_p with the modification that state 1 is the sole final state. The transducer $T_p^{(1)}$ defined over the alphabet $\{a, b\}$ is shown in Fig. 2. We show in the following lemma that $T_p^{(1)}$ realizes $L_p^{(1)}$.

Lemma 2. *The set of accepting paths of the transducer $T_p^{(1)}$ over Σ corresponds exactly to the language edit strings $L_p^{(1)}$. If π is an accepting path of of $T_p^{(1)}$ and e_π is the corresponding edit string, then the weight of π is $c(e_\pi)$.*

Proof. Consider an accepting path $\pi = \pi_1 \cdots \pi_n$ of $T_p^{(1)}$. Recall from the proof of Lemma 1 the definition of φ and observe that since 0 is not a final state of $T_p^{(1)}$, π must be of the form $\pi = \pi_0 \pi_1$, with $\varphi(\pi_0) \in \mathcal{E}_0^*$ and $\varphi(\pi_1) \in (\mathcal{E} \setminus \mathcal{E}_0)^+$. Thus, $\varphi(\pi)$ must contain at least one non-identity operation and $c(\varphi(\pi)) > 0$.

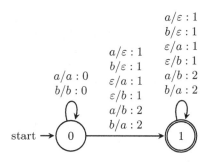

Fig. 2. The transducer $T_p^{(1)}$ over the alphabet $\Sigma = \{a, b\}$.

Now consider an edit string $e \in L_p^{(1)}$. We can decompose e into two parts $e = e_0 e_1$ with $e_0 \in \mathcal{E}_0^*$ and $e_1 \in (\mathcal{E} \setminus \mathcal{E}_0)^+$. Then e_0 corresponds to a computation path that ends in state 0 and e_1 corresponds to a path which begins with a transition from state 0 to state 1 and ends on state 1. Thus any edit string in L_p corresponds to an accepting path in T_p.

By the same argument from the proof of Lemma 1, the weight of an accepting path π of $T_p^{(1)}$ is exactly the cost of the edit string $\varphi(\pi)$.   □

As was the case for T_p, for each pair of words (x, y) with an accepting path in $T_p^{(1)}$, the weight of (x, y) is exactly $d_p(x, y)$ by the same argument as in the proof of Proposition 3. This leads us to the analogue of Theorem 1 for $T_p^{(1)}$.

Theorem 4. *Let L be a regular language recognized by a finite automaton A. Then for $x, y \in L$ with $x \neq y$, (x, y) is an accepting path of $T = A \otimes T_p^{(1)} \otimes A$ and the weight of $(x, y) \in T$ is $d_p(x, y)$.*

Proof. Let $x, y \in L$ and consider the weighted transducer $T = A \otimes T_p^{(1)} \otimes A$. By definition of composition, for any accepting path π of T, the input and output labels must be words recognized by A and the path must correspond to an edit string in $L_p^{(1)}$. Thus, there is an accepting path π of T with label $\ell(\pi) = (x, y)$ which corresponds to an edit string $e_\pi \in L_p^{(1)}$ with $h(e_\pi) = (x, y)$. Furthermore, since $e_\pi \in L_p^{(1)}$, we have $x \neq y$. Thus, by Lemma 2, the weight $w(x, y)$ of T must be $d_p(x, y)$.   □

From this it follows that can compute the inner prefix distance of a regular language by computing the appropriate weighted transducer and finding the minimal weight path from its initial state to one of its final states.

Theorem 5. *For a given NFA A recognizing the language L, the value $d_p(L)$ is computable in polynomial time.*

Proof. By Theorem 4, for $x, y \in L$, the weight of (x, y) in the weighted transducer $T = A \otimes T_p^{(1)} \otimes A$ is $d_p(x, y)$. Then the weight of a minimal weight accepting path in T must be $d_p(L)$ by definition.

Then, as in Theorem 2, the transducer T can be computed in time $O(|A|^2)$ in the worst case and the size of T is $O(|A|^2)$ [7]. Since there are no negative cycles, we can compute the minimal weight path of T in time $O(|E|+|Q|\log|Q|)$, where E is the transition set of T and Q is the state set of T, by using Dijkstra's algorithm [5]. Thus, $d_p(L)$ can be computed in polynomial time.   □

We have shown how to compute the inner prefix distance of a regular language. We can make similar modifications to the edit systems L_s and L_f and construct transducers which model those edit systems. Such an edit system can be defined for the suffix distance by

$$L_s^{(1)} = (\mathcal{E} \setminus \mathcal{E}_0)^+ \mathcal{E}_0^*.$$

The case of subword distance is slightly more complicated, as we require at least one edit operation with nonzero weight. For an edit string e, such an operation can occur as either a prefix or a suffix but we cannot require that there is a symbol with nonzero weight in both the prefix and suffix. Thus, we can define the edit system $L_f^{(1)}$ by

$$L_f^{(1)} = ((\mathcal{E} \setminus \mathcal{E}_0)^* \mathcal{E}_0^* (\mathcal{E} \setminus \mathcal{E}_0)^+) \cup ((\mathcal{E} \setminus \mathcal{E}_0)^+ \mathcal{E}_0^* (\mathcal{E} \setminus \mathcal{E}_0)^*).$$

Then using an analogous approach as for the prefix distance, it is possible to compute the inner suffix and subword distances of a regular language in polynomial time.

Theorem 6. *For a given NFA A recognizing the language L,*

1. *$d_s(L)$ is computable in polynomial time, and*
2. *$d_f(L)$ is computable in polynomial time.*

6   Conclusion

We have shown how to compute the prefix distance of two regular languages in polynomial time by using weighted transducers. We have also used this algorithm to compute the inner prefix distance of a regular language in polynomial time. These algorithms can also be applied to compute the suffix and subword distances for regular languages.

One direction for future research is computing prefix, suffix, and subword distances between non-regular languages. It is known that computing distances between context-free languages is undecidable [4]. However, Han et al. [8] gave an algorithm for computing the edit distance between a regular language and a context-free language. For prefix, suffix, and subword distances, the problem of computing the distance between a regular language and a context-free language or subclasses of context-free languages remains open.

References

1. Benedikt, M., Puppis, G., Riveros, C.: Bounded repairability of word languages. J. Comput. Syst. Sci. **79**(8), 1302–1321 (2013)
2. Benedikt, M., Puppis, G., Riveros, C.: The per-character cost of repairing word languages. Theor. Comput. Sci. **539**, 38–67 (2014)
3. Bruschi, D., Pighizzini, G.: String distances and intrusion detection: bridging the gap between formal languages and computer security. RAIRO Informatique Théorique et Appl. **40**, 303–313 (2006)
4. Choffrut, C., Pighizzini, G.: Distances between languages and reflexivity of relations. Theor. Comput. Sci. **286**(1), 117–138 (2002)
5. Cormen, T.H., Leiserson, C.E., Rivest, R.L., Stein, C.: Introduction to Algorithms, 2nd edn. MIT Press, Cambridge (2001)
6. Deza, M.M., Deza, E.: Encyclopedia of Distances. Springer, Berlin, Heidelberg (2009)
7. Droste, M., Kuich, W., Vogler, H.: Handbook of Weighted Automata. Springer, Berline, Heidelberg (2009)
8. Han, Y.S., Ko, S.K., Salomaa, K.: The edit-distance between a regular language and a context-free language. Int. J. Found. Comput. Sci. **24**(07), 1067–1082 (2013)
9. Kari, L., Konstantinidis, S.: Descriptional complexity of error/edit systems. J. Autom. Lang. Comb. **9**(2/3), 293–309 (2004)
10. Kari, L., Konstantinidis, S., Perron, S., Wozniak, G., Xu, J.: Computing the hamming distance of a regular language in quadratic time. WSEAS Trans. Inf. Sci. Appl. **1**(1), 445–449 (2004)
11. Konstantinidis, S.: Computing the edit distance of a regular language. Inf. Comput. **205**(9), 1307–1316 (2007)
12. Konstantinidis, S., Silva, P.V.: Computing maximal error-detecting capabilities and distances of regular languages. Fundam. Inform. **101**, 257–270 (2010)
13. Kutrib, M., Meckel, K., Wendlandt, M.: Parameterized prefix distance between regular languages. In: Geffert, V., Preneel, B., Rovan, B., Štuller, J., Tjoa, A.M. (eds.) SOFSEM 2014. LNCS, vol. 8327, pp. 419–430. Springer, Heidelberg (2014)
14. Mohri, M.: Edit-distance of weighted automata: general definitions and algorithms. Int. J. Found. Comput. Sci. **14**(6), 957–982 (2003)
15. Pighizzini, G.: How hard is computing the edit distance? Inf. Computat. **165**(1), 1–13 (2001)
16. Shallit, J.: A Second Course in Formal Languages and Automata Theory. Cambridge University Press, Cambridge (2009)
17. Yu, S.: Regular languages. In: Rozenberg, G., Salomaa, A. (eds.) Handbook of Formal Languages, pp. 41–110. Springer, Berlin, Heidelberg (1997)

Complexity of Sets of Two-Dimensional Patterns

Daniel Průša[(⊠)]

Faculty of Electrical Engineering, Czech Technical University,
Karlovo náměstí 13, 121 35 Prague 2, Czech Republic
prusapa1@fel.cvut.cz

Abstract. We study the two-dimensional pattern matching implemented using the two-dimensional on-line tessellation automaton, which
is a restricted type of the cellular automaton able to simulate the Baker-
Bird algorithm, proposed as the first algorithm for the two-dimensional
pattern matching. We further explore capabilities of this automaton to
carry out the matching task against an arbitrary set of equally sized
patterns. To measure amount of resources needed to accomplish it, we
introduce the pattern complexity of a picture language. We show that
this complexity spreads in a wide range. It is demonstrated by giving
examples, deriving general techniques and proving some lower bounds.

Keywords: Two-dimensional pattern matching · Two-dimensional
on-line tessellation automaton · Picture languages · Descriptional
complexity

1 Introduction

The task of the exact two-dimensional pattern matching is to detect occurrences
of a two-dimensional pattern of symbols in a two-dimensional text array. The
interest in this natural extension of the well known one-dimensional pattern
matching problem dates back to the seventies of the past century. The first efficient algorithm, reducing the problem to the one-dimensional pattern matching,
was independently found by Bird [5] and Baker [4]. Different types of algorithms
further improving worst case or average time complexity have appeared during
the intensive research in this area [2,3,8].

At about the same time when Bird and Baker published their results, Inoue
and Nakamura presented the two-dimensional on-line tessellation automaton for
picture recognition [7]. The computation of this model follows the principle of
dynamic programming – it fills a two-dimensional table where a value of a field
depends on values of the left and top neighboring fields and the input symbol
at the field's position. This process is suitable for implementing the Baker-Bird
algorithm, as it has been shown by Toda et al. [11]. Moreover, a usage of the
nondeterministic version of the automaton for the exact and approximate two-
dimensional pattern matching was studied by Polcar and Melichar [10].

In this paper, we consider the two-dimensional pattern matching in a broader
sense. Instead of one pattern, we perform matching against an arbitrary set of

© Springer International Publishing Switzerland 2016
Y.-S. Han and K. Salomaa (Eds.): CIAA 2016, LNCS 9705, pp. 236–247, 2016.
DOI: 10.1007/978-3-319-40946-7_20

equally sized patterns. In real applications, such a set can be generated by various formalism like templates or formulas specifying contents of the searched patterns. For example, one can use wildcards allowing any symbol at some positions in a fixed pattern. The pattern matching in this setting can be implemented efficiently if there is a two-dimensional on-line tessellation automaton with a reasonable number of states detecting matches. Patterns of different size but the same nature can be further grouped into a picture language. In accordance with these thoughts, we define the pattern complexity of a picture language as a function $\sigma(n)$ giving the number of states of a state-minimal deterministic two-dimensional on-line tessellation automaton which detects in a text array all subarrays $n \times n$ that belong to the picture language. We study properties of the defined measure.

The paper is structured as follows. In Sect. 2 we give basics on picture languages, the two-dimensional on-line tessellation automaton and Baker-Bird algorithm. The pattern complexity of a picture language is introduced in Sect. 3. It is demonstrated on several examples, including those cases handled by the Baker-Bird algorithm. Some techniques allowing to prove lower bounds on the complexity are exploited in Sect. 4. Effect of operations over picture languages on the pattern complexity is explored in Sect. 5. Finally, a short summary and open problems are given in Sect. 6.

2 Preliminaries

We use the common notation and terms on pictures and picture languages (see, e.g., [6]). If Σ is a finite alphabet, then a picture P over Σ is a two-dimensional array of symbols from Σ. If P has m rows and n columns, we say it is of size $m \times n$, and we write $\mathrm{rows}(P) = m$ and $\mathrm{cols}(P) = n$. If P is a square picture of size $n \times n$, we shortly say P is of size n. Rows of P are indexed from 1 to m, columns of P are indexed from 1 to n, $P(i,j)$ denotes the symbol of P in the i-th row and j-th column. In the graphical visualizations of pictures, the top-left corner is associated with position $(1,1)$. A subpicture $P[r,c;k,\ell]$ of size $k \times \ell$ is defined iff $1 \le r \le m+1-k$ and $1 \le c \le n+1-\ell$. Then, $P[r,c;k,\ell](i,j) = P(r-1+i, c-1+j)$ for all $(i,j) \in \{1,\dots,k\} \times \{1,\dots,\ell\}$. The set of all pictures of size $m \times n$ over Σ is denoted $\Sigma^{m,n}$. Moreover, $\Sigma^{+,+} = \bigcup_{i \ge 1, j \ge 1} \Sigma^{i,j}$ is the set of all non-empty pictures over Σ. If a picture is of size $1 \times n$ or $m \times 1$, we treat it as a string.

$\mathbb{N}$ denotes the set of positive integers.

Let $\mathcal{A} = (Q, \Sigma, \delta, q_0, F)$ be a deterministic finite automaton (DFA), where Q is a set of states, Σ is an input alphabet, $\delta : Q \times \Sigma \to Q$ is a transition function, $q_0 \in Q$ is the initial state and $F \subseteq Q$ is a set of accepting states. The extended transition function $\hat{\delta} : Q \times \Sigma^* \to Q$ is defined by $\hat{\delta}(q, \lambda) = q$, $\hat{\delta}(q, a) = \delta(q, a)$ and $\hat{\delta}(q, aw) = \hat{\delta}(\delta(q, a), w)$ for all $a \in \Sigma$, $w \in \Sigma^*$ and $q \in Q$.

Let $P \in \Sigma^{m,n}$. Define strings $w_{i,j} = P[i,1;1,j]$ for all $i = 1, \dots, m$ and $j = 1, \dots, n$ (each $w_{i,j}$ is thus a prefix of a row of P). For a DFA $\mathcal{A} = (Q, \Sigma, \delta, q_0, F)$, define $P_{\mathrm{rows}}^{\mathcal{A}}$ as the picture of size $m \times n$ over Q where $P_{\mathrm{rows}}^{\mathcal{A}}(i,j) = \hat{\delta}(q_0, w_{i,j})$.

Analogously, define strings $v_{i,j} = P[1, j; i, 1]$ (prefixes of columns of P) and $P^{\mathcal{A}}_{\text{cols}}$ as the picture of size $m \times n$ over Q where $P^{\mathcal{A}}_{\text{cols}}(i, j) = \hat{\delta}(q_0, v_{i,j})$.

Let $P \in \Sigma^{+,+}$ be an input picture and $R \in \Sigma^{+,+}$ a pattern whose occurrences in P have to be detected. The Baker-Bird algorithm works in two phases. During the first phase, rows of P are scanned for occurrences of rows of R. This task is solved by the well-known Aho-Corasick algorithm [1]. A so called dictionary-matching finite automaton $\mathcal{A}$ is constructed for this purpose. The second phase deals with a series of one-dimensional string matchings as it searches columns of $P^{\mathcal{A}}_{\text{rows}}$ for a sequence of final states of $\mathcal{A}$ encoding all the rows of the pattern R in their order. An illustration of the whole algorithm is given in Fig. 1.

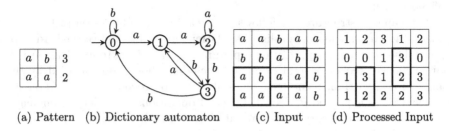

(a) Pattern   (b) Dictionary automaton   (c) Input   (d) Processed Input

Fig. 1. The Baker-Bird algorithm creates a dictionary-matching automaton (b) for rows of a pattern (a). The automaton processes rows of an input picture (c). Then it searches columns of the processed input for substrings 32 encoding the rows of the pattern. This results in a detection of the two highlighted matches.

Informally, the two-dimensional on-line tessellation automaton consists of cells covering the input picture $P \in \Sigma^{m,n}$. Each cell at a position (i, j) performs a transition exactly once, depending on final states of the left and top neighbor and the symbol $P(i, j)$. If there is no left or right neighbor, the blank symbol $\#$ is supplied to inform the cell that it is located at the left or top border, respectively. Transitions of the cells can thus be computed when the cells are processed e.g. row by row or column by column, using $\mathcal{O}(\max(m, n))$ space.

Definition 1. A *deterministic two-dimensional on-line tessellation automaton* (2DOTA) $\mathcal{A}$ is a tuple (Q, Σ, δ, F), where

- Q is a finite set of states
- Σ is an input alphabet
- $\delta : (Q \cup \{\#\}) \times \Sigma \times (Q \cup \{\#\}) \rightarrow Q$ is a transition function
- $F \subseteq Q$ is a set of accepting states

Let $P \in \Sigma^{m,n}$ be an input picture. We define $P^{\mathcal{A}}$ as the picture of size $m \times n$ representing final states of the $\mathcal{A}$'s cells as follows.

$$P^{\mathcal{A}}(i, j) = \delta(q_L, P(i, j), q_U) \text{ where}$$

$$q_L = \begin{cases} \# & \text{if } j = 1, \\ P^{\mathcal{A}}(i, j - 1) & \text{otherwise,} \end{cases} \text{ and } q_U = \begin{cases} \# & \text{if } i = 1, \\ P^{\mathcal{A}}(i - 1, j) & \text{otherwise.} \end{cases}$$

3   Pattern Complexity

Definition 2. Let L be a set of square pictures of size $n \geq 1$ over Σ and $\mathcal{A} = (Q, \Sigma, \delta, F)$ be a 2DOTA. We say that $\mathcal{A}$ *searches patterns of L* if for any input $P \in \Sigma^{+,+}$ it holds $P^{\mathcal{A}}(i, j) \in F$ if and only if $i, j \geq n$ and $P[i - n + 1, j - n + 1; n, n] \in L$.

For $k, \ell \in \mathbb{N}$ and a picture language L over Σ, let $L^{k,\ell} = L \cap \Sigma^{k,\ell}$.

Definition 3 (Pattern complexity of a picture language). Let L be a picture language over Σ. For each $n \in \mathbb{N}$, let $\mathcal{A}_n = (Q_n, \Sigma_n, \delta_n, F_n)$ be a state-minimal 2DOTA that searches patterns of $L^{n,n}$. *Pattern complexity of L* is a function $\sigma_L : \mathbb{N} \to \mathbb{N}$ fulfilling $\sigma_L(n) = |Q_n|$ for all $n \in \mathbb{N}$.

For simplicity, all the patterns we consider are squares. We thus ignore all non-square pictures in the definition. There is a straightforward extension of the pattern complexity to rectangular patterns, which does not result in principally different properties.

A simulation of the Baker-Bird algorithm by 2DOTA was described by Toda et al. [11]. Here we present it in a more concise way.

Theorem 4. *Let L be a picture language over Σ such that $|L^{n,n}| = 1$ for all $n \in \mathbb{N}$. Then, $\sigma_L(n) = \mathcal{O}(n^2)$.*

Proof. For $n \in \mathbb{N}$, let R be the only picture in $L^{n,n}$. Consider a dictionary-matching finite automaton $\mathcal{A}_1 = (Q_1, \Sigma, \delta_1, \#, F)$ for rows of R. Since there are at most n such rows, it holds $|Q| \leq n^2 + 1$. Let $w = f_1 f_2 \ldots f_n$ be the word over F where f_i is the final state of $\mathcal{A}_1$ representing the i-th row of R. Next, consider a dictionary-matching finite automaton $\mathcal{A}_2 = (Q_2, F, \delta_2, q_0, \{q_A\})$ for the only word w. It holds $|Q_2| = n + 1$. Assume that $Q_1 \cap Q_2 = \emptyset$. For $q \in Q_2 \setminus \{q_0\}$, let $s(q)$ denote the symbol of F whose reading is needed to reach q by the control unit of $\mathcal{A}_2$ (i.e., there is $q' \in Q_2$ such that $\delta_2(q', s(q)) = q$).

Construct a 2DOTA $\mathcal{A} = (Q, \Sigma, \delta, \{q_A\})$ searching patterns of $L^{n,n} = \{R\}$. Let $Q = Q_1 \cup (Q_2 \setminus \{q_0\})$. Define mappings $\pi : Q \to Q_1$ and $\tau : Q \to Q_2$ where

$$\pi(q) = \begin{cases} q & \text{if } q \in Q_1, \\ s(q) & \text{if } q \in Q_2 \setminus \{q_0\}, \end{cases} \qquad \tau(q) = \begin{cases} q_0 & \text{if } q \in Q_1, \\ q & \text{if } q \in Q_2 \setminus \{q_0\}. \end{cases}$$

States of Q are designed to represent progress of $\mathcal{A}_1$ as well as $\mathcal{A}_2$. For $q_L, q_U \in Q$ and $a \in \Sigma$, transitions of $\mathcal{A}$ are defined as follows.

$$\delta(q_L, a, q_U) = \begin{cases} \delta_1(\pi(q_L), a) & \text{if } \delta_1(\pi(q_L), a) \notin F \text{ or} \\ & \qquad \delta_2(\tau(q_U), \delta_1(\pi(q_L), a)) = q_0, \\ \delta_2(\tau(q_U), \delta_1(\pi(q_L), a)) & \text{otherwise.} \end{cases}$$

The obtained 2DOTA $\mathcal{A}$ reaches its accepting state q_A at those positions of the input where the Baker-Bird algorithm finds a match. Moreover, $|Q| = \mathcal{O}(n^2)$ as it has been required. $\square$

Example 5. Define $L = \{a^{n,n} \mid n \in \mathbb{N}\}$ over $\Sigma = \{a, b\}$. Then, $\sigma_L(n) = \mathcal{O}(n)$.

For a pattern $a^{n,n}$, denote the rows and columns dictionary-matching finite automata as $\mathcal{A}_1 = (Q_1, \Sigma, \delta_1, q_0, \{q_n\})$ and $\mathcal{A}_2 = (Q_2, Q_1, \delta_2, p_0, \{p_n\})$ where $Q_1 = \{q_0, q_1, \ldots, q_n\}$, $Q_2 = \{p_0, p_1, \ldots, p_n\}$ and, for convenience, $q_0 = \#$. An example is given in Fig. 2. Following Theorem 4, we obtain 2DOTA $\mathcal{A} = (Q_1 \cup Q_2 \setminus \{p_0\}, \Sigma, \delta, \{p_n\})$ searching patterns of $L^{n,n} = \{a^{n,n}\}$ where

- $\delta(q_L, b, q_U) = q_0$ for any q_L and q_U,
- $\delta(q_i, a, q_U) = q_{i+1}$ for $i = 0, \ldots, n-2$ and any q_U,
- $\delta(q_L, a, q_i) = p_1$ for $q_L \in \{q_{n-1}\} \cup Q_2$ and $i = 0, \ldots, n-1$,
- $\delta(q_L, a, p_i) = p_{i+1}$ for $q_L \in \{q_{n-1}\} \cup Q_2$ and $i = 1, \ldots, n-1$,
- $\delta(q_L, a, p_n) = p_n$ for $q_L \in \{q_{n-1}\} \cup Q_2$.

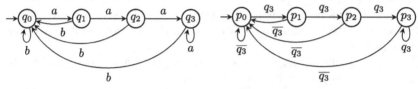

(a) Dictionary automaton $\mathcal{A}_1$ (b) Dictionary automaton $\mathcal{A}_2$

Fig. 2. (a) Rows and (b) columns dictionary-matching finite automaton for pattern $a^{3,3}$ where $\overline{q_3}$ represents states of $Q_1 \setminus \{q_3\} = \{q_0, q_1, q_2\}$.

The pattern complexity of the empty picture language is trivially $\Theta(1)$. The following lemma describes it for the other corner case.

Lemma 6. *Let $L = \Sigma^{+,+}$ for some alphabet Σ. Then, $\sigma_L(n) = \mathcal{O}(\log n)$.*

Proof. Assume $n \geq 2$. To accept all matches of patterns of $\Sigma^{n,n}$ requires to detect all positions (i, j) in the input where $i, j \geq n$. We show, how to construct a 2DOTA $\mathcal{A}$ with $\mathcal{O}(\log n)$ states which localizes the cell at position (n, n).

Assume first that $n = 2^k \geq 2$. Let $Q = \{c, d, y, \$\} \cup \{e_i, o_i, x_i \mid i = 1, \ldots, k\}$ be the set of states of $\mathcal{A}$ where y is the only accepting state. The automaton fills the cells in the following manner.

1. In the first row, $\mathcal{A}$ marks each odd position by state o_1 and each even position by state x_1. In the i-th row, $1 < i \leq k$, $\mathcal{A}$ marks by x_i each position which is a multiple of 2^i. To achieve this, $\mathcal{A}$ checks the previous row for cells in state x_{i-1} and counts their number modulo 2 using states o_i and e_i (where o_i or e_i indicates an odd or even number of occurrences of x_{i-1}, respectively). Moreover, this computation is performed only in the cells above the diagonal.
2. The first $2^k - 1$ cells of the diagonal enter state d. Moreover, all the cells bellow the diagonal enter state $\$$.
3. The cell at position $(k, 2^k)$ is the first one entering state x_k. This state is copied to all cells below (in the same column).

4. The diagonal and the 2^k-th column meet at position $(2^k, 2^k)$. When the cell at this position detects this configuration, it enters the accepting state y. This state is recursively taken over by cells at positions (i, j) where $i, j \geq 2^k$.
5. The remaining cells above the diagonal and bellow the k-th row enter state c.

An example of the computation is depicted in Fig. 3.

Now, let n be an arbitrary integer written as $n = \sum_{i=0}^{k} a_i 2^i$ where each $a_i \in \{0, 1\}$ and $a_k = 1$. Modify the set of states Q by removing states d, y and adding states $d_0, d_1, \ldots, d_k$ and $y_0, y_1, \ldots, y_k$. Let $I = \{i \mid a_i = 1\}$. In this case, the automaton computes position $(2^k, 2^k)$ using the same procedure as described above, however, it fills the diagonal by d_k instead of d (this indicates that k rows have to be processed using markers $x_1, \ldots, x_k$), and fills the cell at position $(2^k, 2^k)$ by y_k. Position $(2^k + 1, 2^k + 1)$ is then filled by d_ℓ where ℓ is the second greatest element of I. It is done using the following transitions. If the left neighbor of a cell is in state y_k, the cell enters state y_k. If the top neighbor of a cell is in state y_k and its left neighbor is not in state \$, the cell enters state d_ℓ. Then, $\mathcal{A}$ performs again the same procedure to compute position $(2^\ell, 2^\ell)$ relative to position $(2^k + 1, 2^k + 1)$, locating thus position $(2^k + 2^\ell, 2^k + 2^\ell)$. And this is repeated for all elements of I in the descending order, entering finally state y_m at position (n, n) where m is the minimal element of I.   □

d	x_1	o_1	x_1	o_1	x_1	o_1	x_1	o_1
\$	d	o_2	x_2	e_2	o_2	o_2	x_2	e_2
\$	\$	d	o_3	o_3	o_3	o_3	x_3	e_3
\$	\$	\$	d	c	c	c	x_3	c
\$	\$	\$	\$	d	c	c	x_3	c
\$	\$	\$	\$	\$	d	c	x_3	c
\$	\$	\$	\$	\$	\$	d	x_3	c
\$	\$	\$	\$	\$	\$	\$	y	y

Fig. 3. Computation of position $(2^3, 2^3)$ in a picture of size 8×9.

The next lemma generalizes the technique of DFAs simulation from Theorem 4.

Lemma 7. *Let $\mathcal{A}_1 = (Q_1, \Sigma, \delta_1, q_0^1, F_1)$ and $\mathcal{A}_2 = (Q_2, Q_1, \delta_2, q_0^2, F_2)$ be DFAs. There is a 2DOTA $\mathcal{A} = (Q, \Sigma, \delta, F)$ with $\mathcal{O}(|Q_1| \cdot |Q_2|)$ states such that, for any input $P \in \Sigma^{+,+}$, $P^{\mathcal{A}}(i, j) \in F \Leftrightarrow (P_{\text{rows}}^{\mathcal{A}_1})_{\text{cols}}^{\mathcal{A}_2}(i, j) \in F_2$.*

Proof. Define 2DOTA $\mathcal{A} = (Q, \Sigma, \delta, F)$ where $Q = Q_1 \times Q_2$, $F = Q_1 \times F_2$ and δ simulates computation of both, $\mathcal{A}_1$ and $\mathcal{A}_2$, as follows. Let $\pi_1 : Q \cup \{\#\} \to Q_1$, $\pi_2 : Q \cup \{\#\} \to Q_2$ be mappings such that $\pi_1((q_1, q_2)) = q_1$, $\pi_1(\#) = q_0^1$, $\pi_2((q_1, q_2)) = q_2$ and $\pi_2(\#) = q_0^2$ for all $(q_1, q_2) \in Q$. For $q_L, q_U \in Q \cup \{\#\}$ and $a \in \Sigma$, define

$$\delta(q_L, a, q_U) = (\delta_1(\pi_1(q_L), a), \delta_2(\pi_2(q_U), \delta_1(\pi_1(q_L), a))).$$ □

Example 8. For $c \in \Sigma$ and $P \in \Sigma^{+,+}$, let $|P|_c = |\{(i,j) \mid P(i,j) = c\}|$ denote the number of occurrences of c in P. Let $\Sigma = \{a,b\}$. For $k \in \mathbb{N}$, define $L_k = \{P \in \Sigma^{+,+} \mid |P|_b = k\}$. Using Lemma 7, we show that $\sigma_{L_k}(n) = \mathcal{O}(n^{2(k+1)} \log n)$.

Consider square patterns in L_k of size n where $n > k$. Construct a finite automaton $\mathcal{A}_1$ with states of the form $(x_1, \ldots, x_{k+1})$ where each $x_i \in \{1, \ldots, n+1\}$ and $x_1 \le x_2 \le \ldots \le x_{k+1}$. These vectors serve to record relative positions of the last $k+1$ occurrences of b in the (at most) n previously read symbols of a row. If the i-th component is of value $n+1$, then there are no more than $i-1$ occurrences of b in the (at most) n previously read symbols. See Fig. 4(a) for an example. Take $(n+1, \ldots, n+1)$ as the initial state. For $x \in \{1, \ldots, n+1\}$, define

$$\overline{x} = \begin{cases} x+1 & \text{if } x < n+1, \\ x & \text{if } x = n+1. \end{cases}$$

Transitions of $\mathcal{A}_1$ are defined as follows:

$$\delta_1((x_1, \ldots, x_{k+1}), a) = (\overline{x_1}, \ldots, \overline{x_{k+1}})$$
$$\delta_1((x_1, \ldots, x_{k+1}), b) = (1, \overline{x_1}, \ldots, \overline{x_k}).$$

Analogously, construct a finite automaton $\mathcal{A}_2$ with states $(y_1, \ldots, y_{k+1})$ where $y_i \in \{1, \ldots, n+1\}$ and $y_1 \le y_2 \le \ldots \le y_{k+1}$. For each position (i,j), such states are used to record row positions of the last $k+1$ occurrences of b in the subpicture $R = P[\max(1, i-n+1), \max(1, j-n+1); n, n]$, as it is illustrated in Fig. 4(a). Hence, the transition function is defined by

$$\delta_2((y_1, \ldots, y_{k+1}), (x_1, \ldots, x_{k+1})) = (1, \ldots, 1, \overline{y_1}, \ldots, \overline{y_{k+1-s}})$$

where $s = |\{x_i \mid x_i < n+1, i \in \{1, \ldots, n+1\}\}|$. A vector $(y_1, \ldots, y_{k+1})$ is a final state of $\mathcal{A}_2$ iff $y_1, \ldots, y_k < n+1$ and $y_{k+1} = n+1$, indicating that there are exactly k occurrences of b in the considered subpicture R.

Lemma 7 applied to $\mathcal{A}_1$ and $\mathcal{A}_2$ gives a 2DOTA $\mathcal{A}$ with $\mathcal{O}(n^{2(k+1)})$ states which correctly detects pattern matches at positions (i,j) where $i,j \ge n$, however, it might also enter a final state at positions (i,j) for which $i < n$ or $j < n$. To suppress these false detections, it suffices to combine $\mathcal{A}$ with the automaton constructed in the proof of Lemma 6 – a product automaton is created, resulting in an automaton searching patterns of $L_k^{n,n}$ with $\mathcal{O}(n^{2(k+1)} \log n)$ states.

Theorem 9. *For any $L \in \Sigma^{+,+}$, where $|\Sigma| \ge 2$, it holds $\sigma_L(n) = \mathcal{O}(|\Sigma|^{n^2})$.*

Proof. For $n \in \mathbb{N}$, construct a 2DOTA $\mathcal{A}_n = (Q_n, \Sigma, \delta_n, F_n)$ whose states represent an enumeration of all pictures of size up to $n \times n$ so that, for an input $P \in \Sigma^{+,+}$, it holds $P^{\mathcal{A}}(i,j) = P[\max(1, i-n+1), \max(1, j-n+1); n, n]$ for all positions (i,j). This is achieved when $Q_n = \bigcup_{1 \le i,j \le n} \Sigma^{i,j}$ and $\delta_n(P_L, a, P_U)$ produces a picture of size $\min(n, \text{rows}(P_U) + 1) \times \min(n, \text{cols}(P_L) + 1)$ specified in Fig. 4(b).

a	a	b	b	b
a	a	a	a	a
a	a	b	a	a
b	b	a	b	a

(a)

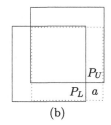

(b)

Fig. 4. (a) Assuming we search for subpictures 4×4 containing $k = 3$ symbols b, the bottom-right field is assigned by vectors $(2, 4, 5, 5)$ and $(1, 1, 2, 4)$ to represent positions of the closest $k + 1$ b's in the last row suffix of length 4 and the subpicture 4×4 ending at the position $(4, 5)$, respectively. (b) The square subpictures P_L, P_U and the symbol a determine uniquely the dotted square subpicture.

To estimate size of Q_n, derive

$$|Q_n| = \sum_{1 \leq i,j \leq n} |\Sigma^{i,j}| = \sum_{i=1}^{n} \sum_{j=1}^{n} (|\Sigma|^i)^j = \sum_{i=1}^{n} \frac{|\Sigma|^i}{|\Sigma|^i - 1} \left(|\Sigma|^{in} - 1 \right) <$$

$$< 2 \sum_{i=1}^{n} (|\Sigma|^n)^i = 2 \frac{|\Sigma|^n}{|\Sigma|^n - 1} \left(|\Sigma|^{n^2} - 1 \right) < 4|\Sigma|^{n^2} = \mathcal{O}\left(|\Sigma|^{n^2} \right).$$

Finally, F_n equals the set of patterns to be detected, hence $F_n = L^{n,n}$.   □

4   Lower Bounds

For a picture language L over Σ, let $\theta_{L,n} : \Sigma^{+,+} \rightarrow \{0,1\}^{+,+}$ be a mapping where $\theta_{L,n}(P)$ is of the same size as P, and $\theta_{L,n}(P)(i,j) = 1$ iff $i,j \geq n$ and $P[i - n + 1, j - n + 1; n, n] \in L^{n,n}$. Such a mapping assigns 1 to each field of P which is the bottom-right corner of a $n \times n$ match against a pattern of $L^{n,n}$ (otherwise it assigns 0). The mapping extends to a picture language $L' \in \Sigma^{+,+}$ as follows: $\theta_{L,n}(L') = \{\theta_{L,n}(P) \mid P \in L'\}$.

Lemma 10. *Let L be a picture language over Σ. For $n \in \mathbb{N}$, let $L_n \subseteq \Sigma^{n,2n-1}$ where for any $P, R \in L_n$, $(i,j) \in (\{1,\ldots,n-1\} \times \{1,\ldots,2n-1\}) \cup (\{n\} \times \{n+1,\ldots,2n-1\}) \Rightarrow P(i,j) = R(i,j)$. Then, $\sigma_L(n) \geq |\theta_{L,n}(L_n)|$.*

Proof. Let $\mathcal{A} = (Q, \Sigma, \delta, F)$ be a 2DOTA searching patterns of $L^{n,n}$. Pictures in L_n are of the form depicted in Fig. 5(a). After processing a picture $P \in L_n$, $\mathcal{A}$ can enter an accepting state only at positions $(n, n), \ldots, (n, 2n - 1)$. Since the content of rows $1, \ldots, n - 1$ is fixed for all pictures in L_n, $\mathcal{A}$ enters the same states $r_1, \ldots, r_{2n-1}$ in the $n - 1$-st row (see Fig. 5(b)). Moreover, the suffix of the last row is also fixed, hence states $q_{n+1}, \ldots, q_{2n-1}$ are determined solely by state q_n. Since all elements of $\theta_{L,n}(L_n)$ have to be distinguished by q_n, it holds $|Q| \geq |\theta_{L,n}(L_n)|$.   □

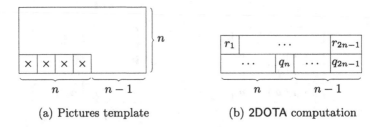

(a) Pictures template   (b) 2DOTA computation

Fig. 5. (a) Pictures in L_n of size $n \times (2n - 1)$ can differ only at the crossed positions. (b) The last two rows of 2DOTA's cells processing a picture from L_n.

We apply the lemma to show $\sigma_L(n) = \Omega(n)$ for the picture language L from Example 5. For $n \in \mathbb{N}$, define L_n consisting of all pictures $P \in \{a, b\}^{+,+}$ where $\text{rows}(P) = n$, $\text{cols}(P) = 2n - 1$, $|P|_b = 1$ and the only symbol b is located at a position (n, i) with $1 \leq i \leq n$. Then, $|\theta_{L,n}(L_n)| = n$, hence $\sigma_L(n) = \Omega(n)$.

Theorem 11. *There is a picture language L over $\{0, 1\}$ with $\sigma_L(n) = 2^{\Omega(n^2)}$.*

Proof. For a binary string w, let $K(w)$ be its Kolmogorov complexity [9]. It is known that, for each $n \in \mathbb{N}$, there is a string w_n of length n such that $K(w_n) \geq n$.

For each $m \in \mathbb{N}$, create a set L_m of selected square pictures of size $m \times m$ as follows. Consider, w.l.o.g., m is even. For $n = \frac{m^2}{2} \cdot 2^{\frac{m^2}{2}}$, split w_n into $\ell = \frac{m^2}{2}$ strings of length $2^{\frac{m^2}{2}}$. I.e., write $w_n = v_1 \ldots v_\ell$ where $|v_i| = 2^{\frac{m^2}{2}}$ for all i. For each $i \in \{1, \ldots, \ell\}$, create a picture $P_{m,i}$ of size $m \times m$ by filling the first half of rows by the binary representation of $i - 1$ of length $\frac{m^2}{2}$ and the second half of rows by v_i (both the considered strings are split into $\frac{m}{2}$ rows, each containing m bits). The set L_m consists of all pictures $P_{m,i}$ and $L = \bigcup_{j \geq 1} L_j$.

Let $\mathcal{A} = (Q, \Sigma, \delta, F)$ be a 2DOTA searching patterns of L_m. This also means $\mathcal{A}$ can be directly used to recognize whether a square picture $P \in \Sigma^{m,m}$ is in L_m or not. Construct a (one-dimensional) deterministic Turing machine $\mathcal{M}$ outputting w_n (where $n = \frac{m^2}{2} \cdot 2^{\frac{m^2}{2}}$) when started over a blank tape by executing procedure

    for $i = 0$ **to** $2^{\frac{m^2}{2}} - 1$ **do**
        for all $v \in \{0, 1\}^{m^2/2}$ **do**
            if $\mathcal{A}$ accepts $S(\text{str}(i)v)$ **then**
                print v
            end if
        end for
    end for

where $\text{str}(i)$ is the binary string of length $\frac{m^2}{2}$ representing i and $S(w)$ is the square of size m filled row by row by bits of a binary string w of length m^2. It remains to describe how $\mathcal{M}$ simulates $\mathcal{A}$. It uses m^2 blocks of size $\lceil \log_2 |Q| \rceil$ tape fields. Blocks represent cells of $\mathcal{A}$, collected row by row. To perform an $\mathcal{A}$'s cell transition requires to memorize states of the left and top neighbor cell.

This requires $\mathcal{O}(|Q|^2)$ states of $\mathcal{M}$. Furthermore, it is necessary to track the position of $\mathcal{M}$'s head with respect to the simulated two-dimensional array of cells of $\mathcal{A}$. The position is fully described by the row and column index and offset within the scanned block. This itself can be represented using $\mathcal{O}(m^2 \log |Q|)$ states. Printing out v can be realized by copying a bit by bit of v to the first empty $\mathcal{M}$'s tape cell to the right. This involves counting a position in v and a constant number of states to perform the transfer. In total, $\mathcal{M}$ needs $\mathcal{O}(|Q|^2 m^2 \log |Q|) = \mathcal{O}(|Q|^3 m^2)$ states. Hence, there is $c \in \mathbb{N}$ such that

$$c|Q|^3 m^2 \geq K(w_n) \geq \frac{m^2}{2} \cdot 2^{\frac{m^2}{2}} \quad \text{implying} \quad |Q| \geq (2c)^{-\frac{1}{3}} \cdot 2^{\frac{m^2}{6}} = 2^{\Omega(m^2)}. \quad \square$$

5   Operations over Picture Languages

Theorem 12. *Let L_1 and L_2 be picture languages over Σ. Define $L_\cup = L_1 \cup L_2$ and $L_\cap = L_1 \cap L_2$. Then $\sigma_{L_\cup} = \mathcal{O}(\sigma_{L_1}\sigma_{L_2})$ as well as $\sigma_{L_\cap} = \mathcal{O}(\sigma_{L_1}\sigma_{L_2})$. Moreover, there are picture languages L_1, L_2 such that $\sigma_{L_1}(n) = \Omega(n)$, $\sigma_{L_2}(n) = \Omega(n)$ and $\sigma_{L_\cup} = \Omega(\sigma_{L_1}\sigma_{L_2})$.*

Proof. For $n \in \mathbb{N}$, let $\mathcal{A}_1 = (Q_1, \Sigma, \delta_1, F_1)$ and $\mathcal{A}_2 = (Q_2, \Sigma, \delta_2, F_2)$ be a state-minimal 2DOTA searching for patterns of L_1 and L_2, respectively. We apply the product automaton construction to obtain 2DOTA with the set of states $Q_1 \times Q_2$ and the set of final states $(F_1 \times Q_2) \cup (Q_1 \times F_2)$ or $F_1 \times F_2$ searching for patterns of $L_\cup$ or $L_\cap$, respectively.

Let $\Sigma = \{a_1, b_1\} \times \{a_2, b_2\}$. Define $L_1 = \{P \mid P \in \{(a_1, a_2), (a_1, b_2)\}^{+,+}\}$ and $L_2 = \{P \mid P \in \{(a_1, a_2), (b_1, a_2)\}^{+,+}\}$. These picture languages over Σ are extensions of the picture language of uniform pictures from Example 5. For $P \in L_1$ or $P \in L_2$, the first or second component of each $P(i,j)$ equals a_1 or a_2, respectively. It is thus obvious that $\sigma_{L_1}(n) = \Theta(n)$ and $\sigma_{L_2}(n) = \Theta(n)$. We will show that $\sigma_{L_\cup}(n) = \Omega(n^2)$.

Apply Lemma 10. For $n \in \mathbb{N}$, $k = 1, \ldots, \lfloor \frac{n}{2} \rfloor$ and $\ell = \lfloor \frac{n}{2} \rfloor + 1, \ldots, n$, construct an input picture $P_{k,\ell}$ over Σ of size $n \times (2n-1)$ where the first $n-1$ rows contain only symbol (a_1, a_2) and the last row fulfills

$$P(n,j) = \begin{cases} (b_1, a_2) & \text{if } j = k \text{ or } j = \lfloor \frac{3}{2}n \rfloor, \\ (a_1, b_2) & \text{if } j = \ell, \\ (a_1, a_2) & \text{otherwise.} \end{cases}$$

As demonstrated in Fig. 6, the picture $P_{k,\ell}$ contains bottom-right corners of searched matches in its last row at column positions $k + n, \ldots, \lfloor \frac{3}{2}n \rfloor - 1$ and $\ell + n, \ldots, 2n - 1$. This means that $\theta_{L_\cup, n}(P_{k,\ell})$ is unique for all considered $\Omega(n^2)$ pictures $P_{k,\ell}$, which implies $\sigma_{L_\cup}(n) = \Omega(n^2)$. $\quad \square$

Theorem 13. *For every picture language L, it holds $\sigma_{\overline{L}}(n) = \mathcal{O}(\sigma_L(n) \log n)$ and $\sigma_{\overline{L}}(n) = \Omega(\sigma_L(n)/\log n)$.*

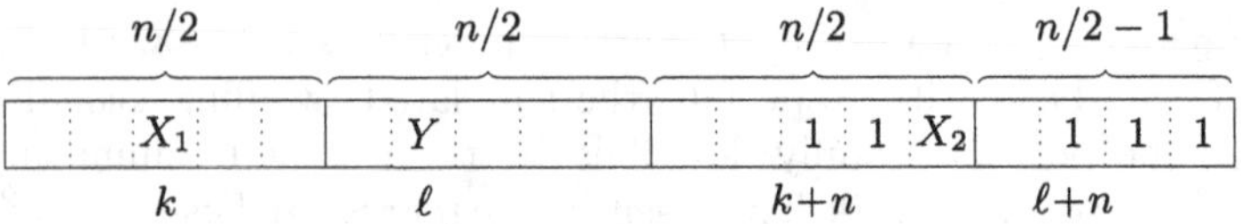

Fig. 6. A row of length $2n - 1$ where n is assumed to be even. Fields X_1 and X_2 at positions k and $\frac{3}{2}n$, respectively, contain symbol (b_1, a_2), field Y at position ℓ contains (a_1, b_2), the other fields contain (a_1, a_2). Ends of substrings of length n containing at most one of the fields X_1, Y, X_2 are marked by 1.

Proof. For $n \in \mathbb{N}$, let $\mathcal{A} = (Q, \Sigma, \delta, F)$ be a state-minimal 2DOTA searching for patterns of $L^{n,n}$. Moreover, let $\mathcal{A}' = (Q', \Sigma, \delta', F')$ be the 2DOTA searching for patterns of $\Sigma^{n,n}$ constructed following the proof of Lemma 6. The product 2DOTA with the set of states $Q \times Q'$ and the set of final states $(Q \setminus F) \times F'$ searches for patterns of $\overline{L}^{n,n} = \Sigma^{n,n} \setminus L^{n,n}$.

For $f, g : \mathbb{N} \to \mathbb{N}$, $f = \mathcal{O}(g)$ implies $g = \Omega(f)$, hence $\sigma_L(n) = \mathcal{O}(\sigma_{\overline{L}}(n) \log n)$ implies $\sigma_{\overline{L}}(n) \log n = \Omega(\sigma_L(n))$ and $\sigma_{\overline{L}}(n) = \Omega(\sigma_L(n)/\log n)$. $\square$

6   Conclusion

We have presented basic results on the defined pattern complexity of picture languages. It has been demonstrated that this complexity ranges from $\Theta(1)$ to $2^{\Theta(n^2)}$. In Lemma 7, we have shown that the idea of reducing the task to the one-dimensional pattern matching is useful even for sets of patterns. Lemma 6 suggests that the dynamic programming approach carried out by the two-dimensional on-line tessellation automaton has benefits over processing rows and tables separately in two phases. It is a question whether this advantage can be utilized to a larger extent.

We have studied, how the pattern complexity is affected when performing basic operations over picture languages. This can be understood as combining templates producing patterns. Analysis of the complexity of more operations over picture languages like rotation or mirroring could be a subject to future work. This could even cover the task of the approximate pattern matching since e.g. the Hamming distance can work as an operator over a picture language L producing a set of those pictures whose Hamming distance to a picture from L is bounded by a chosen k.

The pattern complexity has been based on 2DOTA, as it seems to be a suitable choice due to its ability to simulate the Baker-Bird algorithm. However, it could be worth to explore the pattern complexity based on other computational models.

Acknowledgement. This work was supported by the Czech Science Foundation under grant no. 15-04960S.

References

1. Aho, A.V., Corasick, M.J.: Efficient string matching: An aid to bibliographic search. Commun. ACM **18**(6), 333–340 (1975). http://doi.acm.org/10.1145/360825.360855
2. Amir, A., Benson, G., Farach, M.: Alphabet independent two dimensional matching. In: Proceedings of the Twenty-Fourth Annual ACM Symposium on Theory of Computing, STOC 1992, pp. 59–68. ACM, New York (1992). http://doi.acm.org/10.1145/129712.129719
3. Baeza-Yates, R., Régnier, M.: Fast two-dimensional pattern matching. Inf. Process. Lett. **45**(1), 51–57 (1993). http://www.sciencedirect.com/science/article/pii/002001909390250D
4. Baker, T.P.: A technique for extending rapid exact-match string matching to arrays of more than one dimension. SIAM J. Comput. **7**(4), 533–541 (1978). http://dx.doi.org/10.1137/0207043
5. Bird, R.S.: Two dimensional pattern matching. Inf. Process. Lett. **6**(5), 168–170 (1977). http://dx.doi.org/10.1016/0020-0190(77)90017-5
6. Giammarresi, D., Restivo, A.: Two-dimensional languages. In: Rozenberg, G., Salomaa, A. (eds.) Handbook of Formal Languages, vol. 3, pp. 215–267. Springer, New York (1997)
7. Inoue, K., Nakamura, A.: Some properties of two-dimensional on-line tessellation acceptors. Inf. Sci. **13**(2), 95–121 (1977)
8. Kärkkäinen, J., Ukkonen, E.: Two and higher dimensional pattern matching in optimal expected time. In: Sleator, D.D. (ed.) Proceedings of the Fifth Annual ACM-SIAM Symposium on Discrete Algorithms. 23–25 January 1994, Arlington, Virginia, pp. 715–723. ACM/SIAM (1994). http://dl.acm.org/citation.cfm?id=314464.314680
9. Li, M., Vitnyi, P.M.: An Introduction to Kolmogorov Complexity and its Applications, 3rd edn. Springer Publishing Company, Incorporated, New York (2008)
10. Polcar, T., Melichar, B.: Two-dimensional pattern matching by two-dimensional online tessellation automata. In: Domaratzki, M., Okhotin, A., Salomaa, K., Yu, S. (eds.) CIAA 2004. LNCS, vol. 3317, pp. 327–328. Springer, Heidelberg (2005). http://dx.doi.org/10.1007/978-3-540-30500-2_38
11. Toda, M., Inoue, K., Takanami, I.: Two-dimensional pattern matching by two-dimensional on-line tessellation acceptors. Theor. Comput. Sci. **24**, 179–194 (1983). http://dx.doi.org/10.1016/0304-3975(83)90048-8

The Complexity of Fixed-Height Patterned Tile Self-assembly

Shinnosuke Seki[1] and Andrew Winslow[2]([⊠])

[1] The University of Electro-Communications, Tokyo, Japan
s.seki@uec.ac.jp
[2] Université Libre de Bruxelles, Brussels, Belgium
awinslow@ulb.ac.be

Abstract. We characterize the complexity of the PATSproblem for patterns of fixed height and color count in variants of the model where seed glues are either chosen or fixed and identical (so-called *non-uniform* and *uniform* variants). We prove that both variants are NP-complete for patterns of height 2 or more and admit $O(n)$-time algorithms for patterns of height 1. We also prove that if the height and number of colors in the pattern is fixed, the non-uniform variant admits a $O(n)$-time algorithm while the uniform variant remains NP-complete. The NP-completeness results use a new reduction from a constrained version of a problem on finite state transducers.

Keywords: Tile self-assembly · DNA computing · Finite state transducer

1 Introduction

Winfree [13] introduced the *abstract tile assembly model (aTAM)* to capture nanoscale systems of DNA-based particles aggregating to form intricate crystals, leading to an entire field devoted to understanding the theoretical limits of such systems (see surveys by Doty [3] and Patitz [10]). Ma and Lombardi [9] introduced the *patterned self-assembly tile set synthesis (PATS)* problem, of designing a tile set of minimum size that assembles into a given $n \times h$ colored pattern by attaching to an L-shaped seed.

Czeizler and Popa [2] were the first to provide a proof that the PATS problem is NP-hard, thus establishing the problem as NP-complete. Subsequent work studied the hardness of the constrained version where the patterns have at most c colors, called the c-PATS problem. This line of work proved the 60-PATS [11], 29-PATS [6], 11-PATS [7], and finally the 2-PATS [8] problems NP-complete.

A full version of this paper can be found at http://arxiv.org/abs/1604.07190

S. Seki—Work supported in part by JST Program to Disseminate Tenure Tracking System, MEXT, Japan, No. 6F36 and by JSPS Grant-in-Aid for Research Activity Start-up No. 15H06212.

© Springer International Publishing Switzerland 2016
Y.-S. Han and K. Salomaa (Eds.): CIAA 2016, LNCS 9705, pp. 248–259, 2016.
DOI: 10.1007/978-3-319-40946-7_21

Here we study the complexity of parameterized *height-h* PATS and *c*-PATS problems where patterns have a specified fixed height h and increasing width n. We consider both *uniform* and *non-uniform* model variants, where the glues along the seed are fixed and identical or chosen in tandem with the tile set, respectively. We characterize the computational complexity of these problems via the following results:

- The height-2 PATS problem is NP-complete in both models (Sect. 4).
- The uniform height-2 3-PATS problem is NP-complete (Sect. 5).
- The non-uniform height-*h* *c*-PATS problem and uniform height-1 PATS problems admit $c^{c^{O(h)}} n$-time and $O(n)$-time algorithms, respectively (Sect. 6).

The NP-completeness results also apply to patterns of height greater than 2. Thus the complexity of the PATS problem for all combinations of height, color, and uniformity are characterized, except uniform height-2 2-PATS.

The NP-hardness reductions are based on a reduction for a new variant of the minimum-state finite state transducer problem, originally proved NP-hard by Angluin [1] and by Vazirani and Vazirani [12]. In this variant, any solution transducer is also promised to satisfy additional constraints on its transitions. The reduction is also substantially simpler than the reduction given in [1] and uses input and output strings of just two symbols, rather than the three of [12].

2   Preliminaries

Patterns, Tiles, Assemblies, and Seeds. Define $\mathbb{N}_k = \{1, 2, \ldots, k\}$. A *pattern* is a partial function $P : \mathbb{N}^2 \to C$, i.e. a function that maps a rectangular region of lattice points to a set of colors. If $\mathrm{dom}(P) = \mathbb{N}_w \times \mathbb{N}_h$, then P is a *width-w height-h* pattern. The codomain of P, i.e. the colors seen in the pattern, is denoted color(P). A pattern P is *c-color* provided $|\mathrm{color}(P)| \leq c$.

A *tile type* t is a colored unit square with each edge labeled; these labels are called *glues*. A tile type's color is denoted color(t). For a direction $d \in \{\mathtt{N}, \mathtt{W}, \mathtt{S}, \mathtt{E}\}$, $t[d]$ denotes the glue assigned to side d of t. A tile type is non-rotatable, and thus is uniquely identified by its color and four glues. Instances of tile types, called *tiles*, are placed with their centers in $\mathbb{N}^2$.

An *assembly* is an arrangement of tiles from a set of tile types T; formally a partial function $A : \mathbb{N}^2 \to T \cup \{\varnothing\}$. A *seed* is an "L-shaped" assembly with domain $\{(0,0)\} \cup \{(x,0) : x \in \mathbb{N}_w\} \cup \{(0,y) : y \in \mathbb{N}_h\}$ for some $w, h \in \mathbb{N}$. The *pattern* of an assembly A is defined as $P_A((x,y)) = \mathrm{color}(A((x,y)))$ for $(x, y) \in \mathrm{dom}(A) \cap \mathbb{N}^2$, i.e. the color pattern of A, *excluding* the seed.

RTASs. A *rectilinear tile assembly system (RTAS)* is a pair $\mathcal{T} = (T, \sigma)$, where T is a set of tile types and σ is a seed. An assembly A *yields* an assembly A' with $\mathrm{dom}(A') = \mathrm{dom}(A) \cup \{(x,y)\}$ provided $(x - 1, y), (x, y - 1) \in \mathrm{dom}(A)$ and $A((x - 1, y))[\mathtt{E}] = A'((x, y))[\mathtt{W}]$, $A((x, y - 1))[\mathtt{N}] = A'((x, y))[\mathtt{S}]$. The set of *producible assemblies* of an RTAS are those that can be yielded, starting with the seed assembly σ. That is:

RTAS Tiling Rule: A tile of type t can be added to an assembly A at location (x, y) provided $(x - 1, y), (x, y - 1) \in \text{dom}(A)$ and the east and north glues of the tiles at $(x - 1, y)$ and $(x, y - 1)$ are the same as the west and south glues of t, respectively.

As a result, tiling proceeds from southwest to northeast, i.e., a tile is first placed at $(1, 1)$, then at either $(1, 2)$ or $(2, 1)$, etc. The *terminal assemblies* of a RTAS are the producible assemblies that do not yield other (larger) assemblies. If every terminal assembly of the system has pattern P, the system is said to *uniquely self-assemble* P. An RTAS (T, σ) is *directed*, i.e. deterministic, provided that for any distinct tile types $t_1, t_2 \in T$, either $t_1[\text{W}] \neq t_2[\text{W}]$ or $t_1[\text{S}] \neq t_2[\text{S}]$.

Uniform RTASs. We also define a practical variant of a RTAS called a *uniform* RTAS. An RTAS (T, σ) is *uniform* provided there exist two glues $\ell_{\text{E}}, \ell_{\text{N}}$ such that $\sigma((x, 0))[\text{E}] = \ell_{\text{E}}$ for all $x \in \mathbb{N}_w$ and $\sigma((0, y))[\text{N}] = \ell_{\text{N}}$ for all $y \in \mathbb{N}_h$. In other words, the seed glues cannot be programmed and are generic.

The PATS Problem. The *pattern self-assembly tile set synthesis problem (PATS)* [9] asks for the minimum-size RTAS that uniquely self-assembles a given rectangular color pattern, where the size of an RTAS (T, σ) is $|T|$, the number of tile types. Bounding the number of colors or height of the input pattern yields the following practically motivated special cases of PATS:

Problem 1 (c-colored PATS or c-PATS). Given a c-colored pattern P and integer t, does there exist an RTAS of size $\leq t$ that uniquely self-assembles P?

Problem 2 (Height-h PATS). Given a height-h pattern P and integer t, does there exist an RTAS of size $\leq t$ that uniquely self-assembles P?

Restricting the system to be uniform gives rise to *uniform* variants as well, contrasting with the conventional *non-uniform* variants (Fig. 1).

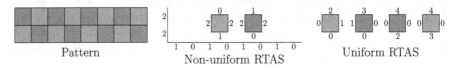

Pattern   Non-uniform RTAS   Uniform RTAS

Fig. 1. A height-2 2-color pattern and minimum-size RTASs uniquely assembling the pattern in the non-uniform and uniform models.

3   Minimum-State Finite State Transducer is NP-hard

The reduction is from 3-PARTITION, a well-known NP-hard problem on integers, to a problem on *finite state transducers* or *FSTs*: finite automata where each transition is augmented with an output symbol and thus *transduces* an input string into an output string of equal length.

Problem 3 (3-PARTITION). Given a multiset of integers[1] $A = \{a_1, a_2, \ldots, a_{3n}\}$ with $\sum_{a_i \in A} a_i/n = p$ and $p/4 < a_i < p/2$, does there exist a partition of A into n sets, each with sum p?

Theorem 1 ([4]). 3-PARTITION *is* NP-*hard*.

Formally, a FST is a 4-tuple $T = \langle \Sigma, Q, s_0, \delta \rangle$, where Σ is the *alphabet*, Q is a finite set of *states* of T, $s_0 \in Q$ is the *start state* of T, and $\delta : Q \times \Sigma \to Q \times \Sigma$ is the *transition function* of T. The *size* of T is equal to $|Q|$. An input-output quadruple $\delta(s_i, b) = (s_j, b')$ is a *transition*, i.e., a (b, b')-*transition* or *b-transition*.

Problem 4 (ENCODING BY FST). Given two strings S, S' and integer K, does there exist a FST with at most K states that transduces S to S'?

Lemma 1. ENCODING BY FST *is* NP-*hard*.

Proof. We borrow from [12] the approach of constructing S and S' by concatenating *segments*: pairs of input and output substrings of equal length that enforce specific structure in a solution FST. An input string A and output string B paired as a segment is denoted $A \to B$.

The integer output by the reduction is $K = 3pn + n + 1$, where n is the number of parts in the partition and p the size of each part. The first segment is $0^{K-1}00^{K-1} \to 0^{K-1}10^{K-1}$. This segment enforces that a solution FST must have K states; label them $s_1, s_2, \ldots, s_K$. Then for all $i < n$, $\delta(s_i, 0) = (s_{i+1}, 0)$ and $\delta(s_K, 0) = (s_1, 1)$.

The problem of partitioning integers of A into groups of size p is implemented in the collection of 1-transitions that leave each state. Each state has a 1-transition that either points to itself (a *fixed singleton*) or is one edge in a

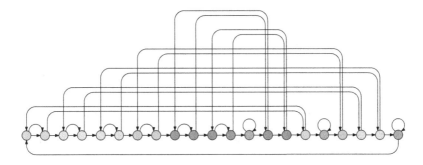

Fig. 2. A solution FST for a toy reduction from an (invalid) 3-PARTITION instance with integers $1, 2, 1, 2$ to ENCODING BY FST. The left-to-right states are s_1 to s_K, colored by their half-fixed interval (fixed singletons are gray). Transitions above the states are $(1, 1)$-transitions. All others are $(0, 0)$-transitions except the lowermost, a $(0, 1)$-transition.

---

[1] 3-PARTITION is *strongly* NP-hard, meaning that the problem is NP-hard when the elements of A are given in unary.

3-cycle formed by two consecutive specified states and an unspecified third state (a *half-fixed triple*). Half-fixed triples are further organized into *half-fixed intervals*, each consisting of a group of $2a_i$ consecutive specified states and a group of a_i consecutive unspecified states for some distinct a_i. The states are partitioned into three groups:

– States s_1 through s_{2pn} are the specified halves of the half-fixed intervals.
– $n + 1$ equally-spaced fixed singletons in states $s_{2pn+1}, \ldots, s_K$.
– The remaining pn states in states $s_{2pn+1}, \ldots, s_K$ partitioned into n sets of p consecutive states.

See Fig. 2 for a toy example of the reduction.

The unspecified halves of the half-fixed intervals can be assigned to the third group of states if and only if the input 3-PARTITION instance has a solution. All that remains is to describe the segments that force the construction of a fixed singleton, half-fixed triple, and half-fixed interval.

Fixed Singleton. The fixed singleton segment ensures that a given state s_i has $\delta(s_i, 1) = (s_i, 1)$. This is done by moving the current state to s_i, transducing a 1 to a 1, and checking whether the current state is still s_i (see Fig. 3).

$$0^{i-1}10^{K-i}0 \rightarrow$$
$$0^{i-1}10^{K-i}1$$

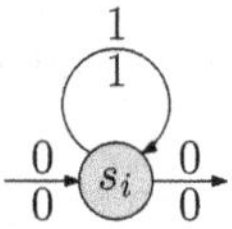

Fig. 3. The fixed single segment and corresponding FST structure enforced.

Half-fixed Triple. The half-fixed triple segment forces two specified *fixed* states s_i, s_{i+1} and an unspecified *free* third state s_j to have $\delta(s_i, 1) = (s_{i+1}, 1)$, $\delta(s_{i+1}, 1) = (s_j, 1)$, and $\delta(s_j, 1) = (s_i, 1)$ (see Fig. 4).

$$0^{i-1}10^{K-i-1}00^i110^{K-i-1}0 \rightarrow$$
$$0^{i-1}10^{K-i-1}10^i110^{K-i-1}1$$

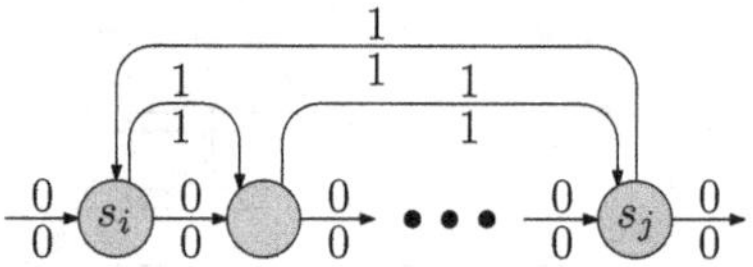

Fig. 4. The half-fixed triple segment and corresponding FST structure enforced.

The segment consists of two subsegments that each ensures a portion of the structure. The first, $0^{i-1}10^{K-i-1}0 \rightarrow 0^{i-1}10^{K-i-1}1$, and ensures that $\delta(s_i, 1) = (s_{i+1}, 1)$. The second, $0^i110^{K-i-1}0 \rightarrow 0^i110^{K-i-1}1$, ensures that $\delta(s_{i+1}, 1) = (s_j, 1)$ and $\delta(s_j, 1) = (s_i, 1)$. The state s_j cannot be in a fixed state of another half-fixed triple segment with fixed states s_i', s_{i+1}' and free state s_j', as then either:

– $s_j = s_i'$ and thus $\delta(s_j, 1) = (s_{i+1}', 1) \neq (s_i, 1)$ (and thus the segment $0^{i-1}10^{K-i-1}0 \rightarrow 0^{i-1}10^{K-i-1}1$ is not transduced).
– $s_j = s_{i+1}'$ and $(s_j, 1) = \delta(s_{i+1}', 1) = (s_i, 1)$, so $\delta(s_j', 1) = (s_{i+1}, 1) \neq (s_i', 1)$ (and thus the segment $0^{i'-1}10^{K-i'-1}0 \rightarrow 0^{i'-1}10^{K-i'-1}1$ is not transduced).

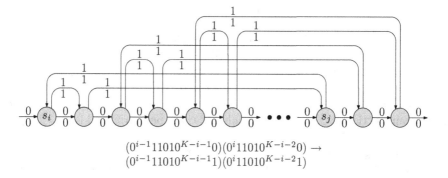

$$(0^{i-1}11010^{K-i-1}0)(0^i11010^{K-i-2}0) \rightarrow$$
$$(0^{i-1}11010^{K-i-1}1)(0^i11010^{K-i-2}1)$$

Fig. 5. The half-fixed interval segment for three consecutive free states and corresponding FST structure enforced.

Half-fixed Interval. The half-fixed interval forces a collection of half-fixed triples with consecutive fixed states to also have consecutive free states. It does so by a simple traversal of the free states, checking that each has the expected pair of consecutive fixed states (see Fig. 5). □

Any solution transduction for the previous reduction uses an FST where each state has at most one incoming 1-transition and 0-transition, since every transition lies on a cycle (of length 1, 3, or K). Also, any solution transduction by an FST with K states traverses $2K$ distinct transitions (with $K-1$ $(0,0)$-transitions, 1 $(0,1)$-transition, and K $(1,1)$-transitions). Any other solution FST must have at least $2K$ states and traverse at least $2K+1$ distinct transitions: $2K$ 0-transitions and at least one 1-transition. Thus the following problem is also NP-hard by the prior reduction:

Problem 5 (PROMISE ENCODING BY FST). Given two strings S, S' and an integer K with the following promises about any FST T with at most K states transducing S to S', does such a T exist?

- Each state of T has at most one incoming 0-transition.
- Each state of T has at most one incoming 1-transition.
- When transducing S to S':
  - $K-1$ distinct $(0,0)$-transitions are used.
  - K distinct $(1,1)$-transitions are used.
  - 1 distinct $(0,1)$-transition is used.
  - The transitions are traversed in a unique specified order given as part of the input.

Corollary 1. *The* PROMISE ENCODING BY FST *problem is* NP-*hard.*

4   Height-2 PATS is NP-complete

Göös and Orponen [5] establish that all the variations of the PATS problem considered here are in NP. So we need only consider their NP-hardness.

Theorem 2. *The non-uniform height-2 PATS problem is NP-hard.*

Proof. The pattern output by the reduction consists of a bottom row encoding S and a top row encoding the sequence of transitions traversed when transducing S to S' (provided as part of the PROMISE ENCODING BY FST instance). The bottom row encoding uses two colors, pink and red, corresponding to the two symbols in S. The top row encoding uses $2K$ colors, one for each transition used in the transduction of S to S'. The number of tile types permitted is $T = 2K+2$: one type per color.

The north glues of the bottom row either encode S (distinct north glues for the pink and red tile types) or $0^{|S|}$ (same glue). The latter is impossible, since then the leftmost $|S|$ locations of the top row are filled by many repetitions of the same K transitions. So the north glues of the bottom row encode S.

A set of $2K$ tile types that assemble the top row is equivalent to a set of $2K$ transitions transducing S to S', with source and destination states corresponding to west and east glues. So the top row can be assembled using $2K$ tile types exactly when S can be transduced to S' using $2K$ transitions of the specified types traversed in the specified order. Thus the pattern can be assembled using a tile set of at most $2K$ types exactly when the corresponding instance of PROMISE ENCODING BY FST has a solution transducer.   □

Theorem 3. *The uniform height-2 PATS problem is NP-hard.*

The addition of more rows with a new common color and increasing T by 1 suffices to prove both the uniform and non-uniform variants NP-hard for greater heights.

5   Uniform Height-2 3-PATS is NP-complete

Problem 6 (MODIFIED PROMISE ENCODING BY FST). Given two strings S, S' and an integer $K \not\equiv 0 \pmod 3$ with the following promises about any FST T with at most K states transducing S to S', does such a T exist?

– The first and last symbols of S' are 2.
– Each state of T has at most one incoming 0-transition.
– Each state of T has at most one incoming 1-transition.
– Every $(1,1)$-transition lies on a 1-cycle or 3-cycle of $(1,1)$-transitions.
– When transducing S to S':
    • $K - 1$ distinct $(0,0)$-transitions are used.
    • $K - 1$ distinct $(1,1)$-transitions are used.
    • 1 distinct $(0,1)$-transition is used.
    • 1 distinct $(1,2)$-transition is used.
    • The transitions are traversed in a unique specified order given as part of the input.

Lemma 2. *The MODIFIED PROMISE ENCODING BY FST problem is NP-hard.*

Theorem 4. *The uniform height-2, 3-PATS problem is* NP-*hard.*

Proof. Let P be the following width-$(1+|S'|+K^2)$, height-2 pattern over 3 colors $\{\blacksquare, \blacksquare, \blacksquare\}$:

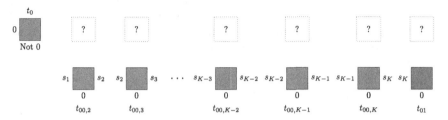

where, for $1 \leq i \leq K-1$,

$$w_i = \begin{cases} \blacksquare^{(3i \bmod K)-1} \blacksquare\blacksquare^{K-3-(3i \bmod K)} & \text{if } 3i \leq K-3 \pmod{K} \\ \blacksquare^{K-3} & \text{otherwise.} \end{cases}$$

Notice that, for any $1 \leq i < j \leq K$, w_i and w_j differ in the position of 1. Split the pattern P into the leftmost $|S'|+1$ columns and the remainder, called the *transduction* and *FST-constructor* gadgets, respectively. The FST-constructor gadget is further partitioned into K rectangular subpatterns of width K.

Next, consider the constraints on RTASs with at most $|S'|+2K+2$ tile types that uniquely self-assemble P. Lemma 1 of Göös and Orponen [5] states that any smallest RTAS that uniquely self-assembles a pattern is directed. As we will prove, directed RTASs uniquely self-assembling P have size at least $|S'|+2K+2$ tile types; thus we need only consider directed systems.

Let the north and east glues of the seed be 0. The leftmost $|S'|+1$ locations in the bottom row of P are orange, with a cyan location following. So these positions must be tiled with orange tiles of pairwise-distinct type; the need for $|S'| + 1$ distinct orange tile types thus arises. Similarly, the leftmost $K-1$ cyan locations in the bottom row must use $K-1$ distinct cyan tile types. These tile types share the south glue 0, and since the system is directed, their west glues are pairwise distinct. Label these $K-1$ cyan tile types left-to-right $t_{00,2}, t_{00,3}, \ldots, t_{00,K}$ and the gray tile type immediately right t_{01}, as seen below.[2] The cyan tile in the northwest corner of P cannot have the same type as any of these $K-1$ types, since otherwise this tile can also appear in the southwest corner of P. Call this type t_0. There are K tile types to be colored yet (illustrated as a dotted square).

These K tile types will turn out to be necessary, implying $(|S'|+1)+(K-1)+2+K-1+1 = S'+2K+2$ types total with $K-1$ colored gray and one colored

---

[2] In these later labels, the first subscript indicates the kind of transition of the FST that the tile type will be shown to simulate, e.g., $t_{00,i}$ is a $(0,0)$-transition, t_{01} and $(0,1)$-transition, etc.

orange. For this, we claim that the bottom row of all blocks but the first assemble identically by establishing that the gray tiles attaching to the southeast corner of the first two blocks are identical. Suppose not. Then the bottom row of the second block cannot reuse cyan tile types used in the bottom row of the first block. So the uncolored K tile types must be one gray and $K-1$ cyan types with south glue 0. Thus the complete tile set includes only two gray tile types with the south glue 0.

Consider the gray tile attaching at the northeast corner of the first block. Its south glue is 0 and its west glue is equal to the east glue of the gray tile attaching to its immediate left. This contradicts the directedness of the system, since a cyan tile is provided with the same pair of west and south glues. Indeed, both gray tile types appear at the southeast corner of a block and to their east are cyan tiles attaching.

The verified claim brings following properties for all but the first block:

Property 1: For any $1 \leq i \leq K-1$, tiles attaching at the i-th top-row position of any two blocks but the first one have the same south glue; tiles attaching at the K-th top-row position (northeast corner) of any two blocks including the first one have the same south glue.

Property 2: Any such pair of tiles have pairwise-distinct east glues (and types).

Property 3: The assembly of the bottom row is provided with at least two different kinds of north glues.

Property 2 holds since a orange tile is placed in the northeast corner of only the last block. Thus without Property 3, $o(K^2)$ tile types would be necessary to place the orange tile. Observe that for each $1 \leq i \leq K-3$, the i-th position of exactly one block is gray and the counterpart of all other blocks are cyan; for each $K-2 \leq i \leq K$, the i-th position of only the last block is orange and the counterpart of all others are gray. Thus, Properties 1 and 2 imply that the tile type set must contain one orange and $K-1$ gray tile types whose south glue is equal to the north glue of t_{01} and one gray and $K-2$ cyan tile types with a common south glue.

We claim these requirements enforce that the north glue of t_{01} is not 0. Suppose otherwise. Then the former requirement implies $K-2$ extra gray tile types with south glue 0. So at most 3 tile types, including t_0, have the non-0 south glue, and Property 3 cannot be satisfied. Thus, the north glue of t_{01} is not 0; call it 1. Tiles attaching at the northeast corner of the blocks must all have distinct types due to Property 2, and now also their south glues must be 1. The K uncolored tile types thus have south glue 1, and one is colored orange and all the others are colored gray.

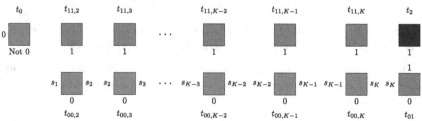

Note that a t_0 tile cannot attach anywhere in the blocks. Indeed, it causes glue mismatch with the seed being placed on the bottom row, and in order for it to attach on the top row, it must share its south glue with $K-3$ cyan tile types due to Properties 1 and 2. In summary, any minimum tile set uniquely assembling P consists of K cyan tile types, K gray ones, and $|S'|+2$ orange ones.

Now we prove constraints on the glues of these types. With only $K-1$ cyan tile types with south glue 0, even the first block must assemble its bottom row as other blocks do. That is, the bottom row of all blocks assemble as $t_{00,2}t_{00,3}\cdots t_{00,K}t_{01}$. Thus, the east glue of t_{01} is equal to the west glue of $t_{00,2}$, that is, s_1. Since t_0 does not appear in any block, Property 2 implies that the north glues of $t_{00,2}, t_{00,3}, \ldots, t_{00,K-2}$ are 0 and that the north glues of $t_{00,K-1}$ and $t_{00,K}$ are 1.

The top row of the last block is w_K■■■ $=$ ■$^{K-4}$■■■■ and its last four positions ■■■■ are assembled as $t_{01}t_2t_2t_2$. This imposes that both the east and west glues of t_2 must be equal to the east glue of t_{01}, that is, s_1. Since S' begins with 2, the east glue of t_0 is s_1.

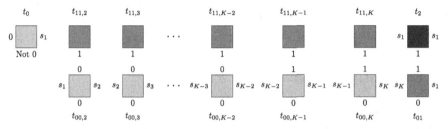

Since S' ends with 2, no tile (necessarily of type $t_{11,2}, t_{11,3}, \ldots, t_{11,K}$ by Properties 1 and 2) appearing at the northeast corner of a block has east glue s_1. Moreover, tiles attaching to their east are of type $t_{00,2}, \ldots, t_{00,K}$ or t_{01}, thus their east glues are in $\{s_2, s_3, \ldots, s_K\}$. Without loss of generality, assign them as follows:

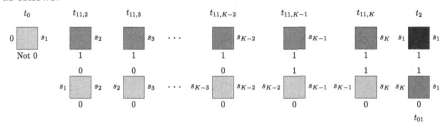

So the east glues of all tile types in the FST-construction gadget are in $\{s_1, \ldots, s_K\}$. The east glues of $t_{11,2}, \ldots, t_{11,K}$ are distinct and selected from $\{s_1, \ldots, s_K\}$. Since $t_{11,2}, \ldots, t_{11,K}$ share the south glue 1 with t_2, the west glue of t_2 is s_1, and the system is directed, the west glues are distinct and from $\{s_2, s_3, \ldots, s_K\}$.

The only remaining flexibility in the design of the tile set is assigning west glues to $t_{11,2}, \ldots, t_{11,K}$, corresponding to the assignment of $(1,1)$-transition sources in MODIFIED PROMISE ENCODING BY FST. All that remains is to prove this correspondence indeed holds.

The glue 0 is not in $\{s_1, \ldots, s_K\}$, as otherwise a cyan or gray tile could appear in the southwest corner of P. So none of the cyan, gray, or t_2 tile types has east glue 0 and thus a t_0 tile cannot attach anywhere but the northwest corner of P. Also, observe that the tile set constraints imply that the north glue of a tile in the bottom row can be discerned by examining the color of the tile north and (possibly) northeast (a cyan tile north implies a 0 glue, a gray tile north implies a 0 or 1 glue if the color of the northeast tile is cyan or orange, respectively).

Finally, consider the tile types, excluding t_0 and all orange tile types except t_2, as a set of transitions of an FST, with $t_{00,i}$ types as $(0,0)$-transitions, $t_{11,i}$ types as $(1,1)$-transitions, t_{01} a $(0,1)$-transition, and t_2 a $(1,2)$-transition. The constraints induced on the resulting transitions (e.g. that there are $K - 1$ $(1,1)$-transitions because there are $K - 1$ color-1 tile types with south glue 1) is found as a constraint on the transitions in the statement of Lemma 2. In particular, the choice of w_i's in consecutive blocks requires that every $(1,1)$-transition lies on a 1-cycle or 3-cycle of $(1,1)$-transitions (the last constraint of Lemma 2). Thus there exists a solution FST to the MODIFIED PROMISE ENCODING BY FST instance if and only if there exists a solution tile set.   □

6   Efficiently Solvable PATS Problems

The non-uniform height-1 PATS problem is trivially solvable using one tile type for each color. This idea can be generalized for all patterns of fixed height:

Theorem 5. *The non-uniform height-h c-PATS problem can be solved in $c^{c^{O(h)}} n$ time.*

As established in Sect. 5, a similar algorithm for the uniform model is impossible unless P = NP. Nevertheless, the uniform height-1 PATS problem can be solved in linear time using a pigeonhole argument and a DFS-based search for the longest repetitive suffix of a given height-1 pattern:

Theorem 6. *The uniform height-1 PATS problem can be solved in $O(n)$ time.*

7   Conclusion

Our work here extends the extensive prior work on the parameterized c-PATS problem to also incorporate pattern height and uniformity, and finds a more delicate complexity landscape: limited height and colors do not make the PATS problem tractable, except when combined in the non-uniform model, or in degenerate cases (height-1 or 1-PATS). A single combination of parameters and model remains unresolved; we conjecture the following:

Conjecture 1. The uniform height-2 2-PATS problem is NP-hard.

We encourage further parameterized analysis of problems in tile self-assembly in support of recent efforts in developing a more complete understanding of the structural complexity of tile self-assembly (see [14]).

Acknowledgements. We thank Yo-Sub Han for very fruitful discussions about finite automata and tile self-assembly, and anonymous reviewers for comments that improved the paper.

References

1. Angluin, D.: On the complexity of minimum inference of regular sets. Inf. Control **39**, 337–350 (1978)
2. Czeizler, E., Popa, A.: Synthesizing minimal tile sets for complex patterns in the framework of patterned DNA self-assembly. Theor. Comput. Sci. **499**, 23–37 (2013)
3. Doty, D.: Theory of algorithmic self-assembly. Commun. ACM **55**(12), 78–88 (2012)
4. Garey, M.R., Johnson, D.S.: Complexity results for multiprocessor scheduling under resource constraints. SIAM J. Comput. **4**(4), 397–411 (1975)
5. Göös, M., Orponen, P.: Synthesizing minimal tile sets for patterned DNA self-assembly. In: Sakakibara, Y., Mi, Y. (eds.) DNA 16 2010. LNCS, vol. 6518, pp. 71–82. Springer, Heidelberg (2011)
6. Johnsen, A.C., Kao, M.-Y., Seki, S.: Computing minimum tile sets to self-assemble color patterns. In: Cai, L., Cheng, S.-W., Lam, T.-W. (eds.) Algorithms and Computation. LNCS, vol. 8283, pp. 699–710. Springer, Heidelberg (2013)
7. Johnsen, A.C., Kao, M.-Y., Seki, S.: A manually-checkable proof for the NP-hardness of 11-color pattern self-assembly tile set synthesis. J. Comb. Optim. (2015) (In press)
8. Kari, L., Kopecki, S., Meunier, P.É., Patitz, M.J., Seki, S.: Binary pattern tile set synthesis is NP-hard. In: Halldórsson, M.M., Iwama, K., Kobayashi, N., Speckmann, B. (eds.) ICALP 2015. LNCS, vol. 9134, pp. 1022–1034. Springer, Heidelberg (2015)
9. Ma, X., Lombardi, F.: Synthesis of tile sets for DNA self-assembly. IEEE Trans. Comput.-Aided Des. Integr. Circuits Syst. **27**(5), 963–967 (2008)
10. Patitz, M.J.: An introduction to tile-based self-assembly. In: Durand-Lose, J., Jonoska, N. (eds.) UCNC 2012. LNCS, vol. 7445, pp. 34–62. Springer, Heidelberg (2012)
11. Seki, S.: Combinatorial optimization in pattern assembly. In: Mauri, G., Dennunzio, A., Manzoni, L., Porreca, A.E. (eds.) UCNC 2013. LNCS, vol. 7956, pp. 220–231. Springer, Heidelberg (2013)
12. Vazirani, U.V., Vazirani, V.V.: A natural encoding scheme proved probabilistic polynomial complete. Theor. Comput. Sci. **24**(3), 291–300 (1983)
13. Winfree, E.: Algorithmic self-Assembly of DNA. Ph.D. thesis, Caltech (1998)
14. Woods, D.: Intrinsic universality and the computational power of self-assembly. Philos. Trans. R. Soc. A **373**(2046) (2015)

Derivative-Based Diagnosis of Regular Expression Ambiguity

Martin Sulzmann[1](✉) and Kenny Zhuo Ming Lu[2]

[1] Faculty of Computer Science and Business Information Systems,
Karlsruhe University of Applied Sciences,
Moltkestrasse 30, 76133 Karlsruhe, Germany
martin.sulzmann@hs-karlsruhe.de

[2] School of Information Technology, Nanyang Polytechnic, 180 Ang Mo Kio Ave 8,
Singapore 569830, Singapore
luzhuomi@gmail.com

Abstract. Regular expressions are often ambiguous. We present a novel method based on Brzozowski's derivatives to aid the user in diagnosing ambiguous regular expressions. We introduce a derivative-based finite state transducer to generate parse trees and minimal counter-examples. The transducer can be easily customized to either follow the POSIX or Greedy disambiguation policy and based on a finite set of examples it is possible to examine if there are any differences among POSIX and Greedy.

Keywords: Regular expressions · Derivatives · Ambiguity · POSIX · Greedy

1 Introduction

A regular expression is *ambiguous* if a string can be matched in more than one way. For example, consider the expression $x^* + x$ where input string x can either be matched against x^* or x. Hence, this expression is ambiguous.

Earlier Works. There exist well-established algorithms to check for regular expression ambiguity. However, most works report ambiguity in terms of an automata which results from an ambiguity-preserving translation of the original expression, e.g. see the work by Book, Even, Greibach and Ott [2]. From a user perspective, it is much more useful to report ambiguity in terms of the original expression. We are only aware of two works which like us perform the ambiguity analysis on the original expression.

Brabrand and Thomsen [4] establish a structural relation to detect ambiguity based on which they can provide minimal counter examples. They consider some disambiguation strategies but do not cover the POSIX interpretation.

Borsotti, Breveglieri, Crespi-Reghizzi and Morzenti [3] show how to derive parse trees based on marked regular expressions [8] as employed in the Berry-Sethi algorithm [1]. They establish criteria to identify ambiguous regular expressions. Like ours, their approach can be customized to support either the

© Springer International Publishing Switzerland 2016
Y.-S. Han and K. Salomaa (Eds.): CIAA 2016, LNCS 9705, pp. 260–272, 2016.
DOI: 10.1007/978-3-319-40946-7_22

POSIX [7] or Greedy [10] disambiguation policy. However, for POSIX/Greedy disambiguation, their approach requires tracking of dynamic data based on the Okui-Suzuki method [9]. Our approach solely relies on derivatives, no dynamic tracking of data is necessary.

Our Work. Brzozowski's derivatives [5] support the symbolic construction of automata where expressions represent automata states. In earlier work [12], we have studied POSIX matching based on derivatives. In this work, we show how to adapt and extend the methods developed in [12] to diagnose ambiguous expressions.

Contributions and Outline. In summary, our contributions are:

- We employ derivatives to compute all parse trees for a large class of (non-problematic) regular expressions (Sect. 3).
- We can build a finite state transducer to compute these parse trees (Sect. 4).
- We can easily customize the transducer to either compute the POSIX or greedy parse tree (Sect. 5).
- We can identify simple criteria to detect ambiguous expressions and to derive a finite set of minimal counter-examples. Thus, we can statically verify if there are any differences among POSIX and Greedy (Sect. 6).

The online version of this paper contains further details including proofs.[1] Next, we introduce our notion of regular expression, parse trees and ambiguity.

2   Regular Expressions, Parse Trees and Ambiguity

The development largely follows [4,6]. We assume that symbols are taken from a fixed, finite alphabet Σ. We generally write x, y, z for symbols.

Definition 1 (Words and Regular Expressions). *Words are either empty or concatenation of words and defined as follows:* $w ::= \epsilon \mid x \in \Sigma \mid w \cdot w$.

We denote regular expressions by r, s, t. Their definition is as follows: $r ::= x \in \Sigma \mid r^* \mid r \cdot r \mid r + r \mid \epsilon \mid \phi$ *The mapping to words is standard.* $\mathcal{L}(x) = \{x\}$. $\mathcal{L}(r^*) = \{w_1 \cdot \ldots \cdot w_n \mid n \geq 0 \land w_i \in \mathcal{L}(r) \land i \in \{1, .., n\}\}$. $\mathcal{L}(r \cdot s) = \{w_1 \cdot w_2 \mid w_1 \in \mathcal{L}(r) \land w_2 \in \mathcal{L}(s)\}$. $\mathcal{L}(r + s) = \mathcal{L}(r) \cup \mathcal{L}(s)$. $\mathcal{L}(\epsilon) = \{\epsilon\}$. $\mathcal{L}(\phi) = \{\}$.

We say an expression r is nullable *iff $\epsilon \in \mathcal{L}(r)$.*

As it is common, we assume that $+$ and $\cdot$ are right-associative. That is, $x + y + x \cdot y \cdot z$ stands for $x + (y + (x \cdot (y \cdot z)))$.

A parse tree explains which subexpressions match which subwords. We follow [6] and view expressions as types and parse trees as values.

---

[1] http://arxiv.org/abs/1604.06644.

Definition 2 (Parse Trees). *Parse tree values are built using data constructors such as lists, pairs, left/right injection into a disjoint sum etc. In case of repetitive matches such as in case of Kleene star, we make use of lists. We use Haskell style notation and write* $[v_1, ..., v_n]$ *as a short-hand for* $v_1 : ... : v_n : []$.

$$v ::= () \mid x \mid (v,v) \mid L\ v \mid R\ v \mid vs \qquad vs ::= [] \mid v : vs$$

The valid relations among parse trees and regular expressions are defined via a natural deduction style proof system.

$$\vdash [] : r^* \qquad \frac{\vdash v : r \quad \vdash vs : r^*}{\vdash (v : vs) : r^*} \qquad \frac{\vdash v_1 : r_1 \quad \vdash v_2 : r_2}{\vdash (v_1, v_2) : r_1 \cdot r_2}$$

$$\frac{\vdash v_1 : r_1}{\vdash L\ v_1 : r_1 + r_2} \qquad \frac{\vdash v_2 : r_2}{\vdash R\ v_2 : r_1 + r_2} \qquad \vdash () : \epsilon \qquad \frac{x \in \Sigma}{\vdash x : x}$$

Definition 3 (Flattening). *We can flatten a parse tree to a word as follows:*

$$|()| = \epsilon \quad |x| \qquad = x \qquad |L\ v| = |v| \quad |v : vs| = |v| \cdot |vs|$$
$$|[]| = \epsilon \quad |(v_1, v_2)| = |v_1| \cdot |v_2| \quad |R\ v| = |v|$$

Proposition 4 (Frisch/Cardelli [6]). *Let* r *be a regular expression. If* $w \in \mathcal{L}(r)$ *for some word* w, *then there exists a parse tree* v *such that* $\vdash v : r$ *and* $|v| = w$. *If* $\vdash v : r$ *for some parse tree* v, *then* $|v| \in \mathcal{L}(r)$.

Example 5. We find that $x \cdot y \in \mathcal{L}((x \cdot y + x + y)^*)$ where $[L\ (x,y)]$ is a possible parse tree. Recall that $+$ is right-associative and therefore we interpret $(x \cdot y + x + y)^*$ as $(x \cdot y + (x + y))^*$.

An expression is ambiguous if there exists a word which can be matched in more than one way. That is, there must be two distinct parse trees which share the same underlying word.

Definition 6 (Ambiguous Regular Expressions). *We say a regular expression* r *is ambiguous iff there exist two distinct parse trees* v_1 *and* v_2 *such that* $\vdash v_1 : r$ *and* $\vdash v_2 : r$ *where* $|v_1| = |v_2|$.

Example 7. $[L\ (x,y)]$ and $[R\ (L\ x), R\ (R\ y)]$ are two distinct parse trees for expression $(x \cdot y + x + y)^*$ and word $x \cdot y$.

Our ambiguity diagnosis methods will operate on arbitrary expressions. However, formal results are restricted to a certain class of 'non-problematic' expressions.

Definition 8 (Problematic Expressions). *We say an expression* r *is problematic iff it contains some sub-expression of the form* s^* *where* $\epsilon \in \mathcal{L}(s)$.

For problematic expressions, the set of parse trees is infinite, otherwise finite.

Example 9. Consider the problematic expression ϵ^* where for the empty input word we find the following (infinite) sequence of parse trees $[], [()], [(), ()], \ldots$

Proposition 10 (Frisch/Cardelli [6]). *For non-problematic expressions, the set of distinct parse trees which share the same underlying word is always finite.*

Next, we consider computational methods based on Brzozowski's derivatives to compute parse trees.

3   Computing Parse Trees via Derivatives

Derivatives denote left quotients and they can be computed via a simple syntactic transformation.

Definition 11 (Regular Expression Derivatives). *The* derivative *of expression* r *w.r.t. symbol* x, *written* $d_x(r)$, *is computed by induction on* r:

$$d_x(\phi) = \phi \quad d_x(\epsilon) = \phi \quad d_x(r_1 + r_2) = d_x(r_1) + d_x(r_2) \quad d_x(r^*) = d_x(r) \cdot r^*$$
$$d_x(y) = \begin{cases} \epsilon \ \text{if } x = y \\ \phi \ \text{otherwise} \end{cases} \quad d_x(r_1 \cdot r_2) = \begin{cases} d_x(r_1) \cdot r_2 + d_x(r_2) \ \text{if } \epsilon \in \mathcal{L}(r_1) \\ d_x(r_1) \cdot r_2 \qquad\qquad \text{otherwise} \end{cases}$$

The extension to words is as follows: $d_\epsilon(r) = r$. $d_{x \cdot w}(r) = d_w(d_x(r))$.

A descendant *of* r *is either* r *itself or the derivative of a descendant. We write* $r \preceq s$ *to denote that* s *is a descendant of* r. *We write* $d(r)$ *to denote the set of descendants of* r.

Proposition 12 (Brzozowski [5]). *For any expression* r *and symbol* x *we find that* $\mathcal{L}(d_x(r)) = \{w \mid x \cdot w \in \mathcal{L}(r)\}$.

Thus, we obtain a simple word matching algorithm by repeatedly building the derivative and then checking if the final derivative is nullable. That is, $w \in \mathcal{L}(r)$ iff $\epsilon \in \mathcal{L}(d_w(r))$. Nullability can easily be decided by induction on r. We omit the straightforward details.

Example 13. Consider expression $(x+y)^*$ and input $x \cdot y$. We find $d_x((x+y)^*) = (\epsilon + \phi) \cdot (x+y)^*$ and $d_y(d_x((x+y)^*)) = (\phi + \phi) \cdot (x+y)^* + (\phi + \epsilon) \cdot (x+y)^*$. The final expression is nullable. Hence, we can conclude that $x \cdot y \in \mathcal{L}((x+y)^*)$.

Based on the derivative method, it is surprisingly easy to compute parse trees for some input word w. The key insights are as follows:

1. Build all parse trees for the final (nullable) expression.
2. Transform a parse tree for $d_x(r)$ into a parse tree for r by injecting symbol x into $d_x(r)$'s parse tree. Injecting can be viewed as reversing the effect of the derivative operation.

Definition 14 (Empty Parse Trees). *Let r be an expression. Then, allEps$_r$ yields a set of parse trees. The definition of allEps$_r$ is by induction on r.*

$$allEps_\epsilon = \{()\} \quad allEps_\phi = \{\} \quad allEps_x = \{\}$$
$$allEps_{r*} = \{[]\} \quad allEps_{r_1 \cdot r_2} = \{(v_1, v_2) \mid v_1 \in allEps_{r_1} \wedge v_2 \in allEps_{r_2}\}$$
$$allEps_{r_1+r_2} = \{L\ v_1 \mid v_1 \in allEps_{r_1}\} \cup \{R\ v_2 \mid v_2 \in allEps_{r_2}\}$$

If the expression is not nullable it is easy to see that we obtain an empty set. For nullable expressions, allEps$_r$ yields empty parse trees.

Proposition 15 (Empty Parse Trees). *Let r be a nullable expression. Then, for any $v \in allEps_r$ we have that $\vdash v : r$ and $|v| = \epsilon$.*

Example 16. For the final (nullable) expression from Example 13 we find that $allEps_{(\phi+\phi)\cdot(x+y)^*+(\phi+\epsilon)\cdot(x+y)^*} = R\ (R\ (), [])$.

For nullable, non-problematic expressions r, we can state that allEps$_r$ yields all parse trees v for r where $|v| = \epsilon$.

Proposition 17 (All Empty Non-problematic Parse Trees). *Let r be a non-problematic expression such that $\epsilon \in \mathcal{L}(r)$. Let v be a parse tree such that $\vdash v : r$ where $|v| = \epsilon$. Then, we find that $v \in allEps_r$.*

The non-problematic assumption is necessary. Recall Example 9.

What remains is to describe how to derive parse trees for the original expression. We achieve this by injecting symbol x into $d_x(r)$'s parse tree.

Definition 18 (Injecting Symbols into Parse Trees). *Let r be an expression and x be a symbol. Then, $injs_{d_x(r)}$ is a function[2] which maps $d_x(r)$'s parse tree to a set of parse trees of r. The definition is by induction on r.*

$$injs_{d_x(\epsilon)}\ \text{-} = \{\} \quad injs_{d_x(\phi)}\ \text{-} = \{\} \quad injs_{d_x(x)}\ () = \{x\} \quad injs_{d_x(y)}\ \text{-} = \{\}$$
$$injs_{d_x(r^*)}\ (v, vs) = \{v' : vs \mid v' \in injs_{d_x(r)}\ v\}$$
$$injs_{d_x((r_1\cdot r_2))} =$$
$$\lambda v.\text{case } v \text{ of}$$
$$(v_1, v_2) \rightarrow \{(v, v_2) \mid v \in injs_{d_x(r_1)}\ v_1\}$$
$$L\ (v_1, v_2) \rightarrow \{(v, v_2) \mid v \in injs_{d_x(r_1)}\ v_1\}$$
$$R\ v_2 \rightarrow \{(v, v') \mid v \in allEps_{r_1} \wedge v' \in injs_{d_x(r_2)}\ v_2\}$$
$$injs_{d_x((r_1+r_2))} =$$
$$\lambda v.\text{case } v \text{ of}$$
$$L\ v_1 \rightarrow \{L\ v \mid v \in injs_{d_x(r_1)}\ v_1\}$$
$$R\ v_2 \rightarrow \{R\ v \mid v \in injs_{d_x(r_2)}\ v_2\}$$

---

[2] Additional arguments are x and r but we use notation $injs_{d_x(r)}$ to highlight that the definition is defined by pattern match over the various cases of the derivative operation.

In the above, we use Haskell style syntax such as lambda-bound functions etc. The first couple of cases are straightforward. For brevity, we use the 'don't care' pattern _ and make use of a non-linear pattern in the third equation. In case of Kleene star, the parse tree is represented by a sequence. We call the injection function of the underlying expression on the first element. In case of concatenation $r_1 \cdot r_2$, we observe the shape of the parse tree of $d_x(r_1 \cdot r_2)$. For example, if we encounter $R\ v_2$, the left component r_1 must be nullable. Hence, we apply $allEps_{r_1}$.

Via a straightforward inductive proof on r, we can verify that the injection function yields valid parse trees.

Proposition 19 (Soundness of Injection). *Let r be an expression, x be a symbol and v be a parse tree such that $\vdash v : d_x(r)$. Then, for any $v' \in injs_{d_x(r)}$ we find that $\vdash v' : r$.*

Example 20. Consider our running example where $\vdash R\ (R\ (),[]) : d_y(d_x((x + y)^*))$. Then, $injs_{d_y(d_x((x+y)^*))}\ (R\ (R\ (),[])) = \{(L\ (),[y])\}$ where $\vdash (L\ (),[y]) : d_x((x + y)^*)$ and $d_x((x + y)^*) = (\epsilon + \phi) \cdot (x + y)^*$.

As in case of $allEps_r$, we can only guarantee completeness for non-problematic expressions.

Proposition 21 (Completeness of Non-problematic Injection). *Let r be a non-problematic expression and v a parse tree such that $\vdash v : r$ where $|v| = x \cdot w$ for some letter x and word w. Then, there exists a parse tree v' such that (1) $\vdash v' : d_x(r)$ and (2) $v \in injs_{d_x(r)}\ v'$.*

Definition 22 (Parse Tree Construction). *Let r be an expression. Then, the derivative-based procedure to compute all parse trees is as follows.*

$$allParse\ r\ \epsilon = allEps_r$$
$$allParse\ r\ x \cdot w = \{v \mid v \in injs_{d_x(r)}\ v' \wedge v' \in (allParse\ d_x(r)\ w)\}$$

Proposition 23 (Valid Parse Trees). *Let r be an expression. Then, for each $v \in allParse\ r\ |v|$ we find that $\vdash v : r$.*

For non-problematic expressions, we obtain a complete parse tree construction method.

Proposition 24 (All Non-problematic Parse Trees). *Let r be a non-problematic expression and v a parse tree such that $\vdash v : r$. Then, we find that $v \in allParse\ r\ |v|$.*

In case of a fixed expression r, calls to $allParse\ r$ repeatedly build the same set of derivatives. We can be more efficient by constructing a finite state transducer (FST) for a fixed expression r where states are descendants of r. The outputs are parse tree transformation functions. This is what we will discuss next.

4   Derivative-Based Finite State Transducer

The natural candidate for FST states are derivatives. That is, $\delta(r, x) = d_x(r)$. In general, descendants (derivatives) are not finite. Thankfully, Brzozowski showed that the set of dissimilar descendants is finite.

Definition 25 (Similarity).   *We say two expressions r and s are similar, written $r \approx s$, if one can be transformed into the other by application of the following rules.*

$$(Idemp)\ r + r \approx r \quad (Comm)\ r_1 + r_2 \approx r_2 + r_1$$

$$(Assoc)\ (r_1 + r_2) + r_3 \approx r_1 + (r_2 + r_3) \quad (Ctxt)\ \frac{s \approx t}{R[s] \approx R[t]}$$

The (Ctxt) rules assumes expressions with a hole. We write $R[s]$ to denote the expression where the hole $[]$ is replaced by s.

$$(Hole\ Expressions)\ R[] :: = [] \mid R[] \cdot s \mid s \cdot R[] \mid R[] + s \mid s + R[]$$

There is no hole inside Kleene star because the derivative operation will only ever be applied on unfoldings of the Kleene star but never within a Kleene star expression.

We write $d(r)/\approx$ to denote the set of equivalence classes of $d(r)$ w.r.t. the equivalence relation $\approx$.

Proposition 26 (Brzozowski [5]). $d(r)/\approx$ *is finite for any expression r.*

Based on the above, we build an automata where the set of states consists of a canonical representative for all descendants of some expression r. A similar approach is discussed in [13].

Definition 27 (Canonical Representative).   *For each expression r we compute an expression $\mathcal{C}(r)$ by systematic application of the similarity rules: (1) Put alternatives in right-associative normal form via rule (Assoc). (2) Remove duplicates via rules (Idemp) where via rule (Comm) we push the right-most duplicates to the left. (3) Repeat until there are no further changes.*

Proposition 28 (Canonical Normal From).   *Let r be an expression. Then, $\mathcal{C}(r)$ represents a canonical normal form of r.*

Furthermore, alternatives keep their relative position. For example, $\mathcal{C}(r + s + s_1 + ... + s_n + s + t) = r + s + s_1 + ... + s_n + t$. This is important for the upcoming construction of POSIX and Greedy parse trees.

Proposition 29 (Finite Dissimilar Canonical Descendants).   *Let r be an expression. Then, the set $\mathcal{D}(r) = \{\mathcal{C}(s) \mid r \preceq s\}$ is finite.*

Like in case of *injs*, we need to maintain information how to transform parse trees among similar expressions. Hence, we attach parse tree transformation functions to the similarity rules.

Definition 30 (Similarity with Parse Tree Transformation). *We write* $r \overset{f}{\gg} s$ *to denote that expressions r and s are similar and a parse tree of s can be transformed into a parse tree of r via function f. In case the function returns a set of parse trees we write $r \overset{fs}{\gg} s$. We write $r \gg s$ if the parse tree transformation is not of interest.*

$$(Idemp) \; \frac{fs(u) = \{L\ u, R\ u\}}{r + r \overset{fs}{\gg} r} \qquad (Comm) \; \frac{\begin{array}{l} f(L\ u) = R\ u \\ f(R\ u) = L\ u \end{array}}{r_1 + r_2 \overset{f}{\gg} r_2 + r_1}$$

$$(Assoc) \; \frac{\begin{array}{ll} f(L\ u_1) & = L\ (L\ u_1) \\ f(R\ (L\ u_2)) = L\ (R\ u_2) \\ f(R\ (R\ u_3)) = R\ u_3 \end{array}}{(r_1 + r_2) + r_3 \overset{f}{\gg} r_1 + (r_2 + r_3)} \qquad (Lift) \; \frac{r \overset{f}{\gg} s}{\quad fs(u) = \{f(u)\} \quad} \\ \frac{}{r \overset{fs}{\gg} s}$$

$$(C1) \; \frac{gs(u_r, u_t) = \{(u_r, u_s) \mid u_s \in fs(u_t)\}}{r \cdot s \overset{gs}{\gg} r \cdot t} \qquad (C2) \; \frac{gs(u_t, u_r) = \{(u_s, u_r) \mid u_s \in fs(u_t)\}}{s \cdot r \overset{gs}{\gg} t \cdot r}$$

$$(C3) \; \frac{\begin{array}{l} s \overset{fs}{\gg} t \\ gs(L\ u_r) = \{L\ u_r\} \\ gs(R\ u_t) = \{R\ u_s \mid u_s \in fs(u_t)\} \end{array}}{r + s \overset{gs}{\gg} r + t} \qquad (C4) \; \frac{\begin{array}{l} s \overset{fs}{\gg} t \\ gs(L\ u_t) = \{L\ u_s \mid u_s \in fs(u_t)\} \\ gs(R\ u_r) = \{R\ u_r\} \end{array}}{s + r \overset{gs}{\gg} t + r}$$

The above rules are derived from the ones in Definition 25 by providing the appropriate parse tree transformations. Due to the similarity rule (Idemp) we may obtain a set of parse trees. Rules (C1-4) cover all the cases described by rule (Ctxt). The attached (transformation) functions yield valid parse trees (soundness) and every parse tree of a similar expression can be obtained (completeness).

Proposition 31 (Soundness of Transformation). *Let r and s be two expressions and fs a function such that $r \overset{fs}{\gg} s$. Then, we find that (1) $r \approx s$ and (2) for any parse tree v where $\vdash v : s$ we have that $\vdash v' : r$ for any $v' \in fs(v)$.*

Proposition 32 (Completeness of Transformation). *Let r and s be two expressions and v be a parse tree such that $\vdash v : r$ and $r \approx s$. Then, $r \overset{fs}{\gg} s$ where $v \in fs(v')$ for some v' such that $\vdash v' : s$.*

The FST to compute parse trees for some expression r consists of states $\mathcal{D}(r)$. Each state transition from s to $\mathcal{C}(d_x(s))$ yields as output a parse tree transformer function which is a composition of $injs_{d_x(s)}$ and fs where $d_x(s) \overset{fs}{\gg} \mathcal{C}(d_x(s))$.

Definition 33 (FST Construction). *Let r be an expression. We define $\mathcal{FST}(r) = (Q, \Sigma, \delta, q_0, F)$ where $Q = \mathcal{D}(r)$, $q_0 = r$, $F = \{s \in Q \mid \epsilon \in \mathcal{L}(s)\}$ and for each $s \in Q$ and $x \in \Sigma$ we set $\delta(s, x) = (\mathcal{C}(d_x(s)), gs)$ where $d_x(s) \overset{fs}{\gg} \mathcal{C}(d_x(s))$ and $gs(u) = \{u_2 \mid u_1 \in fs(u) \wedge u_2 \in injs_{d_x(s)} u_1\}$.*

The transition relation δ is inductively extended to words as follows. We define $\delta(s, \epsilon) = (s, \lambda u.\{u\})$ and $\delta(s, x \cdot w) = (r, fs)$ where $\delta(s, x) = (t, gs)$ and $\delta(t, w) = (r, hs)$ where $fs(u) = \{u_2 \mid u_1 \in hs(u) \wedge u_2 \in gs(u_1)\}$.

Proposition 34 (All Non-problematic Parse Trees via FST). *Let r be a non-problematic expression and v a parse tree such that $\vdash v : r$. Let $\mathcal{FST}(r) = (Q, \Sigma, \delta, q_0, F)$. Then, we find that $v \in fs(allEps_{r'})$ where $\delta(r, |v|) = (r', fs)$.*

5   Computing POSIX and Greedy Parse Trees

Based on our earlier work [12] we can immediately conclude that the 'first' (left-most) match obtained by executing $\mathcal{FST}(r)$ is the POSIX match.[3] The use of derivatives guarantees that the longest left-most (POSIX) parse tree is computed.

Proposition 35 (POSIX). *Let r be an expression and w be a word such that $w \in \mathcal{L}(r)$. Let $\mathcal{FST}(r) = (Q, \Sigma, \delta, q_0, F)$. Let $\delta(r, w) = (r', fs)$ for some expression r' and transformer fs. Then, $fs(allEps_{r'}) = \{v_1, ..., v_n\}$ for some parse trees v_i where v_1 is the POSIX match.*

With little effort it is possible to customize our FST construction to compute Greedy parse trees. The insight is to normalize derivatives such that they effectively correspond to partial derivatives. Via this normalization step, we obtain as the 'first' result the Greedy (left-most) parse tree. This follows from our earlier work [11] where we showed that partial derivatives naturally yield greedy matches.

We first define partial derivatives which are a non-deterministic generalization of derivatives. Instead of a single expression, the partial derivative operation yields a set of expressions.

Definition 36 (Partial Derivatives). *Let r be an expression and x be a symbol. Then, the partial derivative of r w.r.t. x is computed as follows:*

$$pd_x(\phi) = \{\} \qquad pd_x(y) = \begin{cases} \{\epsilon\} & \text{if } x = y \\ \{\} & \text{otherwise} \end{cases}$$

$$pd_x(\epsilon) = \{\}$$

$$pd_x(r_1 + r_2) = pd_x(r_1) \cup pd_x(r_2) \qquad pd_x(r^*) = \{r' \cdot r^* \mid r' \in pd_x(r)\}$$

$$pd_x(r_1 \cdot r_2) = \begin{cases} \{r_1' \cdot r_2 \mid r_1' \in pd_x(r_1)\} \cup pd_x(r_2) & \text{if } \epsilon \in \mathcal{L}(r_1) \\ \{r_1' \cdot r_2 \mid r_1' \in pd_x(r_1)\} & \text{otherwise} \end{cases}$$

Let $M = \{r_1, ..., r_n\}$ be a set of expressions. Then, we define $+M = r_1 + ... + r_n$ and $+\{\} = \phi$.

To derive partial derivatives via derivatives, we impose the following additional similarity rules.

---

[3] Technically, we treat the set of parse trees like a list. Recall that *allEps* and the simplification rule (Idemp) favor the left-most match. Alternatives keep their relative position in an expression.

Definition 37 (Partial Derivative Similarity Rules).

$$(Dist) \ \frac{f(L\ (u_r, u_t)) = (L\ u_r, u_t)}{f(R\ (u_s, t_t)) = (R\ u_s, u_t)} \quad (ElimPhi) \ \phi \cdot r \overset{\perp}{\gg} \phi$$
$$\frac{}{(r+s)\cdot t \overset{f}{\gg} r\cdot t + s\cdot t}$$

Rule (Dist) mimics the set-based operations performed by $pd.(\cdot)$ in case of concatenation and Kleene star. Rule (ElimPhi) covers cases where the set is empty. We use $\perp$ to denote the undefined parse tree transformer function. As there is no parse tree for ϕ this function will never be called.

Proposition 38 (Partial Derivatives as Normalized Derivatives). *Let r be an expression and x be a symbol. Then, we have that $+pd_x(r)$ is syntactically equal to some expression s such that $d_x(r) \gg s$. We ignore the transformer function which is not relevant here.*

Based on the above and our earlier results in [11] we can immediately conclude the following.

Proposition 39 (Greedy). *Let r be an expression and w be a word such that $w \in \mathcal{L}(r)$. Let $\mathcal{FST}(r) = (Q, \Sigma, \delta, q_0, F)$ where we additionally apply the similarity rules in Definition 37 such that canonical representatives satisfy the property stated in Proposition 38. Let $\delta(r, w) = (r', fs)$ for some expression r' and transformer fs. Then, $fs(allEps_{r'}) = \{v_1, ..., v_n\}$ for some parse trees v_i where v_1 is the Greedy parse tree.*

6 Ambiguity Diagnosis

We can identify three situations where ambiguity of r arises during the construction of $\mathcal{FST}(r)$. The first situation concerns nullable expressions. If we encounter multiple empty parse trees for a nullable descendant (accepting state) then we end up with multiple parse trees for the initial state. Then, the initial expression is ambiguous.

The second situation concerns the case of injecting a symbol into the parse tree of a descendant. Recall that the $injs$ function from Definition 18 possibly yields a set of parse trees. This will only happen if we apply the derivative operation on some subterm $t_1 \cdot t_2$ where t_1 is a nullable expression with multiple empty parse trees.

The final (third) situation ambiguous situation arises in case we build canonical representatives. Recall Definition 30. We end up with multiple parse trees whenever we apply rule (Idemp).

These are the only situations which may give rise to multiple parse trees. That is, if none of these situations arises the expression must be unambiguous. We summarize these observations in the following result.

Definition 40 (Realizable State). *We say that $s \in \mathcal{D}(r)$ is realizable, if there exists a path in $\mathcal{FST}(r)$ such that (1) we reach s and (2) along this path all states (expressions) including s do not describe the empty language.*

Proposition 41 (Ambiguity Criteria). *Let r be a non-problematic expression. Then, r is ambiguous iff there exists a realizable $s \in \mathcal{D}(r)$ and some symbol x where one of the following conditions applies:*

A1. $|allEps_s| > 1$, *or*
A2. $s = R[t_1 \cdot t_2]$ *where* $|allEps_{t_1}| > 1$, *or*
A3. $\mathcal{L}(\mathcal{C}(d_x(s))) \neq \{\}$ *and* $d_x(s) \overset{fs}{\gg} \mathcal{C}(d_x(s))$ *with rule* $(Idemp)$ *applied.*

The above criteria are easy to verify. In terms of the FST generated, criteria **A1** is always connected to a final state whereas criteria **A2** and **A3** are always connected to transitions.

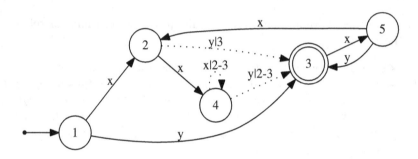

Fig. 1. $\mathcal{FST}((x \cdot x^* + y \cdot x + x \cdot y \cdot x)^* \cdot y)$

Consider the following example taken from [3]. See Fig. 1. We find ambiguous transitions due to **A2** and **A3**. Such transitions are represented as dotted arrows with labels to indicate **A2** and **A3**. Ambiguity due to **A1** does not arise for this example.

Let us investigate the ambiguous transition from state 2 to state 3. We carry out the constructions of states starting with the initial expression $r \cdot y$ where $r = (x \cdot x^* + y \cdot x + x \cdot y \cdot x)^*$. For brevity, we make use of additional similarity rules such as $\epsilon \cdot s \approx s$ to keep the size of descendants manageable. In the following, we write $r \overset{x}{\rightarrow} s$ if $s = d_x(r)$.

$$r \cdot y \overset{x}{\rightarrow} ((x^* + y \cdot x) \cdot r) \cdot y \overset{y}{\rightarrow} (x \cdot r) \cdot y + (x \cdot r) \cdot y + \epsilon \approx (x \cdot r) \cdot y + \epsilon$$

In the last step, we apply rule (Idemp). Hence, the ambiguous transition from state 2 to state 3.

State 3 is final, however, $x \cdot y$ is not yet a full counter-example to exhibit ambiguity. In essence, $x \cdot y$ is a prefix of the full counter-example $x \cdot y \cdot x \cdot y$. For this example, we obtain parse trees $([L\ (x, []), L\ (R\ (x, y))], y)$ and $(R\ (R\ (x, (y, x))), y)$. The first one is obtained via Greedy and the second one via POSIX.

To summarize, from the FST it is straightforward to derive minimal prefixes of counter-examples. To obtain actual counter-examples, minimal prefixes need to be extended so that a final state is reached. Based on the FST, we could perform a breadth-first search to calculate all such minimal counter-examples. Alternatively, we can built (minimal) counter-examples during the construction of the FST.

There is clearly much scope for more sophisticated ambiguity diagnosis based on the information provided by the FST. An immediate application is to check (statically) any differences among Greedy and POSIX. We simply check both methods against the set of minimal counter-examples. It is clear that there are only finitely many (minimal) counter-examples as there are a finite number of states and transitions. Obtaining more precise bounds on their size is something to consider in future work.

Acknowledgments. We thank Peter Thiemann and the reviewers for their comments.

References

1. Berry, G., Sethi, R.: From regular expressions to deterministic automata. Theoret. Comput. Sci. **48**(1), 117–126 (1986). http://dl.acm.org/citation.cfm?id=39528. 39537
2. Book, R., Even, S., Greibach, S., Ott, G.: Ambiguity in graphs and expressions. IEEE Trans. Comput. **20**(2), 149–153 (1971). http://dx.doi.org/10.1109/T-C.1971.223204
3. Borsotti, A., Breveglieri, L., Crespi Reghizzi, S., Morzenti, A.: From ambiguous regular expressions to deterministic parsing automata. In: Drewes, F. (ed.) CIAA 2015. LNCS, vol. 9223, pp. 35–48. Springer, Heidelberg (2015)
4. Brabrand, C., Thomsen, J.G.: Typed and unambiguous pattern matching on strings using regular expressions. In: Proceedings of PPDP 2010, pp. 243–254. ACM (2010)
5. Brzozowski, J.A.: Derivatives of regular expressions. J. ACM **11**(4), 481–494 (1964)
6. Frisch, A., Cardelli, L.: Greedy regular expression matching. In: Díaz, J., Karhumäki, J., Lepistö, A., Sannella, D. (eds.) ICALP 2004. LNCS, vol. 3142, pp. 618–629. Springer, Heidelberg (2004)
7. Institute of Electrical, Electronics Engineers (IEEE): Standard for information technology - Portable Operating System Interface (POSIX) - Part 2 (Shell and utilities), Section 2.8 (Regular expression notation), IEEE Standard 1003.2, New York (1992)
8. McNaughton, R., Yamada, H.: Regular expressions and finite state graphs for automata. IRE Trans. Electron. Comput. EC **9**(1), 38–47 (1960)
9. Okui, S., Suzuki, T.: Disambiguation in regular expression matching via position automata with augmented transitions. In: Domaratzki, M., Salomaa, K. (eds.) CIAA 2010. LNCS, vol. 6482, pp. 231–240. Springer, Heidelberg (2011)
10. PCRE - Perl Compatible Regular Expressions. http://www.pcre.org/
11. Sulzmann, M., Lu, K.Z.M.: Regular expression sub-matching using partial derivatives. In: Proceedings of PPDP 2012, pp. 79–90. ACM (2012)

12. Sulzmann, M., Lu, K.Z.M.: POSIX regular expression parsing with derivatives. In: Codish, M., Sumii, E. (eds.) FLOPS 2014. LNCS, vol. 8475, pp. 203–220. Springer, Heidelberg (2014)
13. Watson, B.W.: A taxonomy of finite automata minimization algorithmes. Computing Science Note 93/44. Eindhoven University of Technology, The Netherlands (1993)

Regular Approximation of Weighted Linear Nondeleting Context-Free Tree Languages

Markus Teichmann$^{(\boxtimes)}$

Department of Computer Science,
Technische Universität Dresden, 01062 Dresden, Germany
markus.teichmann@mailbox.tu-dresden.de

Abstract. We show how to train a weighted regular tree grammar such that it best approximates a weighted linear nondeleting context-free tree grammar concerning the Kullback-Leibler divergence between both grammars.

1 Introduction

In the field of natural language processing (NLP), formal grammars play a central role to model sentences of a natural language. There is the classical compromise between modeling capabilities and computation costs. In the existing literature, the connection between computationally favorable regular string grammars (REGs) and the more expressive context-free string grammars (CFGs) is studied. It has been shown how to approximate a CFG by a REG [8] and how to approximate weight structures in the weighted case [9].

Since context-free string grammars are not sufficient to capture all phenomena of natural languages [15], more powerful extensions are investigated. Tree grammars include additional information about the structure of sentences into the formal representation of a natural language. One well-established class of tree grammars is the class of tree adjoining grammars (TAGs) [5]. The approximation result from the string case has been extended in [10] showing how to approximate a weighted TAG by a weighted regular tree grammar (wRTG). Since TAGs are expressively equivalent to linear monadic context-free tree grammars [3,6], it seems worthwhile to lift the approximation result to the more general case of linear nondeleting context-free tree grammars (lnCFTGs, cf. [14]).

We show how a given weighted lnCFTG (wlnCFTG) can be approximated by a wRTG where the underlying RTG-structure is given. For this, we extend the string case, the result for TAGs, and use similar concepts as in the approximation of weighted CFG given infinite tree corpora [1]. As a technical tool, we use the intersection of the given wlnCFTG and an unambiguous RTG, which is done similarly to the case for TAGs [10] and for synchronous lnCFTGs [11].

In detail, we proceed as follows. Given a wlnCFTG (G, p_G) and an unambiguous RTG H, we will intersect (G, p_G) and H to obtain a wlnCFTG (K, p_K). We then use (K, p_K) to obtain an optimal weight assignment p_H for H such that the Kullback-Leibler divergence between (G, p_G) and (H, p_H) is minimized.

M. Teichmann—Financially supported by DFG Graduiertenkolleg 1763 (QuantLA).

Y.-S. Han and K. Salomaa (Eds.): CIAA 2016, LNCS 9705, pp. 273–284, 2016.
DOI: 10.1007/978-3-319-40946-7_23

2   Preliminaries

Mathematical Notions. The *set of natural numbers* $\{0, 1, \ldots\}$ is denoted by $\mathbb{N}$ and $\mathbb{N}_+ = \mathbb{N} \setminus \{0\}$. The set of finite sequences over $\mathbb{N}_+$ is denoted by $\mathbb{N}_+^*$ (including the empty sequence ε). For $n \in \mathbb{N}$, we let $[n] = \{1, \ldots, n\}$; hence $[0] = \emptyset$. An *alphabet* is a nonempty and finite set. The *set of words* over the alphabet Σ is denoted by Σ^* with ε being the *empty word*. Let I and A be sets and $f \colon I \to A$ a function. We call f an *I-indexed family* over A (for short: family), denoted by $f = (f_i \mid i \in I)$ with $f_i = f(i) \in A$. The powerset of A is denoted by $\mathcal{P}(A)$.

We fix an infinite list $x_1, x_2, \ldots$ of pairwise distinct *variables*. We write $X = \{x_1, x_2, x_3, \ldots\}$ and $X_k = \{x_1, \ldots, x_k\}$. We abbreviate $x_1, \ldots, x_k$ to $x_{1..k}$ and apply this abbreviation also to sequences of other objects.

In this paper, we use weights over $\mathbb{R}$ with usual sum and product. We denote the logarithm in base 2 by $\log(\cdot)$ and assume $0 \cdot \log \frac{0}{0} = 0 \cdot \log 0 = 0$.

Trees. A *ranked alphabet* is a pair $(\Delta, \mathrm{rk}_\Delta)$, where Δ is an alphabet and $\mathrm{rk}_\Delta \colon \Delta \to \mathbb{N}$ is a function. For every $\delta \in \Delta$, we call $\mathrm{rk}_\Delta(\delta)$ the *rank* of δ. We abbreviate the set $\mathrm{rk}_\Delta^{-1}(k)$ to $\Delta^{(k)}$ and $(\Delta, \mathrm{rk}_\Delta)$ to Δ assuming that rk_Δ is the rank function. In this paper, Δ denotes an arbitrary ranked alphabet. We assume that $\Delta \cap X = \emptyset$ and write α instead of $\alpha()$ for each nullary symbol $\alpha \in \Delta^{(0)}$.

Let U be a set. We denote the set of *trees over Δ and U* by $\mathrm{T}_\Delta(U)$ and write T_Δ for $\mathrm{T}_\Delta(\emptyset)$. Positions in trees are identified by finite sequences over $\mathbb{N}_+$. Formally, for each $\xi \in \mathrm{T}_\Delta(U)$, the *set of positions* of ξ, denoted by $\mathrm{pos}(\xi)$, is defined inductively as follows: (i) if $\xi \in \Delta^{(0)} \cup U$, then $\mathrm{pos}(\xi) = \{\varepsilon\}$, and (ii) if $\xi = \delta(\xi_1, \ldots, \xi_k)$ for some $\delta \in \Delta^{(k)}$, $k \geq 1$, and $\xi_1, \ldots, \xi_k \in \mathrm{T}_\Delta(U)$, then $\mathrm{pos}(\xi) = \{\varepsilon\} \cup \{iv \mid i \in [k], v \in \mathrm{pos}(\xi_i)\}$. For a position $w \in \mathrm{pos}(\xi)$, the *label of* ξ *at* w and the *subtree of* ξ *at* w are denoted by $\xi(w)$ and $\xi|_w$, respectively. For every $V \subseteq \Delta \cup U$, we denote the *set of positions of* ξ *labeled by an element of* V by $\mathrm{pos}_V(\xi)$; if V is a singleton $\{v\}$, then we simply write $\mathrm{pos}_v(\xi)$. We define the *yield of* ξ, denoted by $\mathrm{yield}(\xi)$, as the string over $\Delta^{(0)} \cup U$ of all leaf nodes read from left to right.

Let U be a finite set with $\Delta \cap U = \emptyset$. A *context over Δ and U* is a tree in $\mathrm{T}_\Delta(U)$ in which each element $u \in U$ occurs exactly once. The set of all such contexts is denoted by $\mathrm{C}_\Delta(U)$.

Let $k \in \mathbb{N}$, $\xi \in \mathrm{T}_\Delta(X)$, and $\xi_{1..k} \in \mathrm{T}_\Delta(X)$. The *tree concatenation of* ξ *with* $\xi_{1..k}$ *at* $x_{1..k}$, denoted by $\xi[\xi_{1..k}]$, is obtained from ξ by simultaneously replacing each x_i by ξ_i for each $i \in [k]$.

3   Weighted Context-Free Tree Languages and Weighted Regular Tree Languages

A *linear nondeleting context-free tree grammar*[1] (lnCFTG) is a tuple $G = (N, \Delta, A_0, R)$, where N and Δ are ranked alphabets (of *nonterminals* and *terminals*, respectively) such that $N \cap \Delta = \emptyset$, $A_0 \in N^{(0)}$ (*initial nonterminal*),

---

[1] Sometimes called simple context-free tree grammars in the literature.

and R is a finite set of *rules* of the form $A(x_{1..k}) \rightarrow \xi$ with $k \in \mathbb{N}$, $A \in N^{(k)}$, and $\xi \in C_{N \cup \Delta}(X_k)$. In a rule $r \in R$ of the form $A(x_{1..k}) \rightarrow \zeta$ the *left-hand side nonterminal of* r is A, denoted by $\text{lhn}(r)$. For each $A \in N$, we abbreviate $A(x_{1..\text{rk}_N(A)})$ by $A(\overline{x})$ and we define the *A-restriction of* R, denoted by $R|_A$, as $R|_A = \{r \in R \mid \text{lhn}(r) = A\}$.

In the following, let $G = (N, \Delta, A_0, R)$ be an arbitrary lnCFTG. The *derivation relation* $\Rightarrow$ is defined as follows. For trees $\xi, \xi' \in T_{N \cup \Delta}(X)$ and a rule r of the form $A(x_{1..k}) \rightarrow \zeta$ in R, we have $\xi \Rightarrow_r \xi'$ if there is a position $w \in \text{pos}_A(\xi)$ such that ξ' is obtained from ξ by replacing the subtree at position w by $\zeta[\xi|_{w1}, \ldots, \xi|_{wk}]$. Thus if ξ is a context, then so is ξ'. We write $\xi \Rightarrow \xi'$ if $\xi \Rightarrow_r \xi'$ for some $r \in R$, i.e., $\Rightarrow = \bigcup_{r \in R} \Rightarrow_r$. We denote the reflexive, transitive closure of $\Rightarrow$ by $\Rightarrow^*$. For $v \in R^*$, we inductively define $\Rightarrow_v$ as follows. If $v = \varepsilon$, then $\Rightarrow_\varepsilon$ is the identity. If $v = ru$ for some $r \in R$ and $u \in R^*$, then $\Rightarrow_v = \Rightarrow_r \circ \Rightarrow_u$.

Let $k \in \mathbb{N}$ and $A \in N^{(k)}$. A *derivation of* G *starting in* A is a sequence $\xi_0 \Rightarrow_{r_1} \xi_1 \Rightarrow_{r_2} \ldots \Rightarrow_{r_n} \xi_n$ where $n \in \mathbb{N}$, $\xi_0 = A(\overline{x})$, and for $i \in [n]$, we have $\xi_i \in C_{N \cup \Delta}(X_k)$ and $\xi_{i-1} \Rightarrow_{r_i} \xi_i$. We call a derivation *complete*, if $\xi_n \in C_\Delta(X_k)$, i.e., there are no nonterminals in ξ_n. We call a derivation *leftmost outermost* if always the leftmost and outermost (considering a tree as a term) nonterminal occurrence is derived.

For $k \in \mathbb{N}$ and $\zeta \in C_{N \cup \Delta}(X_k)$, the *tree language induced by* ζ *on* G is

$$\mathcal{L}(G, \zeta) = \{\xi \in C_\Delta(X_k) \mid \zeta \Rightarrow^* \xi\}.$$

The *tree language of* G, denoted by $\mathcal{L}(G)$, is defined as $\mathcal{L}(G) = \mathcal{L}(G, A_0)$.

Tree-Shaped Derivations. Let $G = (N, \Delta, R, A_0)$ be a lnCFTG. We consider R as a ranked alphabet. Let $r \in R$ be of the form $A(\overline{x}) \rightarrow \zeta$. Then $\text{rk}_R(r) = |\text{pos}_N(\zeta)|$. The *nonterminal word of* r, denoted by $\text{nt}(r)$, is the word of nonterminal occurrences in ζ ordered lexicographically by their position. Note that $\text{nt}(r) \in N^*$ and that, considering ζ as a term, the order of $\text{nt}(r)$ is leftmost outermost.

We simultaneously define an N-indexed family over $\mathcal{P}(T_R)$, denoted by $(\mathcal{D}_G(A) \mid A \in N)$ as follows. Let $A \in N$, $d \in T_R$, and $r = d(\varepsilon)$. Then $d \in \mathcal{D}_G(A)$ if (i) $\text{lhn}(r) = A$ and (ii) for each $i \in [\text{rk}_R(r)]$ we have $d|_i \in \mathcal{D}_G(A_i)$ where $A_i = \text{nt}(r)(i)$. We call each element in $\mathcal{D}_G(A)$ a *tree-shaped derivation starting in* A.

Observation 1. *Let* $A \in N$. *There is a one-to-one correspondence between* $\mathcal{D}_G(A)$ *and the set of complete leftmost outermost derivations starting in* A.

Let $d \in \mathcal{D}_G(A)$ and d' be the corresponding complete leftmost outermost derivation starting in A according to Observation 1. We write $A \Rightarrow_d \xi$ instead of $A \Rightarrow_{d'} \xi$.

We abbreviate $\mathcal{D}_G(A_0)$ to $\mathcal{D}_G$. Furthermore, for each $\xi \in T_\Delta$ and $r \in R$, we define the sets

- $\mathcal{D}_G(\xi)$ of *derivations ending in* ξ as $\mathcal{D}_G(\xi) = \{d \in \mathcal{D}_G \mid A_0 \Rightarrow_d \xi\}$ and
- $\mathcal{D}_G(r)$ of *derivations starting with* r as $\mathcal{D}_G(r) = \{d \in \mathcal{D}_G(\text{lhn}(r)) \mid r = d(\varepsilon)\}$.

We are also interested in derivations where one occurrence of a nonterminal is not derived. For $A, B \in N$, we define the *set of B-partial derivations starting in A* denoted by $\mathcal{D}_G^B(A)$ which are defined in the following. Let $d \in C_R(X_1)$. Then $d \in \mathcal{D}_G^B(A)$ if there is $d_B \in \mathcal{D}_G(B)$ such that $d[d_B] \in \mathcal{D}_G(A)$.

We say that the lnCFTG G is *unambiguous* if, for each $\xi \in T_\Delta$, we have that $|\mathcal{D}_G(\xi)| \in \{0,1\}$, i.e., if ξ is in $\mathcal{L}(G)$, then there is exactly one derivation for ξ.

A *regular tree grammar* (RTG) is a lnCFTG in which each nonterminal has rank 0. A tree language $L \subseteq T_\Delta$ is *regular over* Δ if there is a RTG $G = (N, \Delta, A_0, R)$ such that $\mathcal{L}(G) = L$. We may say that a tree language is *regular* if it is regular over Δ for some Δ.

Weighted lnCFTGs. A *weighted linear nondeleting context-free tree grammar over* $\mathbb{R}$ *(wlnCFTG)* is a tuple (G, p_G) where $G = (N, \Delta, A_0, R)$ is a lnCFTG and p_G is a mapping $p_G \colon R \to \mathbb{R}$, which we call *weight assignment for G*.

Let (G, p_G) be a wlnCFTG, $A, B \in N$, and $d \in \mathcal{D}_G(A) \cup \mathcal{D}_G^B(A)$. Then the *weight of d in* (G, p_G), denoted by $p_G(d)$, is defined as $p_G(d) = \prod_{w \in \mathrm{pos}_R(d)} p_G(d(w))$. For each $\xi \in T_\Delta$, we define the *weight of ξ in* (G, p_G), denoted by $p_G(\xi)$, as $p_G(\xi) = \sum_{d \in \mathcal{D}_G(\xi)} p_G(d)$.

We call p_G *proper* if for each $A \in N$ we have $\sum_{r \in R|_A} p_G(r) = 1$. Furthermore, p_G is called *consistent* if $\sum_{\xi \in T_\Delta} p_G(\xi) = 1$. We call (G, p_G) *proper* (*consistent*) if p_G is proper (consistent). The *support of* (G, p_G), denoted by $\mathrm{supp}(G, p_G)$, is defined as $\mathrm{supp}(G, p_G) = \{\xi \in T_\Delta \mid p_G(\xi) \neq 0\}$.

A *weighted regular tree grammar* (wRTG) is a wlnCFTG (H, p_H) where H is a RTG.

Normal Forms Let $G = (N, \Delta, A_0, R)$ be a lnCFTG. We say that G is in *nonterminal form* if every rule is of one of the following two forms:

– Type I: $A(x_{1..k}) \to \xi$ with $\xi \in C_N(X_k)$, or
– Type II: $A(x_{1..k}) \to \delta(x_{1..k})$ for some $\delta \in \Delta$.

A wlnCFTG (G, p_G) is in *nonterminal form* if G is. The following lemma is proven in [13, p. 113]. The weighted version follows from a simple extension.

Lemma 2. *For every lnCFTG G, there is a lnCFTG G' in nonterminal form such that $\mathcal{L}(G) = \mathcal{L}(G')$.*

For every wlnCFTG (G, p_G), there is a wlnCFTG $(G', p_{G'})$ in nonterminal form such that, for each $\xi \in T_\Delta$, we have $p_G(\xi) = p_{G'}(\xi)$. The construction preserves properness and consistency.

Let $H = (N_H, \Delta, B_0, R_H)$ be a RTG. We say that H is *producing*[2] if each rule is of the form $B \to \delta(B_{1..k})$ for some $\delta \in \Delta$ and $B, B_1, \ldots, B_k \in N_H$. A wRTG (H, p_H) is *producing* if H is.

Lemma 3. *[4, Chapter II, Lemma 3.4] For each RTG H, there is a producing RTG H' such that $\mathcal{L}(H) = \mathcal{L}(H')$.*

---

[2] In the existing literature about RTG, producing is called *normal form*. In the literature about CFTG, the nonterminal form is also called *normal form*. Thus, the term *normal form* is avoided in this paper.

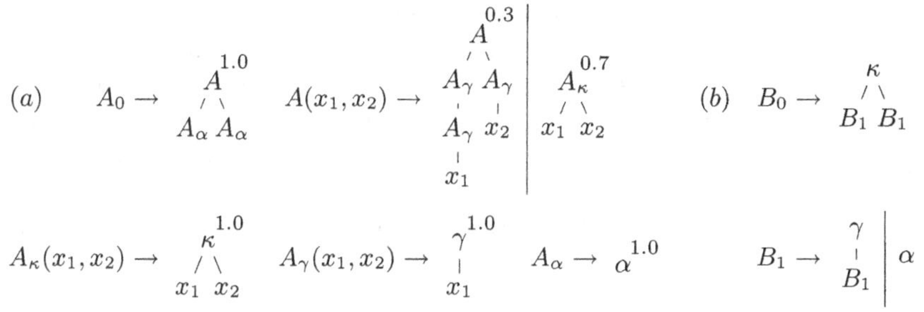

Fig. 1. (a) Example wlnCFTG (G, p_G).   (b) Example RTG H.

4 Approximation of a WlnCFTG by a wRTG

As main result, we show how to approximate a wlnCFTG (G, p_G) by a wRTG (H, p_H). We will assume that we are given the unambiguous and producing RTG H, and find the best possible proper weight assignment p_H according to the Kullback-Leibler divergence. Note that each RTG can be made unambiguous by the following steps. First, construct the associated nondeterministic finite tree automaton. Then determinize it using standard techniques [4, Chapter II, Theorem 2.6] and finally transform it back to an RTG.

The approximation is similar to the approach of [9,10] and we will need two intermediate concepts. First, we show how to intersect a (w)lnCFTG with a (w)RTG. Second, we define the notion of the expected frequency of a rule in a wlnCFTG. Then, the intersection of (G, p_G) with $(H, \mathbb{1})$ ($\mathbb{1}$ is a trivial weight assignment) is used to define expected frequencies for the rules in H. These frequencies are used to obtain a weight assignment p_H such that (H, p_H) best approximates (G, p_G).

As a running example, we consider the wlnCFTG (G, p_G) from Fig. 1(a) and the RTG from Fig. 1(b). Note that G is in nonterminal form and H is producing. By considering the rules, it can be seen that $\mathcal{L}(G) = \{\kappa(\gamma^{(2n)}(\alpha), \gamma^n(\alpha)) \mid n \in \mathbb{N}\}$ and $\mathcal{L}(H) = \{\kappa(\gamma^n(\alpha), \gamma^m(\alpha)) \mid n, m \in \mathbb{N}\}$ and thus, H is a superset approximation of G. Note that the rules of H cannot ensure that the number of γ's in the left subtree of κ is larger than in the right subtree.

4.1 Intersection of a lnCFTG and a RTG

The class of context-free tree languages is closed under intersection with regular tree languages [13, p. 114]. In the original proof, Theorem 7 of [14] is applied which introduces copies and deletion. It is possible to exchange Theorem 7 of [14] by a construction which preserves linearity and nondeletion (cf., e.g., [12, Lemma 3] for the more general case of linear and nondeleting one-state weighted pushdown-extended tree transducers). A complete construction for the intersection is described in [11, p. 60], which, given a (synchronous) lnCFTG G and RTG H, yields a (synchronous) lnCFTG K such that the language of K is the intersection of the languages of G and H.

Here, we show a slightly modified version of the construction in [11] and identify properties of the constructed grammar which we will use in the rest of the paper. The main idea is, as in both [11,12], to annotate nonterminals of G with nonterminals of H. In contrast to [11] and since we require G to be in nonterminal form and H to be producing, we can describe the intersection similar to the string case [9]. Each nonterminal in G is annotated with all possible combinations of nonterminals from H. Terminal symbols can only be generated if the guess was correct. In [11] this checking is partly done using additional RTGs.

Let $G = (N_G, \Delta, A_0, R_G)$ be a lnCFTG in nonterminal form and $H = (N_H, \Delta, B_0, R_H)$ be a producing RTG. We let

$$N' = \bigcup_{k \in \mathbb{N}} N_H \times N_G^{(k)} \times (N_H)^k$$

be a ranked alphabet where $\mathrm{rk}_{N'}((B, A, B_1 \ldots B_k)) = \mathrm{rk}_{N_G}(A) = k$.

For each right hand side ζ of a rule from G of Type I, we represent the choice of corresponding nonterminals by an assignment τ that assigns one state to each position in ζ. Formally, for every $\zeta \in T_{N_G}(X)$ and function $\tau \colon \mathrm{pos}(\zeta) \to N_H$, we define $\zeta_\tau \in T_{N'}(X)$ as follows. For each $w \in \mathrm{pos}(\zeta)$ we have

$$\zeta_\tau(w) = \begin{cases} \big(\tau(w), \zeta(w), \tau(w1) \ldots \tau(w\ell)\big) & \text{if } \zeta(w) \in N_G^{(\ell)}, \\ \zeta(w) & \text{if } \zeta(w) \in X. \end{cases}$$

We define the lnCFTG $K = (N', \Delta, A_0', R')$, where $A_0' = (B_0, A_0, \varepsilon)$ and R' is defined as follows. For each Type I rule $r_G \in R_G$ of the form $A(x_{1..k}) \to \zeta$, $B, B_1, \ldots, B_k \in N_H$, and $\tau \colon \mathrm{pos}(\zeta) \to N_H$ such that (i) $\tau(\varepsilon) = B$ and (ii) for each $i \in [k]$ and $w_i \in \mathrm{pos}_{x_i}(\zeta)$, it holds that $\tau(w_i) = B_i$, we let

$$r' \colon (B, A, B_1 \ldots B_k)(x_{1..k}) \to \zeta_\tau \tag{1}$$

be in R'. Note that r' is a rule of Type I. We call r_G the *corresponding rule* to r', denoted by $\mathrm{cor}(r') = r_G$.

For each Type II rule $r_G \in R_G$ of the form $A(x_{1..k}) \to \delta(x_{1..k})$ and each $r_H \in R_H$ of the form $B \to \delta(B_{1..k})$, we let

$$r' \colon (B, A, B_1 \ldots B_k)(x_{1..k}) \to \delta(x_{1..k}) \tag{2}$$

be in R'. Note that r' is a rule of Type II. We call r_G and r_H the *corresponding rules* to r', denoted by $\mathrm{cor}(r') = (r_G, r_H)$.

Note that the function $\mathrm{cor}(\cdot)$ formally consists of two functions: One mapping Type-I-rules of R' to R_G, the other mapping Type-II-rules of R' to $R_G \times R_H$. We do not distinguish between the two, since the choice is clear from the context.

Figure 2 depicts some of the useful rules of K which is obtained as the intersection of G and H. Note that there are also some useless rules from which no terminal symbol can be created, e.g., the rule $(B_0, A, B_0 B_1)(x_1, x_2) \to (B_0, A_\kappa, B_0 B_1)(x_1, x_2)$.

We proof $\mathcal{L}(K) = \mathcal{L}(G) \cap \mathcal{L}(H)$ by relating the derivations of K, G, and H in the following two lemmas (we omitted the proofs due to space limitations).

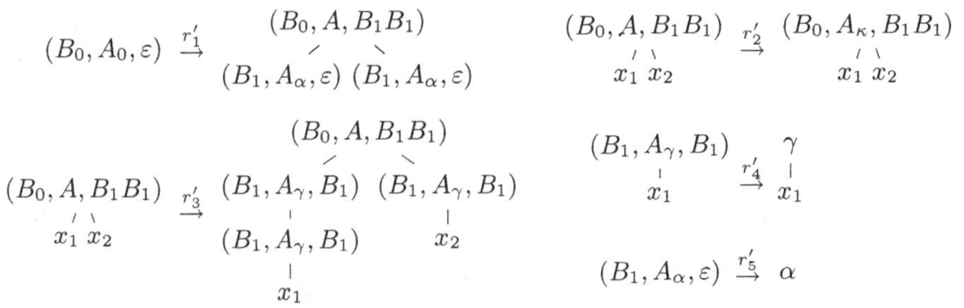

$$(B_0, A_0, \varepsilon) \xrightarrow{r_1'} \begin{array}{c} (B_0, A, B_1 B_1) \\ \diagup \quad \diagdown \\ (B_1, A_\alpha, \varepsilon) \ (B_1, A_\alpha, \varepsilon) \end{array} \qquad (B_0, A, B_1 B_1) \xrightarrow{r_2'} \begin{array}{c} (B_0, A_\kappa, B_1 B_1) \\ \diagup \ \diagdown \qquad\qquad \diagup \ \diagdown \\ x_1 \ x_2 \qquad\qquad x_1 \ x_2 \end{array}$$

$$(B_0, A, B_1 B_1) \xrightarrow{r_3'} \begin{array}{c} (B_0, A, B_1 B_1) \\ \diagup \qquad \diagdown \\ (B_1, A_\gamma, B_1) \ (B_1, A_\gamma, B_1) \end{array}$$

$$\begin{array}{c} (B_0, A, B_1 B_1) \\ \diagup \ \diagdown \\ x_1 \ x_2 \end{array} \xrightarrow{r_3'} \begin{array}{c} (B_1, A_\gamma, B_1) \ (B_1, A_\gamma, B_1) \\ | \qquad\qquad | \\ (B_1, A_\gamma, B_1) \qquad x_2 \\ | \\ x_1 \end{array}$$

$$\begin{array}{c} (B_1, A_\gamma, B_1) \qquad \gamma \\ | \qquad\quad \xrightarrow{r_4'} \quad | \\ x_1 \qquad\qquad x_1 \end{array}$$

$$(B_1, A_\alpha, \varepsilon) \xrightarrow{r_5'} \alpha$$

Fig. 2. Some rules of the lnCFTG K which is the intersection of G and H.

Lemma 4. *Let $\xi \in T_\Delta$ and $n = |\xi|$. For each $d \in \mathcal{D}_K(\xi)$ there are unique derivations $d_G \in \mathcal{D}_G(\xi)$ and $d_H \in \mathcal{D}_H(\xi)$ such that the following holds:*

- *$\mathrm{pos}(d) = \mathrm{pos}(d_G)$,*
- *for each $w \in \mathrm{pos}(d)$ where w is not a leaf, we have that $\mathrm{cor}(d(w)) = d_G(w)$,*
- *$n = |\xi| = |d_H| = |\mathrm{yield}(d)|$, and*
- *there is a one-to-one correspondence φ between the positions in $\mathrm{yield}(d)$ and positions in d_H such that for each $i \in [n]$, we have that*

$$\mathrm{cor}\big(\mathrm{yield}(d)(i)\big) = \big(\mathrm{yield}(d_G)(i), d_H(\varphi(i))\big).$$

(The formal proof of this lemma centers around the construction of K.)

Let $d \in \mathcal{D}_K$, $d_G \in \mathcal{D}_G$, and $d_H \in \mathcal{D}_H$ be the unique derivations according to Lemma 4. We denote this fact by $\mathrm{cor}(d) = (d_G, d_H)$.

Lemma 5. *Let $\xi \in T_\Delta$, $d_G \in \mathcal{D}_G(\xi)$, and $d_H \in \mathcal{D}_H(\xi)$. There is a unique $d \in \mathcal{D}_K(\xi)$ such that $\mathrm{cor}(d) = (d_G, d_H)$. (The proof relies on the fact that the labels in d_G and d_H uniquely determine the labels in d).*

Lemma 6. *For each lnCFTG G and RTG H, there is a lnCFTG K in nonterminal form such that $\mathcal{L}(K) = \mathcal{L}(G) \cap \mathcal{L}(H)$.*

Proof. By Lemmas 2 and 3 we can assume that G is in nonterminal form and H is producing. Then, we apply the construction of this section. $\mathcal{L}(K) \subseteq \mathcal{L}(G) \cap \mathcal{L}(H)$ follows from Lemma 4. $\mathcal{L}(K) \supseteq \mathcal{L}(G) \cap \mathcal{L}(H)$ is a consequence of Lemma 5. Considering (1) and (2), it can be seen that K is in nonterminal form. $\qquad \square$

Weighted Intersection. We extend the result to the weighted case. Let (G, p_G) be a wlnCFTG in nonterminal form and (H, p_H) be a producing wRTG. We let K be the lnCFTG obtained by intersecting G and H according to Lemma 6. We extend K to a wlnCFTG (K, p_K) as follows. For each $r \in R'$, we define

$$p_K(r) = \begin{cases} p_G(\mathrm{cor}(r)) & \text{if } r \text{ is of Type I,} \\ p_G(r_G) \cdot p_H(r_H) & \text{if } r \text{ is of Type II and } \mathrm{cor}(r) = (r_G, r_H). \end{cases}$$

Lemma 7. *For each wlnCFTG (G, p_G) and producing wRTG (H, p_H), there is a wlnCFTG (K, p_K) in nonterminal form such that $p_K(\xi) = p_G(\xi) \cdot p_H(\xi)$ for each $\xi \in T_\Delta$.*

Proof. By Lemma 2, we can assume that (G, p_G) is in nonterminal form. As described in this section, we let (K, p_K) be the wlnCFTG obtained by intersecting (G, p_G) and (H, p_H). By Lemma 6, we have $\mathcal{L}(K) = \mathcal{L}(G) \cap \mathcal{L}(H)$ and K is in nonterminal form. For each $\xi \in T_\Delta \setminus \mathcal{L}(K)$, it holds that $p_G(\xi) = 0$. Hence, it remains to prove that, for each $\xi \in \mathcal{L}(K)$, we have $p_K(\xi) = p_G(\xi) \cdot p_H(\xi)$.

From Lemmas 4 and 5, we get that there is a one-to-one connection between each $d \in \mathcal{D}_K(\xi)$ and (d_G, d_H) where $d_G \in \mathcal{D}_G(\xi)$ and $d_H \in \mathcal{D}_H(\xi)$ such that $\text{cor}(d) = (d_G, d_H)$. From the proof of Lemma 5, we get the connection between the rule occurrences in the three derivations. Since we have commutativity, the order of multiplication does not matter. Hence, the lemma holds. $\square$

Corollary 8. *For each $\xi \in T_\Delta$ and each $d \in \mathcal{D}_K(\xi)$ such that $\text{cor}(d) = (d_G, d_H)$ we have $p_K(d) = p_G(d_G) \cdot p_H(d_H)$.*

4.2   Expected Frequencies

Let (G, p_G) be a consistent wlnCFTG with $G = (N, \Delta, A_0, R)$. For each $r \in R$ where $\text{lhn}(r) = A$, we define the *expected rule frequency of* r as

$$E(r) = \sum_{\substack{d_1 \in \mathcal{D}_G^A \\ d_2 \in \mathcal{D}_G(r)}} p_H(d_1[d_2]). \tag{3}$$

For each $r' \in R$ and $d \in \mathcal{D}_G$, we let $\sharp_{r'}(d)$ denote the number of occurrences of r' in d, or formally, $\sharp_{r'}(d) = |\text{pos}_{r'}(d)|$. Then, we have

$$E(r) = \sum_{d \in \mathcal{D}_G} p_G(d) \cdot \sharp_r(d). \tag{4}$$

Another way of representing the expected rule frequency is using inner and outer values. Assume that $\text{nt}(r) = A_1 \ldots A_k$. Then it holds that

$$E(r) = \text{outer}(A) \cdot p_G(r) \cdot \prod_{i \in [k]} \text{inner}(A_i), \tag{5}$$

where, for each $A' \in N$, we define

$$\text{outer}(A') = \sum_{d \in \mathcal{D}_G^{A'}} p_G(d) \quad \text{and} \quad \text{inner}(A') = \sum_{d \in \mathcal{D}_G(A')} p_G(d).$$

We define, for every $A' \in N$, the value $d(A' = A_0)$ to be 1 if $A' = A_0$ and 0 otherwise. Then, we rephrase inner and outer values as

$$\text{outer}(A') = d(A' = A_0) + \sum_{\substack{r' \in R^{(\ell)}, \, j \in [\text{rk}_R(r')] \\ \text{nt}(r') = A_1' \ldots A_\ell', \, A' = A_j'}} \text{outer}(\text{lhn}(r')) \cdot p_G(r')$$

$$\cdot \prod_{i \in ([\ell] \setminus \{j\})} \text{inner}(A_i') \quad \text{and} \tag{6}$$

$$\text{inner}(A') = \sum_{\substack{r' \in R|_{A'}^{(\ell)} \\ \text{nt}(r') = A_1' \ldots A_\ell'}} p_G(r') \cdot \prod_{i \in [\ell]} \text{inner}(A_i').$$

In the string case [9], the initial nonterminal may not occur in any right-hand side and, by definition, $\text{outer}(A_0) = 1$. We do not impose this restriction in the tree case and allow for A_0 to occur in right-hand sides. To account for A_0 as the initial nonterminal, we add the value $d(A' = A_0)$. Hence, if A_0 does not occur in any right-hand side, then $\text{outer}(A_0) = 1$. Otherwise, $\text{outer}(A_0) \geq 1$.

As in the string case (cf. [9, p. 5]), the values for $\text{inner}(A)$ and $\text{outer}(A)$ can be approximated by fixed-point iteration.

4.3   Approximation

In [1] it is shown how to train a weighted CFG based on an infinite set of derivation trees or an infinite set of strings. We extend the result to the realm of trees similar to [10] as follows. Given an unambiguous and producing RTG H, we will approximate a wlnCFTG (G, p_G) by means of the wRTG (H, p_H). We construct the intersection wlnCFTG (K, p_K) of (G, p_G) and $(H, \mathbb{1})$. Then, the training data is the set of derivations from K from which we will calculate expected frequencies for the rules in K and use them to obtain an optimal p_H.

In the following, we let (G, p_G) be a consistent wlnCFTG where $G = (N_G, \Delta, A_0, R_G)$. Furthermore, we let $H = (N_H, \Delta, B_0, R_H)$ be an unambiguous and producing RTG such that $\text{supp}(G, p_G) \cap \mathcal{L}(H) \neq \emptyset$. As described in Sect. 4.1, we choose (K, p_K) to be the wlnCFTG obtained by intersecting (G, p_G) and $(H, \mathbb{1})$, where $\mathbb{1}(r_H) = 1$ for every $r_H \in R_H$. We denote the ingredients of K as $K = (N_K, \Delta, A_0', R')$ and according to Lemma 7, we have, for each $\xi \in \mathcal{L}(G) \cap \mathcal{L}(H)$, that $p_K(\xi) = p_G(\xi) \cdot \mathbb{1}(\xi)$, i.e., $p_K(\xi) = p_G(\xi)$. Since not all trees of $\mathcal{L}(G)$ must occur in $\mathcal{L}(H)$, we normalize p_G such that only trees in $\mathcal{L}(H)$ are assigned a non-null weight. We define $p_G|_H$ for each $\xi \in T_\Delta$ as

$$p_G|_H(\xi) = \begin{cases} \dfrac{p_G(\xi)}{\sum_{\xi' \in \mathcal{L}(H)} p_G(\xi')} & \text{if } \xi \in \mathcal{L}(H), \\ 0 & \text{otherwise.} \end{cases}$$

We note that $p_G|_H = p_G$ if $\text{supp}(G, p_G) \subseteq \mathcal{L}(H)$.

Lemma 9. *For each $\xi \in T_\Delta$ we have* $p_G|_H(\xi) = \dfrac{p_K(\xi)}{\sum_{\xi' \in T_\Delta} p_K(\xi')}$.

Proof. We observe that $\sum_{\xi' \in T_\Delta} p_K(\xi') > 0$ since in this section we required that $\text{supp}(G, p_G) \cap \mathcal{L}(H) \neq \emptyset$ holds. We show the following for each $\xi \in \mathcal{L}(H)$.

$$\begin{aligned} p_G(\xi) &= \sum_{d_G \in \mathcal{D}_G(\xi)} p_G(d_G) \\ &= \sum_{d_G \in \mathcal{D}_G(\xi), d_H \in \mathcal{D}_H(\xi)} p_G(d_G) \cdot \mathbb{1}(d_H) && (H \text{ is unambiguous}) \\ &= \sum_{d \in \mathcal{D}_K(\xi)} p_K(d) && (\text{Lemma 5, Corollary 8}) \\ &= p_K(\xi) \end{aligned}$$

Let $\xi \in T_\Delta$. If $\xi \notin \mathcal{L}(H)$, then we have $p_{G|H}(\xi) = 0$ and, since $p_H(\xi) = 0$, we also have $\frac{p_K(\xi)}{\sum_{\xi' \in T_\Delta} p_K(\xi')} = 0$. If $\xi \in T_{\mathcal{L}(H)}$, then we have

$$p_{G|H}(\xi) = \frac{p_G(\xi)}{\sum_{\xi' \in T_{\mathcal{L}(H)}} p_G(\xi')} = \frac{p_K(\xi)}{\sum_{\xi' \in T_{\mathcal{L}(H)}} p_K(\xi')} = \frac{p_K(\xi)}{\sum_{\xi' \in T_\Delta} p_K(\xi')}.$$

Note that since (G, p_G) is consistent, Lemma 9 implies that $p_{G|H}$ is consistent.

□

In the following we obtain p_H such that (H, p_H) best approximates (G, p_G). We measure the quality of such an approximation using the *Kullback-Leibler (KL) divergence*. Note that the notion of cross-entropy of two distributions is closely connected to the KL divergence. In general, the KL divergence of two distributions p_G and p_H, denoted by $KL(p_G \parallel p_H)$, is given by (cf. [7, Eq. 2.4])

$$KL(p_G \parallel p_H) = \sum_{\xi \in T_\Delta} p_G(\xi) \cdot \log \frac{p_G(\xi)}{p_H(\xi)}.$$

Formally, we let P_H be the set of all proper weight assignments for H, i.e., $P_H = \{p'_H \mid p'_H \colon R_H \to \mathbb{R}, \ p'_H \text{ is proper}\}$ and determine p_H such that we have $p_H = \operatorname{argmin}_{p'_H \in P_H} KL(p_{G|H} \parallel p'_H)$. We abbreviate $\sum_{\xi' \in T_\Delta} p_K(\xi')$ to Z.

$$\operatorname{argmin}_{p'_H \in P_H} KL(p_{G|H} \parallel p'_H) = \operatorname{argmin}_{p'_H \in P_H} \sum_{\xi \in T_\Delta} p_{G|H}(\xi) \cdot \log \frac{p_{G|H}(\xi)}{p'_H(\xi)}$$

$$= \operatorname{argmin}_{p'_H \in P_H} \sum_{\xi \in T_\Delta} \frac{p_K(\xi)}{Z} \cdot \log \frac{p_K(\xi)}{p'_H(\xi) \cdot Z} \qquad \text{(Lemma 9)}$$

$$= \operatorname{argmin}_{p'_H \in P_H} \left(\sum_{\xi \in T_\Delta} \frac{p_K(\xi)}{Z} \cdot \log \frac{p_K(\xi)}{Z} \right)$$
$$- \left(\frac{1}{Z} \cdot \sum_{\xi \in T_\Delta} p_K(\xi) \cdot \log p'_H(\xi) \right)$$

$$= \operatorname{argmin}_{p'_H \in P_H} - \sum_{\xi \in T_\Delta} p_K(\xi) \cdot \log p'_H(\xi) \qquad (7)$$

Equation (7) holds, since Z and p_K are independent of p'_H.

For each $r_H \in R_H$, we define the *subset of R' corresponding to r_H* as

$$R'|_{r_H} = \{r' \in R' \mid \operatorname{cor}(r') = (r_G, r_H) \text{ for some } r_G \in R_G\}.$$

In [1] it is shown how to use a weighted CFG to approximate an infinite corpus of trees. The corpus is regarded as set of derivations of the weighted CFG and is given as a probability distribution p_T over the derivations. Using the technique of Lagrange Multipliers, it is shown how to choose a weight assignment $p_{\min}$ such that the cross-entropy between p_T and $p_{\min}$ is minimized. The cross-entropy corresponds to (7). Informally, $p_{\min}$ is defined for every rule r of the CFG as (cf. Eq. (9) of [1])

$$p_{\min}(r) = \frac{\sum_{\text{derivation tree } d} p_T(d) \cdot \#_r(d)}{\sum_{\text{derivation tree } d} \sum_{r' \text{ with lhn}(r') = \text{lhn}(r)} p_T(d) \cdot \#_{r'}(d)}. \qquad (8)$$

This result cannot be straightforwardly applied to our scenario for two reasons. First, we are not given a distribution over derivations of H and second, in contrast to (8), our setting would require an unsupervised training (similar to [1, Sect. 7.2]). However, using the intersection grammar (K, p_K), we can infer derivations for H. This approach corresponds to the string case [9] and the case of TAGs [10]. For the trick to work, we consider for each rule $r_H \in R_H$ the set of corresponding rules $R'|_{r_H}$. This allows the reconstruction of the Lagrange multipliers in [1, Sect. 3] for our scenario as follows.

Lemma 10. *We define p_H, for each $r_H \in R_H$, as*

$$p_H(r_H) = \frac{\sum_{r' \in R'|_{r_H}} \sum_{d \in \mathcal{D}_K} p_K(d) \cdot \sharp_{r'}(d)}{\sum_{r \in R_H|_{\mathrm{lhn}(r_H)}} \sum_{r' \in R'|_r} \sum_{d \in \mathcal{D}_K} p_K(d) \cdot \sharp_{r'}(d)}. \tag{9}$$

Then it holds that $p_H = \mathrm{argmin}_{p'_H \in P_H} \mathrm{KL}(p_{G|H} \| p'_H)$.

Using expected frequencies from Sect. 4.2, we rewrite (9) and get

$$p_H(r_H) = \frac{\sum_{r' \in R'|_{r_H}} E(r')}{\sum_{r \in R_H|_{\mathrm{lhn}(r_H)}} \sum_{r' \in R'|_r} E(r')}. \tag{10}$$

Since R_H and R' are finite sets, we can calculate p_H based on $E(r')$ for each $r_H \in R_H$ and $r' \in R'|_{r_H}$. Since r' is of Type II, we have $E(r') = \mathrm{outer}(\mathrm{lhn}(r')) \cdot p_K(r')$. Furthermore, because inner and outer values can be approximated to an arbitrary precision (cf. Sect. 4.2), we can effectively obtain an approximation of p_H.

Theorem 11. *For each consistent wlnCFTG (G, p_G) and each unambiguous and producing RTG H such that $\mathcal{L}(G) \cap \mathcal{L}(H) \neq \emptyset$, the weight assignment p_H from (10) is such that $p_H = \mathrm{argmin}_{p'_H \in P_H} \mathrm{KL}(p_{G|H} \| p'_H)$ holds.*

Proof. By Lemma 7, we obtain (K, p_K) as intersection of (G, p_G) and $(H, \mathbb{1})$. Then Lemma 10 applies and defines the minimal p_H.   □

Consider some rules of K depicted in Fig. 2. It is easy to see that r'_1 and r'_2 occur exactly once in each derivation and thus $E(r'_1) = E(r'_2) = 1$. The rule r'_3 occurs n times in a derivation of $\kappa(\gamma^{(2n)}(\alpha), \gamma^n(\alpha))$ and we get $E(r'_3) = \sum_{n \in \mathbb{N}} 0.3^n \cdot 0.7 \cdot n = \frac{3}{7}$. Since r'_4 occurs three times as often as r'_3, we get $E(r'_4) = 3 \cdot E(r'_3) = \frac{9}{7}$. The rule r'_5 occurs exactly twice, so $E(r'_5) = 2$.

We denote the rules of H from Fig. 1(b) by r_1, r_2, and r_3. We note that r_2 and r_3 correspond to r'_3 and r'_5, respectively. Hence, we obtain an optimal p_H as

$$p_H(r_1) = 1, \quad p_H(r_2) = \frac{\frac{9}{7}}{2 + \frac{9}{7}} = \frac{9}{23} \approx 0.39, \quad p_H(r_3) = \frac{2}{2 + \frac{9}{7}} = \frac{14}{23} \approx 0.61.$$

Although p_H is the optimal weight assignment for H, there is a considerable approximation error since probability mass is lost on trees of shape $\kappa(\gamma^{n_1}(\alpha), \gamma^{n_2}(\alpha))$ where $n_1 \neq 2 \cdot n_2$. Improved approximation results can be obtained by considering other RTGs with a nonterminal structure that better approximates the structure of trees in $\mathcal{L}(G)$. Such RTGs usually have an increased parsing time, since they contain more nonterminals and rules.

5   Further Research

This work motivates to extend the approximation result to the full class of CFTG. A grammar in this class might be nonlinear and deleting, and the derivation mode (cf. [2]) influences the induced language.

References

1. Corazza, A., Satta, G.: Probabilistic context-free grammars estimated from infinite distributions. IEEE Trans. Pattern anal. Mach. Intell. **29**(8), 1379–1393 (2007)
2. Fischer, M.: Grammars with macro-like productions. Ph.D. thesis. Harvard University, Massachusetts (1968)
3. Gebhardt, K., Osterholzer, J.: A direct link between tree-adjoining and context-free tree grammars. In: Proceedings of the 12th International Conference on Finite-State Methods and Natural Language Processing (2015)
4. Gécseg, F., Steinby, M.: Tree automata. In: Kiadó, A. (ed.). Reissued 2015. Akadémiai Kiadó, Budapest (1984). arXiv:1509.06233 [cs.FL]
5. Joshi, A., Schabes, Y.: Tree-adjoining grammars. In: Rozenberg, G., Salomaa, A. (eds.) Handbook of Formal Languages, vol. 3, pp. 69–123. Springer, Heidelberg (1997)
6. Kepser, S., Rogers, J.: The equivalence of tree adjoining grammars and monadic linear context-free tree grammars. J. Logic Lang. Inf. **20**(3), 361–384 (2011)
7. Kullback, S., Leibler, R.A.: On information and sufficiency. Ann. Math. Stat. **22**(1), 79–86 (1951)
8. Nederhof, M.-J.: Regular approximation of CFLs: a grammatical view. In: Bunt, H., Nijholt, A. (eds.) Advances in Probabilistic and Other Parsing Technologies. Text, Speech and Language Technology, vol. 16, pp. 221–241. Springer, Netherlands (2000)
9. Nederhof, M.-J.: A general technique to train language models on language models. Comput. Linguist. **31**(2), 173–186 (2005)
10. Nederhof, M.-J.: Weighted parsing of trees. In: Proceedings of the 11th International Conference on Parsing Technologies, IWPT 2009, pp. 13–24. Association for Computational Linguistics (2009)
11. Nederhof, M.-J., Vogler, H.: Synchronous context-free tree grammars. In: Proceedings of the 11th International Workshop on Tree Adjoining Grammar and Related Formalisms (TAG+11), pp. 55–63 (2012)
12. Osterholzer, J.: Pushdown machines for weighted context-free tree translation. In: Holzer, M., Kutrib, M. (eds.) CIAA 2014. LNCS, vol. 8587, pp. 290–303. Springer, Heidelberg (2014)
13. Rounds, W.C.: Tree-oriented proofs of some theorems on context-free and indexed languages. In: Proceedings of Second Annual ACM Symposium on Theory of Computing, STOC 1970, pp. 109–116. ACM, New York (1970)
14. Rounds, W.C.: Mappings and grammars on trees. Math. Syst. Theor. **4**(3), 257–287 (1970)
15. Shieber, S.: Evidence against the context-freeness of natural language. Linguist. Philos. **8**, 333–343 (1985)

Derivatives for Enhanced Regular Expressions

Peter Thiemann[(✉)]

University of Freiburg, Freiburg im Breisgau, Germany
thiemann@informatik.uni-freiburg.de

Abstract. Regular languages are closed under a wealth of formal language operators. Incorporating such operators in regular expressions leads to concise language specifications, but the transformation of such enhanced regular expressions to finite automata becomes more involved. We present an approach that enables the direct construction of finite automata from regular expressions enhanced with further operators that preserve regularity. Our construction is based on an extension of the theory of derivatives for regular expressions. To retain the standard results about derivatives, we develop a derivability criterion for the compatibility of the extra operators with derivatives.

Some derivable operators do not preserve regularity. Derivatives provide a decision procedure for the word problem of regular expressions enhanced with such operators.

Keywords: Automata and logic · Regular languages · Derivatives

1 Introduction

Brzozowski derivatives [4] and Antimirov's partial derivatives [2] are well-known tools to transform regular expressions to automata and to define algorithms for equivalence and containment on them [1,9]. Brzozowski's automaton construction relies on the finiteness of the set of iterated derivatives when considered up to similarity (commutativity, associativity, and idempotence for union). Derivatives had quite some impact on the study of algorithms for regular languages on finite words and trees [5,12].

While derivative-based algorithms have been deprecated for performance reasons [16], there has been renewed interest in the study of derivatives and partial derivatives. On the practical side, Owens and coworkers [11] report a functional implementation that revives many features. Might and coworkers [10] implement parsing for context-free languages using derivatives.

A common theme on the theory side is the study of derivative structures for enhancements of regular expressions. While Brzozowski's original work covered extended regular expressions, partial derivatives were originally limited to simple expressions without intersection and complement. It is a significant effort to define partial derivatives for extended regular expressions [5].

Derivatives have also been used to study various shuffle operators for applications in modeling concurrent programs [13]. Later extensions consider forkable expressions with a new operator that abstracts process creation [14].

© Springer International Publishing Switzerland 2016
Y.-S. Han and K. Salomaa (Eds.): CIAA 2016, LNCS 9705, pp. 285–297, 2016.
DOI: 10.1007/978-3-319-40946-7_24

Caron and coworkers [6] study derivatives for multi-tilde-bar expressions. The tilde (bar) operator adds (removes) ε from a language. Multi-tilde-bar applies to a list of languages and (roughly) defines a selective concatenation operation that can be configured to include or exclude certain languages of the list.

Champarnaud and coworkers [8] consider derivatives of approximate regular expressions (ARE). AREs extend regular expressions with a family of unary operators $\mathbb{F}_k$, for $k \in \mathbb{N}$, which enhance their argument language L with all words u such that $d(u, w) \leq k$, for some word $w \in L$. Here, d is a suitable distance function, for example, Hamming distance or Levenshtein distance.

Traytel and Nipkow [15] obtain decision procedures for MSO using a suitably defined derivative operation on regular expressions with a projection operation.

The general framework of Caron and coworkers [7] generalizes the syntactic structure underlying a derivative construction to a *support*. A support generalizes expressions (for constructing Brzozowski derivatives), sets of expressions (for Antimirov's partial derivatives), and sets of clausal forms over sets of regular expressions, and thus yields an encompassing framework in which different kinds of derivative constructions can be formalized and compared. The authors give a sufficient criterion for a support to generate a finite number of iterated derivatives from a given expression along with automata constructions for deterministic, nondeterministic, and alternating finite automata. Their work applies to extended regular expressions with arbitrary boolean functions.

Contributions

In this work, we identify a pattern in the definition of (standard) derivatives for enhancements of regular expressions that go beyond boolean functions. Concretely, we consider regular expressions enhanced with further operators on languages (e.g., shuffle, homomorphism, approximation). Then we propose *left derivability* and *ε-testability* as a sufficient condition for the set of operators such that a syntactic derivative operation is definable for enhanced expressions. This condition gives rise to a decision procedure for the membership test for enhanced expressions via expression derivation.

A refinement, *linear left derivability*, is a sufficient condition to guarantee finiteness of the set of dissimilar derivatives of an enhanced expression. The finiteness condition enables the direct construction of a deterministic finite automaton. We show that every linear left derivable operator can be defined by a rational finite state transducer and thus preserves regularity.

A technical report with proofs and further examples will be available on arxiv https://arxiv.org/abs/1605.00817.

2   Preliminaries

We write $\mathbb{N}$ for the set of natural numbers, $\mathbb{B} = \{0, 1\}$ for the set of booleans, and $X \uplus Y$ for the disjoint union of sets X and Y. We sometimes write $(\overline{E_k}^{k=1,\ldots,n})$ for the tuple $(E_1, \ldots, E_n)$ where E_k is some entity depending on k.

An alphabet Σ is a finite set of symbols. The set Σ^* denotes the set of finite words over Σ, $\varepsilon \in \Sigma^*$ stands for the empty word, and $\Sigma^+ = \Sigma^* \setminus \{\varepsilon\}$. For $u, v, w \in \Sigma^*$, we write $|u| \in \mathbb{N}$ for the length of u, $u \cdot v$ (or just uv) for the concatenation of words, and $w \succ v$ if v is a proper suffix of w, that is, $\exists u \in \Sigma^+$ such that $w = u \cdot v$.

Given languages $U, V, W \subseteq \Sigma^*$, concatenation extends to languages as usual: $U \cdot V = \{u \cdot v \mid u \in U, v \in V\}$. The Kleene closure is defined as the smallest set $U^* \subseteq \Sigma^*$ such that $U^* = \{\varepsilon\} \cup U \cdot U^*$. We write the *left quotient* as $U \backslash W = \{v \mid v \in \Sigma^*, \exists u \in U : uv \in W\}$ and the *right quotient* as $W/U = \{v \mid v \in \Sigma^*, \exists u \in U : vu \in W\}$. For a singleton language $U = \{u\}$, we write $u \backslash W$ (W/u) for the left (right) quotient.

A *ranked alphabet* $\mathcal{F}$ is a finite set of *operator symbols* with a function $\# : \mathcal{F} \to \mathbb{N}$ that determines the *arity* of each symbol. We write $\mathcal{F}^{(n)} = \{F \in \mathcal{F} \mid \#(F) = n\}$ for the symbols of arity n. The set $T_{\mathcal{F}}(X)$ of $\mathcal{F}$-*terms over a set* X is defined inductively. If $x \in X$, then $x \in T_{\mathcal{F}}(X)$. If $n \in \mathbb{N}$, $F \in \mathcal{F}^{(n)}$, and $t_1, \ldots, t_n \in T_{\mathcal{F}}(X)$, then $F(t_1, \ldots, t_n) \in T_{\mathcal{F}}(X)$.

An $\mathcal{F}$-*algebra* consists of a carrier set M and an interpretation function $\mathcal{I} : (n : \mathbb{N}) \to \mathcal{F}^{(n)} \to M^n \to M$. Given a function $\mathcal{I}_0 : X \to M$, the *term interpretation* $\hat{\mathcal{I}}(t)$, for $t \in T_{\mathcal{F}}(X)$, is defined inductively as follows. If $x \in X$, then $\hat{\mathcal{I}}(x) = \mathcal{I}_0(x)$. If $F \in \mathcal{F}^{(n)}$ and $t_1, \ldots, t_n \in T_{\mathcal{F}}(X)$, then $\hat{\mathcal{I}}(F(t_1, \ldots, t_n)) = \mathcal{I}(n)(F)(\hat{\mathcal{I}}(t_1), \ldots, \hat{\mathcal{I}}(t_n))$. We often write $T_{\mathcal{F}}$ in place of $T_{\mathcal{F}}(\emptyset)$.

To avoid notational clutter, we fix an arbitrary alphabet Σ.

Definition 1. *The* regular alphabet *is defined by* $\mathcal{R} = \Sigma \uplus \{\mathbf{0}, \mathbf{1}, \cdot, +, *\}$ *with arities* $\#(x) = 0$, *for* $x \in \Sigma$, $\#(\mathbf{1}) = \#(\mathbf{0}) = 0$, $\#(*) = 1$, *and* $\#(\cdot) = \#(+) = 2$.

Similarity is defined as the smallest equivalence relation $\equiv \subseteq T_{\mathcal{R}} \times T_{\mathcal{R}}$ *that enforces left and right unit, idempotence, commutativity, and associativity for the* $+$ *operator. For all* $r, s, t \in T_{\mathcal{R}}$, *the relation* $\equiv$ *contains the pairs:*

$$r + \mathbf{0} \equiv r \quad \mathbf{0} + s \equiv s \quad r + r \equiv r \quad r + s \equiv s + r \quad (r + s) + t \equiv r + (s + t)$$

The set R of regular expressions over Σ *is defined as the quotient term algebra* $R = T_{\mathcal{R}}/(\equiv)$.

The language of $r \in R$ *is defined by* $\mathcal{L}(r) = \hat{\mathcal{I}}(r)$, *that is, the interpretation of the term in the* $\mathcal{R}$-*algebra with carrier set* $\wp(\Sigma^*)$ *and interpretation function*

$$
\begin{aligned}
\mathcal{I}(0)(\mathbf{0}) &= \{\} & \mathcal{I}(1)(*) &= U \mapsto U^* \\
\mathcal{I}(0)(\mathbf{1}) &= \{\varepsilon\} & \mathcal{I}(2)(\cdot) &= (U, V) \mapsto U \cdot V \\
\mathcal{I}(0)(x) &= \{x\} & \mathcal{I}(2)(+) &= (U, V) \mapsto U \cup V
\end{aligned}
$$

The interpretation function $\mathcal{I}$ is compatible with the definition of R as a quotient term algebra because the interpretation of $+$ maps equivalent expressions to the same language. We usually work with a unique representative for each equivalence class computed by a function nf (see [9]). We use parenthesized infix notation for the binary operators $\cdot$ and $+$ and postfix superscript for the unary $*$. We adopt the convention that $\cdot$ binds stronger than $+$ to omit parentheses. The overloading of $\mathbf{0}$ and $\mathbf{1}$ as regular expressions and boolean values is deliberate.

Definition 2. *The operations* $\odot, \oplus : R \times R \to R$ *are* smart concatenation *and* union *constructors for regular expressions. Operator* $\odot$ *binds stronger than* $\oplus$.

$$r \odot s = \begin{cases} 0 & r = 0 \vee s = 0 \\ r & s = 1 \\ s & r = 1 \\ (r \cdot s) & \text{otherwise.} \end{cases} \qquad r \oplus s = \mathsf{nf}(r + s)$$

Lemma 3. *For all* r, s: $\mathcal{L}(r \odot s) = \mathcal{L}(r \cdot s)$; $\mathcal{L}(r \oplus s) = \mathcal{L}(r + s)$.

Definition 4. *A regular expression* r *is* nullable *if* $\varepsilon \in \mathcal{L}(r)$. *The function* $N :$ $R \to \{0, 1\}$ *detects nullable expressions:* $N(1) = 1$. $N(0) = 0$. $N(x) = 1$. $N(r \cdot s) = N(r) \odot N(s)$. $N(r + s) = N(r) \oplus N(s)$. $N(r^*) = 1$.

Lemma 5. *For all* $r \in R$. $N(r) = 1$ *iff* $\varepsilon \in \mathcal{L}(r)$.

Definition 6. *The* Brzozowski derivative *[4] is a function* $D : \Sigma \times T_{\mathcal{R}} \to T_{\mathcal{R}}$ *defined inductively for* $a \neq b \in \Sigma$ *and* $r, s \in T_{\mathcal{R}}$.

$$\begin{aligned} D(a, 0) &= 0 & D(a, r + s) &= D(a, r) \oplus D(a, s) \\ D(a, 1) &= 0 & D(a, r \cdot s) &= D(a, r) \odot s \oplus N(r) \odot D(a, s) \\ D(a, a) &= 1 & D(a, r^*) &= D(a, r) \odot r^* \\ D(a, b) &= 0 \end{aligned}$$

It extends to a function on words and languages $D : \Sigma^* \times T_{\mathcal{R}} \to T_{\mathcal{R}}$ *and* $D :$ $\wp(\Sigma^*) \times T_{\mathcal{R}} \to \wp(T_{\mathcal{R}})$ *as usual* ($a \in \Sigma$, $w \in \Sigma^*$, $U \subseteq \Sigma^*$):

$$D(\varepsilon, r) = r \quad D(a \cdot w, r) = D(w, D(a, r)) \quad D(U, r) = \{D(w, r) \mid w \in U\}.$$

Theorem 7 ([4]). *For all* $w \in \Sigma^*$, $r \in T_{\mathcal{R}}$, $\mathcal{L}(D(w, r)) = w \setminus \mathcal{L}(r)$.

Theorem 8 ([4]). *For all* $r \in T_{\mathcal{R}}$, $\mathcal{L}(r) = \mathcal{L}\Big(N(r) + \sum_{a \in \Sigma} D(a, r)\Big)$.

Definition 9. *A (nondeterministic)* finite automaton *(NFA) is a tuple* $\mathcal{A} = (Q, \Sigma, \delta, q_0, F)$ *where* Q *is a finite set of states,* Σ *an alphabet,* $\delta : Q \times \Sigma \to \wp(Q)$ *the transition function,* $q_0 \in Q$ *the initial state, and* $F \subseteq Q$ *the set of final states.*

Let $n \in \mathbb{N}$. *A* run *of* $\mathcal{A}$ *on* $w = a_0 \ldots a_{n-1} \in \Sigma^*$ *is a sequence* $q_0 \ldots q_n \in Q^*$ *such that, for all* $0 \leq i < n$, $q_{i+1} \in \delta(q_i, a_i)$. *The run is* accepting *if* $q_n \in F$. *The language* $\mathcal{L}(\mathcal{A}) = \{w \in \Sigma^* \mid \exists$ *accepting run of* $\mathcal{A}$ *on* $w\}$ *is* recognized *by* $\mathcal{A}$.

The automaton $\mathcal{A}$ *is* total deterministic *if* $|\delta(q, a)| = 1$, *for all* $q \in Q$, $a \in \Sigma$.

3   Enhanced Derivatives

An operation on languages takes one or more languages as arguments and yields another language. In this section, we enhance the syntax and semantics of regular expressions with extra operations and consider conditions for the existence of a syntactic derivative for such enhanced expressions. Many examples can be drawn from the closure properties of regular languages.

Definition 10. *A function $f : (\Sigma^*)^n \to \Sigma^*$ is regularity-preserving if for all regular languages $R_1, \ldots, R_n$ the image $f(R_1, \ldots, R_n)$ is a regular language.*

Example 11. We give a range of examples for operators on languages. All operators, except shuffle closure, are regularity-preserving. Proofs may be found in textbooks on formal languages unless otherwise indicated. We let $U, V, L \subseteq \Sigma^*$ range over regular languages; $a, b \in \Sigma$ range over symbols.

1. The intersection $U \cap V$ and the complement $\neg U$ of regular languages are regular.
2. The shuffle of two regular languages is defined by $U \| V = \bigcup \{u \| v \mid u \in U, v \in V\}$ where $\varepsilon \| v = \{v\}$, $u \| \varepsilon = \{u\}$, and $au \| bv = \{a\} \cdot (u \| bv) \cup \{b\} \cdot (au \| v)$, is regular.
   The shuffle closure operation $L^{\|} = \{\varepsilon\} \cup L \cup (L \| L) \cup (L \| L \| L) \cup \ldots$ does **not** preserve regularity.
3. The inverse homomorphism, i.e., $h^{-1}(U) = \{w \in \Sigma^* \mid h(w) \in U\}$ is regular for a function $h : \Sigma \to \Sigma^*$ that is extended homomorphically to a function $\Sigma^* \to \Sigma^*$ (for simplicity, we do not consider homomorphisms between different alphabets, which can be simulated by using the disjoint union of the alphabets).
   The non-erasing homomorphism $h(L) = \{h(w) \mid w \in L\}$ is regular where $h : \Sigma \to \Sigma^+$.
4. The language of every k-th symbol starting from position i from words in a regular language L is regular: for $k > 0$ and $0 < i \leq k$
   $f_{i,k}(L) = \{a_i a_{i+k} a_{i+2k} \cdots a_{i+k \lfloor (n-i)/k \rfloor} \mid n \in \mathbb{N}, a_1 \ldots a_n \in L\}$.
5. The left quotient $\backslash$ and the right quotient $/$ of regular languages are regular.
6. Functions $suffixes(L) = \Sigma^* \backslash L$ and $prefixes(L) = L / \Sigma^*$ preserve regularity.
7. The function $reverse(L) = \{a_n \cdots a_1 \mid n, i \in \mathbb{N}, 1 \leq i \leq n, a_i \in \Sigma, a_1 \ldots a_n \in L\}$ preserves regularity.
8. For each $k \in \mathbb{N}$, the function $\mathbb{H}_k(L) = \{v \mid v \in \Sigma^*, \exists u \in L.d(u, v) \leq k\}$ is regularity preserving where the *Hamming distance* of words $a_1 \cdots a_n$ and $b_1 \cdots b_m$ is defined by $h = d(a_1 \cdots a_n, b_1 \cdots b_m)$. If $m = n$, then $h = |\{i \mid 1 \leq i \leq n, a_i \neq b_i\}|$. Otherwise $h = \infty$.
   Analogously, $\mathbb{L}_k(L)$ is a regularity preserving approximation that uses the Levenshtein distance (see [8]).
9. The tilde and bar operators defined by $\tilde{L} = L \cup \{\varepsilon\}$ and $\bar{L} = L \backslash \{\varepsilon\}$ preserve regularity (they are the primitive building blocks of multi-tilde-bar expressions [6], which we do not consider to save space).

The notion of a nullable expression is an important ingredient in the definition of the derivative (Definition 6). Nullability can be computed by induction on a regular expression because each regular operator corresponds to a boolean function on the nullability of the operator's arguments. The following definition imposes exactly this condition on the extra operators in regular expressions.

Definition 12. *A function $f : (\Sigma^*)^n \to \Sigma^*$ is ε-testable, if there is a boolean function $B_f : \mathbb{B}^n \to \mathbb{B}$ such that $\varepsilon \in f(L_1, \ldots, L_n)$ iff $B_f((\varepsilon \in L_1), \ldots, (\varepsilon \in L_n))$.*

Example 13. Some of the functions from Example 11 are ε-testable.

1. intersection, complement: $B_\cap = \wedge$, $B_\neg = \neg$;
2. shuffle: $B_\| = \wedge$; the shuffle closure operation $L^\|$ is ε-testable using $B^\|(b) = 1$;
3. inverse homomorphism: $B_{h^{-1}}(b) = b$, for $b \in \mathbb{B}$; homomorphism h: if h is non-erasing, then $B_h(b) = b$; erasing homomorphism is not ε-testable: consider $L_1 = \{a\}$, $L_2 = \{b\}$, and an erasing homomorphism h defined by $h(a) = \varepsilon$ and $h(b) = b$. Thus, $h(L_1) = \{\varepsilon\}$ and $h(L_2) = \{b\}$. If there was a boolean function f_h to vouch for ε-testability of h, then L_1 shows that $f_h(0) = 1$ and L_2 yields $f_h(0) = 0$, a contradiction.
4. k-th letter extraction: $\varepsilon \in f_{i,k}(L)$ if $\exists w \in L$ such that $|w| < i$, so $f_{i,k}$ is not ε-testable. To see this let $i = k = 2$, $L_1 = \{a\}$, and $L_2 = \{aa\}$ and assume that B_f is the boolean function required for ε-testability. Now $f_{2,2}(L_1) = \{\varepsilon\}$ and $f_{2,2}(L_2) = \{a\}$, so that $B_f(0) = 1$ (by L_1) and $B_f(0) = 0$ (by L_2), a contradiction.
5. The left quotient is not ε-testable because $\varepsilon \in U \backslash W$ iff $U \cap W \neq \emptyset$: consider $U = \Sigma^*$ with $a \in \Sigma$, $W_1 = \emptyset$, and $W_2 = \{a\}$ so that $U \backslash W_1 = \emptyset$ and $U \backslash W_2 = \{\varepsilon, a\}$. A binary boolean function $B_\backslash$ for ε-testability would have to satisfy $B_\backslash(0,0) = 0$ (for W_1) and $B_\backslash(0,0) = 1$ (for W_2), a contradiction. The same reasoning applies, mutatis mutandis, to the right quotient.
6. The *suffixes* function is not ε-testable by the proof for the left quotient. The proof for *prefixes* is analogous to the one for the right quotient.
7. The *reverse* function is ε-testable: $B_{reverse}(b) = b$.
8. The approximation for Hamming distance is ε-testable by $B_{\mathbb{H}_k}(b) = b$. The approximation for Levenshtein distance $\mathbb{L}_k$ is not ε-testable for $k > 0$. The argument here is similar as for erasing homomorphism because a word at distance k from a given word w may be up to k symbols shorter than w.
9. The tilde and bar operators are obviously ε-testable with the constants 1 and 0, respectively.

Definition 14 (Enhanced Regular Expression). *Let $\mathcal{F} \supseteq \mathcal{R}$ be a ranked alphabet, an* enhanced regular alphabet. *Let further $\mathcal{J}$ be an interpretation function for $\mathcal{F}$ on the carrier $\wp(\Sigma^*)$ extending the regular interpretation $\mathcal{I}$ from Definition 1. The set of $\mathcal{F}$-regular expressions over a set X is the set of terms $T_{\mathcal{F}}(X)$. For $t \in T_{\mathcal{F}}(X)$ we define its language $\mathcal{L}(t) = \hat{\mathcal{J}}(t)$. The resulting $\mathcal{F}$-algebra $(\wp(\Sigma^*), \mathcal{J})$ is a* regular enhancement *if every symbol $F \in \mathcal{F}^{(n)}$ is interpreted by a regularity-preserving function $\mathcal{J}(n)(F)$.*

Example 15. To extend regular expressions with a shuffle operator, consider $\mathcal{F}^\| = \mathcal{R} \cup \{\|\}$ with $\#\| = 2$.

To extend expressions with kth-letter extraction, we consider $\mathcal{F}^{x-k} = \mathcal{R} \cup \{f_{i,k} \mid 0 < i \leq k\}$ with $\#f_{i,k} = 1$.

Lemma 16. *If $\mathcal{J}(F)$ is ε-testable, for each $F \in \mathcal{F}$, then the nullability function N can be extended to $\mathcal{F}$.*

To obtain syntactic derivability for an enhanced regular expression, it must be possible to express the derivative of an operator in terms of a regular expression that applies the derivative to the arguments of the operator. We first define a suitable property semantically as an algebraic property of a regular enhancement.

Definition 17. *Let $\mathcal{F}$ be an enhanced regular alphabet and $\mathcal{J}$ an extension of the regular interpretation $\mathcal{I}$. The $\mathcal{F}$-algebra $(\wp(\Sigma^*), \mathcal{J})$ is left derivable if, for each $F \in \mathcal{F}^{(k)}$ and $a \in \Sigma$, there exists a finite subset $X \subset \{x_{v,j} \mid v \in \Sigma^*, 1 \leq j \leq k\}$ and an $\mathcal{F}$-regular expression $r \in T_{\mathcal{F}}(X)$ such that, for all $L_1, \ldots, L_k \subseteq \Sigma^*$ the left quotient $a \backslash (\mathcal{J}(F)(L_1, \ldots, L_k))$ can be expressed as $\hat{\mathcal{J}}(r)$ using the interpretation $\mathcal{J}_0(x_{v,j}) = v \backslash L_j$.*

Example 18. We revisit the previous examples of functions on languages and examine them for being left derivable.

1. Intersection is left derivable: $a \backslash (L_1 \cap L_2) = \hat{\mathcal{J}}(x_{a,1} \cap x_{a,2}) = a \backslash L_1 \cap a \backslash L_2$. For negation $\neg$, the pattern is the same.
2. Shuffle is left derivable:
$a \backslash (L_1 \| L_2) = (a \backslash L_1) \| L_2 \cup L_1 \| (a \backslash L_2) = \hat{\mathcal{J}}(x_{a,1} \| x_{\epsilon,2} + x_{\epsilon,1} \| x_{a,2});$
shuffle closure is also left derivable:

$$a \backslash L^\| = (a \backslash L) \| L^\| = \hat{\mathcal{J}}(x_{a,1} \| x_{\epsilon,2} + x_{\epsilon,1}^\|)$$

3. Inverse homomorphism is left derivable:
$a \backslash (h^{-1}(L)) = h^{-1}(h(a) \backslash L) = \hat{\mathcal{J}}(h^{-1}(x_{h(a),1})).$
Non-erasing homomorphism is left derivable:
$a \backslash (h(L)) = \bigcup_{b \in \Sigma, h(b) = av} v \cdot h(b \backslash L) = \hat{\mathcal{J}}(\sum_{b \in \Sigma, h(b) = av} v \cdot h(x_{b,1})).$
4. For $k > 1$, the set $\{f_{i,k} \mid 0 < i \leq k\}$ is left derivable.
$a \backslash (f_{i,k}(L)) = f_k(\bigcup_{|w|=i-1} wa \backslash L) = \hat{\mathcal{J}}(\sum_{|w|=i-1} f_k(x_{wa,1})).$
5. The left and right quotients are left derivable.
$a \backslash (L_1 \backslash L_2) = (L_1 \cdot a) \backslash L_2 = \hat{\mathcal{J}}((x_{\epsilon,1} \cdot a) \backslash x_{\epsilon,2}).$
$a \backslash (L_1 / L_2) = (a \backslash L_1) / L_2 = \hat{\mathcal{J}}(x_{a,1} / x_{\epsilon,2}).$
6. The function *suffixes* is not left derivable because $a \backslash suffixes(L) = \{w \mid \exists u.uaw \in L\} = (\Sigma^* \cdot a) \backslash L$ cannot be finitely expressed using just derivatives, the *suffixes* function, and the regular operators.
To see this, consider the family of languages $L_n = w_n^*$ where $w_n = (abab^2 \cdot ab^n)^*$, for all $n \in \mathbb{N}$, and find that

$$L_n' = a \backslash suffixes(L_n) = bab^2 \cdot ab^n w_n^* + b^2 \cdot ab^n w_n^* + \cdots + b^n w_n^*$$

Suppose there is a *suffixes*-enhanced regular expression r for $a \backslash L$ that only depends on a and Σ and that refers to finitely many derivatives, say, $v_1 \backslash L, \ldots, v_m \backslash L$. Considering r for L_n', we find that r cannot contain the *suffixes* function because that would introduce words starting with a, which cannot be in L_n' and which cannot be amended by prepending a fixed word without breaking the a-b pattern. There must exist some $v \in w_n^*$ such that each v_j is either a prefix of v that ends with an a or it is not a prefix of v. Now, if we consider L_k where $k = \max(n, |v_1|, \ldots, |v_m|) + 1$ then none of the $v_j \backslash L_k$ can contain $b^k w_k^*$. Note that if v_j is not a prefix of w_n*, then it is not a prefix of w_k^*, for any $k \geq n$, either. Hence, r cannot describe L_k'.

If we assume that $\mathcal{F}$ contains *suffixes* and the left quotient operator, then we could consider $suffixes(L)$ as an abbreviation for $\Sigma^* \backslash L$ and we would regain left derivability. Furthermore, with a suitable variation of Definition 17, *suffixes* is *right derivable*:

$suffixes(L)/a = \{v \mid \exists u.uv \in L\}/a = \{v \mid \exists u.uva \in L\} = suffixes(L/a)$.

The function *prefixes* is left derivable:

$a\backslash prefixes(L) = a\backslash \{v \mid \exists u.vu \in L\} = prefixes(a\backslash L) = \hat{\mathcal{J}}(prefixes(x_{a,1}))$.

7. The function *reverse* is neither left derivable nor right derivable, but swaps between left and right quotients:

$a\backslash reverse(L) = reverse(L/a)$.

To see that *reverse* is no left derivable, consider the language $L = b^*a$. Clearly, $reverse(L) = ab^*$ and $a\backslash reverse(L) = b^*$. Now suppose we can obtain b^* by a regular expression with *reverse* on arbitrary derivatives of L. There are only two distinct derivatives: $a\backslash(b^*a) = \{\varepsilon\}$ and $b\backslash(b^*a) = b^*a$. Hence, for any $w \in \{a,b\}^*$, $w\backslash(b^*a)$ will be either empty, $\{\varepsilon\}$, or b^*a. Now consider a language U constructed from these derivatives by application of regular operators or *reverse*. It can be shown that any word in U is either ε or it contains the symbol a. Thus, U cannot be equal to b^*.

8. The enhancement with the approximation operators $\mathbb{H}_k, \mathbb{H}_{k-1}, \ldots, \mathbb{H}_1, \mathbb{H}_0$ operators (for Hamming distance) is left derivable because

$$a\backslash \mathbb{H}_k(L) = \mathbb{H}_k(a\backslash L) + \sum_{\substack{k>0 \\ x \neq a}} \mathbb{H}_{k-1}(x\backslash L)$$

For approximation with operators $\mathbb{L}_k, \ldots, \mathbb{L}_0$ that rely on Levenshtein distance, we also obtain left closure (assuming that $\mathbb{L}_{-1}(L) = \emptyset$):

$$a\backslash \mathbb{L}_k(L) = \sum_{\substack{w \in \Sigma^* \\ |w| \leq k \\ k>0}} \left(\mathbb{L}_{k-|w|}(wa\backslash L) + \sum_{x \neq a} \mathbb{L}_{k-|w|-1}(wx\backslash L) + \mathbb{L}_{k-|w|-1}(w\backslash L) \right)$$

The terms correspond to the actions "delete w, then match a", "delete w, then substitute a by some x", and "delete w, then insert a".

9. Tilde and bar are trivially left derivable: $a\backslash \tilde{L} = a\backslash L$ and $a\backslash \bar{L} = a\backslash L$.

4 Word Problem

To obtain a decision procedure for the word problem of left derivable enhanced regular expressions, we first define the corresponding syntactic derivative and then extend Brzozowski's result that $w \in \mathcal{L}(r)$ iff $\varepsilon \in \mathcal{L}(D(w, r))$ (which follows from Theorem 7). It is interesting to remark that, for example, we obtain a decision procedure for the word problem for the language of regular expressions enhanced with the shuffle-closure operator is no longer regular.

Theorem 19. *If $(\wp(\Sigma^*), \mathcal{J})$ is a left derivable $\mathcal{F}$-algebra which is ε-testable, then there is a syntactic derivative function $D : \Sigma \times T_{\mathcal{F}} \rightarrow T_{\mathcal{F}}$ such that $\hat{\mathcal{J}}(D(a, t)) = a\backslash \hat{\mathcal{J}}(t)$, for all $a \in \Sigma$ and $t \in T_{\mathcal{F}}$.*

$$D^+(0) = \{0\} \qquad D^+(r+s) = D^+(r) \oplus D^+(s)$$
$$D^+(1) = \{0\} \qquad D^+(r \cdot s) \; = D^+(r) \odot s \oplus \bigoplus D^+(s)$$
$$D^+(a) = \{0,1\} \qquad D^+(r^*) \quad = \bigoplus D^+(r) \odot r^*$$

Fig. 1. Iterated Brzozowski derivatives for $T_\mathcal{R}$

Proof. Define D inductively as an extension of Definition 6 for $F \in \mathcal{F} \setminus \mathcal{R}$:

$$D(a, F(r_1, \ldots, r_n)) = R(F, a)[x_{v,j} \mapsto D(v, r_j) \mid x_{v,j} \in X(F, a)]$$

where N extends to $T_\mathcal{F}$ by Lemma 16 and where D extends to words as before. The statement about the semantics follows by induction on the augmented term using the definition of left derivability. □

Theorem 20. *If $(\wp(\Sigma^*), \mathcal{J})$ is a left derivable $\mathcal{F}$-algebra which is ε-testable, then the word problem for $\hat{\mathcal{J}}(t)$ is decidable, for any $t \in T_\mathcal{F}$.*

Proof. By Theorem 19, there is a nullability function N and a derivative D for $T_\mathcal{F}$. By induction on the length of $w \in \Sigma^*$, we obtain that $w \in \hat{\mathcal{J}}(t)$ iff $\varepsilon \in \hat{\mathcal{J}}(D(w,t))$ iff $N(D(w,t))$. □

5   Finiteness

For classical derivatives on $T_\mathcal{R}$ (cf. Definition 6), Brzozowski showed that the set of iterated derivatives $D(\Sigma^*, r)$ of a given regular expression r is finite, when considered modulo similarity (i.e., associativity, commutativity, and idempotence of union). Hence, we now look for conditions such that the set of dissimilar iterated derivatives is finite for enhanced regular expressions. First, we set up a framework for reasoning about finiteness.

Recent work on determining the number of iterated *partial* derivatives starts with an inductive definition for the set of iterated partial derivatives [3]. We transfer that definition to the classical case and define an upper approximation $D^+(r)$ of the set of iterated derivatives of expression r in Fig. 1 by induction on r. In the definition, we lift $\odot$ and $\oplus$ to sets of expressions (i.e., if $R, S \subseteq T_\mathcal{R}$, then $R \odot S = \{r \odot s \mid r \in R, s \in S\}$ and $R \oplus S = \{r \oplus s \mid r \in R, s \in S\}$). We further write $\bigoplus S$ for the set $\{s_1 \oplus \cdots \oplus s_n \mid n \in \mathbb{N}, s_i \in S\}$ of finite sums of elements drawn from S where the nullary sum stands for 0 and where we assume sums to be identified modulo associativity, commutativity, and idempotence to obtain the following results[1].

Theorem 21. *The set $D^+(r)$ is finite, for all $r \in T_\mathcal{R}$.*

Clearly, the set $D^*(r) = \{r\} \cup D^+(r)$ is also finite for all r.

---

[1] See the technical report for auxiliary lemmas and proofs.

Theorem 22. (Closure Under Derivation).

1. For all r and a, $D(a, r) \in D^+(r)$.
2. For all r and a, if $t \in D^+(r)$, then $D(a, t) \in D^+(r)$.

Corollary 23. *The set $\{D(w, r) \mid w \in \Sigma^+\} \subseteq D^+(r)$, for all r.*

To obtain finiteness for enhanced regular expressions, we strengthen the notion of left derivability. Essentially, we restrict the form of a derivative to a linear combination of enhancement functions applied to derivatives of the arguments.

Definition 24. *Let $\mathcal{F} = \{F_1, \ldots, F_m\}$ be a ranked alphabet. The $\mathcal{F}$-algebra $(\wp(\Sigma^*), \mathcal{J})$ is linear left derivable if, for each $n \in \mathbb{N}$, $F \in \mathcal{F}^{(n)}$, and $a \in \Sigma$, there exists a finite index set J such that, for each $j \in J$, there is a word $v_j \in \Sigma^*$, an index $i_j \in \{1, \ldots, m\}$ of an element of $\mathcal{F}$ with arity $\#(F_{i_j}) = n_j$, and, for $1 \le k \le n_j$, words $w_k^j \in \Sigma^*$ and indexes $\alpha_k^j \in \{1, \ldots, n\}$ of left-hand-side languages, such that for all $L_1, \ldots L_n \subseteq \Sigma^*$, the left quotient can be expressed by:*

$$a \backslash (\mathcal{J}(F)(L_1, \ldots, L_n)) = \bigcup_{j \in J} v_j \cdot \mathcal{J}(F_{i_j})(\overline{w_k^j \backslash (L_{\alpha_k^j})}^{k=1,\ldots,n_j}) \tag{1}$$

Of the standard regular operators, only union (and in fact all boolean functions) is linear left derivable. Concatenation $U \cdot V$ does not fit the pattern because it has a summand which is conditional on $\varepsilon \in U$. The Kleene star does not fit, either, because it concatenates the derivative of the argument with the original term (Definition 6). But many useful operators are linear left derivable (Example 28).

Theorem 25. *Suppose that $\mathcal{F} = \{F_1, \ldots, F_m\} \cup \mathcal{R}$ is an enhanced regular alphabet with interpretation $\mathcal{J}$ such that $(\wp(\Sigma^*), \mathcal{J}_{|\{F_1,\ldots,F_m\}})$ is linear left derivable.*
Then, for all $n \in \mathbb{N}$, $F \in \mathcal{F}^{(n)}$, and $a \in \Sigma$ there exists a finite index set J, for each $j \in J$, there is a word $v_j \in \Sigma^*$, an index $i_j \in \{1, \ldots, m\}$ of an element of $\mathcal{F} \setminus \mathcal{R}$ with arity $\#(F_{i_j}) = n_j$, for each $1 \le k \le n_j$, a word $w_k^j \in \Sigma^*$, and an index $\alpha_k^j \in \{1, \ldots, n\}$ that selects one of the left-hand-side regular expressions as an argument. Then, for each $r_1, \ldots, r_n \in T_{\mathcal{F}}$, the syntactic derivative of $F(r_1, \ldots, r_n)$ by a is given in the form

$$D(a, F(r_1, \ldots, r_n)) = \sum_{j \in J(F,a)} v_j \cdot F_{i_j}(\overline{D(w_k^j, r_{\alpha_k^j})}^{k=1,\ldots,n_j}) \tag{2}$$

In this setting, the set of iterated derivatives of any $\mathcal{F}$-regular expression r is finite. Specifically, in extension of the definition in Fig. 1, we claim that for each $F \in \mathcal{F} \setminus \mathcal{R}$, the set of iterated derivatives

$$D^+(F(r_1, \ldots, r_n)) = \bigoplus \{v \cdot G(\overline{r_i'}) \mid V \succeq v, G \in \mathcal{F} \setminus \mathcal{R}, r_i' \in \bigcup_j D^*(r_j)\} \tag{3}$$

is finite. Here $V = \{v_j \mid j \in J, F \in \mathcal{F}, a \in \Sigma\}$, and we write $V \succeq v$ for $\exists v' \in V. v' \succeq v$.

Corollary 26. *Let $\mathcal{F}$ be an enhanced regular alphabet and $(\wp(\Sigma^*), \mathcal{J})$ be an ε-testable, linear left derivable $\mathcal{F}$-algebra. Then any $\mathcal{F}$-regular expression defines a regular language.*

Proof. Let $r \in T_{\mathcal{F}}$ and let Q_r be the set of dissimilar derivatives of r. As $Q_r \subseteq D^*(r)$, Q_r is finite. Hence $M = (Q_r, \Sigma, D, r, F)$ with $F = \{q \in Q_r \mid N(q)\}$ is a total deterministic finite automaton that recognizes $\mathcal{L}(r)$, which is thus regular. □

Corollary 27. *Let $\mathcal{F}$ be an enhanced regular alphabet and $(\wp(\Sigma^*), \mathcal{J})$ be an ε-testable, linear left derivable $\mathcal{F}$-algebra. Then, for each $F \in \mathcal{F}$, the operation $\mathcal{J}(F)$ preserves regularity.*

Proof. Let $F \in \mathcal{F}^{(n)}$, for some $n \in \mathbb{N}$. Let $R_1, \ldots, R_n$ be regular languages defined by regular expressions $r_1, \ldots, r_n \in T_{\mathcal{R}} \subseteq T_{\mathcal{F}}$. By Corollary 26, $\hat{\mathcal{J}}(F(r_1, \ldots, r_n))$ is regular. Hence $\mathcal{J}(F)$ preserves regularity.   □

Example 28. Many operators are in fact linear left derivable.

1. Intersection and complement are linear left derivable.
2. The shuffle operation is linear left derivable, but the derivative of the shuffle closure contains a nested application of shuffle closure.
3. Inverse and non-erasing homomorphism are linear left derivable.
4. For $k > 0$, the set $\{f_{i,k} \mid 0 < i \leq k\}$ is linear left derivable.
5. The left quotient is not linear left derivable, but the right quotient is linear left derivable.
6. The function *suffixes* is not left derivable; the function *prefixes* is linear left derivable.
7. The function *reverse* is not left derivable.
8. Both, $\mathbb{H}_k$ and $\mathbb{L}_k$ are linear left derivable.
9. Tilde and bar are linear left derivable.

By Corollary 26, regular languages are closed under ε-testable operators that are linear left derivable: $\cap$, $\neg$, $\|$, h^{-1}, non-erasing h, $\mathbb{H}_k$, tilde, bar.

For a set of unary operators, linear left derivability amounts to definability by a rational finite state transducer.

Theorem 29. *Let $\mathcal{F} = \mathcal{F}^{(1)} = \{F_1, \ldots, F_m\}$ be a ranked alphabet of unary operators and $(\wp(\Sigma^*), \mathcal{J})$ be a linear left derivable $\mathcal{F}$-algebra which is ε-testable using the identity function. Then, for each $1 \leq l \leq m$ and $L \subseteq \Sigma^*$, $\mathcal{J}(F_l)(L)$ is equal to $T(L)$ where T is a rational finite state transducer.*

The reverse implication does not hold because transducers may, in general, consume an unbounded amount of input before producing an output. The transducers resulting from Theorem 29 only consume bounded input before producing at least one output symbol.

6   Conclusion

We introduce a framework for constructing derivatives for regular expressions enhanced with new operators. If these operators are left derivable, we obtain an algorithm for the word problem; if they are linear left derivable, we can construct a DFA from an enhanced expression. In fact, unary operators with this property are rational transductions.

Some of the operators considered in this paper are known to be regularity preserving, yet, they fail to be linear left derivable or to be ε-testable. In future work, we plan to address these restrictions by generalizing linear derivability as well as the nullability test.

References

1. Antimirov, V.M.: Rewriting regular inequalities. In: Reichel, H. (ed.) FCT 1995. LNCS, vol. 965, pp. 116–125. Springer, Heidelberg (1995)
2. Antimirov, V.M.: Partial derivatives of regular expressions and finite automaton constructions. Theor. Comput. Sci. **155**(2), 291–319 (1996)
3. Broda, S., Machiavelo, A., Moreira, N., Reis, R.: Study of the average size of Glushkov and partial derivative automata, October 2011
4. Brzozowski, J.A.: Derivatives of regular expressions. J. ACM **11**(4), 481–494 (1964)
5. Caron, P., Champarnaud, J.-M., Mignot, L.: Partial derivatives of an extended regular expression. In: Dediu, A.-H., Inenaga, S., Martín-Vide, C. (eds.) LATA 2011. LNCS, vol. 6638, pp. 179–191. Springer, Heidelberg (2011)
6. Caron, P., Champarnaud, J.-M., Mignot, L.: Multi-tilde-bar derivatives. In: Moreira, N., Reis, R. (eds.) CIAA 2012. LNCS, vol. 7381, pp. 321–328. Springer, Heidelberg (2012)
7. Caron, P., Champarnaud, J., Mignot, L.: A general framework for the derivation of regular expressions. RAIRO Theor. Inf. Appl. **48**(3), 281–305 (2014)
8. Champarnaud, J.-M., Jeanne, H., Mignot, L.: Approximate regular expressions and their derivatives. In: Dediu, A.-H., Martín-Vide, C. (eds.) LATA 2012. LNCS, vol. 7183, pp. 179–191. Springer, Heidelberg (2012)
9. Grabmayer, C.: Using proofs by coinduction to find "Traditional" proofs. In: Fiadeiro, J.L., Harman, N.A., Roggenbach, M., Rutten, J. (eds.) CALCO 2005. LNCS, vol. 3629, pp. 175–193. Springer, Heidelberg (2005)
10. Might, M., Darais, D., Spiewak, D.: Parsing with derivatives: a functional pearl. In: Proceedings of ICFP 2011, pp. 189–195. ACM (2011)
11. Owens, S., Reppy, J., Turon, A.: Regular-expression derivatives reexamined. J. Funct. Program. **19**(2), 173–190 (2009)
12. Roşu, G., Viswanathan, M.: Testing extended regular language membership incrementally by rewriting. In: Nieuwenhuis, R. (ed.) RTA 2003. LNCS, vol. 2706, pp. 499–514. Springer, Heidelberg (2003)
13. Sulzmann, M., Thiemann, P.: Derivatives for regular shuffle expressions. In: Dediu, A.-H., Formenti, E., Martín-Vide, C., Truthe, B. (eds.) LATA 2015. LNCS, vol. 8977, pp. 275–286. Springer, Heidelberg (2015)
14. Sulzmann, M., Thiemann, P.: Forkable regular expressions. In: Dediu, A.-H., Janoušek, J., Martín-Vide, C., Truthe, B. (eds.) LATA 2016. LNCS, vol. 9618, pp. 194–206. Springer, Heidelberg (2016). doi:10.1007/978-3-319-30000-9_15

15. Traytel, D., Nipkow, T.: Verified decision procedures for MSO on words based on derivatives of regular expressions. J. Funct. Program. **25**, e18 (2015)
16. Watson, B.W.: FIRE lite: FAs and REs in C++. In: Raymond, D.R., Yu, S., Wood, D. (eds.) WIA 1996. LNCS, vol. 1260, pp. 167–188. Springer, Heidelberg (1997)

Weighted Restarting Automata
as Language Acceptors

Qichao Wang and Friedrich Otto[(⊠)]

Fachbereich Elektrotechnik/Informatik, Universität Kassel, 34109 Kassel, Germany
{wang,otto}@theory.informatik.uni-kassel.de

Abstract. We use weighted restarting automata to define classes of
formal languages by combining the acceptance condition of a restarting automaton with a condition on the weight of its accepting computations. Specifically, we consider the tropical semiring $\mathbb{Z}_\infty$ and the semiring
REG(Δ) of regular languages over a finite alphabet Δ. We show that by
using the tropical semiring, we can avoid the use of auxiliary symbols.
Further, a certain type of (word-) weighted restarting automata turns
out to be equivalent to non-forgetting restarting automata, and another
class of languages accepted by (word-) weighted restarting automata is
shown to be closed under intersection.

Keywords: Weighted restarting automaton · Non-forgetting restarting
automaton · Language class · Closure property

1   Introduction

The restarting automaton was introduced as a formal model for the analysis by
reduction, which is a linguistic technique that is used to check the correctness of
sentences of natural languages through sequences of local simplifications [1,2,8].
In order to study quantitative aspects of computations of restarting automata,
the authors introduced *weighted restarting automata* in [9]. These automata are
obtained by assigning an element of a given semiring S as a weight to each
transition of a restarting automaton M. The product (in S) of the weights of
all transitions that are used in a computation then yields a weight for that
computation, and the sum over all weights of all accepting computations of M
for a given input word $w \in \Sigma^*$ yields a value from S. In this way, a partial
function $f : \Sigma^* \to S$ is obtained. By placing a condition T on the value $f(w)$,
we can define a subset $L_T(M)$ of the language $L(M)$ that is accepted by M.

Here we study the case that the semiring S is the *tropical semiring* $\mathbb{Z}_\infty =$
$(\mathbb{Z} \cup \{\infty\}, \min, +, \infty, 0)$ and the case that S is the semiring of regular languages
REG(Δ) = (REG(Δ), $\cup, \cdot, \emptyset, \{\lambda\}$) over a given finite alphabet Δ. In the latter case
we restrict our attention to the *word-weighted* restarting automata introduced
in [10], that is, the weight of each transition is taken to be a singleton. In [10]
these automata were studied as transducers, while here we use them as language
acceptors.

© Springer International Publishing Switzerland 2016
Y.-S. Han and K. Salomaa (Eds.): CIAA 2016, LNCS 9705, pp. 298–309, 2016.
DOI: 10.1007/978-3-319-40946-7_25

We present the following results. First we show that, for each type of restarting automaton, the use of auxiliary symbols can be replaced by a condition on the weights of accepting computations in the semiring $\mathbb{Z}_\infty$, that is, each restarting automaton of type RWW or RRWW can be simulated by a weighted restarting automaton of type RW or RRW with weights in $\mathbb{Z}_\infty$. Then we prove that, for each type X of restarting automata, the word-weighted restarting automata of type X are equivalent to the non-forgetting restarting automata of the same type, where a non-forgetting restarting automaton is not required to reset its state to its initial state on executing a restart operation [4,5,7]. Finally, we present some closure properties for the classes of languages that are defined by weighted restarting automata. In particular, we prove that the class of languages that are defined by a certain type of word-weighted restarting automata is closed under intersection, which is only the second time that closure under intersection is established for a class of languages accepted by restarting automata that does not coincide with the class of regular languages [6].

2   Definitions and Examples

We assume that the reader is familiar with the standard notions and concepts of theoretical computer science, such as monoids, finite automata, and semirings. Throughout the paper we use $|w|$ to denote the length of a word w and λ to denote the empty word. Further, $\mathbb{P}(X)$ denotes the power set of a set X, and $\mathbb{P}_{\text{fin}}(X)$ denotes the set of all finite subsets of X.

A *restarting automaton* (or RRWW-automaton) is a nondeterministic machine with a finite-state control, a flexible tape with endmarkers, and a read/write window [1,2]. Formally, it is described by an 8-tuple $M = (Q, \Sigma, \Gamma, \mathcal{c}, \$, q_0, k, \delta)$, where Q is a finite set of states, Σ is a finite input alphabet, Γ is a finite tape alphabet containing Σ, the symbols $\mathcal{c}, \$ \notin \Gamma$ are used as markers for the left and right border of the work space, respectively, $q_0 \in Q$ is the initial state, $k \geq 1$ is the size of the *read/write window*, and δ is the (partial) *transition relation* that associates finite sets of transition steps to pairs of the form (q, u), where q is a state and u is a possible content of the read/write window. There are four types of transition steps. A *move-right step* (MVR) causes M to shift its read/write window one position to the right and to change the state. A *rewrite step* causes M to replace the content u of the read/write window by a shorter string v, thereby reducing the length of the tape, and to change the state. Further, the read/write window is placed immediately to the right of the string v. However, occurrences of the delimiters $\mathcal{c}$ and $\$$ can neither be deleted nor newly created by a rewrite step. A *restart step* causes M to place its read/write window over the left end of the tape, so that the first symbol it sees is the left sentinel $\mathcal{c}$, and to reenter the initial state q_0, and, finally, an *accept step* causes M to halt and accept. A *non-forgetting* restarting automaton M has extended restart steps, which are combined with a change of state just like the move-right and rewrite operations [5,7]. The prefix nf- is used to denote types of non-forgetting restarting automata.

If $\delta(q,u)$ is undefined for some pair (q,u), then M necessarily halts in a corresponding situation, and we say that M *rejects*. Finally, if each rewrite step is combined with a restart step into a joint rewrite/restart operation, then M is called an *RWW-automaton*. An RRWW-automaton is called an *RRW-automaton* if its tape alphabet Γ coincides with its input alphabet Σ, that is, if no auxiliary symbols are available. It is an *RR-automaton* if it is an RRW-automaton for which the right-hand side v of each rewrite step $(q',v) \in \delta(q,u)$ is a scattered subword of the left-hand side u. Analogously, we obtain the *RW-automaton* and the *R-automaton* from the RWW-automaton.

A *configuration* of M is a string $\alpha q \beta$, where $q \in Q$, and either $\alpha = \lambda$ and $\beta \in \{¢\} \cdot \Gamma^* \cdot \{\$\}$ or $\alpha \in \{¢\} \cdot \Gamma^*$ and $\beta \in \Gamma^* \cdot \{\$\}$; here q is the current state, and $\alpha\beta$ is the current content of the tape, where it is understood that the window contains the first k symbols of β or all of β when $|\beta| \leq k$. A *restarting configuration* is of the form $q_0 ¢ w \$$. If $w \in \Sigma^*$, then $q_0 ¢ w \$$ is an *initial configuration*.

Any computation of M consists of certain phases. A phase, called a *cycle*, starts in a restarting configuration, the head moves along the tape performing move-right operations and a single rewrite operation until a restart operation is performed and thus a new restarting configuration is reached. If no further restart operation is performed, the computation necessarily finishes in a halting configuration – such a phase is called a *tail*. It is required that in each cycle M performs *exactly one* rewrite step. A word $w \in \Sigma^*$ is accepted by M, if there is an accepting computation which starts from the initial configuration $q_0 ¢ w \$$. By $L(M)$ we denote the language consisting of all (input) words that are accepted by M.

For studying quantitative aspects of computations of restarting automata, the weighted restarting automaton has been introduced in [9]. A *weighted restarting automaton* of type X, a wX-automaton for short, is a pair (M,ω), where M is a restarting automaton of type X, and ω is a *weight function* from the transitions of M into a semiring S, that is, ω assigns an element $\omega(t) \in S$ as a weight to each transition t of M. The product (in S) of the weights of all transitions that are used in a computation then yields a weight for that computation, and the sum over all weights of all accepting computations of M for a given input word $w \in \Sigma^*$ yields a value from S. In this way, a partial function $f_\omega^M : \Sigma^* \to S$ is obtained. Here we use weighted restarting automata to define sublanguages of the language that is accepted by the underlying (unweighted) restarting automaton.

Definition 1. *Let* $M = (Q,\Sigma,\Gamma,¢,\$,q_0,k,\delta)$ *be a restarting automaton, let* ω *be a weight function from* M *into a semiring* S*, and let* $\mathcal{M} = (M,\omega)$*. For a subset* T *of* S*,* $L_T(\mathcal{M}) = \{ w \in L(M) \mid f_\omega^M(w) \in T \}$ *is the language* accepted by $\mathcal{M}$ relative to T, *that is, a word* $w \in \Sigma^*$ *belongs to the language* $L_T(\mathcal{M})$ *iff* $w \in L(M)$ *and* $f_\omega^M(w) \in T$.

Definition 2. *Let* X *be a type of restarting automaton, let* S *be a semiring, and let* $\mathbb{H}$ *be a family of subsets of* S*. Then*

$$\mathcal{L}(\mathsf{X},S,\mathbb{H}) = \{ L_T(\mathcal{M}) \mid \mathcal{M} \text{ is a weighted restarting automaton of type } \mathsf{X}, \text{ and } T \in \mathbb{H} \}$$

is the class of languages that are accepted by weighted restarting automata of type X *relative to* $\mathbb{H}$.

We continue with an example that illustrates our definitions.

Example 3. Let $M_1 = (Q, \Sigma, \Gamma, \text{\textcent}, \$, q_0, k, \delta)$ be the R-automaton that is defined by taking $Q := \{q_0, q_r\}$, $\Gamma := \Sigma := \{a, b\}$, and $k := 4$, where δ is defined as follows:

$$
\begin{array}{llll}
t_1: & (q_0, \text{\textcent}aaa) \to (q_0, \text{MVR}), & t_7: & (q_0, abb\$) \to (q_r, b\$), \\
t_2: & (q_0, aaaa) \to (q_0, \text{MVR}), & t_8: & (q_0, abb\$) \to (q_r, \$), \\
t_3: & (q_0, aaab) \to (q_0, \text{MVR}), & t_9: & (q_0, \text{\textcent}ab\$) \to \text{Accept}, \\
t_4: & (q_0, aabb) \to (q_0, \text{MVR}), & t_{10}: & (q_0, \text{\textcent}\$) \to \text{Accept}, \\
t_5: & (q_0, abbb) \to (q_r, bb), & t_{11}: & (q_0, \text{\textcent}aab) \to (q_0, \text{MVR}), \\
t_6: & (q_0, abbb) \to (q_r, b), & t_{12}: & (q_0, \text{\textcent}abb) \to (q_0, \text{MVR}), \\
\end{array}
$$
$$
t_{13,x}: (q_r, x) \quad \to \text{Restart for all admissible } x.
$$

It is easily seen that $L(M_1) = \{\, a^m b^n \mid 0 \le m \le n \le 2m \,\}$. Further, for $a^n b^n$ and for $a^n b^{2n}$, M_1 has just a single accepting computation.

Let $(\text{REG}(\Delta), \cup, \cdot, \emptyset, \{\lambda\})$ be the semiring of regular languages over $\Delta = \{c, d\}$, let ω_1 be the weight function that assigns the set $\{c\}$ to the transitions t_5, t_7 and t_9, that assigns the set $\{d\}$ to the transitions t_6 and t_8, and that assigns the set $\{\lambda\}$ to all other transitions. Finally, let $\mathcal{M}_1 = (M_1, \omega_1)$, and let

$$
T_1 = \{\, \{c^m\} \mid m \ge 0 \,\} \cup \{\, \{d^n\} \mid n \ge 0 \,\}.
$$

Then $f^{\mathcal{M}_1}_{\omega_1}(w) \in T_1$ iff $w \in L(M_1)$, and $|w|_a = |w|_b$ or $2 \cdot |w|_a = |w|_b$, which yields

$$
L_{T_1}(\mathcal{M}_1) = \{\, a^n b^n \mid n \ge 0 \,\} \cup \{\, a^n b^{2n} \mid n \ge 0 \,\}.
$$

It is known that the language $L_{T_1}(\mathcal{M}_1)$ is not even accepted by any RW-automaton [2]. Hence, we see that the notion of relative acceptance increases the expressive power of R-automata.

3   On the Classes of Languages Accepted Relative to the Tropical Semiring $\mathbb{Z}_\infty$

Here we consider the languages that are accepted by weighted restarting automata relative to subsets of the tropical semiring $\mathbb{Z}_\infty = (\mathbb{Z} \cup \{\infty\}, \min, +, \infty, 0)$. Our first result states that by using the notion of acceptance relative to the family $\mathbb{H}_0 = \{\{0\}\}$, we can avoid auxiliary symbols.

Theorem 4. *For all* $\mathsf{X} \in \{\mathsf{RRW}, \mathsf{RW}\}$, $\mathcal{L}(\mathsf{XW}) \subseteq \mathcal{L}(\mathsf{X}, \mathbb{Z}_\infty, \mathbb{H}_0)$.

For proving this result, we need the following technical lemma.

Lemma 5. *For each $R(R)WW$-automaton M, there exists an $R(R)WW$-automaton M' accepting only on empty tape such that $L(M) = L(M')$.*

Proof. Let $M = (Q, \Sigma, \Gamma, \text{\textcent}, \$, q_0, k, \delta)$ be an R(R)WW-automaton. We construct an R(R)WW-automaton M' that simulates M as follows. In each cycle M' first guesses whether to simulate a cycle of M or whether M has already accepted.

– In the former case, another cycle is simulated, in which M' performs the same move-right, rewrite, and restart steps as M. However, each accept transition of M is simulated by a rewrite transition of M' that replaces the content of the window by a special symbol @, which indicates that the corresponding computation of M has accepted. If during the simulation of a cycle of M, the symbol @ is encountered by M', then M' halts without accepting.
– In the latter case, on seeing the symbol @, M' simply erases its left-hand neighbour and restarts - if, however, the symbol @ is not encountered, then M' halts without accepting.

This process is repeated until a tape content of the form ¢@$a$$ is reached, which M' then deletes symbol by symbol from right to left. After deleting the last symbol, M' halts and accepts. It should be clear that $L(M) = L(M')$.   □

Proof of Theorem 4. Let $M = (Q, \Sigma, \Gamma, ¢, \$, q_0, k, \delta)$ be an R(R)WW-automaton with input alphabet Σ. In order to prove the above inclusion, we construct an R(R)W-automaton M' and a weight function ω from M' to $\mathbb{Z}_\infty$ such that $L_{\{0\}}((M', \omega)) = L(M)$. By Lemma 5, we can assume without loss of generality that M always accepts on empty tape. Let $M' = (Q, \Gamma, \Gamma, ¢, \$, q_0', k', \delta')$ be the R(R)W-automaton that is obtained from M by simply taking all symbols as input symbols. Now we define the weight function ω as follows. To each rewrite transition of the form $(q', v) \in \delta(q, u)$, we assign the weight $|v|_{\Gamma \setminus \Sigma} - |u|_{\Gamma \setminus \Sigma}$, where $|x|_{\Gamma \setminus \Sigma}$ denotes the number of occurrences of symbols from the set $\Gamma \setminus \Sigma$ in x, that is, the number of occurrences of auxiliary letters in x, and we take $\omega(t) = 0$ for all other transitions t. Then, for each accepting computation AC of M' on input $w \in \Gamma^+$, $\omega(AC)$ is the number of auxiliary symbols that are written onto the tape during this computation minus the number of auxiliary symbols that are removed from the tape during this computation. Since M' (just as M) always accepts on empty tape, we see that $\omega(AC) = -|w|_{\Gamma \setminus \Sigma}$. Hence, $f_\omega^{M'}(w) = 0$ iff w does not contain any auxiliary symbols, that is, $L_{\{0\}}((M', \omega)) = L(M)$.   □

It remains open whether the inclusion in Theorem 4 is a proper one. We complete this section with a closure property for the language classes of the form $\mathcal{L}(\text{RRWW}, \mathbb{Z}_\infty, \mathbb{H})$.

Corollary 6. *The language class $\mathcal{L}(\text{RRWW}, \mathbb{Z}_\infty, \mathbb{H})$ is closed under the operation of reversal for each family $\mathbb{H}$ of subsets of $\mathbb{Z}_\infty$.*

This result follows immediately from the proof that the class of languages that are accepted by RRWW-automata is closed under the operation of reversal [3] and the fact that the tropical semiring $\mathbb{Z}_\infty$ is commutative.

4   On the Classes of Languages Accepted by Word-Weighted Restarting Automata

Now we study the classes of languages that are accepted by weighted restarting automata relative to subsets of a semiring of regular languages

$REG(\Delta) = (REG(\Delta), \cup, \cdot, \emptyset, \{\lambda\})$. In general, the weight of a transition of a restarting automaton M can be any regular language over Δ. However, some more restricted types of weighted restarting automata were introduced in [10]. Here we only consider the so-called *word-weighted restarting automata* that are defined as follows.

Definition 7. *A weighted restarting automaton $\mathcal{M} = (M, \omega)$ is called a word-weighted restarting automaton of type* X *(a* w_{word}X*-automaton for short), if M is a restarting automaton of type* X *and ω is a weight function from M into a semiring of the form* $REG(\Delta)$ *such that the weight $\omega(t)$ of each transition t of M is a singleton set, that is, it is of the form $\omega(t) = \{v\}$ for some $v \in \Delta^*$.*

For these word-weighted restarting automata, we define the following notion of relative acceptance.

Definition 8. *Let $\mathcal{M} = (M, \omega)$ be a* w_{word}X*-automaton with input alphabet Σ, where ω maps the transitions of M to singleton sets over Δ.*

(a) *For a set $T \in REG(\Delta)$, $\hat{L}_T(\mathcal{M}) = \{w \in L(M) \mid f_\omega^M(w) \cap T \neq \emptyset\}$ is the language accepted by $\mathcal{M}$ relative to the set T, that is, a word $w \in \Sigma^*$ belongs to the language $\hat{L}_T(\mathcal{M})$ iff $w \in L(M)$ and $f_\omega^M(w)$ contains at least one element of T.*

(b) *Let $\mathbb{H}$ be a family of subsets of $REG(\Delta)$. Then*

$$\hat{\mathcal{L}}(X, REG(\Delta), \mathbb{H}) = \{\hat{L}_T(\mathcal{M}) \mid \mathcal{M} \text{ is a } w_{word}X \text{-automaton and } T \in \mathbb{H}\}$$

is the class of languages that are accepted by w_{word}X*-automata relative to $\mathbb{H}$.*

For word-weighted restarting automata we have the following inclusion result.

Lemma 9. *For all types* $X \in \{R, RR, RW, RRW, RWW, RRWW\}$,

$$\mathcal{L}(nf\text{-}X) \subseteq \hat{\mathcal{L}}(X, REG(\Delta), REG(\Delta)).$$

Proof. Let $M = (Q, \Sigma, \Gamma, \mathbb{c}, \$, q_0, k, \delta)$ be a non-forgetting restarting automaton of type X. In order to prove the above inclusion, we construct a w_{word}X-automaton $\mathcal{M}' = (M', \omega)$ and we define a set $T \in REG(\Delta)$ such that $L(M) = \hat{L}_T(\mathcal{M}')$. The main problem in simulating M is the fact that, when executing a restart step, M can enter any state, while M' must return to its initial state q_0. To overcome this problem, for each restart transition $t : (q, \text{Restart}) \in \delta(p, u)$ of M, the automaton M' will have a restart transition $t' : \text{Restart} \in \delta'(p, u)$ with associated weight $\omega(t') = \{q\}$. Further, when starting from a restarting configuration, M' guesses the state q of M with which M begins the current cycle, and it then proceeds to simulate the next cycle of M starting in this state. In addition, the corresponding transition is given the weight $\{q\}$. If in each cycle, M' guesses the correct state of M, then the weight of the resulting accepting computation of M' is of the form $\{q_0 q_1 q_1 q_2 q_2 \ldots q_n q_n\}$ for some $q_1, q_2, \ldots, q_n \in Q$ and $n \geq 0$. Accordingly, we take $\Delta = Q$ and $T = \{q_0 q_1 q_1 q_2 q_2 \ldots q_n q_n \mid q_1, q_2, \ldots, q_n \in Q, n \geq 0\}$. To realize the above simulation, we take $M' = (Q', \Sigma, \Gamma, \mathbb{c}, \$, q_0, k, \delta')$, where $Q' = Q \cup \{(q, q_1, q_2) \mid q, q_1, q_2 \in Q\}$, and the transition relation δ' and the weight function ω are defined as shown below:

1. First, in order to allow M' to guess the state with which M begins the current cycle, δ' contains the transition $t' : ((p, q, q'), op) \in \delta'(q_0, \mathrevq{c}u)$ with associated weight $\{q\}$ for each transition $t : (p, op) \in \delta(q, \mathrevq{c}u)$ of M and each $q, q' \in Q$. Here the first state component is the current state of M, the second component is the guessed state with which the current cycle of M begins, and the third component is the guessed state that M will enter through the next restart step (if any).

2. In a state of the form (p, q, q'), M' proceeds just as M proceeds in state p, leaving state components 2 and 3 untouched, until it reaches a restart transition. All these move-right, rewrite and accept steps of M' have weight $\{\lambda\}$.

3. Finally, for each restart transition of the form $(q', \mathsf{Restart}) \in \delta(q, u)$, δ' contains the transitions $t_{q_1} : \mathsf{Restart} \in \delta'((q, q_1, q'), u)$ for all $q_1 \in Q$, which all have weight $\{q'\}$.

For an input $w \in \Sigma^*$, M' may have many more accepting computations than M. In fact, in general, $L(M')$ will be a proper superset of $L(M)$. However, $f_\omega^{M'}(w) \cap T \neq \emptyset$, iff M' has an accepting computation AC on input w such that $\omega(AC) \in T$. This means that within this computation, M' always guesses the correct state after each restart step, which shows that AC is the correct simulation of an accepting computation of M. It follows that $\hat{L}_T(\mathcal{M}) = L(M)$.   □

In fact, also the converse inclusions hold.

Lemma 10. *For all types* $\mathsf{X} \in \{\mathsf{R}, \mathsf{RR}, \mathsf{RW}, \mathsf{RRW}, \mathsf{RWW}, \mathsf{RRWW}\}$,

$$\hat{\mathcal{L}}(\mathsf{X}, \mathsf{REG}(\Delta), \mathsf{REG}(\Delta)) \subseteq \mathcal{L}(\mathsf{nf\text{-}X}).$$

Proof. Let Δ be a finite alphabet, let $\mathcal{M} = (M, \omega)$ be a $\mathsf{w_{word}}\mathsf{X}$-automaton, where $M = (Q, \Sigma, \Gamma, \mathrevq{c}, \$, q_0, k, \delta)$ is a restarting automaton of type X, and ω is a weight function that assigns to each transition of M a subset of Δ^* of cardinality one, and let $T \in \mathsf{REG}(\Delta)$. In order to prove the above inclusion we provide a non-forgetting restarting automaton M' of type X such that $L(M) = \hat{L}_T(\mathcal{M})$. For each $w \in \Sigma^*$, we have $w \in \hat{L}_T(\mathcal{M})$ iff $w \in L(M)$ and $f_\omega^M(w) \cap T \neq \emptyset$. As the set T is regular, there exists a deterministic finite automaton (DFA) $A = (Q_A, \Delta, \delta_A, q_0^A, F_A)$ such that $L(A) = T$, where Q_A is a finite set of states, Δ is the input alphabet for A, $\delta_A : Q_A \times \Delta \to Q_A$ is the transition function, q_0^A is the initial state, and F_A is a set of accepting states. For an input $w \in \Sigma^*$, M' has to check whether $w \in L(M)$ and whether $f_\omega^M(w) \cap L(A) \neq \emptyset$. Therefore, M' needs to simulate both M and A simultaneously. Accordingly, each state of M' is a pair (p, q), where $p \in Q$ and $q \in Q_A$, and when simulating a step of M, M' needs to ensure that A has a transition that is applicable to the weight of this step. As it is non-forgetting, M' can always remember the actual state of A, even after executing a restart step. If M accepts, then M' also accepts, provided that A reaches a final state by reading the weight of the current accept step.   □

Together, Lemmas 9 and 10 yield the following characterization.

Theorem 11. *For all types* $X \in \{R, RR, RW, RRW, RWW, RRWW\}$,

$$\mathcal{L}(\mathsf{nf}\text{-}X) = \hat{\mathcal{L}}(X, REG(\Delta), REG(\Delta)).$$

Based on this result a characterization of the class CFL of context-free languages in terms of word-weighted restarting automata can be derived from a corresponding result for non-forgetting restarting automata [7]. Also this section closes with a look at some closure properties.

Theorem 12. *The class* $\hat{\mathcal{L}}(X, REG(\Delta), REG(\Delta))$ *is closed under the operation of union for each* $X \in \{R, RR, RW, RRW, RWW, RRWW\}$.

Proof. Let $\mathcal{M}_1 = (M_1, \omega_1)$ and $\mathcal{M}_2 = (M_2, \omega_2)$ be $\mathsf{w}_{\mathsf{word}}X$-automata, and let $T_1, T_2 \in REG(\Delta)$. By Theorem 11, there exist non-forgetting restarting automata M_1' and M_2' of type X such that $L(M_1') = L_{T_1}(\mathcal{M}_1)$ and $L(M_2') = L_{T_2}(\mathcal{M}_2)$. Thus, in order to prove the above closure property, it suffices to construct a non-forgetting restarting automaton M of type X such that $L(M) = L(M_1') \cup L(M_2')$. At the start, M nondeterministically chooses an index $i \in \{1, 2\}$, and then it simply works exactly like M_i'. As M is non-forgetting, it can store its guess within its finite-state control. If M_i' accepts, then M also accepts. Thus, M accepts on input w iff at least one of M_1' or M_2' accepts on input w. It follows that $L(M) = L(M_1') \cup L(M_2')$. □

In [3] it is shown that the language classes $\mathcal{L}(RWW)$ and $\mathcal{L}(RRWW)$ are closed under the operation of concatenation. This result also holds for the class of languages that are defined by word-weighted restarting automata.

Theorem 13. *The class* $\hat{\mathcal{L}}(X, REG(\Delta), REG(\Delta))$ *is closed under the operation of concatenation for each* $X \in \{RWW, RRWW\}$.

Proof. The proof for RWW- and RRWW-automata given in [3] proceeds as follows. On input a word w, a factorization $w = uv$ is guessed such that u is accepted by the first automaton and v is accepted by the second. To fix this guess, the last symbol a of u and the first symbol b of v are rewritten into a special symbol $[a, b]$, and then the first automaton is simulted on u. If and when it accepts, then the second automaton is simulated on v. In the same way, we can proceed for non-forgetting RWW- and RRWW-automata. By Theorem 11 this yields the intended closure property for $\hat{\mathcal{L}}(X, REG(\Delta), REG(\Delta))$. □

Finally, we return to the operation of reversal. In [3] it is shown that the class of languages that are accepted by RRWW-automata is closed under reversal. As the proof carries over to non-forgetting RRWW-automata, we immediately obtain the following result from Theorem 11.

Corollary 14. *The class* $\hat{\mathcal{L}}(RRWW, REG(\Delta), REG(\Delta))$ *is closed under the operation of reversal.*

5   A Stronger Restriction for Word-Weighted Restarting Automata

A word $w \in \Sigma^*$ is an element of the language $\hat{L}_T(\mathcal{M})$ for a word-weighted restarting automaton $\mathcal{M} = (M, \omega)$ and a set $T \in \mathsf{REG}(\Delta)$, if $w \in L(M)$ and the weight $\omega(AC)$ is an element of T for at least one accepting computation AC of M on input w. Thus, there may be other accepting computations of M on this very input that have an associated weight that does not belong to the set T. The following definition requires that $\omega(AC)$ must belong to T for *each* accepting computation AC of M on input w.

Definition 15. *Let $\mathcal{M} = (M, \omega)$ be a $\mathsf{w}_{\mathsf{word}}X$-automaton with input alphabet Σ, where ω maps the transitions of M to singleton sets over Δ. For a set $T \subseteq \mathbb{P}_{\mathsf{fin}}(\Delta^*)$, $L_T(\mathcal{M}) = \{\, w \in L(M) \mid f_\omega^M(w) \in T \,\}$ is the language strongly accepted by $\mathcal{M}$ relative to the set T, that is, a word $w \in \Sigma^*$ belongs to the language $L_T(\mathcal{M})$ iff $w \in L(M)$ and $f_\omega^M(w)$ coincides with an element of T.*

Actually, this definition is exactly in the spirit of Definition 2. Indeed, let S be the semiring of regular languages over Δ. For a $\mathsf{w}_{\mathsf{word}}X$-automaton $\mathcal{M} = (M, \omega)$ and an input $w \in \Sigma^*$, the value $f_\omega^M(w)$ is a finite subset of Δ^*. Thus, it suffices to consider subsets of S that consist of finite languages. Accordingly, if T is a collection of finite subsets of Δ^*, then an input word $w \in \Sigma^*$ belongs to the language $L_T(\mathcal{M})$ iff $w \in L(M)$ and $f_\omega^M(w)$ is an element of T.

Concerning the classes of languages that are defined using this stronger notion, we have the following important closure property.

Theorem 16. *Let $\mathcal{M}_1 = (M_1, \omega_1)$ and $\mathcal{M}_2 = (M_2, \omega_2)$ be $\mathsf{w}_{\mathsf{word}}X$-automata, where $X \in \{\mathsf{RWW}, \mathsf{RRWW}\}$ and ω_1 and ω_2 map the transitions of M_1 and M_2 to singleton sets over Δ, and let $T_1, T_2 \subseteq \mathbb{P}_{\mathsf{fin}}(\Delta^*)$. Then there are an alphabet Δ', a $\mathsf{w}_{\mathsf{word}}X$-automaton $\mathcal{M} = (M, \omega)$, where ω maps the transitions of M to singleton sets over Δ', and a set $T \subseteq \mathbb{P}_{\mathsf{fin}}(\Delta'^*)$ such that $L_T(\mathcal{M}) = L_{T_1}(\mathcal{M}_1) \cap L_{T_2}(\mathcal{M}_2)$.*

Proof. For $i = 1, 2$, let $\mathcal{M}_i = (M_i, \omega_i)$, where $M_i = (Q_i, \Sigma, \Gamma_i, \mathbb{c}, \$, q_0^{(i)}, k_i, \delta_i)$ is an RRWW-automaton, ω_i is a weight function that maps the transitions of M_i to singleton sets over Δ, and let $T_i \subseteq \mathbb{P}_{\mathsf{fin}}(\Delta^*)$. We construct an alphabet Δ', a word-weighted RRWW-automaton $\mathcal{M} = (M, \omega)$, where ω maps the transitions of M to singleton sets over Δ', and a subset $T \subseteq \mathbb{P}_{\mathsf{fin}}(\Delta'^*)$ such that $L_T(\mathcal{M}) = L_{T_1}(\mathcal{M}_1) \cap L_{T_2}(\mathcal{M}_2)$, that is, for all $w \in \Sigma^*$, $w \in L_T(\mathcal{M})$ iff $w \in L_{T_1}(\mathcal{M}_1)$ and $w \in L_{T_2}(\mathcal{M}_2)$.

On input a word $w \in \Sigma^*$, the automaton M will be able to simulate M_1 as well as M_2. Essentially, the simulation of an accepting computation of M_1 on input w should give the same weight as the corresponding computation of M_1, and analogously, the simulation of an accepting computation of M_2 on input w should give the same weight as the corresponding computation of M_2. However, as the elements of T_1 and T_2 are subsets of Δ^*, it could happen that $f_{\omega_1}^{M_1}(w)$ is an element of T_2, although $w \notin L_{T_1}(M_1)$. Thus, we must ensure that the set of weights of the simulations of all accepting computations of M_1 for an input word w cannot be an element of T_2, and analogously for simulations of accepting computations of M_2 and T_1.

For this purpose, we define the following sets and mappings. Let $\Delta_1 = \Delta \cup \{@\}$, $\hat{\Delta} = \{\hat{a} \mid a \in \Delta\}$, $\Delta_2 = \hat{\Delta} \cup \{\hat{@}\}$, and let $\Delta' = \Delta_1 \cup \Delta_2$. Further, let σ_1, σ_2, and σ' be the mappings that are given through

$$\begin{aligned}
\sigma_1(w) &= w@ && \text{for } w \in \Delta^*, \\
\sigma_2(w) &= \hat{a}_1 \hat{a}_2 \ldots \hat{a}_n \hat{@} \text{ for } w = a_1 a_2 \ldots a_n \in \Delta^*, \\
\sigma'(\lambda) &= \lambda, \\
\sigma'(w) &= \hat{a}_1 \hat{a}_2 \ldots \hat{a}_n && \text{for } w = a_1 a_2 \ldots a_n \in \Delta^+,
\end{aligned}$$

which are extended to sets by simply applying them to all elements of a given set. Finally, let ω_1' and ω_2' be the weight functions that are defined as follows:

$$\begin{aligned}
\omega_1'(t) &= \{u@\} && \text{for each accept transition } t \in \delta_1, \text{where } \omega_1(t) = \{u\}, \\
\omega_1'(t) &= \omega_1(t) && \text{for all other transitions } t \in \delta_1, \\
\omega_2'(t) &= \{\sigma'(u)\hat{@}\} \text{ for each accept transition } t \in \delta_2, \text{where } \omega_2(t) = \{u\}, \\
\omega_2'(t) &= \sigma'(\omega_2(t)) \text{ for all other transition } t \in \delta_2.
\end{aligned}$$

Now let $\mathcal{M}_1' = (M_1, \omega_1')$ and $\mathcal{M}_2' = (M_2, \omega_2')$, and let $T_1' = \{\sigma_1(V) \mid V \in T_1\}$ and $T_2' = \{\sigma_2(V) \mid V \in T_2\}$. Then $L_{T_1}(\mathcal{M}_1) = L_{T_1'}(\mathcal{M}_1')$ and $L_{T_2}(\mathcal{M}_2) = L_{T_2'}(\mathcal{M}_2')$. It is easily seen that $f_{\omega_1'}^{M_1}(w) \subseteq \Delta_1^+$ and $f_{\omega_2'}^{M_2}(w) \subseteq \Delta_2^+$ for each $w \in \Sigma^*$.

The RRWW-automaton M and the weight function ω are defined as follows, where $M = (Q, \Sigma, \Gamma, \mathfrak{c}, \$, q_0, k, \delta)$. If $\max\{k_1, k_2\} = 1$, we take $k = 2$; otherwise, we take $k = \max\{k_1, k_2\}$. Starting from the initial configuration on input $w \in \Sigma^*$, M first guesses whether to simulate M_1 or M_2. In order to remember its guess, δ contains some transitions that allow M to combine the first two symbols a_1 and a_2 of w into a special auxiliary symbol of the form $[a_1, a_2, i]$, where $i \in \{1, 2\}$ is the above guess. The weight function ω assigns the set $\{\lambda\}$ to these transitions. In the subsequent cycles, M simulates the machine M_i on seeing the symbol $[a_1, a_2, i]$. Of course, the symbol $[a_1, a_2, i]$ leads to some adjustments in the construction of the transitions of M that simulate M_1 and M_2. However, this technique has already been presented in detail in the proof of Theorem 6 of [9]. The simulation of an accepting computation AC of M_i on input w will yield an accepting computation of M that has exactly weight $\omega_i'(AC)$. Finally, let $T = \{V_1 \cup V_2 \mid V_1 \in T_1' \text{ and } V_2 \in T_2'\}$. Then, for each $w \in \Sigma^*$, $w \in L_T(\mathcal{M})$ iff $w \in L(M_1) \cup L(M_2)$ and it holds that $f_\omega^M(w) \in T$. The latter means that there exist a subset $V_1 \in T_1'$ and a subset $V_2 \in T_2'$ such that $f_\omega^M(w) = V_1 \cup V_2$. This, however, implies that $f_{\omega_1'}^{M_1}(w) = V_1$ and $f_{\omega_2'}^{M_2}(w) = V_2$, which means in particular that both, M_1 and M_2, accept on input w. If follows that $L_T(\mathcal{M}) = L_{T_1'}(\mathcal{M}_1') \cap L_{T_2'}(\mathcal{M}_2')$. For RWW-automata, the result can be proved in exactly the same way. □

This is the first result that shows that a class of languages defined in terms of a quite general class of restarting automata is closed under the operation of intersection. Using essentially the same technique also the following closure property can be derived.

Theorem 17. *Let $\mathcal{M}_1 = (M_1, \omega_1)$ and $\mathcal{M}_2 = (M_2, \omega_2)$ be $\mathsf{w}_{\mathsf{word}}\mathsf{X}$-automata, where $\mathsf{X} \in \{\mathsf{RWW}, \mathsf{RRWW}\}$ and ω_1 and ω_2 map the transitions of M_1 and M_2 to singleton sets over Δ, and let $T_1, T_2 \subseteq \mathbb{P}_{\mathrm{fin}}(\Delta^*)$. Then there are an alphabet Δ', a $\mathsf{w}_{\mathsf{word}}\mathsf{X}$-automaton $\mathcal{M} = (M, \omega)$, where ω maps the transitions of M to singleton sets over Δ', and a set $T \subseteq \mathbb{P}_{\mathrm{fin}}(\Delta'^*)$ such that $L_T(\mathcal{M}) = L_{T_1}(\mathcal{M}_1) \cup L_{T_2}(\mathcal{M}_2)$.*

In fact, we also have the following result.

Theorem 18. *Let $\mathcal{M}_1 = (M_1, \omega_1)$ and $\mathcal{M}_2 = (M_2, \omega_2)$ be $\mathsf{w}_{\mathsf{word}}\mathsf{X}$-automata, where $\mathsf{X} \in \{\mathsf{RWW}, \mathsf{RRWW}\}$ and ω_1 and ω_2 map the transitions of M_1 and M_2 to singleton sets over Δ, and let $T_1, T_2 \subseteq \mathbb{P}_{\mathrm{fin}}(\Delta^*)$. Then there are an alphabet Δ', a $\mathsf{w}_{\mathsf{word}}\mathsf{X}$-automaton $\mathcal{M} = (M, \omega)$, where ω maps the transitions of M to singleton sets over Δ', and a set $T \subseteq \mathbb{P}_{\mathrm{fin}}(\Delta'^*)$ such that $L_T(\mathcal{M}) = L_{T_1}(\mathcal{M}_1) \cdot L_{T_2}(\mathcal{M}_2)$.*

Proof. It is known that the language classes $\mathcal{L}(\mathsf{RWW})$ and $\mathcal{L}(\mathsf{RRWW})$ are closed under the operation of concatenation [3]. The central idea of the proof is to guess a factorization $w = uv$ for an input w, to combine the last symbol a of u and the first symbol b of v into a special symbol $[a, b]$ in order to fix this guess, and to simulate the first automaton on u and the second on v. Using the alphabets and mappings from the proof of Theorem 16, and by taking $T = \{ V_1 \cdot V_2 \mid V_1 \in T_1' \text{ and } V_2 \in T_2' \}$, the simulation technique from [3] can be used. □

6   Conclusions

We have introduced the notion of acceptance relative to a subset of a semiring to use the weight function of a weighted restarting automaton to specify a language. This language is obtained from the language that is accepted by the underlying (unweighted) restarting automaton by restricting the weight associated to a given input word through an additional requirement. Here we have only considered the case of the tropical semiring $\mathbb{Z}_\infty$ and that of the semiring of regular languages $\mathsf{REG}(\Delta)$ over a finite alphabet Δ. We have seen that by using the semiring $\mathbb{Z}_\infty$, we can simulate the computations of a restarting automaton with auxiliary symbols by an automaton without auxiliary symbols. For the notion of acceptance relative to a regular language, we have shown that our notion of relative acceptance just corresponds to the non-forgetting restarting automata. Finally, in a more restricted setting we have even presented a class of languages that is specified by general restarting automata, but which is nevertheless closed under the operation of intersection.

Actually, it can be shown that also the latter two types of relative acceptance can be used to replace restarting automata with auxiliary symbols by automata without auxiliary symbols. However, many problems remain open. Most importantly, we do not yet have a characterization for the classes of languages that are accepted by the various types of weighted restarting automata relative to subsets of the associated semiring.

References

1. Jancar, P., Mráz, F., Plátek, M., Vogel, J.: Restarting automata. In: Reichel, H. (ed.) FCT 1995. LNCS, vol. 965, pp. 283–292. Springer, Heidelberg (1995)
2. Jancar, P., Mráz, F., Plátek, M., Vogel, J.: On monotonic automata with a restart operation. J. Auto. Lang. Comb. **4**(4), 287–312 (1999)
3. Jurdziński, T., Lorys, K., Niemann, G., Otto, F.: Some results on RWW- and RRWW-automata and their relation to the class of growing context-sensitive languages. J. Autom. Lang. Comb. **9**(4), 407–437 (2004)
4. Messerschmidt, H., Otto, F.: Cooperating distributed systems of restarting automata. Int. J. Found. Comput. Sci. **18**, 1333–1342 (2007)
5. Messerschmidt, H., Stamer, H.: Restart-Automaten mit mehreren Restart-Zuständen. In: Bordihn, H. (ed.) Workshop "Formale Methoden in der Linguistik" und 14. Theorietag "Automaten und Formale Sprachen". pp. 111–116. Institut für Informatik, Universität Potsdam, Potsdam (2004)
6. Messerschmidt, H.: CD-Systems of restarting automata. Ph.D. thesis, Fachbereich Elektrotechnik/Informatik, Universität Kassel (2008)
7. Messerschmidt, H., Otto, F.: On nonforgetting restarting automata that are deterministic and/or monotone. In: Grigoriev, D., Harrison, J., Hirsch, E.A. (eds.) CSR 2006. LNCS, vol. 3967, pp. 247–258. Springer, Heidelberg (2006)
8. Otto, F.: Restarting automata. In: Ésik, Z., Martín-Vide, C., Mitrana, V. (eds.) Recent Advances in Formal Languages and Applications, vol. 25, pp. 269–303. Springer, Heidelberg (2006)
9. Otto, F., Wang, Q.: Weighted restarting automata. Soft Computing (2016). Accepted. The results of this paper have been announced at WATA 2014 in Leipzig. doi:10.1007/s00500-016-2164-4
10. Wang, Q., Hundeshagen, N., Otto, F.: Weighted restarting automata and pushdown relations. In: Maletti, A. (ed.) Algebraic Informatics. LNCS, vol. 9270, pp. 196–207. Springer, Switzerland (2015)

Enhancing Practical TAG Parsing Efficiency by Capturing Redundancy

Jakub Waszczuk[1]([✉]), Agata Savary[1], and Yannick Parmentier[2]

[1] Laboratoire d'informatique, Université François-Rabelais Tours, Blois, France
{jakub.waszczuk,agata.savary}@univ-tours.fr
[2] LIFO - Université d'Orléans, Orléans, France
yannick.parmentier@univ-orleans.fr

Abstract. The efficiency of parsing with tree adjoining grammars (TAGs) depends not only on the size of the input sentence but also, linearly, on the size of the input TAG, which can attain several thousands of elementary trees. We propose a factorized, finite-state TAG representation to cope with this combinatorial explosion. The associated parsing algorithm shows a substantial performance gain on a real-size French TAG.

Keywords: Parsing · Tree-adjoining grammars · Grammar compression · Finite-state automata · Hypergraphs

1 Introduction

High lexicalization and the so-called *extended domain of locality*[1] of TAGs [9], while beneficial for grammar development, are known to lead to very large grammars with up to several thousands of elementary trees [16]. This poses problems of practical nature – parsing algorithms for TAGs are polynomial in the size of the input sentence but also at least linear in the size of the underlying grammar. While many parsing algorithms for speeding up TAG parsing exist, we propose a novel approach in which redundancy is captured by combining and optimizing several previously proposed techniques: grammar flattening, subtree sharing, rule compression into a unique finite-state automaton, and adaptation of parsing inference rules to this representation. Experiments show that these measures lead to a substantial gain in space and time efficiency.

2 Tree Adjoining Grammars

Let Σ and $\mathcal{N}_0$ be disjoint sets of terminal and non-terminal symbols. An *initial tree* (IT) is a tree with non-terminals in non-leaf nodes and terminals/non-terminals in leaf nodes. An *auxiliary tree* (AT) is similar to an IT but it has one

---

This work has been supported by the PARSEME European COST Action (IC1207) and by the PARSEME-FR French ANR project (ANR-14-CERA-0001-01).

[1] The former meaning that elementary grammar units are typically attached to one or more lexical items, the latter that many syntactic phenomena can be conveniently represented locally, at the level of individual elementary units.

Y.-S. Han and K. Salomaa (Eds.): CIAA 2016, LNCS 9705, pp. 310–321, 2016.
DOI: 10.1007/978-3-319-40946-7_26

distinguished leaf (usually marked with an asterisk), called a *foot*, containing the same non-terminal as the root. For instance, in Fig. 2, t_1, t_5 and t_6 are ITs, while t_2, t_3 and t_4 are ATs. A TAG is defined as a tuple $(\Sigma, \mathcal{N}_0, \mathcal{I}, \mathcal{A}, \mathcal{S})$ where $\mathcal{I}$ is the set of elementary initial trees (EITs), $\mathcal{A}$ is the set of elementary auxiliary trees (EATs), and $\mathcal{S}$ is the start non-terminal.

A *derived tree* is created from EITs and EATs by *substitution* and *adjunction*. Given an IT t, and any tree t', substitution replaces a non-terminal leaf l in t' by t provided that labels in l and in t's root are equal. Given an AT t, and any tree t', adjunction replaces t's foot by a subtree t'' of t' and then inserts this modified t in place of t'' in t', provided that the root non-terminals in t and t'' are identical, as shown in Fig. 1. A *derivation tree* keeps track of the operations and the elementary trees (ETs) involved in the creation of a derived tree.

A sequence of terminals obtained by an in-order traversal of a tree t is called a *projection* of t, written $proj(t)$. We also define $proj_A(t)$, specialized to ATs, as a pair of terminal sequences on the left and on the right of the foot node, respectively.

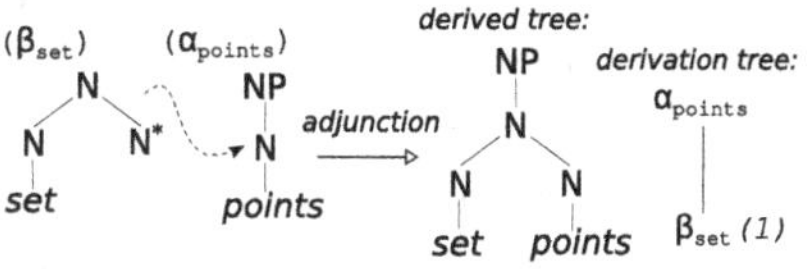

Fig. 1. Adjunction of the tree t_2 to the tree t_5 from Fig. 2

We say that a tree t can be *derived from* a non-trivial subtree[2] t_0 (auxiliary or not) of an ET iff (i) tree t can be derived from the grammar extended with t_0 as an ET and (ii) a derivation tree d of t exists such that t_0 occurs in d's root and, unless t_0 is already part of the grammar, nowhere else in d. We will also say that a non-auxiliary subtree t_0 of an ET is *recognized* over a span (i, l) of the input sentence s iff a tree t can be derived from t_0 such that $proj(t) = s_{(i,l)}$, where $s_{(i,l)}$ is a part of sentence s containing its words between positions i and l. Similarly, we will say that an auxiliary subtree t_0 of an ET is recognized over a span (i, j, k, l) iff a tree t can be derived from t_0 such that $proj_A(t) = (s_{(i,j)}, s_{(k,l)})$.

3   Grammar Factorization

Consider the sentence in example (1) and the toy lexicalized TAG (LTAG) containing trees $t_1, \ldots, t_6$ from Fig. 2 covering several competing interpretations for the two initial words.

(1) **Set points** in tennis belong to official scoring.

The IT t_1 represents *set* as a phrasal verb in imperative mode taking a direct object and a prepositional complement governed by *in*. ATs t_2, t_3 and t_4 consider *set* as a nominal, adjectival and participle modifier of a head noun, respectively. In the IT t_5 *points* is a nominal phrase, while t_6, having two terminals, corresponds to the idiomatic interpretation of *set points* as an NN compound.

---

[2] In the rest of this paper, by *subtree* we mean a non-trivial (of height > 0) subtree, unless explicitly stated otherwise.

3.1   Grammar Flattening with Subtree Sharing

We propose to represent each ET as a set of flat production rules, so that common subtrees are shared (cf. Fig. 2). Each non-terminal from an internal (non-root and non-leaf) node receives a unique index, and each non-leaf node together with its children yields a production rule. E.g., nodes VP and PP with their children in t_1 yield the rules $VP_1 \rightarrow V_2\ NP\ PP_3$, $PP_3 \rightarrow P_4\ NP$, respectively. Additionally, each node on the spine of an AT is marked by an asterisk, e.g., the root of t_2 becomes N^* in the head of the rule $N^* \rightarrow N_5 N^*$.

Note also that the non-terminal N, occurring twice in t_6, yields two different non-terminals N_0 and N_5 in order to prevent non-compatible rule combinations. For instance, we should not admit an NN-compound *points set*, which would be admitted if these two N terminals were not distinguished. Note, however, also that as soon as some subtrees are common for different grammar trees, the indexed non-terminals, and consequently the target rules, can be shared. For example, the nominal interpretations of *set* and *points* common for t_2, t_5 and t_6 can be shared via the common production rules $N_5 \rightarrow set$ and $N_0 \rightarrow points$.

In what follows, we refer to such a grammar conversion as *flattening with subtree sharing* (FSS), and to the conversion result as an *FSS grammar* (FSSG).

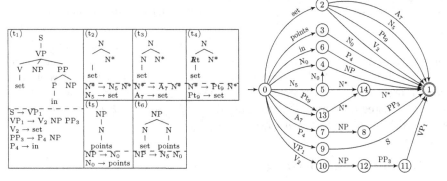

Fig. 2. A toy LTAG grammar and its FSSG.

Fig. 3. Compression of the FSSG from Fig. 2 into an FSSA.

Formally, the FSSG constructed from a TAG $G = (\Sigma, \mathcal{N}_0, \mathcal{I}, \mathcal{A}, \mathcal{S})$ is a set of production rules $\alpha \in \mathcal{N} \times (\mathcal{N} \cup \Sigma)^+$ where the first and the second component represent the head and the non-empty body of the rule, respectively. $\mathcal{N}_0$ is the set of FSSG non-terminals, i.e. triples $X \in \mathcal{N}_0 \times (\mathbb{N} \cup \{-\}) \times \{-, *\}$ where '$-$' indicates that the corresponding value is unbound. Internal nodes are marked with unique identifiers from the set of natural numbers $\mathbb{N}$. A non-terminal $(x, u, a) \in \mathcal{N}$ is alternatively written as x_u^a and unbounded values $(-)$ are ignored. For example, $(N, -, *)$ is equivalent to N^*, $(V, 2, -)$ to V_2 and $(NP, -, -)$ to NP.

The FSS conversion determines a bijection R_0 between non-terminals originating from internal nodes ($X \in \mathcal{N}_0 \times \mathbb{N} \times \{-, *\}$) and proper subtrees of ETs.

A subtree common to several ETs (e.g., the subtree rooted at N dominating *set* in trees t_2 and t_6 in Fig. 2) is represented, in the FSSG, by a single non-terminal (here: N_5). We define a 1-to-many correspondence R between non-terminals ($X = (x, u, a) \in \mathcal{N}$) and TAG subtrees as an extension of this bijection:

$$R(X) = \begin{cases} \{R_0(X)\} & \text{if } u \neq - \\ \mathcal{I}|_x & \text{if } (u, a) = (-, -) \\ \mathcal{A}|_x & \text{if } (u, a) = (-, *) \end{cases} \tag{1}$$

where $\mathcal{I}|_x$ and $\mathcal{A}|_x$ are the sets of all EITs and all EATs, respectively, rooted at $x \in \mathcal{N}_0$. E.g., in Fig. 2, $R(NP) = \{t_5, t_6\}$ and $R(N^*) = \{t_2, t_3, t_4\}$.

3.2   Automaton-Based Grammar Compression

Despite subtree sharing applied to the FSSG in Fig. 2, it still shows some degree of redundancy: the terminal *set* constitutes the body of 4 rules (headed by V_2, N_5, A_7 and Pt_9), the non-terminal NP occurs in the head of 2 rules, and the spine non-terminal N^* appears in the head and in the suffix of 3 rules. This observation leads to the idea of representing the FSSG as a minimal deterministic finite-state automaton (DFSA), called here *FSSA*, as shown in Fig. 3. The FSSA's alphabet consists of terminals and non-terminals of the FSSG rules. Each path represents the right-hand side of a rule followed by its head.[3] For instance, the bottom path, traversing nodes 0, 10, 12, 11 and 1, represents the rule $VP_1 \rightarrow V_2\ NP\ PP_3$. In this representation redundancy is largely avoided: the terminal *set* and the head non-terminals NP and N^*, are represented by unique transitions $(0, set, 2)$, $(4, NP, 1)$ and $(14, N^*, 1)$, respectively. Additionally, transition $(13, N^*, 14)$ is shared by the suffixes of rules $N^* \rightarrow A_7 N^*$ and $N^* \rightarrow Pt_9 N^*$.

In what follows we extend the notion of an FSSA-based grammar compression into the case when the grammar rules are possibly represented as a *set* of FSSAs (with disjoint sets of node identifiers), according to the particular variant of the compression technique. For instance, in [12] all grammar rules having the same head non-terminal are compressed into a separate DFSA. One of the versions of our parser tested in Sect. 5 implements a similar compression idea.

For a grammar represented as a set of FSSAs, and for any state q therein, let $P(q)$ be a set of sequences of labels leading from an initial state to q. For instance, in Fig. 3, $P(14) = \{N_5 N^*, Pt_9 N^*, A_7 N^*\}$. Note that if q is non-final, sequences in $P(q)$ correspond to prefixes of rules' bodies. In particular, $P(q) \in (N \cup T)^*$.

4   Parser

We propose two Earley-style [6] bottom-up TAG parsing algorithms. The first one, called an *FSS parser*, is inspired by [14], and differs from this seminal work in that it uses an FSSG instead of the original TAG and ignores prediction. The other one, called an *FSSA parser* and inspired by [12], is an extension of the FSS

---

[3] Head non-terminals are distinguished from others, which is neglected in Fig. 3.

parser in that it uses the FSSG compressed into FSSAs. In both algorithms parsing can be seen, after [11], as a dynamic construction of a *hypergraph* [8] whose nodes are parsing chart items and whose hyperarcs represent applications of inference rules. The hypergraph representation facilitates comparisons between the two algorithms, and time efficiency estimations (the number of elementary parsing steps can be approximated by the number of hyperarcs). It also provides a compressed representation of all the derived trees for a given input sentence.

4.1   FSS Parser

Figure 4 shows the hypergraph created while parsing the two initial words of sentence (1) by the FSS parser with the FSSG from Fig. 2. Due to space constraints, we do not formally define the inference rules of the FSS parser here. They can be seen as simplified versions of those defined in Sect. 4.4. Each item contains a dotted rule and the span over which the symbols to the left of the dot have been parsed. E.g., the hyperarc leading from $(N_5 \rightarrow \bullet set, 0, 0)$ to $(N_5 \rightarrow set\bullet, 0, 1)$ means that the terminal *set* has been recognized from position 0 to 1. The latter item can be combined with $(NP \rightarrow \bullet N_5 N_0, 0, 0)$ yielding $(NP \rightarrow N_5 \bullet N_0, 0, 1)$, etc. The sentence s has been parsed if a goal item has been reached (spanning from 0 to $|s|$, with a rule headed by $(S, -, -)$ and terminated by a dot).

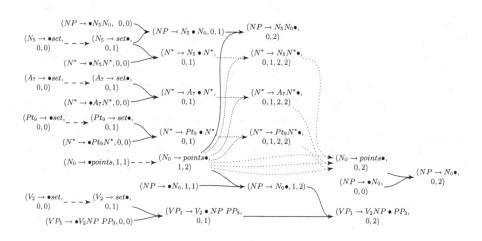

Fig. 4. Hypergraph created by the FSS parser while parsing the substring *set points* with the FSSG from Fig. 2. The dashed and plain hyperarcs roughly correspond to scanner and completer operations in a CFG Earley parser. The densely and loosely dotted hyperarcs represent novel inference rules: foot adjoin and root adjoin.

Items whose spans contain 4 integers (i_1, i_2, i_3, i_4) result from the FSS-based inference rules related to adjunction: i_1 and i_4 represent the whole span of the recognized sequence, while i_2 and i_3 indicate the *gap*, i.e., the part of the sequence matched by the foot node of an AT. For instance, the hyperarc leading from

$(N^* \rightarrow N_5 \bullet N^*, 0, 1)$ and $(N_0 \rightarrow points\bullet, 1, 2)$ to $(N^* \rightarrow N_5N^*\bullet, 0, 1, 2, 2)$ puts forward an adjunction hypothesis. The noun *points* has been recognized over span $(1, 2)$, and *set* recognized over $(0, 1)$ might later be adjoined to it as a modifier. Thus, *points* will fill the gap (from 1 to 2) corresponding to the foot node N^* in the body of rule $N^* \rightarrow N_5N^*$ (stemming from tree t_2). Note further that the combination of items $(N^* \rightarrow N_5N^*\bullet, 0, 1, 2, 2)$ and $(N_0 \rightarrow points\bullet, 1, 2)$ yields $(N_0 \rightarrow points\bullet, 0, 2)$, which corresponds to stage 1 of the adjunction (see Sect. 2). Stage 2 is then represented by the hyperarc leading to $(NP \rightarrow N_0\bullet, 0, 2)$.

4.2   FSSA Parser

The idea behind grammar compression is not only space efficiency but also reducing parsing time [12]. The latter is based on the observation that, whenever bodies of some flat rules share common prefixes and/or suffixes (which is in close relation to sharing sub-paths in the FSSA), partial parsing results can be shared for them. Another related fact is that, for a given position of the dot in a flat dotted rule, the history of the parsing on the left-hand side of the dot does not influence the future parsing on the right-hand side of the dot. Therefore, the position of the dot in a rule can be nicely represented by the FSSA state achieved while parsing the rule, whatever the path which led us to this state.

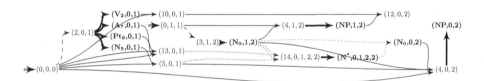

Fig. 5. Hypergraph representing the chart parsing of the substring *set points* with the FSSA from Fig. 3. The double, plain, thick, dashed, densely dotted and loosely dotted hyperarcs represent axioms, pseudo substitution, deactivate, scan, foot adjoin and root adjoin inference rules, respectively (see Sect. 4.4). Passive states are highlighted in bold.

These observations may lead to a substantial reduction of the parsing hypergraph, as shown in Fig. 5. Here, dotted rules in the hypergraph items from Fig. 4 are replaced by states of the FSSA from Fig. 3 (the resulting items are called *active*). Firstly, all 9 initial items (i.e., having the dot at the beginning of their rules' bodies) over span $(0, 0)$ in Fig. 4, e.g., $(N_5 \rightarrow \bullet set, 0, 0)$, $(NP \rightarrow \bullet N_5N_0, 0, 0)$, etc. – are replaced by a unique item $(0, 0, 0)$ in Fig. 5 due to the fact that they all share the same (empty) prefix on the left-hand side of the dot, and the same span. The 10th remaining initial item $(N_0 \rightarrow \bullet points, 1, 1)$ is replaced by $(0, 1, 1)$. Further, rules having dots inside their bodies are replaced by FSSA states, for instance items $(N^* \rightarrow A_7 \bullet N, 0, 1)$ and $(N^* \rightarrow Pt_9 \bullet N, 0, 1)$ are replaced by the unique item $(13, 0, 1)$ since their prefixes A_7 and Pt_9 lead to the same state 13. Finally, complete items (i.e., having the dot at the end of the rule), are replaced by two items, the one containing the arrival state, and the other (called a *passive* item) in which the state is replaced by the head of the fully recognized rule. For instance, items $(N^* \rightarrow N_5 N^* \bullet, 0, 1, 2, 2)$, $(N^* \rightarrow A_7 N^* \bullet, 0, 1, 2, 2)$ and $(N^* \rightarrow Pt_9 N^* \bullet, 0, 1, 2, 2)$ are merged into one active item $(14, 0, 1, 2, 2)$ since they share the same arrival state 14 and span. This item is then followed by a passive item $(N^*, 0, 1, 2, 2)$. The goal item is $(S, 0, |s|)$.

4.3 Items

Let $s = s_0 s_1 \ldots s_{n-1}$ be the input sentence and $Pos(s) = \{0, \ldots, n\}$ the set of positions between the words in s, before s_0 and after s_{n-1}. We define two kinds of items. A **passive item** is a tuple (X, i, j, k, l) where: $X \in \mathcal{N}$, $i, l \in Pos(s)$, $j, k \in Pos(s) \cup \{-\}$, $i \leq l$, and $i \leq j \leq k \leq l$ if $(j, k) \neq (-, -)$. Item (X, i, j, k, l) asserts that X can be *matched* over the span (i, j, k, l), where (i, l) and (j, k) denote the whole span of a matched sequence and the gap, respectively. Formally, a passive item $(X, i, -, -, l)$, or (X, i, l) for short, asserts that an IT $t \in R(X)$, a subtree of an ET in G, can be recognized (cf. Sect. 2) over the span (i, l). E.g., item $(N_0, 1, 2)$ in Fig. 5 indicates that *points* in sentence (1) can be a noun by the subtree rooted at N in t_5 and t_6 in Fig. 2. A passive item (X, i, j, k, l) where $(j, k) \neq (-, -)$ and $X = (x, u, a)$ asserts that (i) an AT $t \in R(X)$, a subtree of some ET in G, can be recognized over (i, j, k, l), and (ii) a subtree t' of an ET[4], with $x \in \mathcal{N}_0$ in its root, can be recognized over (j, k). Thus, the item $(N^*, 0, 1, 2, 2)$ in Fig. 5 means that *set* can be a modifier adjoined to the noun *points*. Here: $t \in \{t_2, t_3, t_4\}$ and t' is the subtree rooted at N in t_5 and t_6. An **active item** is a tuple (q, i, j, k, l), where i, j, k, and l specify the span, as previously, and q is a state in one of the underlying FSSAs. An active item (q, i, j, k, l) asserts that there exists a (not necessarily proper) prefix $\omega \in P(q)$ (of a grammar rule's body) which can be matched over (i, j, k, l), i.e., that the individual elements of ω can be consecutively matched over the adjacent spans of the input sentence, together spanning over (i, l), and that, if $(j, k) \neq (-, -)$, one of the elements of ω, marked with an asterisk, is matched against the item's gap (j, k). E.g., $(12, 0, 2)$ and $(14, 0, 1, 2, 2)$ in Fig. 5 correspond to matching *set points* with $\omega = V_2 NP$ and $\omega \in \{N_5 N^*, Pt_9 N^*, A_7 N^*\}$, respectively.

4.4 Inference Rules

We now formally specify the FSSA parser using the deductive framework [15]. As shown in Table 1, each of the inference rules, whose applications correspond to hyperarcs in the parsing hypergraph, takes zero, one or two chart items on input (*premises*, presented above the horizontal line) and yields a new item (*conclusion*, presented below the line) to be added to the chart if the conditions given on the right-hand side are met. The **axiom** rule (AX, cf. the double hyperarcs with empty inputs leading to $(0, 0, 0)$ and $(0, 1, 1)$ in Fig. 5) fills the initially empty

---

[4] t' must not be an EAT (see the *root adjoin* inference rule in Sect. 4.4 for explanations).

Table 1. Inference rules of the FSSA parser

AX: $\dfrac{}{(q_0,i,-,-,i)}$	$i \in Pos(s)\setminus\{n\}$	PS: $\dfrac{(q,i,j,k,l) \quad (X,l,-,-,l')}{(\delta(q,X),i,j,k,l')}$	$\delta(q,X)$ defined
SC: $\dfrac{(q,i,j,k,l)}{(\delta(q,s_l),i,j,k,l+1)}$	$\delta(q,s_l)$ defined	FA: $\dfrac{(q,i,-,-,l) \quad (X,l,j,k,l')}{(\delta(q,Y),i,l,l',l')}$	$\begin{array}{l}(x,u,a)=X\\(u,a)\neq(-,*)\\Y=(x,-,*)\\\delta(q,Y)\text{ defined}\end{array}$
DE: $\dfrac{(q,i,j,k,l)}{(X,i,j,k,l)}$	$X \in heads(q)$	IA: $\dfrac{(q,i,-,-,l) \quad (X,l,j,k,l')}{(\delta(q,X),i,j,k,l')}$	$\begin{array}{l}\delta(q,X)\text{ defined}\\(j,k)\neq(-,-)\end{array}$
RA: $\dfrac{(X,i,j,k,l) \quad (Y,j,j',k',k)}{(Y,i,j',k',l)}$		$(x,-,*) = X, (y,u,a) = Y, (u,a) \neq (-,*), x = y$	

chart with active items representing the claim that any rule α from the FSSG can be used to parse s starting from any non-final position, for each initial state q_0 of one of the FSSAs. The **scan** rule (SC, cf. the dashed hyperarcs in Fig. 5) matches the FSSAs' terminal symbols with words from the input. **Deactivation** (DE, cf. the thick hyperarcs) transforms an active item into the corresponding passive item, based on the q-outgoing head non-terminals, where $heads(q)$ is the set of symbols over transitions (representing rule heads) from state q to final states of the FSSAs. **Pseudo substitution** (PS, cf. the plain hyperarcs) is similar to scan, but instead of matching FSSA terminals against input words, automaton non-terminals are matched against already inferred non-terminals represented by passive items. Pseudo substitution handles regular TAG substitution, i.e., replacing a leaf non-terminal X by an IT rooted by X (cf. the hyperarc leading from $(10,0,1)$ and $(NP,1,2)$ to $(12,0,2)$), as well as matching two adjacent fragments of the same ET (cf. the hyperarc from $(5,0,1)$ and $(N_0,1,2)$ to $(4,0,2)$). The **foot adjoin** rule (FA, cf. the densely dotted hyperarcs) identifies ranges over which adjunction could possibly occur. It ensures that the resulting item is considered only if an elementary (sub)tree, recognized starting from l, and to which the corresponding AT(s) could be adjoined, exists. For the hyperarc from $(5,0,1)$ and $(N_0,1,2)$ to $(14,0,1,2,2)$, we have $X = N_0 = (N,0,-)$, $Y = (N,-,*) = N^*$, $(j,k) = (-,-)$ and $\delta(5,Y) = 14$. The **internal adjoin** rule (IA, with no instance in Fig. 5) combines an elementary (sub)tree, partially recognized over $(i,-,-,l)$, with its spine subtree, recognized starting from position l. Internal adjoin is similar to pseudo substitution but must be handled by a separate rule because the span of gap in the conclusion stems from the passive rather than the active premise. The **root adjoin** rule (RA, cf. the loosely dotted hyperarcs) represents the actual adjoining of a fully recognized EAT t into the root of a recognized subtree t' of an ET. Information that t' is recognized (with a modified span), is preserved in the conclusion and can be reused in order to recognize the full ET of which t' is a part. E.g., for the hyperarc from $(N^*,0,1,2,2)$ and $(N_0,1,2)$ to $(N_0,0,2)$, we have $X = N^* = (N,-,*)$, $Y = N_0 = (N,0,-)$, $x = y = N$, $(u,a) = (0,-)$, $(j',k') = (-,-)$, $t \in \{t_2,t_3,t_4\}$ and t' is the subtree of t_5 rooted at N.[5]

---

[5] Note that the additional constraint imposed on the modified node is that it must not be a root of an AT $((u,a) \neq (-,*))$. Otherwise, it would be possible to adjoin one AT to a root of another *not yet adjoined* AT. We block this derivation path, so that adjunction can only be carried out on top of an AT which has already been adjoined to some particular IT.

5   Experimental Results

We performed experiments on the FrenchTAG meta-grammar [5] compiled into a set of 9043 non-lexicalized ETs. After removing feature structures (not supported by our parser) 3065 unique trees where obtained. Since no compatible lexicon is available, we lexicalized the grammar with part-of-speech (POS) tags. Namely, to each anchor (i.e., the node meant to receive a terminal from an associated lexicon) in each ET a special terminal, containing the same POS value as the anchor, was attached. Thus, we obtained a grammar which models sentences with regular terminals (e.g., *il* 'it', *de* 'of', *qui* 'who') and POS tags (e.g., *v*, *n*, *adj*) interleaved. Such (inevitably, due to the missing lexicon) artificial lexicalization is not fully satisfactory in the context of TAGs, but it gives us an approximate upper bound on the possible gain from our compression-based approach.

Figure 6(a) shows the total numbers of automaton states and transitions depending on the compression method used to encode the resulting grammar. In the baseline, the grammar is represented as a list of flat rules (encoded as a separate automaton each) but no subtree sharing takes place. With this representation, parsing is roughly equivalent to the Earley-style TAG algorithm [14]. The FSS and FSSA encoding methods were described in Sects. 3.1 and 3.2.

Since treebanks compatible with existing TAGs (especially those generated from metagrammars) are hardly available, parsing evaluation was done on a synthetic corpus. Namely, $\sim$13000 sentences of length 1 to 30, of up to 500 sentences per length, were used to measure performance in terms of the number of hyperarcs explored while parsing a sentence (deactivate operations are ignored). The results are presented in Fig. 6(b), which includes two additional grammar compression methods similar to those in [12] for CFGs: (i) a trie, in which the list of rules is transformed into a prefix tree instead of a DFSA, (ii) a set of FSSAs, where a separate DFSA is constructed for each rule head.

The results show that the baseline version of the parser is only of a theoretical interest. It requires generating on average more than 4×10^4 hyperarcs even for sentences of length 1 (notably due to the POS-based lexicalization). The FSS parser is already a reasonable choice for parsing short sentences. FSSA compression leads, averaging over sentences of length from 1 to 15, to a farther reduction of $\sim$24$\times$ in terms of the number of visited hyperarcs. Using a set of FSSAs instead of a single FSSA is $\sim$2.25 times less efficient on average.

Figure 6(c) compares the FSS and FSSA parsers in terms of speed. In both versions, parsing time is almost linear w.r.t. the number of generated hyperarcs. However, the FSSA version proves more efficient, most likely due to the number of generated hypernodes which is, consistently, significantly higher in the FSS version (e.g., 95666 hypernodes in FSS against 1193 in FSSA for sentences of length 15). This, in turn, is related to the fact that a large number of (trivial) automata is used in the FSS parser, thus a large number of initial states have to be handled by the the **axiom** rule at the very beginning of the parsing process.

Surprisingly, the FSSA compression does not bring significant improvements in comparison to the prefix tree version. This is probably related to the fact that the active/passive distinction already provides a form of suffix sharing – items

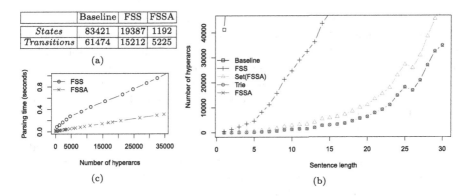

	Baseline	FSS	FSSA
States	83421	19387	1192
Transitions	61474	15212	5225

(a)

(c)

(b)

Fig. 6. (a) Results of the compression experiments, (b) Impact of grammar encoding methods on parsing performance, measured as an average number of hyperacs explored by the parser on ∼13000 sentences randomly generated from the FrenchTAG grammar, (c) Average parsing time as a function of the number of generated hyperarcs.

referring to pre-final states in the prefix tree are automatically transformed into the corresponding passive items. In particular, the number of passive items which can be potentially constructed over a given span equals 1123 in both versions, while the number of potential active items per span diminishes merely from 430 to 301 in the FSSA version. Moreover, due to the left-to-right parsing strategy, prefix sharing impacts parsing performance more visibly than suffix sharing.

6   Related Work

A bottom-up Earley-like algorithm based on flattening is one of the TAG parsing schemata proposed in [2]. While, conversely to our approach, it does not allow multiple adjunctions at the same node, it is similar to our baseline algorithm. Our enhancements of this baseline with subtree sharing and grammar FSA-compression substantially influence space and time efficiency (cf. Sect. 5).

FSA-based grammar encoding considerably speeds up CFG parsing [12] but it is not straightforwardly applicable to TAGs (which consist of trees rather than flat rules). It is, however, enabled by the flattening transformation proposed in this paper. Previous proposals of applying FSA-based compression to TAGs are manifold. [10,13] describe LR parsers for TAGs, in which predictions are pre-compiled off-line into an FSA. Each state of this FSA is a set of dotted production rules closed under prediction. Thus, the FSA represents the parser, while in our approach the FSSA represents the grammar (and the inferences rules of the parser are adapted to this representation).

Another automata-based solution for LTAGs and related lexicalized formalisms has been proposed by [4,7]. The traversal of an ET, starting from its anchor (lexical unit), is represented there as an automaton. Sets of trees attached to common anchors are then converted to automata, merged and minimized using standard techniques. As a result, structure sharing occurs only within tree families, while in our solution all ETs are represented with a single automaton which provides sharing between rules assigned to different lexical units. Another potential advantage of our solution lies in the subtree-sharing it enables, which allows different rules – even when represented by completely different paths in the automaton – to share common middle elements if these middle elements represent common subtrees. Finally, our method can be used for TAGs in general, not only for lexicalized TAGs. [4] report state-level compression ratios equal to 18 for *come*, 18.2 for *break*, and 30 for *give*, over a lexicalized English grammar. We converted the XTAG grammar [1] into an FSSA, obtaining a global, state-level compression of 22.7 (10751 states in the baseline representation vs. 472 in the FSSA). It is, however, difficult to compare these numbers: (i) their grammar is considerably larger than XTAG, (ii) they did not report the compression ratio over the entire grammar, (iii) they use one automaton per input word While they did not measure the impact of their encoding on parsing performance, we believe that our FSSA-based solution is more scalable w.r.t. the input length.

[16] proposes a method of grammar compression directly at stage of its definition. A linguist uses a formal language including factoring operators (e.g., disjunctions over tree fragments, Kleene-star-alike repetitions, optional or shuffled fragments, etc.) and the resulting grammar is then converted into a Logic Push-Down Automaton for parsing. The price to pay for this highly compact resource is its high potential overgeneration. Moreover, grammar description and parsing are not separated, hence large unfactorized TAGs can be hardly coped with. Our solution abstracts away from how the TAG is represented, compression is automatic and the FSSA is strongly equivalent to the original TAG.

Linear indexed grammars (LIGs) compare to our grammar flattening in that they contain flat production rules and are weakly equivalent to TAGs [10]. However, LIGs are more generic than TAGs, thus more specialized and efficient parsers can be potentially designed for TAGs [3]. Also, the TAG-to-LIG conversion does not preserve the extended domain of locality (EDL) ensured by TAGs, which is for us an eliminating criterion. Namely, in future we wish our parser to be driven by the knowledge about possible occurrences of multi-word expressions [17], whose elegant representation in TAGs is precisely due to the EDL property.

7   Conclusions

Our contribution is to design a parsing architecture coping with large TAGs (notably produced from metagrammars). We build on previous work so as to capture redundancy: (i) we flatten TAGs, (ii) we share common subtrees, (iii) we compress the flat grammar into an FSA, (iv) we adapt an Earley-based algorithm to this representation, (v) we show the influence of these steps on the parsing efficiency. To the best of our knowledge this is the first attempt to combine all these steps within one framework. Our parser and evaluation corpus are available under open licenses.[6] This solution does not affect the theoretical complexity of TAG parsing but it greatly improves the practical parsing performance.

---

[6] https://github.com/kawu/partage4xmg/tree/0.1.

References

1. Alahverdzhieva, K.: XTAG using XMG, Master Thesis, Nancy Université (2008)
2. Alonso, M., Cabrero, D., de la Clergerie, E.V., Vilares, M.: Tabular algorithms for TAG parsing. In: EACL 1999, pp. 150–157 (1999)
3. Alonso, M.A., de La Clergerie, É.V., Diaz, V.J., Vilares, M.: Relating tabular parsing algorithms for LIG and TAG. In: Text, Speech and Language Technology, vol. 23, pp. 157–184. Kluwer Academic Publishers (2004)
4. Carroll, J., Nicolov, N., Shaumyan, O., Smets, M., Weir, D.: Grammar compaction and computation sharing in automaton-based parsing. In: Proceedings of the TAPD 1998 Workshop, Paris, France, pp. 16–25 (1998)
5. Crabbé, B.: Représentation informatique de grammaires d'arbres fortement lexicalisées: le cas de la grammaire d'arbres adjoints. Ph.D. thesis, Université Nancy 2 (2005)
6. Earley, J.: An efficient context-free parsing algorithm. Commun. ACM **13**(2), 94–102 (1970)
7. Evans, R., Weir, D.: Automaton-based parsing for lexicalized grammars. In: Proceedings of the IWPT 1997 Workshop, Boston, MA, pp. 66–76 (1997)
8. Gallo, G., Longo, G., Pallottino, S., Nguyen, S.: Directed hypergraphs and applications. Discrete Appl. Math. **42**(2–3), 177–201 (1993)
9. Joshi, A., Schabes, Y.: Tree-adjoining grammars. In: Rozenberg, G., Salomaa, A. (eds.) Handbook of Formal Languages, pp. 69–123. Springer, Heidelberg (1997)
10. Kallmeyer, L.: Parsing Beyond Context-Free Grammars. Springer, Heidelberg (2010)
11. Klein, D., Manning, C.D.: Parsing and hypergraphs. In: Proceedings of the IWPT 2001 Workshop, Tsinghua University Press (2001)
12. Klein, D., Manning, C.D.: Parsing with treebank grammars: empirical bounds, theoretical models, and the structure of the penn treebank. In: Proceedings of ACL 2001, pp. 338–345 (2001)
13. Prolo, C.A.: Fast LR parsing using rich (tree Adjoining) grammars. In: Proceedings of EMNLP 2002, pp. 103–110 (2002)
14. Schabes, Y.: Left to right parsing of lexicalized tree adjoining grammars. Comput. Intell. **10**(4), 506–524 (1994)
15. Shieber, S., Schabes, Y., Pereira, F.: Principles and implementation of deductive parsing. J. Logic Program. **24**(1), 3–36 (1995)
16. de La Clergerie, É.V.: Building factorized TAGs with meta-grammars. In: Proceeding of the TAG+10 Conference (2010)
17. Waszczuk, J., Savary, A.: Towards a MWE-driven A* parsing with LTAGs. In: PARSEME 6th General Meeting, Struga, FYR Macedonia (2016)

Analyzing Matching Time Behavior
of Backtracking Regular Expression Matchers
by Using Ambiguity of NFA

Nicolaas Weideman[1,4]($\boxtimes$), Brink van der Merwe[1], Martin Berglund[2],
and Bruce Watson[3]

[1] Department of Computer Science,
Stellenbosch University, Stellenbosch, South Africa
nhweideman@gmail.com
[2] Department of Computer Science, Umeå University, Umeå, Sweden
[3] FASTAR Research, Information Science,
Stellenbosch University, Stellenbosch, South Africa
[4] Center for AI Research, CSIR, Stellenbosch University, Stellenbosch, South Africa

Abstract. We apply results from ambiguity of non-deterministic finite automata to the problem of determining the asymptotic worst-case matching time, as a function of the length of the input strings, when attempting to match input strings with a given regular expression, where the matcher being used is a backtracking regular expression matcher.

Keywords: Regular expression · Backtracking matcher · Ambiguity

1 Introduction

Catastrophic backtracking is a phenomenon that causes extended matching time, when attempting to match certain input strings with so-called vulnerable regular expressions, when using backtracking regular expression matchers found in programming languages such as Java, Perl and Python. It can be used to launch regular expression denial of service (ReDoS) attacks, and there are numerous online accounts (some listed at [2]) of the occurrence of catastrophic backtracking. Catastrophic backtracking often occurs (although not necessarily or exclusively) when matching with a regular expression R, containing a subexpression S^* (or S^+), where S could match some non-empty input string w in multiple ways. Thus an input string containing w^k (i.e. k copies of w) as substring, may potentially be matched (or attempted to be matched) in exponentially (in k) many ways by R, in cases where the matcher tries most of the possible ways (one after the other, i.e. not using the subset construction) in which the substring w^k can be matched, in an attempt to obtain an overall match. Even though some regular expression matcher implementations do not match input strings in a backtracking fashion [8], these alternative implementations typically do not support all extended regular expression functionality that programmers have become accustomed to, such as back references.

© Springer International Publishing Switzerland 2016
Y.-S. Han and K. Salomaa (Eds.): CIAA 2016, LNCS 9705, pp. 322–334, 2016.
DOI: 10.1007/978-3-319-40946-7_27

Although catastrophic backtracking is typically regarded as being synonymous with exponential worst-case matching time, non-linear polynomial worst-case matching time might still be unsatisfactory from a performance or security point of view. We regard non-linear matching time, vulnerable regular expressions and catastrophic backtracking to be all equivalent. We even point out cases with constant backtracking or matching time, where the constant is so large that the regular expression should be regarded as being vulnerable (from a practical point of view). Non-linear worst-case matching time often occurs when matching with a regular expression R, where R contains one or more occurrences of subexpressions of the form S^*U^* (or more generally S^*TU^*), where S and U (S, T and U) matches some common non-empty input string, say w. Similar to the exponential case, an input string containing w^k as substring, may now be matched (or attempted to be matched) in at least linearly (in k) many ways. The degree of the polynomial describing the worst-case number of ways in which an input string (of a given length) can be matched, depends on the number of occurrences of subexpressions of the form S^*U^* or S^*TU^*, where it is possible to move from one of these subexpressions (in the corresponding non-deterministic finite automaton) to the next while reading some input string. For example, a regular expression containing a subexpression $S_1^*U_1^*S_2^*U_2^*$, where S_i and U_i, for $i = 1, 2$, matches some non-empty input string w_i, could potentially attempt to match an input string containing $w_1^k w_2^k$, as substring, in quadratic (in k) different ways, leading to cubic matching time. To better understand the relationship between polynomial backtracking and matching time, consider a regular expression of the form S^*U^*, where S and U match some common non-empty input string w, but not any string of the form $w^k x$, for some suffix x. The matcher will first try to obtain an overall match by matching all of w^k with S^*, then backtrack and match only w^{k-1} with S^*, and continue this process of attempting to obtain an overall match by matching fewer and fewer of the repetitions of w with S^*, until S^* matches only the empty string. Thus since U^* first matches the empty string, then the last repetition of w, then the last two repetitions of w, etc., until it matches all of w^k, the matching time is quadratic in k, and thus in the length of $w^k x$.

A necessary condition to have exponential worst-case matching time is that the non-deterministic finite automaton (NFA), corresponding to the regular expression under consideration, contains a state with at least two loops that can be followed while processing the same substring (in a given input string). This condition is necessary and sufficient (under the additional assumptions that the NFA is trim and does not contain ε-loops) for an NFA to be exponentially ambiguous, i.e. to have input strings that can be matched in exponentially many ways in terms of their length [5]. A necessary condition to have non-linear polynomial worst-case matching time, is that the corresponding NFA contains one or more pairs of states, such that for each pair of states p, q, there exists a string $w_{p,q}$ and loops on p and q and a path from p to q, all that can be followed while reading $w_{p,q}$. Let d be the length of the longest sequence of pairs of states, with the above properties, obtained by ordering the pairs of states such that there exists a path from the second state in a pair of states, to the first state in the next pair of states. Then $(d + 1)$ is the maximum degree of the polynomial describing the worst-case matching time. Again, if the NFA under consideration is not exponentially ambiguous, these conditions are necessary and sufficient for the NFA to have polynomial ambiguity of degree d [5].

In the exponential matching time case, we refer to the part of the input string that can be matched in multiple ways while following these loops, as a *pump*, a string prefixed to the pump to ensure that the NFA reaches one of these states with two or more loops, as the *prefix*, and the string that is appended after the pump to ensure that the matcher attempts to match the pump in all possible ways, as the *suffix*. Thus exploit strings will be of the form $pw^k s$, with p the prefix, w the pump and s the suffix. In the non-linear polynomial matching time case, we have a pump for each subexpression of the form S^*U^* (or S^*TU^*). The strings required to move from the second state of a pair of states to the first state of the next pair of states, are referred to as the *pump separators*. Exploit strings will thus be of the form $s_0 w_1^k s_1 w_2^k \ldots s_{n-1} w_n^k s_n$, $k \geq 0$, where the w_i's are the pumps, s_0 the prefix, $s_1, \ldots, s_{n-1}$ the separators, and s_n the suffix. Again, the exploit strings correspond to strings exhibiting ambiguity, of a given form, in the underlying NFA (although strictly speaking, an additional sink accept state, having ε incoming transitions from all states, should be added to the underlying NFA, to make the correspondence between worst-case matching time and ambiguity precise).

As an example of a vulnerable regular expression, consider the following expression used to validate email addresses [4]:

```
R:=^([a-zA-Z0-9_\.\-])+\@(([a-zA-Z0-9\-])+\.)+([a-zA-Z0-9]{2,4})+
```

In the case of R, the subexpression $S := \boxed{(\texttt{[a-zA-Z0-9]\{2,4\})+}}$ can match the input string aaaa in two ways, either by matching aaaa by using the + in S once, or by using + twice by matching each time only aa. Note that this vulnerable regular expression is of a slightly different form than those described earlier, since in this case it is S and not $\boxed{(\texttt{[a-zA-Z0-9]\{2,4\})}}$, that matches some input string in more than one way. We construct an input string capable of exploiting this vulnerability as follows. First, we construct a prefix capable of taking the matcher to the vulnerable subexpression, for example a@a. should suffice. Next, we add multiple repetitions of the pump aaaa. Finally, we force the matcher to reject our specifically crafted string. For this we append, for example, a '$' to the end of the input string. Strings of the form a@a.(aaaa)^k$ can thus be used as exploit strings.

For backtracking regular expression matchers, the different paths which can be traversed to possibly obtain an overall match, are prioritized, and also explored in this prioritized order, one after the other. Also, the matcher will not continue exploring alternative ways of matching the input string, after a match has been found. Consequently, regular expressions that seem very similar, even that match precisely the same language, may have completely different matching time behavior. Consider for example regular expressions of the form $R_1 := S \mid .^*$ and $R_2 := .^* \mid S$, where S has exponential worst-case matching time and '.' is the wild card symbol that matches any single input symbol. These regular expressions are equivalent in terms of languages matched, but not in terms of matching time, due to the fact that in R_1, matching will first be attempted with S, while in R_2, the subexpression $.^*$ will be used first and S will be ignored. A slightly more complicated example is obtained by changing $.^*$ to $.\{m, \}$, i.e. an expression that matches strings of length m or more, with m a positive integer constant, in the regular expressions R_1 and R_2. In $R_2' := .\{m, \} \mid S$, the subexpression S with exponential worst-case matching time will now be reachable (in the corresponding non-deterministic finite automaton), but only for input strings of length shorter than m, leading to a regular expression with linear matching time. A non-trivial example of a similar type is $\boxed{\texttt{(\\\&d[0-9]\{2\}=.*?)+}}$, discussed in Sect. 4.

This paper extends results from [6]. We also consider how to determine the degree of the polynomial describing the worst-case matching time of a regular expression (if worst-case matching time is polynomial), which is listed as future work in [10]. The outline of the paper is as follows. In the next section we give the required definitions. After that, we provide our main results on deciding worst-case matching time behavior of a given regular expression, when using a backtracking regular expression matcher. Finally, we discuss our experimental results and conclude with a discussion on future work.

2 Definitions

In this section we introduce the notation and definitions required for the remainder of the paper. We denote by Σ a non-empty finite alphabet, which is used as input alphabet for automata and also an alphabet over which regular expressions are defined. As usual, ε denotes the empty word, and Σ^ε is used for $\Sigma \cup \{\varepsilon\}$. Also, Σ^* is the Kleene closure applied to Σ, thus the set of finite words over Σ. For $\Sigma_1 \subseteq \Sigma$ and $w = a_1 \ldots a_n \in \Sigma^*$, with $a_i \in \Sigma$, we let $\pi_{\Sigma_1}(w)$ be the word $b_1 \ldots b_n \in \Sigma_1^*$, with $b_i = a_i$ if $a_i \in \Sigma_1$, and $b_i = \varepsilon$ otherwise. For a function $f : A \to B$, and $a \in A$ and $b \in B$, we have that $f_{a \mapsto b} : A \to B$ is the function such that $f_{a \mapsto b}(a) = b$ and $f_{a \mapsto b}(x) = f(x)$ for all $x \in A \setminus \{a\}$. Also, $b^A : A \to B$, with $b \in B$, denotes the constant function with $f(x) = b$ for all $x \in A$. We use $\mathbb{N}$ for the set natural numbers, excluding 0. We denote by $|Q|$ the cardinality of the set Q and $\mathcal{P}(Q)$ the power set of Q.

A regular expression over an alphabet Σ (where $\varepsilon, \emptyset \notin \Sigma$) is either an element of $\Sigma \cup \{\varepsilon, \emptyset\}$ or an expression of one of the forms $(E \mid E')$, $(E \cdot E')$, or (E^*), where E and E' are regular expressions. Some parentheses can be dropped with the rule that * (Kleene closure) takes precedence over $\cdot$ (concatenation), which takes precedence over $\mid$ (union). Further, outermost parentheses can be dropped, and $E \cdot E'$ can be written as EE'. The language of a regular expression E, denoted $\mathcal{L}(E)$, is obtained by evaluating E as usual. When we say that E matches a string w, we mean that $w \in \mathcal{L}(E)$, as opposed to $vwv' \in \mathcal{L}(E)$, for $v, v' \in$

Σ^*. Some of our examples of expressions will use operators other than just union, concatenation and Kleene star, but we will refer to all regular expressions, including the extended expressions, simply as *regexes* in the remainder of the paper.

A *tree* with labels in a set Σ is a function $t: V \to \Sigma$, where $V \subseteq \mathbb{N}^*$ is a non-empty, finite set of vertices (or nodes) which are such that (i) V is prefix-closed, i.e., for all $v \in \mathbb{N}^*$ and $i \in \mathbb{N}$, $vi \in V$ implies $v \in V$, and (ii) V is closed to the left, i.e., for all $v \in \mathbb{N}^*$ and $i \in \mathbb{N}$, $v(i+1) \in V$ implies $vi \in V$. The vertex ε is the root of the tree and vertex vi is the ith child of v. We let $|t| = |V|$ be the size of t. We denote by t/v the tree t' with vertex set $V' = \{w \in \mathbb{N}^* \mid vw \in V\}$, where $t'(w) = t(vw)$ for all $w \in V'$. Given trees $t_1, \ldots, t_n$ and a symbol α, we let $\alpha[t_1, \ldots, t_n]$ denote the tree t with $t(\varepsilon) = \alpha$ and $t/i = t_i$ for all $i \in \{1, \ldots, n\}$.

Next we define non-deterministic finite automata (and runs for them), followed by the prioritized finite automata from [6,7], which are used to model regex matching behaviors exhibited by typical software implementations. In the definition of an NFA below, the transition function δ is defined to allow for parallel transitions on the same symbol between a pair of states. By $\delta(p, \alpha, q) = i > 0$, we indicate that there are i transitions on α between p and q. It is assumed that the transitions (if any) between p and q on α are numbered from 1 to $\delta(p, \alpha, q)$. We indicate by $p \to^{\alpha(j)} q$ (or $p\alpha(j)q$) that the jth-transition on α between p and q is taken. In our investigation, all parallel edges will be on ε, and we simply use $\varepsilon_1, \varepsilon_2, \ldots \varepsilon_n$, instead of $\varepsilon(1), \varepsilon(2), \ldots \varepsilon(n)$. Although parallel transitions do not influence the language accepted by an NFA, they do influence the number of accepting paths of a given input string, and thus play a role in our setting.

Definition 1. *A non-deterministic finite automaton (NFA) is a tuple $A = (Q, \Sigma, q_0, \delta, F)$ where: (i) Q is a finite set of states; (ii) Σ is the input alphabet; (iii) $q_0 \in Q$ is the initial state; (iv) the partial function $\delta : Q \times \Sigma^\varepsilon \times Q \to \mathbb{N}$ is the transition function; and (v) $F \subseteq Q$ is the set of final states.*

Also, $|A|_Q := |Q|$ and $|A|_\delta := \sum_{q_1, q_2 \in Q, \alpha \in \Sigma^\varepsilon} \delta(q_1, \alpha, q_2)$ is the state and transition size respectively.

Definition 2. *For an NFA $A = (Q, \Sigma, q_0, \delta, F)$ and $w \in \Sigma^*$, a run for w is a string $r = s_0 \alpha_1 (j_1) s_1 \cdots s_{n-1} \alpha_n (j_n) s_n$, with $s_0 = q_0$, $s_i \in Q$ and $\alpha_i \in \Sigma^\varepsilon$ such that $\delta(s_i, \alpha_{i+1}, s_{i+1}) \geq j_{i+1}$ for $0 \leq i < n$, and $\pi_\Sigma(r) - w$. A run is accepting if $s_n \in F$. The language accepted by A, denoted by $\mathcal{L}(A)$, is the subset $\{\pi_\Sigma(r) \mid r$ is an accepting run in $A\}$ of Σ^*.*

Definition 3. ([7]). *A prioritized non-deterministic finite automaton (pNFA) is a tuple $A = (Q_1, Q_2, \Sigma, q_0, \delta_1, \delta_2, F)$, where if $Q := Q_1 \cup Q_2$, we have: (i) Q_1 and Q_2 are disjoint finite sets of states; (ii) Σ is the input alphabet; (iii) $q_0 \in Q$ is the initial state; (iv) $\delta_1 : Q_1 \times \Sigma \to Q$ is the deterministic, but not necessarily total, transition function; (v) $\delta_2 : Q_2 \to Q^*$ is the non-deterministic prioritized transition function; and (vi) $F \subseteq Q_1$ are the final states.*

Given a pNFA A, nfa(A) denotes the NFA *associated* with A, which is obtained by ignoring the priorities of the δ_2 transitions of A. Thus for nfa(A),

$\delta(p, a, q) = 1$ if $\delta_1(p, a) = q$ for $p \in Q_1$ and $a \in \Sigma$, and $\delta(p, \varepsilon, q) = j$ for $p \in Q_2$, if $\delta_2(p) = q_1 \ldots q_n$, and q appears $j > 0$ times in the sequence $q_1 \ldots q_n$.

Definition 4. ([7]). *For a pNFA $A = (Q_1, Q_2, \Sigma, q_0, \delta_1, \delta_2, F)$, a path of $w \in \Sigma^*$ in A, is a run $s_0 \alpha_1(j_1) s_1 \cdots s_{n-1} \alpha_n(j_n) s_n$ of w in nfa(A), such that if $\alpha_i = \alpha_{i+1} = \ldots = \alpha_{m-1} = \alpha_m = \varepsilon$, with $i \leq m$, then $(s_{k-1}, j_k, s_k) = (s_{l-1}, j_l, s_l)$, with $i \leq k, l \leq m$, implies $k = l$ – i.e. a path is not allowed to repeat the same transition in a sequence of ε-transitions. For two paths $p = s_0 \alpha_1(j_1) s_1 \cdots s_{n-1} \alpha_n(j_n) s_n$ and $p' = s_0' \alpha_1'(j_1') s_1' \cdots s_{m-1}' \alpha_m'(j_m') s_m'$ we say that p is of higher priority than p', $p > p'$, if $p \neq p'$, $\pi_\Sigma(p) = \pi_\Sigma(p')$ and either p' is a proper prefix of p, or if k is the first index such that $(j_k) s_k \neq (j_k') s_k'$, then $\delta_2(s_{k-1}) = \cdots s_k \cdots s_k' \cdots$ if $s_k \neq s_k'$, or $s_k = s_k'$ and $j_k < j_k'$. An accepting run for A on w is the highest-priority path $p = s_0 \alpha_1(j_1) s_1 \cdots \alpha_n(j_n) s_n$ such that $\pi_\Sigma(p) = w$ and $s_n \in F$. The language accepted by A, denoted by $\mathcal{L}(A)$, is the subset of Σ^* defined by $\{\pi_\Sigma(r) \mid r$ is an accepting run in $A\}$. Note that $\mathcal{L}(A) = \mathcal{L}(nfa(A))$.*

Infinite loops are avoided in backtracking matchers by disallowing the repetition of the same transition in a sequence of ε-transitions, as specified in the definition above. In [6], the input directed depth-first search algorithm, typically used by backtracking regex matchers to find accepting runs in pNFA, was given, and it was observed that the running time of this algorithm can be described by the size of the backtracking run, defined next. It should be noted that although $\mathcal{L}(A) = \mathcal{L}(nfa(A))$ for a pNFA A, the purpose of a pNFA is to associate a run deterministic NFA (i.e. an input string can have at most one accepting run) to a regex, and thus to make it possible to define regex extensions such as capturing groups [7].

Definition 5. ([6]). *Let $A = (Q_1, Q_2, \Sigma, q_0, \delta_1, \delta_2, F)$ be a pNFA, $q \in Q_1 \cup Q_2$, $w = \alpha_1 \cdots \alpha_n \in \Sigma^*$, and $C : Q_2 \to \mathbb{N} \cup \{0\}$. Then the (q, w, C)-backtracking run of A is a tree over $Q_1 \cup Q_2 \cup \{Acc, Rej\}$ ($Acc, Rej \notin Q_1 \cup Q_2$). It succeeds if and only if Acc occurs in it. We denote the (q, w, C)-backtracking run by $btr_A(q, w, C)$ and inductively define it as follows. If $q \in F$ and $w = \varepsilon$ then $btr_A(q, w, C) = q[Acc]$. Otherwise, we distinguish between two cases:*

– *If $q \in Q_1$, then $btr_A(q, w, C)$ equals*

$$\begin{cases} q[btr_A(\delta_1(q, \alpha_1), \alpha_2 \cdots \alpha_n, 0^{Q_2})] & \text{if } n > 0 \text{ and } \delta_1(q, \alpha_1) \text{ is defined,} \\ q[Rej] & \text{otherwise.} \end{cases}$$

– *If $q \in Q_2$ with $\delta_2(q) = q_1 \cdots q_k$, let $i_0 = C(q) + 1$ and $r_i = btr_A(q_i, w, C_{q \mapsto i})$ for $i_0 \leq i \leq k$. Then $btr_A(q, w, C)$ equals*

$$\begin{cases} q[Rej] & \text{if } i_0 > k, \\ q[r_{i_0}, \ldots, r_k] & \text{if } i_0 \leq k \text{ but no } r_i(i_0 \leq i \leq k) \text{ succeeds,} \\ q[r_{i_0}, \ldots, r_i] & \text{if } i \in \{i_0, \ldots, k\} \text{ is the least index such that } r_i \text{ succeeds.} \end{cases}$$

The backtracking run of A on w *is* $btr_A(w) = btr_A(q_0, w, 0^{Q_2})$. *If* $btr_A(w)$ *suc-ceeds, then the accepting run of A on w contains the sequence of states on the right-most path in* $btr_A(w)$.

It should be noted that the argument C, in the definition of $btr_A(q, w, C)$, enforces the condition that a path is not allowed to repeat the same transition in a sequence of ε-transitions in Definition 4. For pNFA without ε-loops, the argument C and corresponding conditions can be removed from the definition of $btr_A(q, w, C)$.

Definition 6. *For a pNFA* $A = (Q_1, Q_2, \Sigma, q_0, \delta_1, \delta_2, F)$, *the matching time of an input string w with A, is defined to be* $|btr_A(w)|$. *Let* $f(n) = \max\{|btr_A(w)| \mid w \in \Sigma^*, |w| \leq n\}$ *for all* $n \in \mathbb{N}$. *We say that A has* exponential worst-case matching time *if* $f \in 2^{\Omega(n)}$ *(or equivalently, if* $f(n) \in 2^{\Theta(n)}$*) and* polynomial matching time of degree k, *for* $k \in \mathbb{N} \cup \{0\}$, *if* $f \in \Theta(n^k)$.

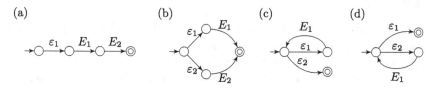

Fig. 1. pNFA corresponding to (a) $E_1 \cdot E_2$, (b) $E_1 \mid E_2$, (c) E_{1*} and (d) $E_{1*?}$

We use a regex to pNFA construction, similar to the one implicitly used in the Java regex matching engine. We denote this pNFA constructed from the regex E, by using inductively the constructions in Fig. 1 (described in [6]), by $J^P(E)$. In Fig. 1(d), $E_1^{*?}$ denotes the reluctant Kleene star operator applied to E_1, which match as few input symbols as possible with E_1, in contrast to greedy Kleene star in Fig. 1(c), which matches as many as possible with E_1. Recall that the subscripts of ε indicates the priority of the transition, with ε_1 having the highest priority.

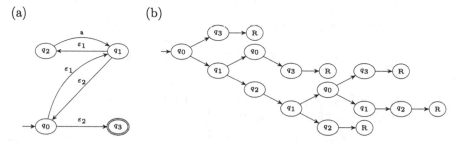

Fig. 2. (a) $J^P((a^*)^*)$, i.e. the Java pNFA for the regex $(a^*)^*$. (b) The backtracking run of $J^P((a^*)^*)$ on input ax. The tree is rotated anticlockwise by 90° and highest priority paths are at the bottom. Leaves are marked with R instead of Rej.

The degree of ambiguity for $w \in \Sigma^*$, with respect to the NFA A, is the number of accepting runs for w in A. The degree of ambiguity of A is the maximum degree of ambiguity over all $w \in \Sigma^*$, which might be infinite, in which case we say A has *infinite degree of ambiguity* (IDA). When A has IDA, we consider the rate at which the maximum number of accepting runs grow in proportion to the length of the input strings, which might be exponential, described by saying A has *exponential degree of ambiguity* (EDA), or polynomial, described as A being *polynomially ambiguous*. Since we determine worst-case matching time of a regex E by using the type of ambiguity of an NFA related to E in a way described in Sect. 3, the next result is of importance to us.

Theorem 1. ([5]). *Let A be a trim ε-loop free finite automaton. Then*

- *It is decidable in time $O((|A|_\delta + |A|_Q^2)^3)$ whether A is infinitely ambiguous, and in time $O(|A|_\delta^2)$ whether A is exponentially ambiguous.*
- *If A is polynomially ambiguous, the degree of polynomial ambiguity of A can be computed in $O(|A|_\delta^3)$.*

3   Deciding Worst-Case Matching Time

We start this section by defining for a pNFA A, potentially with ε-loops, a pNFA flat(A), with matching time behavior very similar to that of A, but without ε-loops. To use ambiguity of NFA to analyze worst-case matching time, we first have to remove ε-loops from an pNFA associated to a regex, and this is the main purpose of defining flat(A). For a pNFA A, let $r_A(Q_2)$ be the subset of Q_2 defined by $Q_2 \cap (\{q_0\} \cup \{\delta_1(q, \alpha) \mid q \in Q_1, \alpha \in \Sigma\})$, i.e. all Q_2 states reachable from a Q_1 state in one transition, and possibly also q_0. A sequence $p_1 j_2 p_2 \cdots p_{n-1} j_n p_n$, with $p_1 \in r_A(Q_2), p_2, \ldots, p_{n-1} \in Q_2, p_n \in Q_1, j_i \in \mathbb{N}$, is a δ_2-path if $\delta_2(p_i) = \cdots p_{i+1} \cdots$, $\delta_2(p_i)$ has at least j_{i+1} occurrences of p_{i+1}, and $(p_i, j_{i+1}, p_{i+1}) = (p_k, j_{k+1}, p_{k+1})$ only if $i = k$. Thus δ_2-paths are maximum length subsequences of ε-transitions only, obtained from paths in a pNFA. For a pNFA A, we define a pNFA flat(A) next, such that the paths for flat(A) are obtained from those for A, by replacing each δ_2-path with a single ε-transition. This ensures that A and flat(A) have the same matching time behavior up to a constant.

Definition 7. *For δ_2-paths $P := p_1 j_2 \cdots j_n p_n$ and $P' := p'_1 j'_2 \cdots j'_m p'_m$, with $p_1 = p'_1$, we define $P > P'$ if the least i such that $j_i p_i \neq j'_i p'_i$ is such that $\delta_2(p_{i-1}) = \cdots p_i \cdots p'_i \cdots$ with $p_i \neq p'_i$, or $p_i = p'_i$ but $j_i < j'_i$. We let flat(A) be $(Q_1, r_T(Q_2), \Sigma, q_0, \delta_1, \delta'_2, F)$, where δ'_2 is defined as follows. For $q \in Q'_2$, let $P_1, \ldots, P_n$ be all δ_2-paths, ordered according to priority, starting at q and ending at a state in Q_1. Then $\delta'_2(q) := q_1 \cdots q_n$, where q_i is the last state in P_i.*

An example of going from a pNFA A to flat(A), is given in Fig. 3.

We now describe two algorithms to analyze worst case matching time of regexes. Due to space limitations, these algorithms are not described in-depth. *Simple analysis* is a procedure for determining an upper bound for the worst-case matching time of a regex. We start with a regex E and turn it into $J^p(E)$, the Java version of the pNFA for E. Next, we remove ε-loops by going from $J^p(E)$ to $J' := \mathrm{flat}(J^p(E))$, and then consider the NFA $N := \mathrm{nfa}(J')$. Finally, the NFA N' is obtained from N by adding an additional sink accept state z to N. We place incoming ε-transitions from all other states to z, and make z the only accept state. Going from N to N', turns the problem of counting all possible transitions that can be taken while attempting to match $w \in \Sigma^*$ with N, into counting the number of accepting paths in N' for w. Thus for a given input string w, the size of the backtracking run of w in J' is bounded by the number of accepting paths of w in N', and we have equality if $w \notin \mathcal{L}(N)$. Thus if $w' \in \Sigma^*$ exists such that $ww' \notin \mathcal{L}(N)$ for all $w \in \Sigma^*$, then the worst-case matching time of J' and thus J, is precisely the ambiguity of N', otherwise the ambiguity for N' is only an upper bound for the worst-case matching time of E.

The unprioritized pNFA (upNFA), $\mathrm{up}(A)$, is an NFA obtained from the pNFA A, by not simply ignoring priorities of ε-transitions of A, but doing a type of subset construction that keeps in a given state also track of the states that are reachable with higher priority paths (on the same input). In the construction of $\mathrm{up}(A)$, we assume A has no ε-loops, otherwise replace A by $\mathrm{flat}(A)$. For an NFA B and Q' a subset of the states of B, let $\overline{Q'}$ be the ε-closure of Q'. For a pNFA A, $\mathrm{up}(A)$ is defined next.

Definition 8. *Let* $A := (Q_1, Q_2, \Sigma, q_0, \delta_1, \delta_2, F)$, *then* $\mathrm{up}(A)$ *is the NFA given by* $(Q', \Sigma, q_0', \delta', F')$, *where:*

(i) $Q' = ((Q_1 \cup Q_2) \times \mathcal{P}(Q_1)) \setminus Q''$, *where* Q'' *is the set of states* (p, P) *such that for all* $w \in \Sigma^*$, *there is a* $p' \in P$, *such that* w *has an accepting path in* $\mathrm{nfa}(A)$ *starting at* p';

(ii) $q_0' = \{(q_0, \emptyset)\}$ *and* $F' = (F \times \mathcal{P}(Q_1)) \cap Q'$;

(iii) *for* $a \in \Sigma$, $\delta'((p, P), a, (p', P')) = 1$ *if* $\delta_1(p, a) = p'$ *and* $\overline{\delta_1(P, a)} \cap Q_1 = P'$, *where* δ_1 *is extended to be defined on sets of states in the obvious way;*

(iv) $\delta'((p, P), \varepsilon, (p_i, \overline{P \cup \{p_1, \ldots, p_{i-1}\}} \cap Q_1)) = i_j$, *for* $1 \leq i \leq n$, *if* $p \in Q_2$ *and* $\delta_2(p) = p_1 \ldots p_n$, *where* i_j *is the number of indices* i' *with* $p_i = p_{i'}$ *and* $\overline{P \cup \{p_1, \ldots, p_{i-1}\}} \cap Q_1 = \overline{P \cup \{p_1, \ldots, p_{i'-1}\}} \cap Q_1$.

Note that the states of $\mathrm{up}(A)$ are of the form (p, P), with $p \in Q$ and $P \subseteq Q_1$. The states from P in (p, P) are those reachable with higher priority paths on the same input. By removing states (p, P) such that for all $w \in \Sigma^*$, there is a $p' \in P$, such that w has an accepting path in $\mathrm{nfa}(A)$ starting at p', we ensure that only paths explored in A on any input w is kept in $\mathrm{up}(A)$. Just as in our simple analysis, we add a sink accept state z to $\mathrm{up}(A)$ to obtain an NFA B', and then perform ambiguity analysis on B'. We refer to the process of constructing $B := \mathrm{up}(\mathrm{flat}(J^p(E)))$ from a regex E, adding the sink accept state z to B to get B', and then determining the ambiguity of B', as doing a *full analysis* of E. At the cost of going form polynomial to exponential time (in the size of the regular expression), full analysis provides a precise answer.

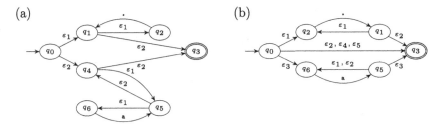

Fig. 3. (a) $J^p(.^* \mid (a^*)^*)$ and (b) flat($J^p(.^* \mid (a^*)^*)$)

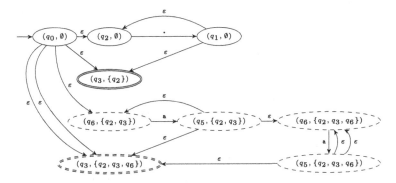

Fig. 4. The unprioritized pNFA for $J^p(.^* \mid (a^*)^*)$. Dashed states indicate the states Q'' in Definition 8, that should be removed.

Example 1. Next we consider the regex E given by $\boxed{.^* \mid (a^*)^*}$, which has EDA, but only linear worst-case matching time, although the subexpression $\boxed{(a^*)^*}$ has exponential worst-case matching time (with input $a \ldots ab$). However, due to priorities, only the subexpression $\boxed{.^*}$ is used during matching. The main steps taken in our full analysis are shown in Figs. 3 and 4. Since up(flat($J^p(E)$)) has constant ambiguity, E has linear worst-case matching time.

4 Experimental Results

All experiments were performed on a machine with a 3.1GHz, 4 cores and a cache size of 6144KB. We performed simple analysis on the Snort rule-set version 2.9.31 ([3]; 12499 expressions) and RegExLib ([1]; 2994 expressions) repositories. Simple analysis only checks for EDA and IDA, yielding one of six results: EDA; IDA (but not EDA); No IDA; whether the regex contains illegal syntax or requires unhandled functionality (indicated as "Skipped"); or if the analysis takes an inordinate (10 s in our experiments) amount of time (indicated as "Timeout in EDA", or "Timeout in IDA"). Both analyses construct exploit strings with properties, as explained in the Introduction. If EDA or IDA is present in simple analysis, full analysis is performed to determine whether a regex indeed has

Table 1. A breakdown of the simple analysis results.

Repository	EDA	IDA	No IDA	Skipped	Timeout in EDA	Timeout in IDA
Snort	11	824	8381	3108	103	72
RegExLib	145	217	1617	912	16	87

Table 2. The matching time behavior, as determined by full analysis, of the 156 and 1041 regexes shown to have EDA and IDA, respectively, by simple analysis.

Simple Analysis	Full Analysis				
EDA	Exponential	Polynomial	Linear	Timeout in EDA	Timeout in IDA
156	122	0	2	32	0
IDA	Exponential	Polynomial	Linear	Timeout in EDA	Timeout in IDA
1041	0	692	24	0	325

exponential or non-linear polynomial matching time (although this is strictly speaking only necessary in cases where the exploit strings obtained in simple analysis do not point out the expected behavior). Simple analysis determined that, in total, 156 regexes have EDA, 1041 have IDA, and 9998 have neither. The remaining 4298 regexes were either skipped, or timed out. Full analysis was performed on the 156 and 1041 regexes with EDA and IDA (in simple analysis), respectively. The results of the full analysis is shown in Table 2, which shows whether the matching time of a regex is exponential, polynomial or linear; or whether the analysis timed out. All exponentially vulnerable regexes were tested against the Java regex matcher with their respective exploit strings, as generated by the full analysis. All but two of these regexes did indeed exhibit exponential matching time. The reason for the full analysis producing faulty exploit strings in these two cases warrants further investigation [2].

As mentioned before, if an NFA for a regex contains EDA, it does not necessarily imply that the regex is vulnerable. The regex $(\backslash\&d[0-9]\{2\}=.*?)+$ from the Snort repository match any input string with a prefix starting with &d, followed by two digits and an equals sign. In order to build an exploit string, we can use ε as prefix and &d00=&d00= as pump. Since $.*?$ matches all strings, two copies of the string &d00= can be matched in two ways – either once with the $\backslash\&d[0-9]\{2\}=$ and once with the $.*?$, or twice with the + operator. But every time the matcher can not match part of the input string with $(\backslash\&d[0-9]\{2\}=)$, it will backtrack and match one character with $.*?$, and thus all strings will be matched in linear time. In the simple analysis, the analyzer detected that the (NFA of the) regex has EDA, but when the full analysis is performed, the analyzer detected that it cannot construct a valid exploit string and therefore classified the regex as not being vulnerable.

Regexes with large constant matching time, might also be regarded as being vulnerable (at least from a practical point of view). Next, we describe an approach that worked well in practice to identify some these regexes. If a regex has a large counted closure, such as $R := (S \mid T)\{0, n\}$, for large n, the regex can be approximated (in terms of language accepted and matching time) with a Kleene star, as in $(S \mid T)^*$. The Snort repository contains an expression of this form, namely $\boxed{\texttt{\textbackslash x20\textbackslash x00([\^{}\textbackslash x00].|.[\^{}\textbackslash x00])\{255\}}}$. Although the counted closure is of the form $\{n\}$, and not $\{0, n\}$, we can still approximate the regex with $\boxed{\texttt{\textbackslash x20\textbackslash x00([\^{}\textbackslash x00].|.[\^{}\textbackslash x00])+}}$ for the purpose of approximating worst-case matching time. By using this approximation approach, our analysis was able to point out that this regex is indeed vulnerable with the exploit string \x20\x00 as prefix, aa as pump and \x00\x00 as suffix.

Our analyzer does not yet support all extensions found in Java regexes. Unsupported extensions include possessive quantifiers and back references.

5   Conclusions and Future Work

We developed an analyzer to identify regexes vulnerable to ReDoS. The analysis was run on two repositories of commonly used regexes, and numerous regexes with non-linear worst-case matching time were discovered. We plan to extend our analysis so that most features found in extended regexes are supported, and also to develop techniques to identify regexes with non-constant worst-case memory usage. One interesting extension to consider is that of possessive quantifiers, allowing the matcher to throw away certain backtracking positions, and creating the possibility to remove some matching time vulnerabilities. To analyze regexes with possessive quantifiers in terms of time and space, we plan to describe the matching of an input string by a pNFA, in terms of a 2-way deterministic pushdown automaton [9]. A further goal is to investigate the complexity of our worst-case matching-time analysis techniques as discussed in Sect. 3.

References

1. Regexlib. http://www.regxlib.com. Accessed 06 Oct 2015
2. Regular expression static analysis project page. http://www.cs.sun.ac.za/~abvdm/regex.html. Accessed 30 Apr 2016
3. Snort. http://www.snort.org. Accessed 06 Oct 2015
4. Adam, B.: Regular expression dos and node.js (2014). https://blog.liftsecurity.io/2014/11/03/regular-expression-dos-and-node.js
5. Allauzen, C., Mohri, M., Rastogi, A.: General algorithms for testing the ambiguity of finite automata. In: Ito, M., Toyama, M. (eds.) DLT 2008. LNCS, vol. 5257, pp. 108–120. Springer, Heidelberg (2008). http://dx.doi.org/10.1007/978-3-540-85780-8_8

6. Berglund, M., Drewes, F., van der Merwe, B.: Analyzing catastrophic backtracking behavior in practical regular expression matching. In: Ésik, Z., Fülöp, Z. (eds.) Proceedings 14th International Conference on Automata and Formal Languages, AFL 2014, EPTCS, vol. 151, Szeged, Hungary, 27–29 May 2014, pp. 109–123 (2014). http://dx.doi.org/10.4204/EPTCS.151.7

7. Berglund, M., van der Merwe, B.: On the semantics of regular expression parsing in the wild. In: Drewes, F. (ed.) CIAA 2015. LNCS, vol. 9223, pp. 292–304. Springer, Heidelberg (2015). http://dx.doi.org/10.1007/978-3-319-22360-5_24

8. Cox, R.: Implementing regular expressions (2007). http://swtch.com/~rsc/regexp/. Accessed 26 Feb 2016

9. Gray, J., Harrison, M.A., Ibarra, O.H.: Two-way pushdown automata. Inf. Control **11**(1/2), 30–70 (1967). http://dx.doi.org/10.1016/S0019-9958(67)90369-5

10. Rathnayake, A., Thielecke, H.: Static analysis for regular expression exponential runtime via substructural logics. CoRR abs/1405.7058 (2014). http://arxiv.org/abs/1405.7058

Author Index

Printed in the United States
By Bookmasters